Fossil Prokaryotes and Protists

Frontispiece. A reconstruction of prokaryotic stromatolites growing in shallow waters lapping onto a volcanically active land mass during the Archaean eon, 3.5 billion years ago. (From a mural, "The Archean Age: Three billion, five hundred million years ago. . ." by Peter Sawyer in "*The Earliest Traces of Life*" hall, Smithsonian Institution, Washington, DC. Used with the permission of the Smithsonian Institution.)

Fossil Prokaryotes and Protists

Edited by

JERE H. LIPPS

Department of Integrative Biology and
Museum of Paleontology
University of California, Berkeley
California

BOSTON

BLACKWELL SCIENTIFIC PUBLICATIONS

OXFORD LONDON EDINBURGH

MELBOURNE PARIS BERLIN VIENNA

Editorial offices:
238 Main Street, Cambridge
Massachusetts 02142, USA
Osney Mead, Oxford OX2 0EL, England
25 John Street, London WC1N 2BL
England
23 Ainslie Place, Edinburgh EH3 6AJ
Scotland
54 University Street, Carlton
Victoria 3053, Australia

Other editorial offices:
Librairie Arnette SA
2, rue Casimir-Delavigne
75006 Paris
France

Blackwell Wissenschafts-Verlag
Meinekestrasse 4
D-1000 Berlin 15
Germany

Blackwell MZV
Feldgasse 13
A-1238 Wien
Austria

First published 1993

Set by Setrite Typesetters Ltd, Hong Kong
Printed in the United States of America
by Braun-Brumfield, Ann Arbor, Michigan

93 94 95 96 4 3 2 1

DISTRIBUTORS

USA
Blackwell Scientific Publications, Inc.
238 Main Street
Cambridge, Massachusetts 02142
(*Orders*: Tel: 800 759–6102
617 876–7000)

Canada
Times Mirror Professional Publishing, Ltd
5240 Finch Avenue East
Scarborough, Ontario M1S 5A2
(*Orders*: Tel: 416 298–1588
800 268–4178)

Australia
Blackwell Scientific Publications
(Australia) Pty Ltd
54 University Street
Carlton, Victoria 3053
(*Orders*: Tel: 03 347–0300)

Outside North America and Australia
Marston Book Services Ltd
PO Box 87
Oxford OX2 0DT
(*Orders*: Tel: 0865 791155
Fax: 0865 791927
Telex: 837515)

Library of Congress
Cataloging-in-Publication Data

Fossil prokaryotes and protists/
editor, Jere H. Lipps.
p. cm.
Includes bibliographical references and index.
ISBN 0–86542–073–4
1. Prokaryotes, Fossil. 2. Protista, Fossil.
I. Lipps, Jere H.,
1939– .
QE719.5.F67 1993
560–dc20

Contents

List of contributors

JOHN A. BARRON
Paleontology and Stratigraphy Branch, US Geological Survey, Menlo Park, California 94025

RICHARD E. CASEY
Marine Studies Program, University of San Diego, San Diego, California 92110

STEPHEN J. CULVER
Department of Geological Sciences, Old Dominion University, Norfolk, Virginia 23508

DAVID W. DEAMER
Department of Zoology, University of California, Davis, California 95616

LUCY E. EDWARDS
Paleontology and Stratigraphy Branch, US Geological Survey, Reston, Virginia 22092

JOHN J. ERNISSEE
Department of Geography and Earth Science, Clarion University, Clarion, Pennsylvania

STJEPKO GOLUBIC
Department of Biology, Boston University, Boston, Massachusetts 02215

JOSEPH E. HAZEL
Department of Geology and Geophysics, Louisiana State University, Baton Rouge, Louisiana 70803

ANDREW H. KNOLL
Botanical Museum, Harvard University, Cambridge, Massachusetts 02138

DAVID R. LINDBERG
Museum of Paleontology, University of California, Berkeley, California 94720

JERE H. LIPPS
Department of Integrative Biology and Museum of Paleontology, University of California, Berkeley, California 94720

KEVIN McCARTNEY
Division of Mathematics and Science, University of Maine, Presque Isle, Maine 04769

CARL V. MENDELSON
Department of Geology, Beloit College, Beloit, Wisconsin 53511

WILLIAM G. SIESSER
Department of Geology, Vanderbilt University, Nashville, Tennessee 37235

HELEN TAPPAN
Department of Earth and Space Sciences, University of California, Los Angeles, California 90024

Preface

Fossil Prokaryotes and Protists is a textbook for an advanced paleontology course dealing with the morphology, systematics, distribution in time and space, and evolution of single-celled organisms, as represented in the fossil record. The book assumes a knowledge of the principles of paleontology or general biology, and students would be well advised to have taken an introductory course in paleontology, biology, botany, or zoology before using this book. No preceding course on systematic paleontology is necessary, because each group considered here is unique and unlike any of the multicellular animals or plants. An understanding of basic geology will be useful, although separate sections on biostratigraphy, paleoceanography, and paleoenvironments are included because that is the chief geologic use of most unicellular fossils.

Traditionally, fossils of protists, although not usually those of prokaryotes, have been taught in courses called *micropaleontology*, but just as traditionally, these courses have covered any tiny fossils that require a microscope for study, including those of many invertebrate, vertebrate, and plant groups. Tiny fossils of higher plants, invertebrates, and vertebrates are excluded from this textbook in order to treat levels of biologic organization together rather than a particular technique of study. Both approaches have merit. Microfossils are found together in the same samples regardless of biologic origin, and the inclusion of the common fossils likely to be encountered in microscopic examination of samples in a single textbook has some practical value. However, modern techniques for study of the very wide variety of microfossils are no longer simple, and nearly every group now requires special techniques. In addition, paleontology in general is making great advances by moving towards a more biologic interpretation of fossils, so textbook treatments of fossils should be usefully organized by systematic affinities. *Fossil Prokaryotes and Protists* does just that by including bacteria and single-celled algae and protozoa in a single volume. The multicellular invertebrate microfossils are considered in the companion textbook to this volume, *Fossil Invertebrates*, edited by Richard S. Boardman, Alan H. Cheetham, and Albert J. Rowell, and also published by Blackwell Scientific Publications in 1987. Vertebrate and plant fossils are considered in any of several modern textbooks.

Some topics in the general field of paleontology have not yet had major impact on the study of fossil prokaryotes and protists. Cladistic analysis, for example, has not been applied to unicellular fossils in any major way, although the method is discussed in this book with the hope that it will be more widely applied by the next generation of protistan paleontologists. Molecular phylogeny has clarified relationships between large groups of prokaryotes and between some protists and other kingdoms, but these techniques have not been applied extensively enough to reveal information useful in the context of this book. So many of the more recent advances in general paleontology and evolutionary biology have great but unrealized potential in prokaryotic and protistan paleontology. Other entire fields are based solidly on unicellular fossils; for example, paleoceanography. While such topics are useful in a textbook on these fossils, in recent years they have grown so much that they too deserve and have textbooks and courses devoted exclusively to them.

The specific goals, then, of *Fossil Prokaryotes and Protists* are:

1 to provide an understanding of the fossil record of unicellular organisms;

2 to provide an understanding of the important role these organisms have played in Earth and life history;

3 to provide an understanding of the usefulness of these fossils in solving certain geologic problems;

4 to include enough information about the biology, morphology, and relationships of prokaryote and protist fossils to enable students to enter the professional literature, to be conversant with specialists on each group, or to proceed to more advanced studies of these groups;

5 to provide a summary of the morphology, systematic paleontology, biology, and paleontology of unicellular fossils sufficiently detailed that advanced students can recognize and develop geologic or biologic research problems in their own programs that can be addressed using these fossils, given advice from their professors about topics amenable to study at their institutions.

The book is organized into two parts—one dealing with general aspects of fossil unicellular organisms, and another describing each major group. Fossil prokaryotes first appear in the fossil record 3.5 billion years ago (Ga), and so they are close to the very origin of life on Earth. A single, brief chapter is included on the origin of life, constrained by reasonable geologic and biochemical evidence, in order to provide an understanding of how the first unicellular fossils themselves may have arisen. Another chapter deals with the problems peculiar to single-celled fossils, and a third covers their applications to geologic problems because protists play an essential role in modern geology. A brief evolutionary history of the prokaryotes and protists and their role in influencing Earth's history provides a summary in which to place succeeding chapters that deal with specific groups of organisms.

The 11 major unicellular contributors to the fossil record are considered in detail in the second part, each written by practising experts. One or two groups that occur very sparsely in the fossil record, for example thecamoebians, are omitted. The chapters are organized similarly: an introduction, history of study, morphology and systematics, biology, paleobiology, biostratigraphy, and an evolutionary history of the group. The level of systematic treatment of each group varies depending on its diversity and complexity. All chapters include a short list of supplementary reading that can provide additional information for the students or instructor.

The paleontology of unicellular organisms has grown so much in the past few decades that no individual can adequately deal with all groups. Well over 40 000 technical papers have appeared on aspects of single-celled fossils in the last two decades alone. Multiauthored texts are now standard. This aproach has the particular problem that chapters vary in treatment and style. Partly that cannot be avoided because each expert has made his or her own judgment about what material should be included within the objectives and outline for each chapter, and each has their own style. Even more important in this book, however, is that each group is quite variable in its complexity, its geologic or ecologic distribution, and in the knowledge accumulated about it. For example, foraminifera have been studied carefully for more than 150 years, have over 60 000 described benthic and planktic species, have perhaps as many as 6000 workers in industrial and research institutions worldwide, range in age from Cambrian to Recent, and live in environments from fresh water to shallow water of all types, to the deep sea, and from polar to equatorial regions. Silicoflagellates, in contrast, have been intensively studied for less than 30 years, have a few hundred species at most, have a dozen or so workers worldwide, are wholly planktic in the surficial waters of the oceans, and range in age from the Cretaceous to the Recent. Naturally, foraminifera will be treated less inclusively than silicoflagellates, although more pages will be devoted to them. Such is the nature of paleontology.

Each chapter has been reviewed by other experts in the field, and the entire book has been reviewed by paleontologists who teach this subject. Nevertheless, the book will surely differ in various respects from most courses. It is the only textbook that treats single-celled fossils exclusively and that includes all the major groups. The editor alone is responsible for this organization, and he hopes that the book marks the initiation of a more unified biologic approach to systematic paleontology in general.

The following colleagues contributed to this book by supplying information, assistance, illustrations, or reviews of particular chapters: David P. Adam, Nairn Albert, George W. Andrews, J. Ashmore, Jack G. Baldauf, F.T. Banner, John Barron, William B.N. Berry, Charles D. Blome, Andre Brack, Timothy J. Bralower, Lloyd H. Burckle, Nicholas J. Butterfield, Martin A. Buzas, Annette Ciccateni, John Dodge, Charles Downie, Paulian Dumitrica, John Firth, Greta Fryxell, Gregory Gaines, A.C. Gary, Rose Graves, Drew Haman, John Hanchar, B.U. Haq, Paul Hargraves, Robin Helby, Carole Hickman, Andrew H. Knoll, Jane Lewis, Hsin Yi Ling, Alfred R. Loeblich Jr, Sigurd Locker, David Loper, Allan Lord, Denis H. Lynn, Al Mahood, Carol Mankiewicz, Erlend Martini, Merrell Miller, Malgorzata Moczydłowska, Harold Morowitz, Shiro Nishida, Yvonne Norpchen, Robert Oscarson, Katharina Perch-Nielsen, J. William Schopf, Eugene Small, Joel Socwell, Lewis Stover, Helen Tappan, Anne Thayer, Gonzalo Vidal, E. Reed Wicander, Y. Wilson, and Sherwood W. Wise. We are particularly grateful to Roger Kaesler and Richard Olsson who provided thorough and useful reviews of the book from the viewpoint of the classroom instructor. All authors associated with academic institutions have benefitted enormously in their knowledge through research supported by various grants from the National Science Foundation, the National Geographic Society, and the Petroleum Research Fund of the American Chemical Society. Authors affiliated with the United States Geological Survey have benefitted from research projects of that agency.

We are especially indebted to Ms Sherrill Herbert for her careful editorial work on this book.

Royalties from the sale of this book, like those of its companion volume *Fossil Invertebrates,* will go to support the science of paleontology with the Paleontological Society and the Cushman Foundation for Foraminiferal Research, Inc., as beneficiaries.

Introduction to fossil prokaryotes and protists

1

Jere H. Lipps

Introduction

Although plants and animals are the most conspicuous organisms on Earth today, they are not the most numerous. Single-celled bacteria, or prokaryotes, and protists far outnumber them and occupy a much wider range of habitats. These microorganisms are abundant from the Antarctic ice deserts to steaming volcanic hot springs and from the highest mountains to the deepest sea. Life on the Earth as we know it today would be impossible without them because they provide so many fundamental necessities for other forms of life such as oxygen, food, and digestion. They also infest other organisms, including humans, and cause many common illnesses (one good example is malaria, caused by a protist, which infects an estimated 350 million people worldwide).

Prokaryotes and protists are also the most abundant organisms preserved in the fossil record, they are the oldest known fossils and, for 85% of the known geologic record of life, they were the only organisms on Earth. They contribute enormously to our understanding of the history of the Earth and of its life. They provide opportunities to develop and test evolutionary models using different **grades** of organization and different ecologic limitations than those provided by multicellular plants and animals, they provide data for paleoecologic interpretations, and they are especially useful in biostratigraphy because of their small size and abundance. The solution to many practical problems in geology, including those of the economically important oil industry, are dependent on the paleontology of protists. For all these reasons and more, prokaryotes and protists are exceptionally important objects of paleontologic study.

Throughout much of the geologic record, rock units sometimes thousands of meters thick are composed of prokaryote and protist fossil remains. Prokaryotic **stromatolites** are common in the Proterozoic and early Phanerozoic, and some of these formed reefs many kilometers in extent and many meters high. Most deep-sea and many freshwater lake deposits consist almost exclusively of carbonate and siliceous skeletons of protists. Some of these rock types even carry protistan names—diatomite or radiolarianite, for example. These rocks may be mined for commercial use. Diatomites from both marine and lake deposits are valuable for use in filters (for beer, swimming pools, and others), carriers of pigments (in lipstick) and chemicals, and as abrasives and polishes. Petroleum, commonly sought with the use of protist fossils, is formed chiefly from the degradation of eukaryotic algae. Prokaryotes and protists have a large impact on our modern lives, even as fossils.

Prokaryotes and protists

The prokaryotes and protists are single-celled organisms, whereas all other organisms are either acellular or multicellular. However, prokaryotes and protists are themselves separated by one of the most significant differences in the organic world. Prokaryotes differ from all other organisms chiefly because their DNA is loosely organized within the cell and not bounded by a membrane into a nucleus, and they lack chromosomes. All other organisms are **eukaryotes** and have chromosomes made of DNA, RNA, and proteins contained within a membrane-bound **nucleus** (Fig. 1.1). Prokaryotes and eukaryotes differ in a number of other significant ways as well (Table 1.1; see Chapter 5). Protists, animals, plants, and fungi are all eukaryotes. Animals, plants, and fungi are multicellular; protists include all other eukaryotes.

Single-celled organisms are often considered simple, but this is hardly true. Each cell must function in most of the same ways that complete higher plants and animals do (Fig. 1.2), yet they do this with just a single cell. Prokaryotes and protists have developed a wide array of **organelles** within their cells that function in marvelous ways. Prokaryotes, protists, fungi, metazoa, and metaphyta are simply different grades of organization, each making a successful living in different ways. Prokaryotes and protists certainly are older and, in this sense, more primitive, but they have also existed for as long as 3.5

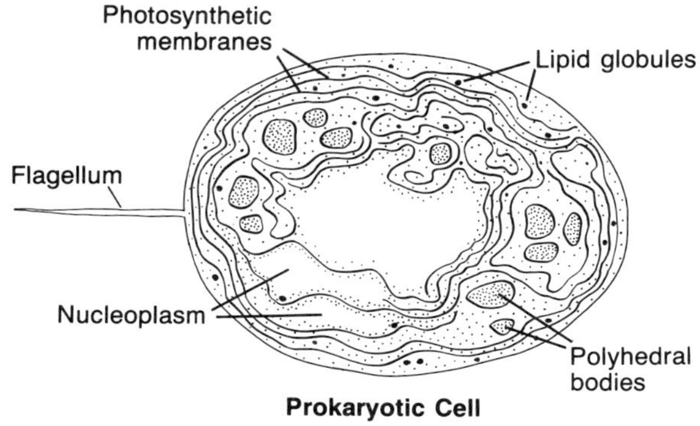

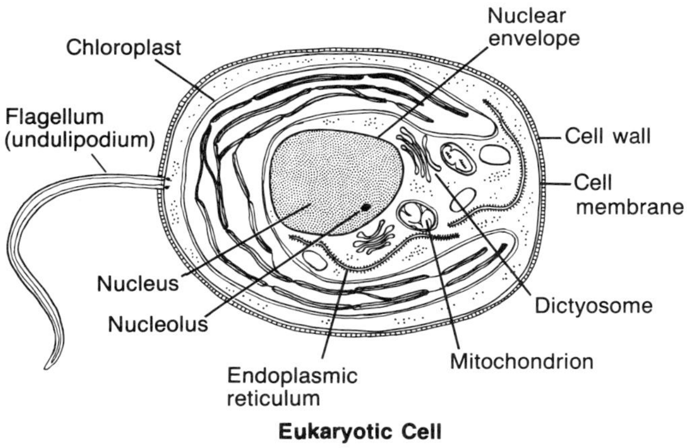

Figure 1.1 Prokaryotic and eukaryotic cells, showing their principal differences (see Table 1.1). The prokaryotic cell lacks a nucleus, chromosomes, mitochondria, and chloroplasts, and generally ranges in size between 0.5 and 15 μm. Eukaryotic cells possess a variety of organelles, one or more nuclei, 2–600 chromosomes, mitochondria, and in some, chloroplasts. They are generally much larger than prokaryotic cells.

Table 1.1 Some primary differences between prokaryotes and eukaryotes (see also Chapter 5)

Prokaryotes	Eukaryotes
Nucleus absent	Nucleus present
Meiosis absent	Meiosis
One basic genome	Chromosome number 2–600
Mitochondria absent	Mitochondria present
Chloroplasts absent	Chloroplasts may be present
Endoplasmic reticulum absent	Endoplasmic reticulum present
Vacuoles absent	Vacuoles present

billion years (By) and, hence, must be considered very successful indeed. During this time, they have also evolved into many different biochemical, physiologic, morphologic, and ecologic types.

Single-celled organisms include two major groups, both of which have an abundant fossil record. These groups are the bacteria or prokaryotes (archaebacteria and eubacteria, including the cyanobacteria) and the protists (eukaryotic unicellular algae and protozoa). Viruses have been included among the single-celled organisms by some, but they have an unusual structure, and are commonly considered by some others to be nonliving or, at least, acellular. They have no fossil record, although they have probably been extant since shortly after the first life appeared on Earth. Some multicellular algae with life histories or morphologies very similar to unicellular algae and protozoa are sometimes grouped with them in a kingdom, the Protoctista, essentially the equivalent of the kingdom Protista as currently used. Because this book considers only **unicellular** organisms, the eukaryotic unicells are referred to as protists.

The fossil record of prokaryotes and protists

Most single-celled organisms are soft bodied and do not fossilize well. Remarkable **body fossils** of prokaryotes have been found in very old Precambrian cherts, and stromatolites, their **trace fossils**, are common from the Archean to the Lower Paleozoic, and less common right up to modern times. Of the 16 major groups of prokaryotes that have been recognized, 10 contain species that mineralize and are capable of preservation in the fossil record. Protists may carry skeletal elements of a wide range of form and composition (Table 1.2). The fossil record of these protists is well known, and they show significant evolutionary development through time. Yet many protists do not have skeletons of any sort and are completely unknown as fossils. Such important and large groups as the ciliates (Fig. 1.2), with the exception of the small group of tintinnids (see Chapter 14) that have shells, are virtually absent from the fossil record. No record exists for the large number of parasitic prokaryotes and protists, although they have probably been infecting other organisms from very early times. Occasionally, an outstanding find of assemblages of abundant, delicate, unicellular fossils is made, such as the intriguing Late Proterozoic scale microfossils found in Arctic Canada that represent fairly highly organized protists.

In spite of a seemingly poor potential for preservation, single cells have, in fact, a remarkable fossil record. Prokaryotes represent the oldest known body and trace fossils, dating from about 3.5 Ga, and they have had a continuous record ever since. The first 3 Ga of the known fossil record contains few fossils other than single cells. They evolved into a wide variety of different kinds, and much of this evolution can be tracked in the fossil record.

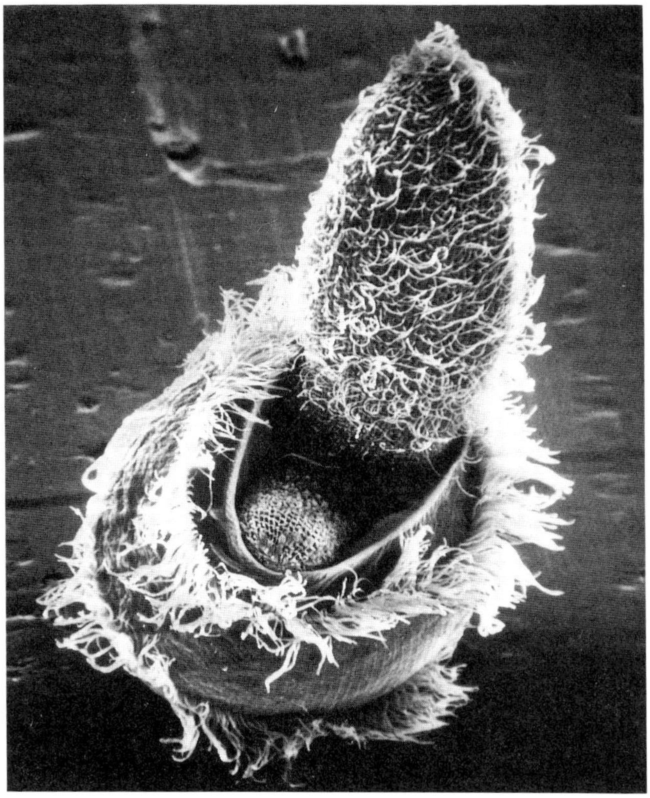

Figure 1.2 Two ciliates, one, *Didinium nasutum*, eating another, *Paramecium caudatum*, showing the complexity of protistan morphology, behavior, and interaction. The *Paramecium* is folded in the gullet of the *Didinium* and its cilia have been removed from that portion of its cell wall. (SEM provided by Eugene B. Small.)

The fossil record of many of the skeletonized protists is better, if less spectacular, than the record of most metazoans and metaphytes.

The completeness of the fossil record is difficult to determine because so many organisms in general simply do not preserve. For the higher plants and animals, estimates of the **species** that are preserved range from 3 to 13.6% of all the species that ever lived. For protists, estimates are in the same range. Among all described species 8.7% are fossils. However, among the species of protists, living and fossil, that have been described, about 50% are fossil because the economic necessity of using protist fossils in the oil industry prompted more thorough search and description. Based on these numbers and the commonly very fine fossil record of skeletonized protists, a high potential exists to understand the course of evolution of many free-living protists. Prokaryotes seem to have undergone little morphologic evolution, and if that morphologic continuity represents genetic continuity as well, then the record of prokaryotes may well be decipherable too. So the record is good, and not at all as hopeless as some have suggested.

Systematics and relationships among organisms

The systematics and classification of prokaryotes and protists have long been confused, multifaceted, and not yet resolved at all levels. For example, among the prokaryotes the cyanobacteria were once believed to be plants and were called blue–green algae. Now their relationship with the bacteria is generally acknowledged although the term blue–green algae is still commonly used. Among the protists, matters are quite complex. These organisms are not strictly comparable with more familiar animals and plants, yet historically their systematics developed in conjunction with either animal or plant systematics. **Protozoologists** traditionally studied the animal-like protists, meaning those that are generally capable of locomotion, and **phycologists** studied plant-like protists, meaning those that generally photosynthesize. "Animal-like" and "plant-like protists" are misnomers in reality, because these protists have nothing in common with either animals or plants—they are not evolutionarily close, they do not develop in the same way, and they do not function or live like animals or plants. Protists represent a great diversity of organisms that deserve to be recognized in their own right.

Protozoologists and phycologists developed their own separate classifications for the overlapping groups of protists they each claimed. Thus protozoologists follow the **International Code of Zoological Nomenclature** and have established a classification in which Protozoa is considered a subkingdom of either the Protista or the Animalia. The most modern classification of the Protozoa recognizes six phyla, each containing a number of classes, orders, families, and **genera**. Phycologists classify the photosynthesizing protists using the **International Code of Botanical Nomenclature** and recognize many of the protists as separate phyla. An order in the protozoologic scheme might well be the equivalent of a division or phylum in the botanical classification. A recent tendency is to recognize many protist groups as phyla within their own kingdom, either the Protoctista or the Protista. *Bacteriologists* have been reluctant even to describe formally higher taxa, and not until 1982 were phyla suggested that might be comparable with phyla in the other kingdoms. Although these nomenclatorial and systematic problems confuse, they are indicative of the complex and **polyphyletic** nature of these groups.

Recent work with sequencing various molecules,

Table 1.2 Fossil prokaryotes and protists, their overall characteristics, composition, habitat, and geologic range

Fossil	Description of fossils	Size	Composition	Habitat	Geologic range
Archaebacteria	Unicells	<10 μm	Organic body fossils, stromatolitic trace fossils	Shallow saline waters; marine	Archean (?) to Recent
Eubacteria	Unicells, filaments	<10 μm	Organic body fossils, stromatolitic trace fossils	All	Archean (?) to Recent
Cyanobacteria	Unicells, filaments	<50 μm	Organic body fossils, calcareous sheaths, stromatolitic trace fossils	Shallow saline waters; marine	Archean to Recent
Prasinophytes	Cysts	<800 μm	Organic	Aquatic	Proterozoic to Recent
Euglenophytes	Unicells	<200 μm	Body fossils	Aquatic	Jurassic Eocene to Recent
Dinoflagellates	Cysts	5–150 μm	Organic	Aquatic	Silurian to Recent
Acritarchs	Cysts of various eukaryotic algae	<100 μm	Organic	Marine	Proterozoic to Recent
Ebridians	Mesh of solid rods	<100 μm	Silica	Marine	Paleocene to Recent
Diatoms	Circular, elongate, or irregular valves	<200 μm	Silica	Aquatic	Jurassic to Recent
Silicoflagellates	Mesh of hollow rods	<100 μm	Silica	Marine	Cretaceous to Recent
Chrysophytes	Cysts	<100 μm	Silica	Aquatic	Proterozoic, Cambrian, Cretaceous to Recent
Calcareous nannofossils	Plates, rods, asters	<50 μm	Calcite	Marine	Triassic to Recent
Amoebas	Cysts	<150 μm	Organic	Parasitic	Quaternary
Arcellinids	Single-chambered tests	0.01–0.5 mm	Organic, agglutinated	Chiefly freshwater or terrestrial in damp places	Mississippian, Tertiary to Recent
Gromians	Single-chambered tests	<5 mm	Organic	Marine	Eocene to Recent
Foraminifera	Tests	0.01–100 mm	Organic, agglutinated, calcium carbonate	Marine	Cambrian to Recent

Table 1.2 *Continued*

Fossil	Description of fossils	Size	Composition	Habitat	Geologic range
Radiolaria	Spherical or conical mesh	0.03–1.5 mm	Silica	Marine	Cambrian to Recent
Acantharia	Spicules	<0.05 mm	Strontium sulfate	Marine	Eocene to Recent
Heliozoans	Spines, scales	<0.05 mm	Silica	Marine or freshwater	Pleistocene to Recent
Heterotrichs	Cysts	<300 μm	Organic	Parasitic, freshwater, marine	Cretaceous, Quaternary
Tintinnids	Single-chambered loricae	<300 μm	Organic, agglutinated, calcium carbonate	Marine	Ordovician to Recent
Hypotrichs	Cysts	<300 μm	Organic	Freshwater, marine	Eocene to Recent

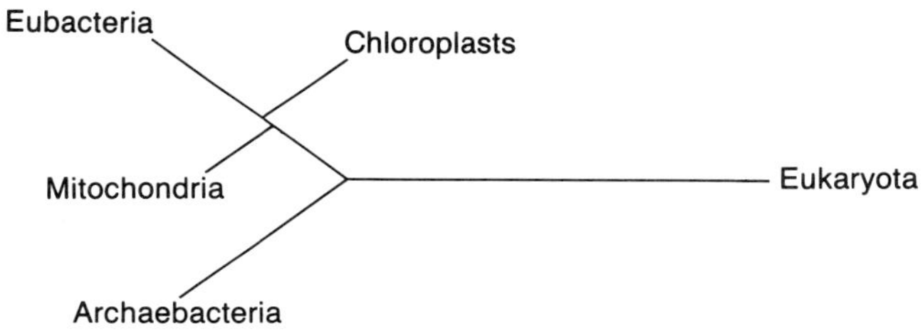

Figure 1.3 RNA phylogeny of life, showing the deep separation between the archaebacteria, eubacteria, and eukaryotes, as well as the close relationship between eukaryotic chloroplasts, mitochondria, and the eubacteria.

especially RNA, contained in prokaryotes and protists indicates deep and separate origins for several groups (Figs 1.3, 1.4). Such divisions, particularly among the protists, may well be the equivalent to the other eukaryote kingdoms Animalia, Plantae, and Fungi, now recognized. These three kingdoms are, in fact, a rather late and minor cluster at the end of a long and complex eukaryote phylogeny (Fig. 1.4).

All life on Earth is related. Seven kinds of life have been recognized:

1 viruses;
2 archaebacteria;
3 eubacteria;
4 protists;
5 animals;
6 plants;
7 fungi.

These seven groups, although not recognized uniformly in biology or paleontology, include in a logical structure all the major types of life. The last six groups are commonly recognized now as kingdoms, while viruses usually are not so recognized because they are most closely related to their own particular hosts, and so are not truly a distinct, separate group.

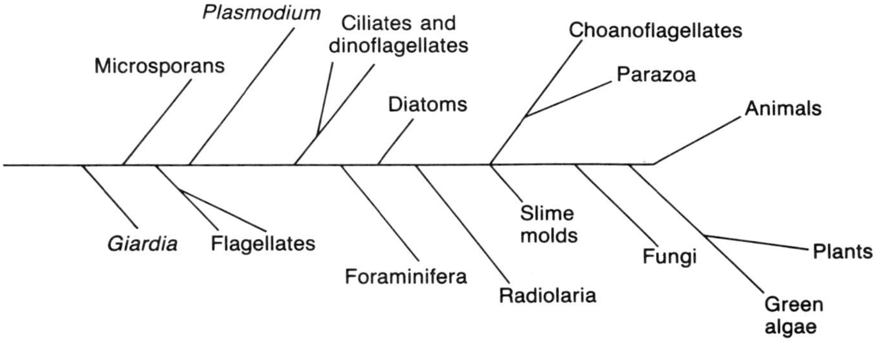

Figure 1.4 A diagrammatic phylogeny of eukaryotes, based on RNA molecular sequencing and similarity of mitochondrial cristae, showing the place of various groups of eukaryotes. The multicellular kingdoms, animals, plants, and fungi, are a much later and less significant development. Indeed, this kind of evidence indicates that many major protist groups are worthy of distinction at the same kingdom level as animals, plants, and fungi.

Viruses

Viruses are basically a molecule of RNA or DNA covered by a protein case or capsid. They range in size from 20 to 300 nm; the largest virus is just about as large as the smallest bacterium. Viruses do not have any of the typical cell characters and, hence, are considered acellular or even nonliving. In fact, some of them can be crystallized, just like a mineral, but can resume activity when conditions are again appropriate.

Viruses reproduce only within the cells of other organisms ranging from bacteria to vertebrates. Some of them may cause diseases such as polio or AIDS. Virologists recognize nearly 50 families of viruses grouped according to the kind of host they infect. Like most parasites, they have no fossil record. Viruses are believed to be nucleic acids that escaped originally from their hosts.

Prokaryotes

These organisms include the kingdoms Archaebacteria and Eubacteria (including the cyanobacteria). Because of the simple morphology and biochemical differences of bacteria, bacterial systematics has been greatly confused, and it did not develop within an evolutionary context. The most commonly used scheme for classifying bacteria implies no relationships whatsoever. Molecular sequencing in these organisms is beginning to indicate relationships, and a reliable scheme may be developed soon.

Molecular sequencing has revealed an exceptional division among the prokaryotes (Fig. 1.3). These data show that one group, the archaebacteria, is separated from another group, the eubacteria, by a difference that is as great as any known elsewhere among organisms. These two groups differ particularly in the sequences in ribosomal RNA. One of these, 16S rRNA, is an especially useful phylogenetic indicator because of its functional constancy and the degree of conservatism shown by various regions of the molecule. Well over 400 bacterial species have now been sequenced.

The archaebacteria are superficially similar to other bacteria but they are also known to differ in significant structural and chemical aspects as well. Furthermore, the split is apparently a very ancient one. This evidence indicates that archaebacteria should be considered yet another distinct kingdom equivalent in status to the other bacteria and eukaryotes in general. The molecular evidence shows that the archaebacteria are closer to the eukaryotes than they are to the eubacteria. They either branched early in their phylogeny from the eubacterial lineage or from the eukaryote lineage shortly after they separated from the eubacteria.

The eubacteria, so named because they contain the most commonly recognized or true bacteria, can now be divided into at least 10 groups on the basis of 16S rRNA sequences. They include the traditionally recognized bacterial categories. Of the free-living bacterial groups, a number of these may be represented by morphologically similar forms among the oldest known fossils some 3500 million years (My) old.

Protista

The Protista includes all the unicellular and fundamentally unicellular protozoa and algae. Multicellular algae have also been grouped together with the protists in the kingdom by some workers because they possess primitive characters, but are excluded by others because they are multicellular. They are not included in this book.

Because the Protista contains organisms that have both animal-like and plant-like characters, no generally accepted, single systematic arrangement has been developed for classifying them. Protozoologists use the most inclusive classification but it omits some important groups such as the diatoms which are important paleontologically. Phycologists who study algal protists use a classification that omits all the protozoan protists. Thus, the most commonly fossilized protist groups may be omitted by one or the other classification so that no single classification serves paleontology well. As a result, some groups of protists may be classified either as algae using the botanical code or as protozoa using the zoologic code.

Protists can be separated into six major groups, if the protists are considered a single kingdom (Table 1.3). Three of these groups contain only parasitic forms and another contains only nonskeletonized forms. Only the flagella- or pseudopodia-bearing Sarcomastigophora and the ciliated Ciliophora possess forms that readily fossilize. Among the Sarcomastigophora, many of the photosynthesizing flagellates (Phytomastigophora) and some of the sarcodines have excellent fossil records. In the Ciliophora, only one order out of 46 has abundant fossils.

Protists arose from prokaryotes probably in the late Archean or early Proterozoic, probably by the incorporation of endosymbiotic prokaryotes inside a single-cell membrane. By at least 1.6 Ga, eukaryotic algal cells were present in the fossil record. Undisputed protozoan protists did not appear as fossils until the Early Cambrian, 550 million years ago (Ma), although peculiar flask-shaped microfossils said to be the **loricae** of **heterotrophic** protists are known from rocks around 850 My old.

Table 1.3 Major groups (phyla) of protists

Group	Characters	Habitats	Fossil record
Sarcomastigophora	Bearing flagella or pseudopodia, nucleus of a single type (except foraminifera)	Marine, freshwater, terrestrial, parasitic	Phytomastigophorans: Proterozoic to Recent. Rhizopods: Cambrian to Recent
Labyrinthomorpha	Sheath-like network with nonameboid cells	Parasitic or saprobic on marine and estuarine algae	None
Apicomplexa	Spherical to elongate cells with a ring of microtubules around one end. Complex life-cycles	Parasites in animals; some causing major problems for vertebrates, including humans (malaria)	None
Microspora	Spherical to plasmodial cells with complex structures and life-cycles specialized for their habitat	Parasitic in the cells of nearly all animal groups	None
Myxozoa	Spherical to elongate cells with complex life-cycles	Parasitic in the organ cavities and tissues of cold-blooded vertebrates and annelids	None
Ciliophora	Spherical to oval cells covered with cilia and containing two types of nuclei	Marine, freshwater, terrestrial, parasitic or commensal with other organisms	One order, the tintinnids, ranges from Ordovician to Recent

These fossils, however, are best interpreted as algal in origin.

Higher kingdoms

Three higher kingdoms are generally recognized: Fungi, Plantae, and Animalia. These kingdoms, all eukaryotes, differ from prokaryotes and protists in their multicellularity and developmental sequences.

Fungi develop from spores that give rise to the structures called hyphae. A large number of hyphae form a mycelium, the **vegetative** form of most fungi. Fungi are heterotrophic, digesting their food outside the cell walls with secreted enzymes and absorbing the molecular remains directly. Their fossil record starts in the early part of the Paleozoic, or maybe even earlier, but it is discontinuous and incomplete, at least in part because few paleontologists study fungi.

Plants are sexually reproducing eukaryotes that develop from an **embryo**. They are all autotrophic, relying on photosynthesis for the manufacture of their food. Most plants are terrestrial, although the marine kelps (rhodophytes and phaeophytes) are usually included in the kingdom but may also be considered with the protists in the Protoctista. The chloroplasts of plants have a close similarity to cyanobacteria, both morphologically and in molecular sequences (16S rRNA), and are believed to have arisen from them (Fig. 1.3). Terrestrial plants first appeared in the Silurian.

Animals are heterotrophic and develop by the fusion of an egg and a sperm to form a **zygote** which then proliferates cells to form a blastula. Most animals are motile or can move their bodies or feeding organs in and out of a protective shell. The oldest animals are the Vendian, or Ediacarian, assemblages of about 580−560 My in age known from many places around the world. None of the known fossils provide a clue to the phylogenetic origins of animals, although they probably arose from a protist perhaps a billion years ago.

Origins

Origin of prokaryotes

The origin of prokaryotes occurred before the time of the earliest record of life on Earth. Although prokaryotes are regarded as the earliest life on Earth, they may have been preceded by an earlier type of primitive organism or self-replicating molecule (see Chapter 2). Prokaryotes were first found 3.5 Ga and no evidence exists for older ones, chiefly because older unaltered sedimentary rocks are unknown. Both the archaebacteria and the eubacteria, and the eukaryotic cytoplasm and nucleus, have very ancient origins as indicated by molecular data. These molecular data indicate that the archaebacteria, the eubacteria, and the eukaryote lineage split very early in Earth history. The first fossils show that prokaryotes had attained local abundance and diversity. Thus their origins and the origin of life itself must have occurred earlier and must be sought in still older rocks. Life may well have evolved as soon as the Earth's surface temperatures allowed, probably more than 4 Ga (see Chapter 2).

The earliest functioning cells and, soon thereafter, prokaryotes, must have been anaerobes because the atmosphere contained no oxygen. When oxygen first became important in life's evolution is, as yet, an unresolved question because of sparse evidence. Such an atmosphere, however, results chiefly from the liberation of free oxygen through photosynthesis. An oxygenated atmosphere was surely present at a billion or half a billion years ago, and evidence from ancient soils indicates it formed about 2 Ga. In any case, the photosynthetic prokaryotes or protists were the only source for it.

Origin of eukaryotes

After the origin of life, the development of the eukaryotic cell is perhaps the most important event in life's history. Like most of the important questions in Earth science, several hypotheses have been proposed to account for it. Two models for the origin of eukaryotic cells have been debated for a long time. One model, for which evidence is mostly lacking, suggests that these cells originated by the invagination of the cell membranes of some prokaryote to form the internal cell structures of eukaryotes. The alternative and most convincing hypothesis is that the basic eukaryotic cell split from the bacterial lineages, but that many of its structures were originally prokaryotic **endosymbionts** living within the cytoplasm. A very large number of eukaryotes today live in extremely close and coordinated association with **symbionts** of various sorts. Perhaps the strongest support for the endosymbiosis hypothesis is the RNA sequence data that show a close relationship between cyanobacteria and the chloroplasts of eukaryotes, and between other eubacteria and the mitochondria of eukaryotes. In this scenario, a preexisting eukaryote-like cell becomes associated first with a mitochondrium-like purple eubacterium and later with a cyanobacterium in endosymbiosis. The endosymbiosis is stabilized in the eukaryote lineage by the transfer of genetic material from the two symbionts to the nucleus, making a fully integrated and functioning eukaryotic cell (Fig. 1.5).

No hypotheses is likely to be negated or supported by studies of the fossil record, at least as it is known now. The postulated events leading to eukaryotic cells were unlikely to have left a fossil record of any kind. At present, the best that paleontologists have to contribute to the question is the minimal age of appearance of eukaryotes. Even this is not without controversy because simple fossil eukaryotic cells are difficult to distinguish from fossil prokaryotic cells. About 1.8−1.6 Ga, larger microfossils appear that are interpreted as eukaryotic algal **cysts**. Such cysts were abundant in some places by 1.4 Ga.

The first eukaryotes were surely protists, a conclusion supported by the sequence of fossils found in Proterozoic rocks. Multicellular organisms arose much later from protists, indicating that protists must have diversified even more than observed in the fossil acritarch algae. **Planktic** and **benthic** protists, without skeletons and of some variety, were likely to have been present as well. A chief problem in determining the fossil record of the evolution from protists to multicellular organisms is that the most probable protistan ancestors, as represented in the modern fauna, had no preservable parts. Nevertheless, protists occupy a central position in the initial development of animals, plants, and fungi.

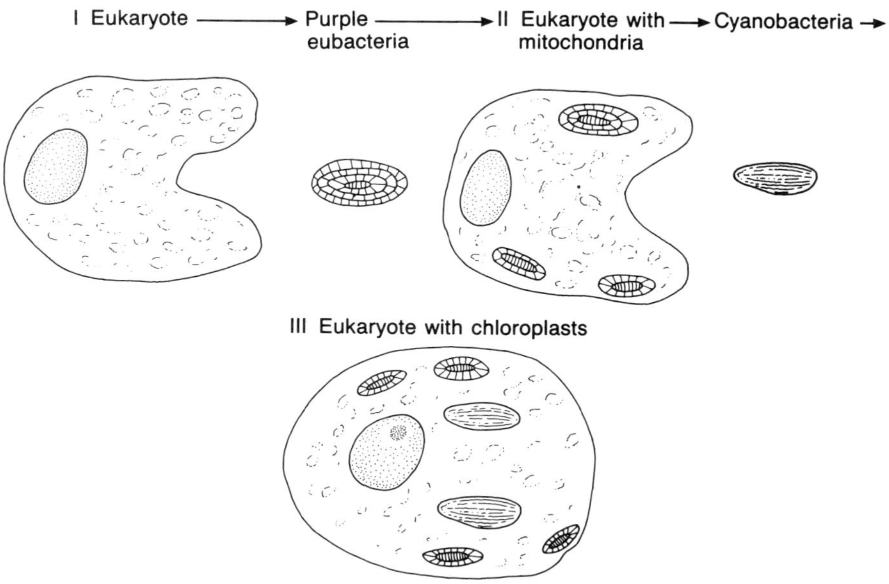

Figure 1.5 Representation of the endosymbiosis theory for the development of the eukaryotic cell. A preexisting eukaryotic lineage with a cell structure and nucleus acquires first a purple bacterium followed by a cyanobacterium as endosymbionts. These transfer genetic material to the nucleus of the cell, thus making an integrated eukaryotic cell.

Origin of animals

Multicellular heterotrophs surely evolved from the protozoan protists through the development of shared functions in colonies or by the cellularization of some multinucleate protozoan. A wide variety of such living colonies and protozoans live today, but again, fossil evidence for this important event is absent. A reasonable model is for increasing specialization of the cells in colonies or around the nuclei of the protozoans until a true animal developed. Specialization for feeding, locomotion, defense, and reproduction should have significant selective advantage and may have required some coordination of cellular function so this evolutionary sequence may well have occurred fairly rapidly in geologic time. Each intermediate type can be envisioned by reference to modern organisms.

Multicellular animals probably arose from protists several times, making the kingdom Animalia polyphyletic. The Porifera or Parazoa likely arose from colonial flagellates, perhaps the choanoflagellates, as their structure is quite similar. The Ctenophora arose before the Cnidaria, perhaps from some flagellate group as well. The exact evolutionary pathway in this case is not clear. The higher **metazoans** probably arose not from a protist but from preexisting metazoans derived from protists. A likely candidate is the Cnidaria, because of the resemblance of the planula larvae to the supposed flatworm ancestral condition.

In general, this sequence of events is supported by the fossil record. The earliest body fossils are those of the unique Vendian and Ediacarian fauna, which is reasonably interpreted as being dominated by cnidarians, although another interpretation considers them as a group that gave rise to no other animals. This assemblage, known worldwide from rocks about 580 My old, precedes most indications of more advanced metazoans. By Cambrian time, 550 Ma, a very large number of skeletonized metazoans appeared, followed by the more commonly known Cambrian brachiopod, trilobite, and archaeocyathid assemblages.

Why metazoans radiated in the earliest Cambrian is the subject of much speculation. Although this event surely postdates the divergence of metazoans from protists, it is of concern to protist paleontology. At the same time that the metazoans developed skeletons, so did a number of protists, including radiolaria, foraminifera, and a number of scale-bearing algae. The algal phytoplankton also experienced major changes at this time. For whatever reasons the metazoans developed skeletons, the protists did likewise. Most likely this radiation resulted from a complex of factors. The one factor that seems to link all of these organisms is trophic interactions, and this may have played a predominant role in structuring these early marine communities.

Origin of fungi

Slime molds and fungi are sometimes grouped together. Slime molds (mycetozoans) are related most closely with the sarcomastigophoran protists (sarcodines and flagellates) because of their ameboid or flagellated locomotion and holozoic nutrition. The group probably did not give rise to other organisms. The origin of true fungi, however, is not obvious. They certainly resemble sarcodine protists in their mode of nutrition and in some aspects of their locomotion, but they have become so specialized that close relationships are difficult to infer.

Origin of plants

Plants, especially the larger, multicellular algae which are sometimes included with the protists because of the closely similar life-cycles, are easily derived from algal protists. The various divisions within the plant kingdom are all rooted in the green flagellates directly, or through ancestors that were.

Summary

Prokaryotes and protists have a remarkable record for groups that, in general, do not fossilize well. Prokaryotes dominate the fossil record for 2 By from 3.5 to near 1.5 Ga when eukaryotic protists joined them. For about another 1 By these two groups remain the only fossils. Protists probably arose from prokaryotes evolving into a wide variety of forms, of which the flagellates gave rise, probably several times, to the multicellular kingdoms (animals, plants, and fungi). Thus, the prokaryotes and the protists play a central and important role in the initial development of higher life on Earth.

Supplementary reading

Barnes RSK, ed. *A Synoptic Classification of Living Organisms.* Oxford: Blackwell Scientific Publications, 1984. A classification of life into kingdoms and phyla.

Lee JJ, Hutner SH, Bovee EC, eds. *An Illustrated Guide to the Protozoa.* Lawrence, Kansas: Society of Protozoologists, 1985. The modern classification of most protists (except diatoms) with illustrations, by a group effort of the Society of Protozoologists.

Margulis L. *Symbiosis in Cell Evolution.* San Francisco: WH Freeman, 1981. Detailed development of the endosymbiosis theory for the origin of eukaryotes and other groups.

Margulis L, Corliss JO, Melkonian M, Chapman DJ. *Handbook of Protoctista.* Boston: Jones and Bartlett, 1990. A rearrangement of all major groups of protoctists, elevating many groups to the level of phylum, together with biologic and geologic information on each.

Margulis L, Schwartz KV. *Five Kingdoms: An Illustrated Guide to Life on Earth.* San Francisco: WH Freeman, 1982. A description with selected illustrations of the phyla of prokaryotes, protists, fungi, plants, and animals.

Parker SP, ed. *Synopsis and Classification of Living Organisms.* New York: McGraw-Hill, 1982. A multivolume classification of all life, including viruses.

Sleigh MA. *Protozoa and other Protists.* London: Routland, Chapman and Hall, 1989. A good review of modern protists.

Sogin ML. Evolution of eukaryotic microorganisms and their small subunit ribosomal RNAs. *Am Zool* 1989; 29:487–99. A summary of molecular sequencing of RNAs and the phylogeny of life that results.

Stackebrandt E, Woese CR. The phylogeny of the prokaryotes. *Microbiol Sci* 1984; 1:117–22. An explanation of the molecular evidence for the relationships of various bacteria, including the separation of archaebacteria from the eubacteria.

Tappan H. *The Paleobiology of Plant Protists.* San Francisco: WH Freeman, 1980. A detailed treatment of the morphology and systematics of living and fossil bacteria and plant protists.

Vidal G. The oldest eukaryotic cells. *Sci Am* 1984; 250:48–57. An illustrated and readable account of the first eukaryotes that appeared about 1.5 Ga.

Woese CR. Bacterial evolution. *Microbiol Rev* 1987; 51:221–71. A summary of recent views of eubacterial and archaebacterial evolution.

Prebiotic conditions and the first living cells 2

David W. Deamer

The **cell** is the unit of life. The origin of cellular life must have involved chemical and physical mechanisms that permitted systems of molecules to self-assemble from organic compounds present on the prebiotic Earth. The life process probably did not begin in a single moment when a tiny portion of inanimate matter suddenly took on the properties of the living state. Instead, life more plausibly arose from countless natural experiments in which mixtures of organic molecules were subjected to such energy sources as light, heat, and electric discharge. Various combinations of the molecules were mixed and recombined into complex interacting systems. Through this process, an imperceptible molecular evolution took place over a period of several hundred million years following the accretion of the Earth, after it had cooled sufficiently for water vapor to condense into oceans. At some point, membrane-bound systems of molecules appeared that could grow and reproduce by using energy and nutrients from the environment. An observer seeing this end-product would conclude that such systems were alive, but would be unable to pinpoint the exact time when the complex mixtures took on the property of life.

To understand this evolutionary process, information is needed about what organic molecules were available, the energy sources, and the physical environment. This chapter will outline a probable set of prebiotic conditions relevant to the origin of cellular life on the early Earth, integrating current knowledge about the primitive atmosphere, the chronology of the Earth's accretion process, and the chemistry of early oceans and land masses. Global considerations include atmospheric composition, temperature ranges, major energy sources, and composition of the hydrosphere. The specific aspects involve the site of origin, the mineral composition of the site, and plausible organic chemical species available to participate in reactions that led to self-assembly of protocellular structures.

The prebiotic environment

The origin of life must have taken place after oceans formed but prior to the appearance of stromatolites, the earliest fossil indicators of prokaryotic cellular activity. During this time, membranous boundary structures appeared, simple replicating molecules began to interact with catalytic molecules, and energy transduction mechanisms evolved to capture light and chemical energy, thereby making it available to drive polymerization reactions. Each of these events presents fundamental questions related to the origin of life, and must be understood in terms of the chemical and physical environment prevailing at that time.

The reconstruction of events and circumstances of this period is speculative at best, arising largely from deductive reasoning, rather than physical evidence. Prebiotic scenarios therefore rely on plausible arguments, very much as circumstantial evidence is argued in a court of law. For the purposes of this chapter, plausibility will be based on three considerations. The first is best described as the principle of continuity: those models which most clearly demonstrate a continuous evolutionary pathway leading from proposed origins to extant life forms will be considered to be more plausible than models requiring a discontinuity. The second principle concerns ubiquity. In general, ubiquitous conditions will be considered more plausible than special cases, unless the special case is necessary in some fundamental way. The third might be called the principle of robustness. A robust model is relatively independent of precisely defined environmental parameters. Plausible simulations should therefore use relatively common organic chemical species that can be added to the prebiotic environment over periods of several hundred million years. Furthermore, plausible models should describe not only synthetic processes, but also concentrating mechanisms and degradation rates. If the earliest forms of life centered around RNA molecules, for instance, conditions under which those molecules could be continuously produced from monomers and maintained at reasonable levels against degradative reactions must be demonstrated.

Chronology of the prebiotic environment

The earliest convincing microfossils of cellular life forms are found in ancient rocks which have been dated at 3.5 By before the present time. The Earth's primary accretion process was essentially completed between 4.6 and 4.5 By, but was followed by several hundred million years of episodic bombardment. Some of this bombardment was of sufficient magnitude to vaporize oceans and heat the Earth's surface to sterilizing temperatures. For this reason, the period from 4.0 to 3.5 By (Late Hadeon and Early Archean) is important for understanding life's origin. Sometime in this interval, the Earth's crust cooled sufficiently to permit oceans to form and remain as a permanent feature, volcanic activity produced the first land masses, and late accretion added the final fraction of the present Earth's mass. The accretion was largely in the form of meteoritic and cometary infall, which may have contributed a significant fraction of the organic carbon available on the Earth's surface.

Site of origin

Plausible sites for the origin of cellular life are environments undergoing physical cycles of temperature and dehydration. These would concentrate potential reactants, and would provide an ubiquitous source of free energy to drive polymerization reactions necessary for the assembly of catalytic and replicating molecules. Examples of such sites are **intertidal zones**, the borders of freshwater lakes, and shallow ponds occasionally filled by rain, then dried by solar heating. Sites that are at physical equilibrium seem less plausible. For example, **polymerization** reactions of dissolved, chemically activated **monomers** could presumably occur in large aqueous volumes (ponds, lakes, oceans) but plausible concentrating mechanisms and sources of chemical free energy are conceptually more difficult to imagine in an open body of water.

Other sites that have been suggested include marine thermal vents and aerosols. However, until these environments can be shown to have properties that are *required* for the origin of life, they must be considered less plausible than the ubiquitous interfaces described above.

Global and local temperatures

Any discussion of temperature must address global mean temperatures, local ambient temperatures, and extreme temperatures. Most plausibly, life's origin followed condensation of the oceans, so that the global mean temperature was below the boiling point of water. Because atmospheric pressure was probably much greater than contemporary values, this may have been considerably higher than 100°C. There was a cooling trend from 4.0 to 3.5 By, when microfossil evidence indicates that prokaryotic organisms were abundant. A reasonable estimate for global mean temperature at the time of life's origin is 40–60°C.

Daily cycles of temperature resulting from solar heating of mineral surfaces would have an amplitude similar to such cycles today, but would range around a higher mean temperature. Therefore the high end of the cycle would be near 100°C. Because of thermal buffering of warm oceans and higher mean global temperatures, freezing temperatures would be an uncommon occurrence, with the possible exceptions of polar regions and volcanic mountain peaks.

Composition of the atmosphere

The first scientists to consider the primitive atmosphere reasoned that the solar system is mostly hydrogen, and that the early atmosphere would therefore be strongly reducing, rather than strongly oxidizing, as it is today. That is, the most common biogenic elements—carbon, oxygen, and nitrogen—would be in their most highly reduced states: CH_4 (methane), H_2O (water), and NH_3 (ammonia).

More recent evidence does not support a strongly reducing atmosphere. Instead, the most abundant component of the early atmosphere was probably carbon dioxide. The argument in favor of a carbon dioxide atmosphere considers that the carbonate now present at the Earth's surface as calcium carbonate sediments must once have been present as carbon dioxide gas. If all of the carbonate could somehow be transformed into carbon dioxide, it would be equivalent to about 60 atmospheres, a significant fraction of which would have composed the Earth's early atmosphere. This idea is given additional support by comparison of the Earth and Venus. The two planets are similar in mass, and probably underwent similar processes of early planetary evolution. However, Venus has retained its original carbon dioxide atmosphere, equivalent to about 100 Earth atmospheres. The reason is that Venus is closer to the sun, and is relatively hot because of a solar greenhouse effect. Therefore water on Venus remained in the vapor phase, rather than condensing as oceans, and carbon dioxide could not be removed from the atmosphere as marine sediments.

Nitrogen on the early Earth would be approximately at the partial pressure of today's atmosphere, and oxygen would have been a rare, short-lived species produced by high energy reactions and quickly consumed by various reduced components of the atmosphere and hydrosphere. Trace amounts of hydrogen, methane, hydrogen cyanide, formaldehyde, and hydrogen sulfide would be continuously added by volcanic outgassing, photochemical reactions in the atmosphere, and cometary infall, so that a modest level of reducing power would be available.

Composition of the hydrosphere

A marine site for the origin of cellular life within the **hydrosphere** seems plausible for two reasons. First, oceans contain most of the Earth's water, and are a preferred site until the chemical and physical properties of a freshwater site can be shown to be advantageous. Second, the earliest fossil stromatolites apparently developed in marine environments, rather than freshwater.

The early ocean was probably more dilute than contemporary seas, but the mineral composition was similar, except that ferrous iron and dissolved carbon dioxide and bicarbonate would be relatively high. The presence of ferrous iron is significant, because it has been shown that ultraviolet (UV) light can drive a photochemical reaction in which ferrous iron reduces carbon dioxide to formaldehyde, a highly reactive molecule that can undergo condensation reactions to form more complex species such as carbohydrates. Ferrous iron also provides a sensitive indicator of oxidation potential of the surrounding environment. For instance, the first appearance of significant atmospheric oxygen is dated by **banded iron formations**, which were produced when soluble ferrous iron was precipitated as insoluble iron oxides.

Mineral phases

The most abundant mineral surface would be igneous rocks and weathered sediments resembling those of contemporary volcanic islands. Thus, lava and volcanic sands are reasonable models. Even today lava surfaces can approach 100°C by solar heating. Any films of organic material on such a surface would be subjected to temperatures sufficient to form covalent chemical bonds through condensation reactions. Volcanic sands would reach similar temperatures at the surface, but would provide protection from such extreme conditions deeper in the interstices between grains. Such environments would also provide protection against UV light, which has sufficient energy to break down chemical bonds. Sand therefore represents a microenvironment in which free energy is available to drive chemical bond formation at the surface, thereby trapping light and heat energy, while a protected environment occurs a few centimeters deeper in which the chemical energy could be used to drive secondary reactions.

Organic components

Organic carbon compounds could have at least three possible sources on the prebiotic Earth: relic carbon remaining from the primary accretion phase, terrestrial synthesis reactions, and meteoritic and cometary infall during late accretion phases. Relic carbon compounds are unlikely to have survived the **pyrolytic** conditions of the primary accretion phase, and the high carbon dioxide content of the primitive atmosphere would not have provided the reduced gas mixture required for high yield, ubiquitous syntheses under Miller–Urey or Fischer–Tropsch conditions. However, volcanic activity releases small amounts of reduced gases such as carbon monoxide and hydrogen sulfide, and elemental iron or ferrous ion in solution are also sources of reducing power. Therefore, some terrestrial synthesis of organic compounds could still occur, as originally suggested by Stanley Miller in 1952.

Another source of reduced organic material is meteoritic and cometary infall during late accretion. Comets contain substantial amounts of organic carbon as hydrogen cyanide and formaldehyde, which under certain conditions undergo secondary reactions that produce amino acids, carbohydrates, and purine bases. We do not yet have access to the organic components of comets, but samples of authentic organic mixtures dating from the origin of the solar system are preserved in **carbonaceous chondrites**. Observed falls of such meteorites occur approximately every 10 years, and several hundred more were collected in Antarctic ice fields along with approximately 10 000 other stony and iron meteorites.

These remarkable meteorites contain hundreds of biologically relevant organic compounds, including **hydrocarbons** and **amino acids** which were probably synthesized under reducing conditions in the early solar system by Miller–Urey and Fischer–Tropsch reactions. Could such infall have provided significant amounts of organic carbon to the prebiotic Earth's surface?

To answer this question, recent theoretical models

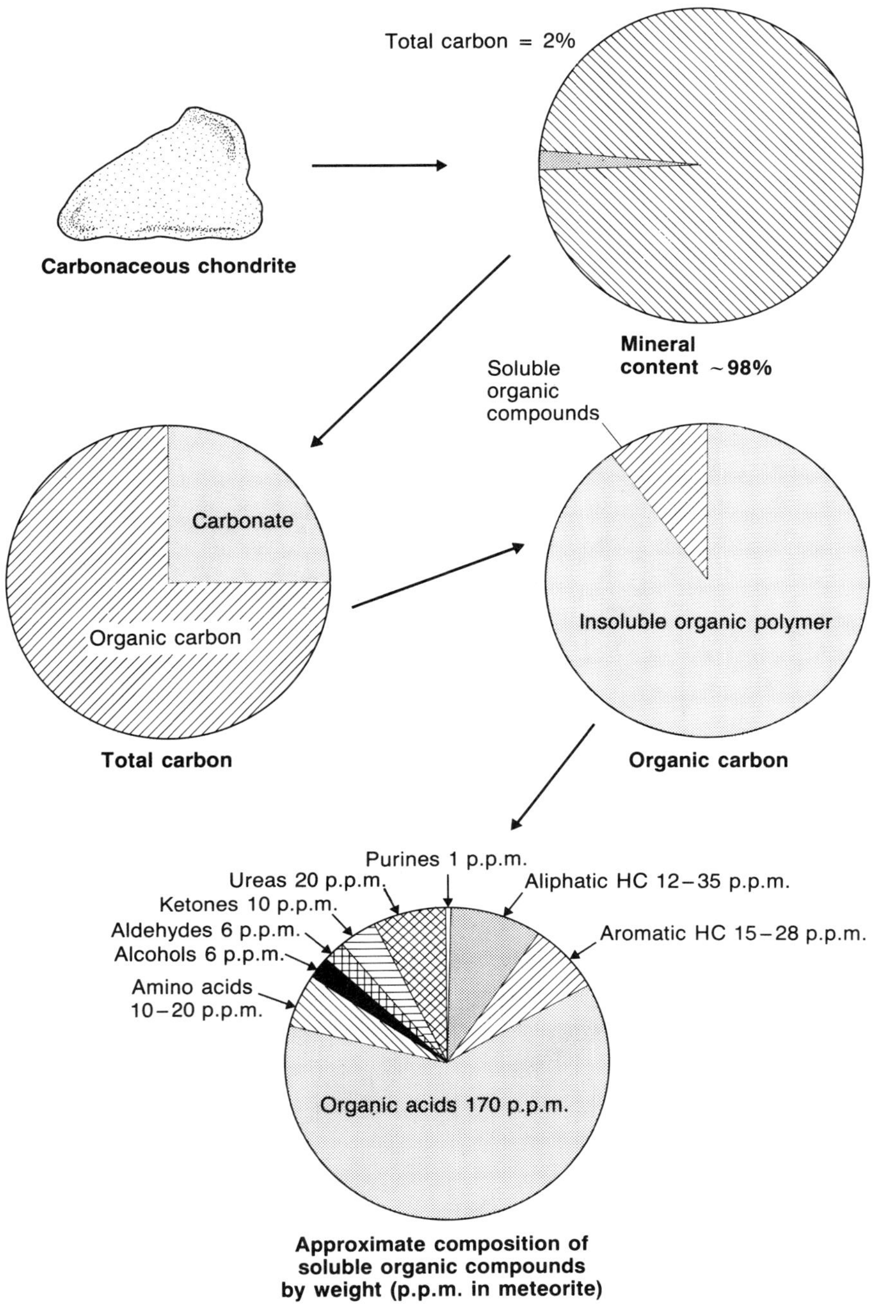

Figure 2.1 A shower of carbonaceous meteorites fell near Murchison, Australia, in September, 1969. Approximately 100 kg of meteoritic material was collected, and because of the relatively large amounts made available, the Murchison meteorite has been thoroughly investigated. The pie graphs show summaries of data compiled from several analyses. About 1.5% of the mass of a typical specimen is organic carbon, mostly in the form of an insoluble, kerogen-like polymer. The soluble organic species include monocarboxylic acids, amino acids, and hydrocarbons. (Data from Wood JA, Chang S. *NASA Publication 476*. Washington, DC: US Government Printing Office, 1985.)

of cometary infall during later stages of accretion of the Earth's mass take into account the increased "aerobraking" of an atmosphere which would be an order of magnitude denser than the contemporary value due to its high carbon dioxide content. Under these conditions, comets in the 100 m size range would provide significant amounts of organic components, sufficient to add 10^{10} g/year, or approximately 10^{18} g over a late accretion period of several hundred million years. Much of the organic material would be in the form of HCN and HCHO, together with **aliphatic** and **aromatic** hydrocarbons derived from **kerogen**-like substances that are presumed to compose nearly 20% of a comet's mass. According to these calculations, larger comets (>1 km) would not be sufficiently slowed to permit survival of organic material, and smaller comets (10 m or less) would provide negligible amounts of organic material, even though they might survive atmospheric entry.

Another approach is to estimate the relative amounts of organic carbon delivered to the Earth's surface by a spectrum of very small micrometeorites (10^{-12} g) to large (10^{18} g) extraterrestrial objects. Surprisingly, most of the intact organic material would be added by microscopic particles in the size range $10^{-6}-10^{-9}$ g, which lose relatively little of their mass during atmospheric entry. The total amount accumulated over 10^8 years has been estimated to be approximately 1.1×10^{20} g.

To give some perspective to this figure, the organic carbon added would be about 0.1% of total crustal carbon, equivalent to a layer of organic material about several centimeters deep spread over the surface of the Earth. It is two orders of magnitude greater than the current organic carbon inventory in the biosphere, estimated to be 6×10^{17} g. Cometary and meteoritic infall therefore seems likely to have been a significant source of organic carbon in the prebiotic environment.

What compounds might be present in a mixture of organic substances derived from such material? The only samples available are the carbonaceous chondrites, in which the most abundant organic substance is a complex aromatic hydrocarbon polymer, followed by monocarboxylic acids, then aliphatic and aromatic hydrocarbons (Fig. 2.1). Amino acids and purines are present in smaller quantities. The major fraction of cometary and meteoritic infall would enter the oceans and presumably release its organic content over a period of time. Water-soluble compounds like amino acids would dissolve, while relatively low-density hydrocarbons and their derivatives would accumulate at the ocean surface, forming a kind of prebiotic oil slick. The action of wind and tide would concentrate the films at intertidal zones, just as it does today.

Because all organic compounds are degraded by several kinds of reactions, particularly photochemical damage caused by UV light, the concentrations that might build up in the oceans are uncertain. The relative amounts of material added by infall and terrestrial synthesis are unknown, as is the rate at which the compounds were degraded. On a global scale, organic concentrations were probably very dilute. Even under the most favorable conditions amino acids would reach little more than millimolar abundance. Therefore local conditions that would concentrate organic compounds were probably essential to the origin of primitive life. There are several possibilities, including hydration–dehydration cycles in tide pools, interfacial concentrating effects on hydrocarbons and **surfactant** species, and **adsorption** to charged surfaces such as those of certain clay minerals.

Energy sources

Light was by far the major energy source available on the early Earth. Therefore the most plausible energy source in prebiotic simulations would be photochemical reactions, rather than such sources as electric discharge, shock, or volcanic heat. At the intertidal site described earlier, a second useful energy source was moderate solar heating and hydration–dehydration cycles, particularly for driving polymerization reactions. Drying cycles would have concentrated reactants, and heating to nearly 100°C would drive simple, low-energy dehydration–polymerization reactions involving amino acids and **nucleotides** (Fig. 2.2).

The first cells

Four physical processes must have been involved in the origin of life: accumulation of simple organic compounds, chemical evolution of such compounds into larger, more complex molecules, self-assembly of the molecules into protocellular structures, and the appearance of a system of molecules which could capture energy and nutrients from the environment to reproduce itself in some manner (Fig. 2.3).

Several examples of molecular self-assembly are shown in Figure 2.4, which illustrates how hydrogen bonding

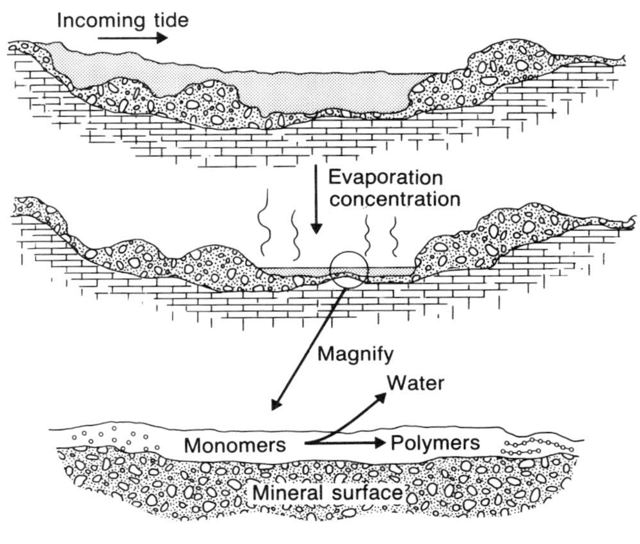

Figure 2.2 Cycling environments. Cycles of heating–cooling and wetting–drying would provide ubiquitous sources of free energy in the prebiotic environment. The energy can be used to concentrate potential reactants from dilute solutions, and to drive relatively low-energy condensation reactions, such as the formation of **ester bonds**. A tide-pool cycle is shown in this figure.

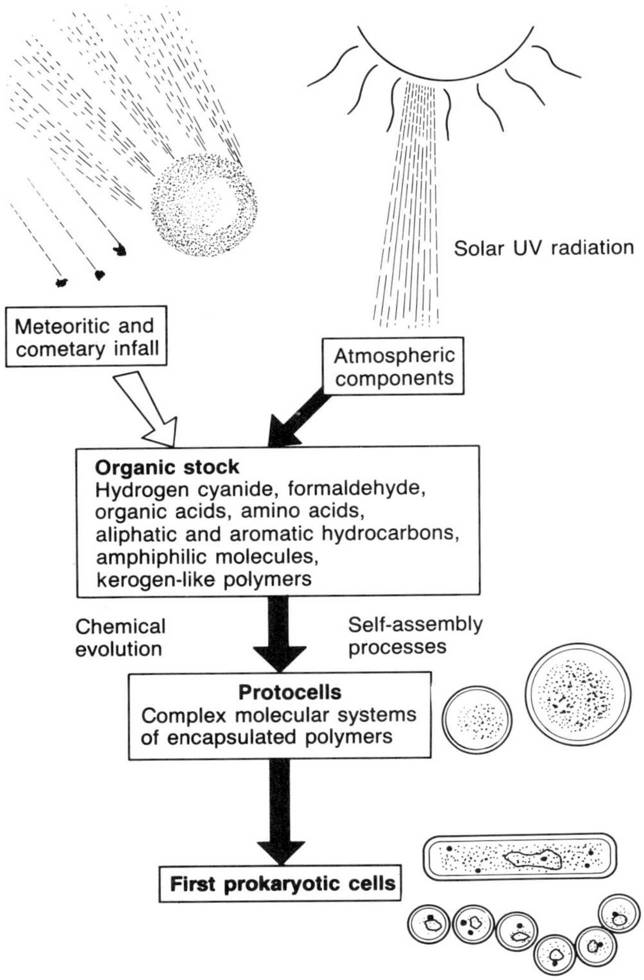

Figure 2.3 Summary of processes proposed to underlie the origin of cellular life. Theoretical and experimental models of the processes leading to the accumulation of organic compounds and self-assembly of protocells are reasonably convincing. The pathway from protocells to the first prokaryotic cells is essentially unknown.

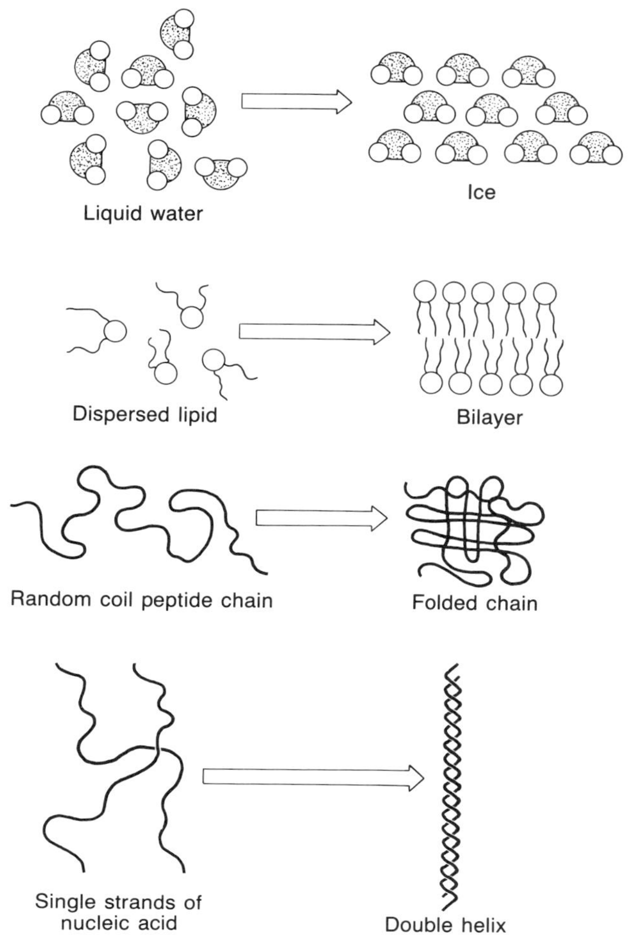

Figure 2.4 Examples of self-assembly processes. In self-assembly, relatively weak physical forces cause certain molecules to group into larger structures. The stabilizing forces are typically hydrogen bonds (ice, secondary, and tertiary protein structure, and the double helix of nucleic acids) or nonpolar entropy-driven effects in the case of lipids.

and nonpolar forces can stabilize orderly arrangements of small and large molecules. A central event in the origin of life was the self-assembly of a molecular system in which catalytic polymers could interact with a second class of polymers having the capacity to store information in a sequence of monomers. That sequence in turn would in some manner determine the sequence of monomers in the catalyst, so that the resulting catalytic–information cycle was able to undergo directed growth. In contemporary cells, the cycle is represented by protein **enzymes** and **nucleic acids**. However, in the protocell, both catalytic and information-containing sites could be present in the same molecule, as suggested by recent studies of RNA ribozymes.

Another important self-assembly process involved **amphiphiles**, lipid-like compounds that formed closed microenvironments. Such boundary structures have the capacity to make energy available in the form of ion gradients, and can provide a selective inward transport of nutrients. Furthermore, membrane vesicles in principle are capable of growth and division, and several possible mechanisms in which amphiphilic molecules could be synthesized from "nutrient" precursors have been proposed.

If a **macromolecular** replicating system could be encapsulated within a vesicular membrane, the components of the system would share the same microenvironment, and the result would be a step toward true cellular function. The encapsulation process is not difficult to imagine, since macromolecules have been shown to be

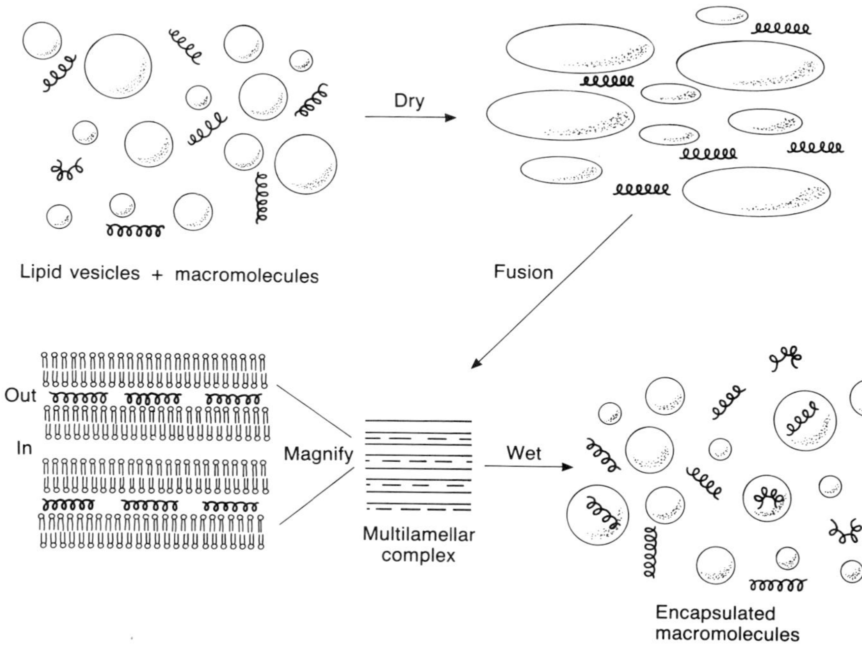

Figure 2.5 Encapsulation processes. Amphiphilic molecules can self-assemble into bimolecular layers that form membrane-bound vesicles. Such vesicles fuse into multilayers upon drying, and if other molecules are present they are trapped between the layers. Vesicles reform during rehydration, thereby encapsulating some of the solute molecules which are otherwise excluded by the membrane barrier. Tide pools undergo similar drying–wetting cycles, and offer a plausible site for the formation of encapsulated replicating macromolecules, one of the primary steps in the origin and evolution of cellular life forms.

readily encapsulated by amphiphilic molecules under simulated prebiotic conditions (Fig. 2.5). An energy-yielding process is also necessary to provide for growth of the encapsulating membrane concurrently with growth of the catalytic–information system.

This system represents a minimal set of basic cellular properties: energy and nutrient capture from the local environment, growth through a catalyzed polymerization mechanism, and replication of an information-storage molecule, all encapsulated within a membranous boundary structure. Would such a system be considered to be a living cell? Probably not. Although it is clearly a protocellular stage in the evolutionary process leading to the first life forms, three primary functions are missing: a method for coding between a nucleotide sequence and a **peptide** sequence (origin of the genetic code), a method for catalyzing the synthesis of peptides from activated amino acids (origin of translation), and a mechanism to introduce random changes in the genome so that populations can respond to their environment and change over time (evolution). Some mechanism for selecting **enantiomeric** (D and L) monomers must also have been available.

None of these questions has been answered even at the most elementary level, and yet the earliest prokaryotes, 3.5 Ga, had advanced to this point when they began to lay down the mineral mass of the stromatolites which are now found in ancient fossil formations. Clearly, the most exciting questions regarding the origin of life still remain as fundamental problems in biology.

Supplementary reading

Anders E. Prebiotic organic matter from comets and asteroids. *Nature* 1989; 342:255–7. A recent estimate of how much organic matter could be delivered to the early Earth by meteoritic infall.

Cairns-Smith G. *Genetic Takeover*. Cambridge: Cambridge University Press, 1982. A stimulating discussion of clay chemistry and its possible relationship to the origin of life.

Chyba CF, Thomas PJ, Brookshaw L, Sagan C. Cometary delivery of organic molecules to the early Earth. *Science* 1990; 249:366–73. Discussion of the amount of organic material surviving impact of comets with the Earth during late accretion.

Deamer DW. Boundary structures are formed by amphiphilic components of the Murchison carbonaceous chondrite. *Nature* 1985; 317:792–4. This paper established that amphiphilic molecules were present in a carbonaceous meteorite, and demonstrated their self-assembly into membrane structures.

Hayatsu R, Anders E. Organic components in meteorites and their origins. *Top Curr Chem* 1981; 99:1–37. This paper reviews the organic compounds found in carbonaceous meteorites.

Holland HD. *The Chemical Evolution of the Atmosphere and Oceans*. Princeton, NJ: Princeton University Press, 1984. An important discussion of the early atmosphere, making a convincing argument for a high carbon dioxide content.

Kasting, J. Earth's earliest atmosphere. *Sci Am* 1988; 258:30–8. Summarizes evidence for a high carbon dioxide content of the early atmosphere.

Kerridge JF, Matthews MF, eds. *Meteorites and the Early Solar System*. Tucson, AZ: University of Arizona Press, 1988. An up-to-date book on meteorites and what they can tell us about the evolution of the solar system.

Levine JS. *The Photochemistry of Atmospheres: Earth, the Outer Planets, and Comets*. New York: Academic Press, 1985. Discussion of how photochemical reactions control the composition of planetary atmospheres.

Miller SL, Orgel LE. *The Origins of Life on the Earth*. Englewood Cliffs, NJ: Prentice Hall, 1974.

Miller SL, Urey HC, Oro J. Origin of organic compounds on the primitive Earth and in meteorites. *J Mol Evol* 1976; 9:59–72. Useful summaries of ideas about the origin of life, written by four investigators who made fundamental contributions to our knowledge.

Sleep N, Zahnle KH, Kasting JF, Morowitz HJ. Annihilation of ecosystems by large asteroid impacts on the early Earth. *Nature* 1989; 342:139–42. This paper attempts to establish a plausible point in time for the origin of life.

Usher DA, McHale AH. Nonenzymatic joining of oligoadenylates on a polyuridylic acid template. *Science* 1976; 192:53–4. Demonstrates how temperatures in the range of solar heating can drive polymerization reactions.

Wood JA, Chang S. The cosmic history of the biogenic elements and compounds. *NASA Publication 476*. Washington DC: US Government Printing Office, 1985. A brief overview of the "big picture," relating cosmic processes to the origin of life on the Earth.

Evolutionary history of prokaryotes and protists 3

Andrew H. Knoll, Jere H. Lipps

Earliest history: the Archean

The geologic record of organic evolution begins about 3.5 Ga (3500 Ma) with geochemical evidence and actual fossils found in weakly metamorphosed sedimentary and volcanic rocks of the greenstone belts of the Early Archean Swaziland Supergroup, South Africa, and Warrawoona Group, Western Australia. Both deposits consist of thick sequences of mafic and ultramafic lavas, subordinate felsic volcanic rocks, and intercalated sedimentary rocks, mostly tuffs, cherts, and carbonates altered to massive chert by early hydrothermal **diagenesis**. Both rock groups are overlain by greywackes and shales, with minor cherts, carbonates, and rocks known as banded iron formations. Many units within these ancient greenstone belts contain significant amounts of organic matter that exhibit carbon **isotopic** ratios indicative of an autotrophically driven carbon cycle. Stromatolites occur in both the South African and Western Australian rocks, further indicating the presence of complex benthic communities containing **phototactic** microorganisms that built the stromatolites. In all probability, these populations were also phototrophic. Microfossils have also been found in both groups of rocks, although some of the reported microstructures are doubtfully biologic. Other assemblages of these objects, however, display size–frequency distributions, divisional cycles, and spatial distributions within the rocks that closely approximate those of younger, unambiguous microfossils. Filaments and cell clusters discovered in cherts of the Warrawoona Group particularly resemble other younger fossils. They clearly indicate that life existed early in Earth history, but **taphonomic** problems of the Archean record make evaluation of early morphologic or physiologic diversity difficult. The Early Archean geologic record yields other evidence of biologic importance: 3500 Ma the Earth's atmosphere and oceans contained only small amounts of oxygen, probably too little to support aerobic metabolism.

In combination with data from molecular phylogeny and comparative physiology, this incomplete geologic record supports the general conclusion that by 3500 Ma, Earth supported complex anaerobic ecosystems energetically sustained by photosynthesis. The fundamental early diversification of anaerobic eubacteria, archaebacteria, and simple eukaryotes indicated by molecular phylogeny was probably already complete by the Early Archean. Quite possibly, the photosynthetic biota included cyanobacteria capable of oxygenic photosynthesis, although most oxygen produced by this process must have been utilized in the oxidation of organic matter, volcanic gases, and minerals.

How much earlier than 3500 Ma life began is unknown. Strongly metamorphosed sedimentary rocks in the 3800 My old Isua supracrustal sequence in Greenland contain graphite and carbonate rocks whose carbon isotopic ratios can be interpreted as the result of a biologic carbon cycle altered by metamorphism. Earlier evidence of life is unlikely to be found on Earth unless older sedimentary rocks are found.

Radiating protists: the Proterozoic

The Proterozoic sedimentary record is more widely and, in many places, better preserved than that of the Archean, and so the fossil record of this eon is correspondingly better and more amenable to confident interpretation. Prokaryotic microfossils are commonly beautifully preserved in cherty iron formations, silicified carbonates, and fine-grained mudstones. The most striking feature of these remains is their close similarity to living prokaryotes found in physically comparable depositional environments. Paleontologic and geochemical evidence indicate that stable, oxygen-containing environments existed by early Proterozoic times, permitting the evolution of **aerobic** metabolisms. With that radiation, the prokaryotic flora assumed its modern character. Although the Phanerozoic evolution of plants and animals has provided continually changing biologic environments to foster continued bacterial evolution, the major features of the archaebacteria and eubacteria appear to have been established at least 2 Ga.

The major evolutionary event of the Early Proterozoic was the rise of eukaryotic cells containing **mitochondria** and, in some cases, **plastids**. Although evolutionary scenarios remain speculative, the spread of oxic environments may have favored the establishment of protomitochondrial endosymbionts within prokaryotic hosts. The fossil record indicates that eukaryotes were significant components of ecosystems at least 1500–1800 Ma. Unornamented phytoplankton (acritarchs) quite similar to younger populations of green algal cysts but unlike the extracellular envelopes produced by colonial chroococcoid cyanobacteria occur abundantly in rocks of this age in China, North America, the Soviet Union, and Australia. At least in Australia, this morphologic record is complemented by abundant **steranes**—biomarker molecules produced predominantly by nucleated cells.

The Late Proterozoic, especially between about 850 and 600 Ma, was a time of major tectonic, climatic, biogeochemical, and biologic change in Earth history. The best known evolutionary development was the Ediacaran or Vendian radiation of large animals, although protists underwent major changes as well, which is important for interpreting the Ediacaran faunas. Middle Proterozoic acritarchs are simple, but by about 1000 Ma, morphologic complexity and diversity began to increase. Late Proterozoic protistan cysts include large, process-bearing acritarchs, double-walled and sculptured **vesicles**, and vase-shaped microfossils. Algal scales, now silicified, occur abundantly in Upper Proterozoic chert nodules in northwest Canada. This occurrence, containing different kinds, indicates that algal protists were quite diverse at that time but many were simply not preserved. These fossils form the basis of a usable Late Proterozoic biostratigraphy, and they indicate that the tempo of evolution in marine environments had increased long before large animals diversified. Macroscopic algae also existed in Middle and Upper Proterozoic rocks, further indicating that large eukaryotic organisms did not originate with macroscopic animals. Large animals radiated into a taxonomically diverse, ecologically complex, and evolutionarily dynamic biologic world. By the time large animals appeared, protists had already undergone at least one protracted episode of diversification and major turnover. Sometime near the end of the Proterozoic most of the characteristic Proterozoic acritarch species became extinct.

Paleozoic radiations and extinctions

Early Paleozoic radiations

The Cambrian and Ordovician were times of sequential radiations of marine, skeletonized, invertebrate animals. At the same time both planktic and benthic protists also diversified in a similar manner. The beginning of the Cambrian is marked by a distinctive radiation of organic-walled phytoplankton. Several genera of morphologically distinct prasinophyte algae appeared at this time, as did a variety of process-bearing acritarchs thought to be the preserved cysts of dinoflagellates, golden-brown algae, or, perhaps, even extinct groups of algae. A greater diversification of acritarchs followed in the Ordovician, with the expansion of baltisphaerid, veryhachid, and other acritarch groups.

Benthic algae, too, radiated in the Cambrian and Ordovician. Benthic calcareous **photoautotrophs** first appear in Upper Proterozoic and expand in Lower Cambrian sedimentary rocks. Many Early Cambrian calcareous algae may in fact be cyanobacteria, although these prokaryotes probably did not precipitate skeletons under active physiologic control in the ways that eukaryotes do. However, rare and commonly fragmentary remains of probable red and green calcareous algae have been found in uppermost Proterozoic and Lower Cambrian rocks. Calcareous red, green, and problematic algae diversified during the Ordovician, becoming sedimentologically and ecologically important.

Heterotrophic protists developed skeletons in the Cambrian as well and became considerably more numerous in the Ordovician. Probable foraminifera and questionable radiolaria have been described from Lower Cambrian rocks, although some workers believe these fossils to be worm tubes or sponges, respectively. By the Middle Cambrian, foraminifera and radiolaria were certainly present in benthic and planktic ecosystems, respectively, but with very few species. In the Ordovician, both groups diversified considerably, and they continued to do so into the Silurian. Early foraminifera had **agglutinated** tests and the radiolaria had siliceous skeletons. Like the metazoans that appeared suddenly in the Cambrian, these protists have no identifiable ancestors preserved in older rocks. As a result, their true age of origin is unknown.

This record of protists complements the history of Early Paleozoic evolution provided by marine invertebrates. It shows that Cambrian and Ordovician animal radiations were part of overall increases in the complexity of ecosystems that affected **primary producers** as well as **consumers**, plankton as well as benthos. Protistan fossils also increase the taxonomic and ecologic complexity of the early evolution of **biomineralization**. Skeletal groups of protists appearing in the Late Proterozoic or Cambrian include radiolaria (SiO_2), benthic foraminifera (agglutinated), green algae (**aragonite**), and red algae (calcite and possibly aragonite), as well as photoautotrophs, possibly cyanobacteria, that may have been calcitized passively. Evolutionary interpretations must accommodate the taxonomic, mineralogic, **trophic**, and environmental heterogeneity encompassed by these early skeletal protists. Such constraints call into question simplistic geochemical, ecologic, or taxon-specific functional hypotheses for the Late Proterozoic–Cambrian evolution of biomineralization and general biotic diversification. No single explanation may be applicable to all organisms.

Late Paleozoic evolution of protists

The later Paleozoic record of protists also resembles that of marine invertebrates. Foraminifera and radiolaria continued to diversify, and they are represented by numerous species, genera, and higher taxa. Agglutinated foraminifera retained a simple tube, either straight or coiled, and in the Devonian developed into the first truly chambered foraminifera. Although agglutinated forms continued to dominate the microbenthos, calcareous foraminifera of the suborder Fusulinina appeared in the Early Silurian. Later—during the Early Mississippian—one superfamily, the Fusulinacea, began a remarkable radiation. An initial complement of small, planispiral forms rapidly evolved into a morphologically varied and ecologically important group of species. They developed large, fusiform **tests** through elongation along the axis of coiling, a complex wall structure, and an enormous increase in **proloculus** diameter. Fusulinacean genera decreased by over 60% in the later Pennsylvanian, increased slightly in the Early Permian, followed by a general decline with a sharp reduction in the Late Permian, and finally, extinction at the end of the period. Some fusulinaceans reached lengths of many centimeters and had very complex tests. Most of them lived in shallow, tropical water, and their large size, wall complexity, and septal fluting indicate that they possessed symbionts in their protoplasm. These symbionts, like those of modern larger foraminifera, were most likely to have been algae that enabled the fusulinaceans to support a large cell volume and deposit large amounts of $CaCO_3$. The rapid diversification and unique features of fusulinaceans make them one of the most useful groups for later Paleozoic biostratigraphy. Two other major groups of foraminifera appeared in the Carboniferous—the suborders Lagenina and Miliolina, both of which became common and diverse in the Mesozoic and Cenozoic.

Among Paleozoic plankton, radiolaria and acritarchs were particularly abundant, although other types such as tintinnids and chrysophytes are found sporadically, indicating that they too may have had a more extensive but unpreserved record. Radiolaria, like the foraminifera, became morphologically more complex through time, with the number of species increasing in the Ordovician and Silurian. During the rest of the Paleozoic, radiolaria were abundant, and were probably a major part of the microzooplankton. In some places their skeletons are the principal component of siliceous rocks. The presence of radiolaria and tintinnids, as well as the phytoplanktic acritarchs, implies that a complex pelagic ecosystem was fully developed. Like the modern oceans, the Paleozoic seas were probably inhabited by abundant soft-bodied larger zooplankton that also would not preserve as fossils.

Paleozoic extinctions

Marine invertebrates underwent major extinctions at the end of the Ordovician, during Late Devonian, and, most dramatically, at the close of the Permian. The protistan record of extinctions is not correspondingly well understood, but, in general, marine protists appear to have undergone major extinctions at about the same times as animals. For example, acritarchs underwent an episode of extinction at the end of the Ordovician, following which they rediversified rapidly at the beginning of the Silurian.

Major extinctions affected both the phytoplankton and benthic calcareous algae near the end of the Devonian, although radiolaria and most foraminifera appear to have been little affected. More detailed stratigraphic work is required before the timing of these extinctions can be compared with confidence to those of invertebrates. Benthic algae rediversified during the Carboniferous as sea level rose to inundate cratonic interiors, but diversity of acritarchs never recovered, remaining low to the present.

The end of the Permian is marked by the greatest extinction known in the fossil record. Perhaps as many as 90% of all marine species disappeared at this time, including many marine protists. Foraminifera, for example, declined from over 100 genera in the Late Permian to fewer than 40 in the Early Triassic. Fusulinids disappeared completely. In the last stages of the Permian, their generic diversity dropped from 50 to nine and finally to zero at the beginning of the Triassic; higher latitude and Andean faunas disappeared well before Tethyan species. Benthic calcareous algae also underwent major extinctions; generic diversity of dasycladacean green algae dropped from 29 in the Late Permian to six (new genera) in the Middle Triassic. No dasycladal algae are known from Lower Triassic rocks. Dasycladacean extinctions have no obvious latitudinal component because all Permian genera lived within 30° of the paleoequator. Radiolaria and acritarchs both underwent extinctions at the end of the Permian, but little detail is available for these groups.

General considerations

The overall similarity of the Paleozoic benthic and planktic protist records to that of the invertebrates indicates that explanations offered for evolutionary events in marine animals may be applicable to a trophically and environmentally wide range of organisms. Unfortunately, few of the mechanisms that have been proposed to account for changes of diversity (sea-level rise and fall, impacts of asteroids or comets, nutrient fluctuations, climatic changes) can be eliminated by this general coincidence of pattern. Events that affected protists will have to be documented in far greater stratigraphic and biogeographic detail before the similarities and differences among fossil records will be useful in constraining theories of evolutionary change in Paleozoic oceans. This presents a significant paleobiologic opportunity for micropaleontologists and paleontologists working on Paleozoic fossils. Important questions regarding the nature of protistan changes during critical evolutionary periods remain to be answered. For most groups, however, the first need is for much additional primary taxonomic, stratigraphic, paleoecologic, and paleogeographic data. Without this, no more than the most obvious evolutionary patterns can be identified and evaluated with confidence.

The Mesozoic oceans: evolution of the modern protistan biota

During the Mesozoic planktic protists underwent remarkable evolutionary innovation. Dinoflagellates, calcareous nannoplankton, diatoms, silicoflagellates, planktic foraminifera, and radiolaria either appeared and radiated for the first time or diversified from preexisting forms into a greater variety of new types. Dinoflagellates may occur in Upper Proterozoic and Paleozoic acritarch assemblages, but unambiguously recognizable dinoflagellate cysts first become diverse and abundant components of plankton assemblages in Upper Triassic rocks. Diversity expanded and retracted several times during the Mesozoic with **species richness** reaching a maximum in the Late Cretaceous (Santonian) times. Calcareous nannoplankton, which likewise have a scarce and questionable Paleozoic record, also diversified in the Late Triassic, became an abundant component of planktic fossils during the Early Jurassic, and fluctuated, but generally increased in species diversity through the later Mesozoic—culminating in a Late Cretaceous (Campanian to Early Maastrichtian) maximum. The siliceous diatoms appeared during the Jurassic and diversified throughout the Cretaceous, and even the radiolaria, the one microplankton group with a long, unambiguous Paleozoic record, produced a number of new families during the Late Mesozoic.

During the Mesozoic, modern ocean basins began to form, with the creation of the Atlantic Ocean basin and beginning of the geographic separation of the Indian and Pacific Oceans. At the same time, subduction around the Pacific Ocean basin destroyed Paleozoic and Triassic sedimentary rocks on the sea floor. Mesozoic and Cenozoic deep-sea sedimentary rocks are widespread, relatively complete, and little altered, and they have been sampled extensively by deep-sea drilling. Undoubtedly, the *appearance* of evolutionary innovation among the phytoplankton is enhanced by these circumstances. After all, delicate coccolith plates or diatom frustrules might well have been destroyed by the diagenetic alteration of many if not most Paleozoic rocks; furthermore, many of the best preserved Paleozoic sedimentary rocks were deposited in shallow, epicontinental seaways where open-ocean plankton might not be present or abundant. These probabilities notwithstanding, the morphologic and diversity trends evident in well-preserved phytoplankton and zooplankton assemblages indicate that a true radiation took place during the Mesozoic.

No simple idea explains why microplankton radiated in the Mesozoic. Although most fossils of organisms that lived in the pelagic environment are protistan, some groups of animals, especially the ammonites and even pelagic reptiles, show stratigraphically similar patterns of radiation and extinction in the Mesozoic oceans. Radiating protists include taxa characterized by both siliceous and calcareous skeletons, as well as unmineralized groups. Thus, a change in ocean chemistry is unlikely to have triggered the evolutionary events of the Mesozoic. Possibly the break up of continents and the partitioning of the world's oceans, including strengthened circulation and upwelling of nutrients, contributed to the ecologic opportunities for diversification of planktic organisms. The causes of the Mesozoic plankton revolution remain uncertain; probably they are as complex as the factors that promoted the radiation of skeletal metazoans and protists at the beginning of the Paleozoic.

Perhaps the best documented example of protistan

radiation into the pelagic realm is that of the foraminifera. All known Paleozoic foraminifera were benthic, and today approximately 4000 (99%) foraminiferal species live on the sea floor or attached to other organisms, while just over 40 (1%) species are planktic. Indeed, the number of planktic foraminiferal species present at any time in geologic history has never been very large. Nonetheless, the planktic foraminifera are present in extraordinary abundance, they are important elements in pelagic ecosystems, they constitute a major sink for calcium carbonate, and their skeletons make up a major proportion of the carbonate oozes that have accumulated on the deep sea floor since the middle of the Mesozoic. Planktic foraminifera first appeared in the Jurassic as tiny, spherical forms with simple trochospiral tests composed of few chambers. Their benthic ancestor remains unknown. These foraminifera continued into the Early Cretaceous, when they gave rise to many species with more complex tests characterized by keels, elongate chambers, apertural plates, and similar structures. The two general morphologic groupings—simple trochospiral and more complex forms—alternate in importance throughout the Mesozoic and Cenozoic history of the group.

Benthic foraminifera are also well preserved in Mesozoic sediments. Following the Permo-Triassic extinctions, benthic foraminifera began a steady increase in diversity throughout the Mesozoic. Starting with about 40 genera in the Early Triassic they rose in diversity to about 300 at the close of the Cretaceous, at which time the foraminiferal faunas of the world began to take on a distinctly modern aspect. Foraminifera appeared in nearly all benthic marine environments during this time, ranging from marginal marine marshes and lagoons to reefs and to the deep sea. They are among the ecologically most important organisms of the Mesozoic oceans.

As was true for the Paleozoic, intervals of diversification and extinction of the plankton generally coincide with those of **nektic** and benthic skeletonized invertebrates, necessitating causal explanations that are broadly applicable to a wide range of ecologically, trophically, and developmentally disparate organisms. Proposed explanations commonly involve changes in **primary productivity**, **paleoceanographic** circulation, sea level, oxygen profiles in the water column, or impacts of asteroids or comets. With the exception of the terminal Cretaceous extinctions (discussed below), little evidence favors extraterrestrial impact. All the other possibilities can be supported by some evidence, and they may not be mutually exclusive. The fluctuations of diversity in planktic protists have been suggested many times to be a function of variations in primary productivity and heterogeneity of **water masses**. During intervals of strong heterogeneity of water masses and high productivity, the oceans can support many specialized species that live effectively, at least for a part of their life-cycles, within a narrow range of environments. When productivity drops and oceanic water masses become less structured and more homogeneous, narrowly adapted species disappear.

Most planktic protists show this pattern, but it is shown best by planktic foraminifera, which radiated at least three times from simple, trochospiral ancestors to produce diverse faunas dominated by morphologically complex species that include flattened, spheroidal, or irregularly shaped forms with keels, elongate chambers, additional apertures, spines, or other derived features. Living planktic foraminifera with complex adult morphologies reproduce at different depths in the water column. Particular morphologies appear to allow foraminifera to take advantage of specific characteristics of water masses as distinct ranges of temperature, salinity, density, and nutrient availability. Isotopic geochemistry of fossil tests indicates that ancient foraminifera with complex morphologies were similarly specialized for different portions of vertically structured oceans.

At three separate times—during the Middle Cretaceous, at the Cretaceous–Tertiary boundary, and near the end of the Eocene—high-diversity planktic foraminiferal faunas rapidly disappeared, with simple trochospiral species surviving to dominate postextinction biotas. Today, morphologically simple species dominate high-latitude oceans characterized by relatively structureless water columns. Geochemical evidence also indicates that ancient oceans dominated by simple, trochospiral forms were similarly relatively homogeneous although much warmer than today's high-latitude seas. In this line of reasoning, the oceans' physical structure provides opportunities for plankton speciation and the biologic interactions of constituent populations. Unfortunately, such biologic interactions are not understood, even for living plankton, and so explanations concentrate on types of oceanographic changes that are known to correlate with stratigraphic changes in diversity.

The Cretaceous–Tertiary boundary

At the end of the Cretaceous mass extinctions occurred in the oceanic plankton and nekton, reef biotas, and some shallow-water benthic biotas, as well as among terrestrial plants and animals. The coincidence of these extinctions with high concentrations of iridium and with shocked quartz indicates to many geologists that they were caused by an impact of an extraterrestrial object like an asteroid or comet with the Earth. Nevertheless, others favor alternative explanations, for example, widespread major volcanic eruptions. The presence of shocked quartz and so much iridium, however, is difficult to explain by any mechanism other than an extraterrestrial impact. In any case, the fossil evidence only indicates that a major extinction took place suddenly, and that ecologic conditions were different before and after the event.

The impact hypothesis must be divided into several independent questions. The first, of course, is whether such an object collided with Earth at the proper time. The next question is whether an impact or any other catastrophic event can be related causally or even correlated stratigraphically with major extinctions. In the popular press (and among a surprising number of paleontologists), this question has been phrased, "Did an extraterrestrial impact kill the dinosaurs?" Unfortunately, the principal conclusion about this issue that can be reached on the basis of dinosaur bones is that the fossil record of large vertebrates is not detailed enough to settle questions whose resolution requires extraordinary stratigraphic control.

The required control is best achieved in marine sedimentary successions containing protistan microfossils. These sections provide a record of continuous sedimentation across the boundary, large population sizes of fossils, and millimeter-scale stratigraphic resolution. Some microplankton groups show rapid and extreme extinctions at the Cretaceous–Tertiary boundary. The diverse, ecologically and biogeographically stable flora of Late Cretaceous calcareous nannoplankton collapsed in a geologic instant at the Cretaceous–Tertiary boundary, although limited populations of some Cretaceous species persisted for up to several tens of thousands of years into the Paleocene. Immediately following this mass extinction, plankton such as the calcareous dinoflagellate *Thoracosphaera* proliferated throughout ocean surface waters. Over a period of about 100 000 years, diversity began to increase rapidly.

Planktic foraminifera had a comparably rapid decline from about 40 to two species. Morphologically complex forms were eliminated, leaving two simple trochospiral species that became widely distributed in the world's oceans. These two species slowly gave rise to other species, but a diverse fauna did not develop for about 3 My. The subsequent evolution of planktic foraminifera in the Tertiary involved completely new families, with the exception of those of the simple survivors.

The record of other microplankton groups at the boundary is less well known than those of the planktic foraminifera and calcareous nannoplankton. Radiolaria apparently underwent major extinctions, with a large percentage of species reportedly perishing. Certainly, higher taxa show marked change from the Cretaceous to the Tertiary. Likewise, silicoflagellates must have undergone systematic reorganization, because the Cretaceous flora contains a number of genera and species that do not appear in Tertiary rocks. Diatoms, however, may have lost only about 23% of their generic diversity, and dinoflagellates lost few species at the boundary. These two groups include many neritic forms that may have been better able to cope with the environmental changes at the end of the Cretaceous. Both dinoflagellates and many diatoms can produce cysts that rest on the sea floor and facilitate survival during periods of ecologic change.

Among benthic foraminifera, those inhabiting shallow-water carbonate environments, particularly reefs, also became extinct at the Cretaceous–Tertiary boundary. These foraminifera were chiefly large species likely to have harbored symbiotic algae much like the larger invertebrates that constructed the reef frameworks. Deep-sea foraminifera were little affected by the Cretaceous–Tertiary boundary event.

Protists, thus, indicate that the shallow-water ecosystems of the world's oceans were disrupted by a large, geologically rapid perturbation at the Cretaceous–Tertiary boundary. Organisms tolerant of variable conditions, such as neritic or cyst-forming species, and deeper water benthic species survived this event well. Although not as complete as data from marine protists, evidence from animals and, especially, land plants is also consistent with the idea of rapid disruption of ecosystems causing widespread but not biologically random extinctions. These paleontologic data are, of course, consistent with the impact hypothesis, but they are also consistent with any of several other postulated mechanisms—as long as their effects are rapid and severe. By themselves, therefore, fossils cannot provide unequivocal evidence on the agency of disruption. That must come from geologic and geochemical evidence, and debate about the Cretaceous–Tertiary boundary continues unabated in these fields. Whether the boundary event was extraterrestrial or terrestrial in origin, a number of extinction mechanisms are possible, including sharp, short-lived temperature change, an oceanographic and climatic restructuring, extremely acid rain, a collapse of primary productivity, or some combination of factors. Whatever the cause, clearly a geologically and ecologically rapid event of catastrophic proportions profoundly altered the composition of oceanic microplankton as well as other ecosystems at the close of the Mesozoic. For many groups, such as terrestrial vertebrates and pelagic animals, the event changed the course of evolutionary history. For protists, groups decimated by the extinction event re-radiated to produce a biota ecologically comparable to the one that was eliminated although composed of new species in new families.

Cenozoic protistan evolution

Paleogene radiations

Protists have been abundant and widespread in the oceans throughout the Cenozoic. Major features of the Cenozoic protistan record can be related to the significant climatic and plate tectonic changes that characterize the era, especially as these changes affected patterns of productivity and circulation of water masses in the oceans. Particularly important to protistan evolution were changes in the structure and properties of oceanic water masses in the Paleocene and at the end of the Eocene, and the general cooling of Earth's polar regions from the Miocene onwards.

As noted earlier, deep-water benthic groups were little affected by the terminal Cretaceous extinction, and they continued to diversify during the Paleogene. Most other groups of benthic foraminifera continued a radiation that began in the Mesozoic and continued throughout the Cenozoic. The most dramatic evolution in the Paleocene occurred among the planktic protists and larger foraminifera inhabiting shallow water, with all groups of plankton regaining the high diversities that existed during the Cretaceous. The Paleocene, then, was characterized by the reestablishment of a diverse planktic and shallow-water tropical biota. These biotas underwent continued evolutionary change resulting in increased diversity until the end of the Eocene.

The Eocene–Oligocene transition

The Eocene–Oligocene boundary has long been recognized as a period of significant biologic, climatologic, and oceanographic change. More recently, it has been suggested as another time of impact-induced mass extinction. Changes of microplankton across the Eocene–Oligocene boundary are well illustrated by the planktic foraminifera. The Eocene was an epoch of pronounced speciation, resulting in diverse, morphologically complex Middle to Late Eocene faunas characterized by marked latitudinal **provinciality**, as well as strong vertical segregation of species. The presence of keels, accessory apertures, secondary calcification, clavate chambers, and similar characters correlates well with isotopic indicators of relative depth, temperature, and surface productivity, implying that these features represent adaptations to strong depth stratification and geographically varying fertility. This contrasts strongly with the Oligocene, a time characterized by low foraminiferal diversity, low morphologic complexity, reduced provinciality, increased homogeneity of surface water masses, and low rates of speciation.

The transition from Eocene to Oligocene faunas was not abrupt but rather took place through several million years of Late Eocene and Early Oligocene time through a series of discrete events that appear to have been climatically and oceanographically induced. Isotopic geochemical data obtained from individual species suggest that evolutionary changes in the foraminiferal fauna were a consequence of oceanographic changes that eliminated a series of cool- to warm-water masses having specific oxygen content and nutrient characteristics — and with them species specialized for such conditions. Prior to their extinction, some taxa initially underwent a contraction of latitudinal range and decreased in importance. Thus times of last appearance of populations of a single species can differ between middle and low latitudes by as much as 500 000 years.

Several of the foraminiferal extinction events and fluctuations of abundance correlate stratigraphically with a series of extraterrestrial impacts recorded in the sedimentary record as microtektite horizons. At each of the two horizons where extinctions and microtektites coincide, three to five foraminiferal species (<15% of standing diversity) disappear; at another microtektite horizon, no foraminifera disappear entirely, but the population structure of assemblages changes markedly, as once important species decrease in abundance while others expand. On the other hand, while some Eocene–Oligocene extinctions and population fluctuations correlate with microtektite horizons, others do not. Also, most of the Eocene extinctions of planktic foraminifera were followed quickly by the first appearances of new species.

Other plankton groups show generally similar patterns of diversity. Calcareous nannoplankton, dinoflagellates, silicoflagellates, and ebridians all have diversities that were high in the Eocene and low in the Oligocene; however, the details of turnover are not identical among taxa. Timings of last appearances, population fluctuations, degree of latitudinal diachroneity in these events, and degree of correlation to microtektite distributions or isotopic events vary within and among major groups. At least in part, these differential responses appear to reflect the biologic and biogeographic differences among major groups of microplankton. As with the Cretaceous–

Tertiary boundary extinction, large air-breathing marine vertebrates and invertebrate nekton also underwent major diversity declines followed by the reradiation of new groups. At the Cretaceous–Tertiary boundary these groups included mosasaurs, plesiosaurs, and ammonites, and at the Eocene–Oligocene transition they included archaeocete whales and nautiloids.

The diversity decline of Late Eocene to Early Oligocene plankton appears related to major climatic, paleoceanographic, and sea-level changes. Indeed, many of the biologic changes that occurred near the Eocene–Oligocene boundary appear to be the culmination of trends heralded by an interval of major plankton turnover at the Middle to Late Eocene boundary. The Eocene–Oligocene changes cannot be characterized as a mass extinction like the terminal Cretaceous event because they do not involve rapid extinction, although the cumulative effect of the several clustered extinction and origination events was almost complete turnover of species. Individual extinction events generally affected only 6–12% of the taxa in any one group, and disappearance of Late Eocene species was often accompanied by the appearance of new species that maintained the levels of diversity. Eocene–Oligocene microplankton, in fact, do not show high rates of extinction. Late Eocene rates of extinction are moderate, and low to moderate rates of extinction continue into the Oligocene. The Early Oligocene is unusual because rates of species origination were low. Reduced rates of **cladogenesis** probably resulted from decreased habitat diversity in the relatively homogeneous Oligocene oceans.

At both the Cretaceous–Tertiary and Eocene–Oligocene boundary events, a diverse microplanktic biota of morphologically complex and ecologically specialized species gave way to a biota of few, broadly adapted species. However, the two episodes have important differences. The Cretaceous–Tertiary boundary is characterized by numerous simultaneous or nearly simultaneous extinctions, high short-term rates of extinction, brief postextinction expansion of surviving species of plankton, and, following that, high rates of cladogenesis. In contrast, the Eocene–Oligocene transition featured moderate rates of extinction that were not simultaneous among appreciable numbers of species, no great expansion of surviving plankton, and low rates of speciation. This is consistent with the fossil records of other groups, especially land plants and vertebrates, which suggests that the dominant controls on evolutionary turnover at these two intervals were different.

Neogene cooling

During the Neogene, Earth's polar regions began to cool markedly, altering oceanographic and atmospheric circulation patterns. Protists and other groups changed in response. Diatoms, in particular, radiated significantly in the cold, nutrient-rich surface waters of high latitudes. Perhaps surprisingly, despite environmental changes that culminated in the great Pleistocene ice sheets, this interval is not marked by major pulses of extinction, although background turnover is essentially continuous through the interval. Major biotic patterns of the Neogene principally involved shifting biogeographic distributions.

In general, both benthic and planktic marine protists became increasingly provincial. Warm-water biotas were increasingly restricted to equatorial latitudes, while new, cool-water biotas developed in the far north and south. Indeed, the Miocene and younger biotas of the Antarctic regions are unique, apparently having evolved in place as a result of the new conditions there. Cool water itself may not have been the most significant factor in the evolution of this biota; the extreme seasonality of primary productivity in high latitudes combined with their oceanographic isolation appears to have allowed the evolution of a completely new biota adapted in many ways to these singular environments. Deep-sea biotas also changed during this period of time as the world's sea floors became bathed in cold water that originated in high latitudes, chiefly near Antarctica.

Although climate and oceanography began to change in response to the general cooling of the Earth's high latitudes in the Miocene, major, rapid changes took place in the Pleistocene as glaciations waxed and waned. For the most part, marine biotas simply migrated in response to changing thermal and oceanographic conditions. Few species disappeared and relatively few have their first appearance during this epoch. Because Pleistocene populations appear to have had environmental tolerances little different from their living descendants, plankton distribution can be useful in the reconstruction of climatic and oceanic circulation during the ice ages. Much of our present knowledge of Pleistocene climate is derived directly or indirectly from such micropaleontologic data. Microfossils and isotopic analyses of their skeletons indicate that the Pleistocene climate has fluctuated between intervals that were warmer or cooler than today's climate. Importantly, this climatic record provides a basis for prediction of the Earth's climate in the next few hundred to thousands of years (Fig. 3.1).

Conclusions

Prokaryote and protist fossils furnish a unique and important evolutionary record that may provide answers to some fundamental questions about rates, processes, and environmental determinants of evolution, as well as paleoenvironmental and paleoceanographic conditions. As with the evolution of plants and animals, microorganisms have not had a single, invariant pace of change; rather they have exhibited a variety of different evolutionary rates related to processes operating at different times and at different time-scales.

Populations of living bacteria evolve rapidly when novel environmental conditions are imposed on them, yet the fossil record suggests that at least major features of the archaebacteria and eubacteria have existed since early in the Earth's history. The three major periods of rapid prokaryotic diversification—the Early Archean, Late Archean to Early Proterozoic, and the Phanerozoic—correspond respectively to the initial exploitation of an anoxic Earth, the spread of stable oxic environments, and the evolution of plants and animals which presented novel environments to bacteria.

Phyletic evolution and cladogenesis have both been documented in protists, and in either case morphologic change can be rapid or slow. Taxonomic rates of evolution are also variable and may be at least a partial function of time-scales of development and decay of water masses. Diachroneity can sometimes be resolved in species' first and last appearances. When extinctions are diachronous, they may indicate either the tempo of paleoceanographic change or the migration rates of competitively successful new species within and between oceans. Diachroneity in first appearances may have resulted from differential migration or, when change is both diachronous and phyletic throughout the range of a species, the rate at which genes move through widely distributed pelagic populations. Although population changes can be resolved on a time-scale of 10^4 to 10^6 years, major bursts of extinction and origination occur only a few times per era, underscoring the importance of extrinsic as well as intrinsic controls on evolutionary rates.

Much of the discussion in this chapter has focused on physical determinants of protistan evolution—climatic and paleoceanographic change, the direct effects of rising and failing sea level, nutrient availability, and impacts of extraterrestrial bodies. Clearly such phenomena have exerted a strong influence on historical patterns of diversity change and morphologic evolution. The **iterative evolution** of planktic foraminifera as specific water masses came and went strongly suggests convergent adaptive evolution in response to changing physical conditions.

However, what about biologic factors; the organism–organism interactions stressed by ecologists working in modern ecosystems? Predation is potentially important in generating patterns of change in the fossil record of protists. Because we know so little about the ecology of living protists, the paleontologic significance of this process has been little explored. Increasing predation pressure has, however, been suggested as the cause of evolutionary patterns in the morphology, diversity, and relative importance of Cenozoic phytoplankton from the Arctic Ocean.

The reported occurrence of character displacement in neo**sympatric sister species** of radiolaria implicates **interspecific competition** in the generation of evolutionary pattern. On a more diffuse level, a Cenozoic trend of decreasing test weight in radiolaria parallels the pronounced radiation of marine diatoms, and the two patterns may be linked through the mutual dependence of radiolaria and diatoms on silica. If the metabolic cost of precipitating a skeleton is at all proportional to ambient SiO_2 concentrations, then the exceptional effectiveness of diatoms in depleting oceanic silica should have exerted a continuing selective pressure in favor of radiolaria able to form functional tests using less SiO_2. One prediction of this hypothesis is that radiolaria living in surface waters should show far more pronounced test diminution than those inhabiting deeper waters, and this appears to be so. Doubtless, other competitive or predatory interactions among species have helped shape protistan evolution, but we still know too little about the nature and scope of such interactions in the modern oceans to frame useful questions about the patterns they might have generated in the fossil record.

One other type of biologic interaction whose effects may be observable in the fossil record is **mutualism**. Some scientists believe that the rise to ecologic prominence of large benthic foraminifera several times during the Phanerozoic has resulted from repeated evolutionary changes in development and the incorporation of symbiotic algae as adaptations to persistent low nutrient levels.

Much has been learned, but much remains to be learned about the paleontologic record of prokaryotic and protistan evolution. In the Cretaceous and Tertiary record, there is an important opportunity to investigate in detail the relationship between physical environmental

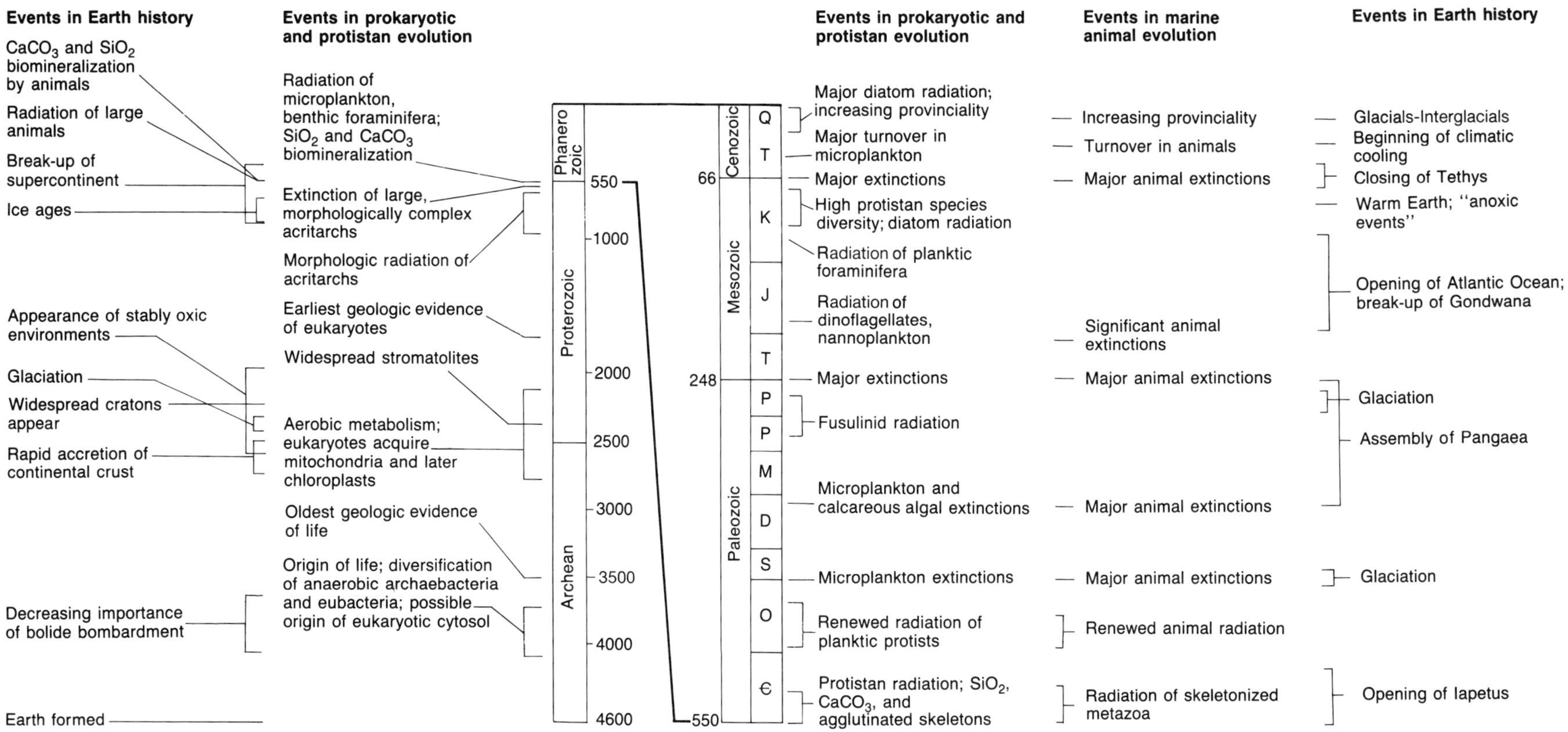

Figure 3.1 Major events in Earth history and in prokaryotic and protistan evolution. Archean and Proterozoic history and evolution (first 4.05 By (4050 My)) is shown on the left and the Phanerozoic (last 550 My) is shown on the right. The time-scale is in millions of years and the letters indicate the periods of the geologic time-scale.

change and patterns of both cladogenesis and **phyletic evolution**. The investigation of multispecies assemblages throughout their geographic ranges will be particularly important in this regard. Paleozoic microfossils also present numerous opportunities for paleobiologically important research. How much can we learn about Paleozoic oceans and the ecologic and biogeographic relationships of the protists that inhabited them? Can isotopic and other physical signals be related to Paleozoic evolutionary patterns as they have been for younger events? How much can biomarker molecular geochemistry and isotopic analyses add to the woefully incomplete picture that conventional paleontology provides of the protistan and bacterial components of ancient ecosystems? Along a different line, how much can the geologic record add to our knowledge of large-scale protistan and prokaryotic phylogeny, at present reconstructed almost exclusively on the basis of comparative ultrastructure, physiology, and molecular phylogeny?

Much of our ability to discriminate among various hypotheses for the explanation of evolutionary change in geologic time will depend on careful and detailed comparisons of evolutionary pattern among trophically, ecologically, reproductively, and developmentally different groups of organisms. Therefore, in addition to its effective use in stratigraphy and paleoenvironmental reconstruction, micropaleontology clearly has much to offer evolutionary paleontology.

Supplementary reading

Bolli HM, Saunders JB, Perch-Nielsen K, eds. *Plankton Stratigraphy*. Cambridge: Cambridge University Press, 1985. Taxonomic and stratigraphic compendium for Mesozoic and Cenozoic planktic protists.

Imbrie J, Berger A, eds. *Milankovitch and Climatic Change*. Dordrecht: Riedel, 1984. A comprehensive treatment of Late Neogene protists and climatic change.

Knoll AH. Patterns of evolution in the Archean and Proterozoic eons. *Paleobiology* 1985; 11:53–64. A comparison of prokaryotic and protistan evolution, as recorded in the Precambrian fossil record.

Lazarus D. Speciation in pelagic Protista and its study in the fossil record: a review. *Paleobiology* 1983; 9:327–40. A thoughtful essay on evolutionary patterns and process in the pelagic realm.

Lipps JH. Extinction dynamics in pelagic ecosystems. In: Elliott DK, ed. *Dynamics of Extinction*. New York: Wiley, 1986: 87–104. An account of mass extinctions in the pelagic realm.

Pomerol CH, Premoli-Silva I, eds. *Terminal Eocene Events*. Amsterdam: Elsevier Science Publishers, 1986. Detailed reviews of Eocene-Oligocene protistan micropaleontology, paleoceanography, and climatic change.

Silver LT, Schultz PH, eds. *Geological Implications of Impacts of Asteroids and Large Comets on the Earth*. Geological Society of America Special Paper 190, 1982. Proceedings of a conference prompted by the asteroid impact hypothesis of extinctions. Useful summaries of extinction patterns in fossil protists and other organisms.

Tappan H. *The Paleobiology of Plant Protists*. San Francisco: WH Freeman, 1980. A superb account of the morphology, biology, and evolution of photosynthetic protists, rich in detail.

Tappan H, Loeblich AR Jr. Evolution of the oceanic plankton. *Earth Sci Rev* 1973; 9:207–40. An early but still useful attempt to gain a quantitative view of the protistan fossil record.

Micropaleontology

4

David R. Lindberg, Jere H. Lipps, Joseph E. Hazel

INTRODUCTION

Jere H. Lipps

For many years, the study of tiny fossils, mostly various protists but also prokaryotes, small invertebrates, pollen, and even some vertebrate bones, has been called micropaleontology. This word was used because the fossils are so small that they must be studied with a microscope. This utilitarian definition is no longer appropriate because all fossils are studied with microscopes, and today even microfossils are studied with a great variety of different kinds of microscopes. The word micropaleontology is still used, however, to designate studies whose principal aim is to decipher geologic problems. Thus, micropaleontology is not defined so much on the kinds of organisms but on the kinds of problems. Most of the problems that micropaleontology deals with are identical to problems solved with larger fossils but, because of their small size and enormous numbers in sedimentary rocks, microfossils are commonly more useful.

A number of fossil groups serve geology well by aiding in the solution of biostratigraphic, paleoenvironmental, including paleoclimatologic, and paleoceanographic problems. Protists have been especially important in these studies because they are abundant, diverse, commonly widespread, and occur in nearly all sedimentary rocks. Oil companies, for example, have used protists extensively in biostratigraphic (age dating and correlation) and paleoenvironmental analyses. In marine geology, fossil protists have provided the basis for much of paleoceanographic, paleoclimatic, and historical geologic interpretations as well as biostratigraphic and paleoenvironmental analysis. In these respects, protists are probably the most utilitarian fossils, and far exceed the usefulness of any other group. In their own right, they provide enormous potential for documenting and understanding evolutionary processes. Fundamental to these studies is proper taxonomy and systematics. This chapter briefly surveys this and some of the applications of prokaryotes and protists in micropaleontology. More detailed information is available in the chapters dealing with particular groups and in the supplementary reading lists.

SYSTEMATIC PALEONTOLOGY

David R. Lindberg

Organismal systematics, according to the *Oxford English Dictionary*, pertains to, follows, or arranges species according to a system of classification. This definition, and the scope of the terms associated with it, however, remains a point of contention with those who practice it. Systematics is often considered to be the study of organismal diversity, although it can also be viewed as a subset of comparative biology. Some systematists restrict the concept of systematics by constraining the scope of diversity under study to that united by genealogy. Like the term systematics, taxonomy has also been interpreted differently by workers. Most have considered taxonomy to be the theory and practice of classification. Others have considered taxonomy and classification to be separate entities that are united under the term systematics which also includes the study of the evolution of the organisms. Often systematics and taxonomy are treated as synonyms, while some workers consider taxonomy and classification to be synonymous.

Examined from the perspective of organizational science, systematics encompasses three interrelated and interacting areas—taxonomy, evolution, and classification, and its aim is to diagnose taxa and discover their ancestry. While the debate and conjecture surrounding the philosophy and methodologies of organismal systematics will undoubtedly continue, the question at hand is, what is the application of systematics to the micropaleontology of prokaryotes and protists?

In protistan paleontology, stratigraphic work provides the framework on which most systematics is done. This is not surprising given the relatively fine resolution that is attainable within samples. However, a strictly stratigraphic approach can lead to circular arguments regarding relationships. For example, the appearance of distinct morphologic sequences through time, combined with the presence of possible morphologic intergrades, is often interpreted as an evolutionary series connecting ancestral and derived taxa. How characters change is also determined from these sequences, but independent, nonstratigraphic evidence is wanting. Interspecific developmental studies, especially those that invoke heterochronic processes, and biogeographic studies, are unequivocally dependent on ancestor–descendant or sister–taxon relationships implied or implicit in classification. For studies such as these, untestable systematic relationships are simply inadequate.

Taxonomy

All taxonomic approaches use features or characters to group individuals. What differs between them is the importance (e.g., *weight*) assigned to characters, how the state of the character is determined (e.g., primitive, intermediate, derived), and the algorithms used to group the taxa based on the characters selected. Character choice in all of these approaches is the key to the practice of rigorous taxonomy. Characters should be homologous between taxa, and not analogs. For example, keels have evolved at least twice in planktic foraminifera, once in the Cretaceous, and at least once in the Tertiary. Outwardly, these structures appear similar but they differ substantially in their internal construction. Thus the keels on these tests are analogous, not homologous structures. Also, character complexes should not be used, but separated into their individual components.

All of these different taxonomic approaches produce nested sets of relationships among the groups analyzed. Typically, they are expressed as branching diagrams in which the different branches are diagnosed by characters, character state changes, or by changes in the degree of similarity. Several distinct terms are used to distinguish these different types of branching diagrams and the kind of information they contain. Phenograms present information on overall similarity of the taxa. Evolutionary trees typically present the supposed relationships between taxa through time, and the width of a given branch often reflects its taxonomic diversity. Cladograms are branching diagrams that present relationships in terms of shared derived characters or **synapomorphies**. Phylograms are constructed on the same criteria as cladograms except branch length is proportional to the number of character state changes (or steps) occurring on a branch. Available software to perform the various taxonomic procedures discussed below are listed after the supplementary reading at the end of this section.

Phenetics

In **phenetics**, taxa are grouped together by overall similarity, and the characters are not weighted. Phenetics is the ancestral form of taxonomy. For example, when Adam named the "fowl of the air" and the "beasts of the field" there is little doubt that he was using a phenetic approach. The birds were grouped together because of their possession of wings and feathers, and the beasts grouped because they had four legs and fur. Modern-day phenetists use as many characters as they can identify, and character data can be represented by presence or absence scores (binary), counts (meristic), or sometimes the variation exhibited by a character (e.g., aperture oval or aperture round).

Numerical taxonomy is phenetics with algorithms. In a numerical taxonomic analysis the phenetic features are scored and a matrix of character values for the taxon entered into a data matrix (Table 4.1). The data matrix is then subjected to one of several possible analysis routines depending on the type of data present in the matrix. The similarity matrix produced by the appropriate analysis is composed of measures of distance between the taxon (Table 4.2). These values are then subjected to a cluster analysis and the resultant values of similarity or distance arrayed in a phenogram (Fig. 4.1).

The information content of the resultant phenogram is nonevolutionary. It only orders taxa based on overall similarity. Assumptions regarding primitiveness or the direction of change of the characters used in the analysis are not made. Moreover, no distinction is made between features that are convergent among taxa and those that result from common ancestry. Phenetic analyses were initially thought to produce more stable and more natural classifications than phylogenetic analyses; however, this is not the case.

Numerical taxonomic studies of microfossils are uncommon; however, ordination of data matrices of microfossil characters is not. Principal component analyses, discriminant function analyses, and other forms of ordi-

Table 4.1 Data matrix for *Anthocyrtidium* species (Radiolaria). (Data modified from Nigrini C, Caulet JR. *Micropaleontology* 1988; 34:343)

Character	*angulare*	*ehrenbergi*	*euryclathrum*	*jenghisi*	*michelinae*	*nosicaae*	*ophirense*	*pliocenica*	*prolatum*	*zanguebaricum*
Length of apical horn	0	0	1	2	0	0	3	2	0	1
Thorns	0	0	0	1	2	0	0	1	0	0
Wings	1	0	0	1	0	0	0	0	0	0
Thoracic shape	1	2	3	1	4	5	2	5	3	5
Terminal teeth	0	0	0	0	1	0	0	0	0	0
Subterminal teeth	0	0	0	0	1	0	0	0	0	0
Thoracic pores	0	0	0	1	1	0	1	0	1	0

Table 4.2 Similarity matrix for *Anthocyrtidium* species

Taxa	*angulare*	*ehrenbergi*	*euryclathrum*	*jenghisi*	*michelinae*	*nosicaae*	*ophirense*	*pliocenica*	*prolatum*	*zanguebaricum*
angulare	1.00									
ehrenbergi	0.71	1.00								
euryclathrum	0.57	0.71	1.00							
jenghisi	0.57	0.29	0.29	1.00						
michelinae	0.14	0.29	0.14	0.14	1.00					
nosicaae	0.71	0.86	0.71	0.29	0.29	1.00				
ophirense	0.43	0.71	0.57	0.43	0.29	0.57	1.00			
pliocenica	0.43	0.57	0.57	0.57	0.14	0.71	0.43	1.00		
prolatum	0.57	0.71	0.71	0.43	0.43	0.71	0.71	0.43	1.00	
zanguebaricum	0.57	0.71	0.86	0.29	0.14	0.86	0.57	0.71	0.57	1.00

nation are commonly used in micropaleontologic studies. Most often they are done to assess the boundaries of identified morphologic groups (i.e., species) or to classify intermediate individuals. They have even been used to explore evolutionary pathways, although no assumptions or statements of ancestry are present in the data.

Stratophenetics is a subdiscipline within phenetics. It is a hierarchical process that first groups taxa based on overall similarity and then establishes ancestor–descendant relationships based on the stratigraphic order of taxa in the fossil record. Character transformations and phylogenetic trends are determined wholly by stratigraphic position (Fig. 4.2) without independent demonstration of primitive vs. derived conditions (e.g., outgroup comparisons) or character transformations (e.g., ontogenetic sequences). Sometimes stratigraphic data are merged with an evolutionary taxonomy approach and the stratigraphic occurrences of morphologic characters that appear to be of "special importance" are used to determine relationships.

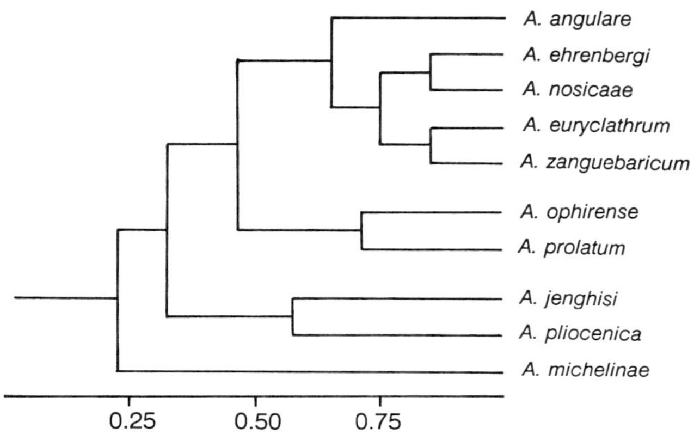

Figure 4.1 Phenogram of members of the radiolarian genus *Anthocyrtidium*. The scale across the bottom denotes degree of similarity. See Table 4.1 for characters used in the analysis.

Evolutionary taxonomy

Often referred to as evolutionary systematics or orthodox classification, evolutionary taxonomy produces groupings of taxa that are brought together by features that were selected based on a worker's previous knowledge of the study organism. These special characters are weighted characters because they are given more importance than other co-occurring features. They are also assumed to be conservative characters that present the worker with an accurate view from which to reconstruct the evolution of the group. Unlike numerical taxonomists and phylogeneticists, evolutionary taxonomists seldom use algorithms. Instead, their data are internally processed to construct relationships and explanatory evolutionary

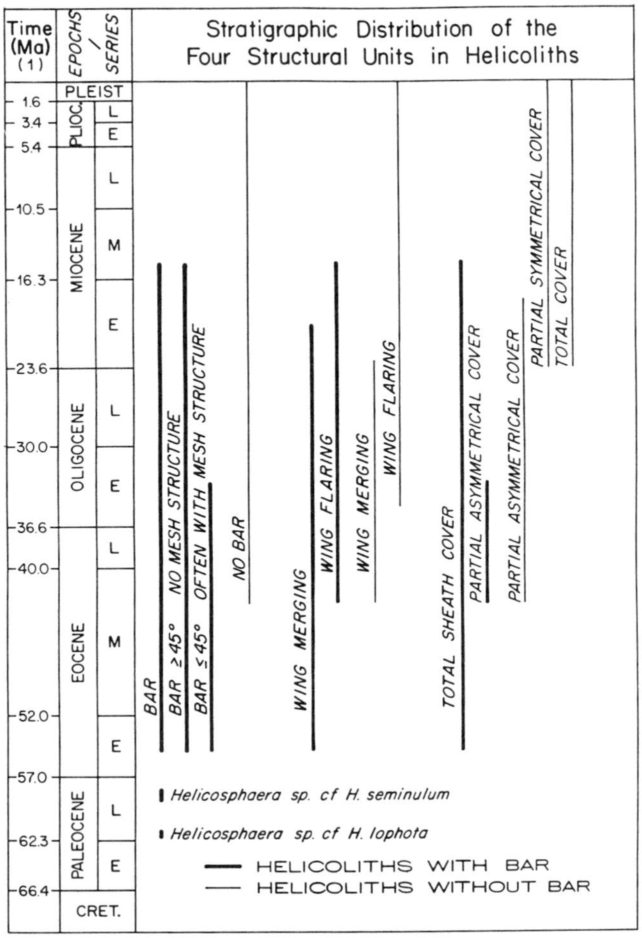

Figure 4.2 Stratigraphic distribution of helicolith character states in the coccolithophorid genus *Helicosphaera*. (Reproduced from Aubry MP. *Paleobiology* 1988; 14:64–80.)

scenarios. Perceived morphologic gaps between taxa are used to construct more inclusive groupings. **Grades** of organization (Fig. 4.3) are also recognized and are often formally named.

Many microfossil classifications are based on an evolutionary taxonomy approach. For example, foraminiferal workers have argued that features of the wall structure of foraminifera are not as subject to convergent evolution as external test characters, and therefore are preferable when constructing relationships among taxa. Clearly, similarity and weighting are the operational units in studies such as these.

Phylogenetics (cladistics)

Taxonomists using **phylogenetics**, or **cladistics**, are interested in discovering nested sets of homologous features. Shared (or similar) characters are still recognized,

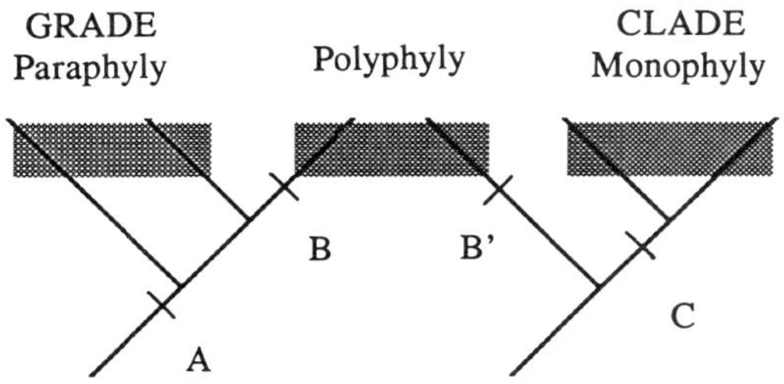

Figure 4.3 Branching diagrams illustrating the concepts of paraphyly, polyphyly, and monophyly. The presence of character state A in two terminal taxa produces a paraphyletic grade outlined by the left stippled pattern. Character states B and B′ are analogous, not homologous characters, and thus produce the polyphyletic taxon demarcated by the middle stippled pattern. Character C is a synapomorphy for the clade highlighted by the right stippled bar which is clearly monophyletic.

but the shared derived characters (synapomorphies) impart information of relationships, not the shared primitive characters (**plesiomorphies**) (Fig. 4.4). **Autapomorphies** (characters that diagnose terminal taxa) also contain no information of relationships (Fig. 4.4). Taxa that included the ancestral taxon and all of its descendants are referred to as **monophyletic**, while groups that include the ancestral taxon and only some of its descendants are **paraphyletic** (Fig. 4.3). **Polyphyletic** groups are those with more than one ancestral taxon (Fig. 4.3).

Like phenetic analyses, cladistic analyses typically begin with a data matrix of the taxa under study, and Table 4.1 can be used to illustrate certain differences between cladistic and phenetic analyses. Whereas the numbers in Table 4.1 represent specific character states present in the taxa, they are not ordered in a particular evolutionary sequence or transformation series. To perform a cladistic analysis the character states should be arranged in a primitive to derived sequence. Thus, for "length of apical horn" the numbers 0, 1, 2, 3 serve not simply to group taxa, but instead represent an evolutionary series of character states for "length of apical horn." The directionality of the transformation (or **polarity**) is determined by outgroup comparisons or ontogeny. In Table 4.1 the species *A. nosicaae* can serve as the **outgroup**. The outgroup is always the **sister** taxon of the taxa under study (the **ingroup**) (Fig. 4.4), and is assumed to exhibit primitive character states relative to the ingroup. In the example, *A. nosicaae* exhibits state 0 for all characters except thoracic shape which is 5 (ovate). Ontogenetic criteria could be useful in determining the polarity of thoracic shape. Cultures of these radiolarian species might show that all of the taxa began with ovate thoraces and through differential growth achieved other morphologies.

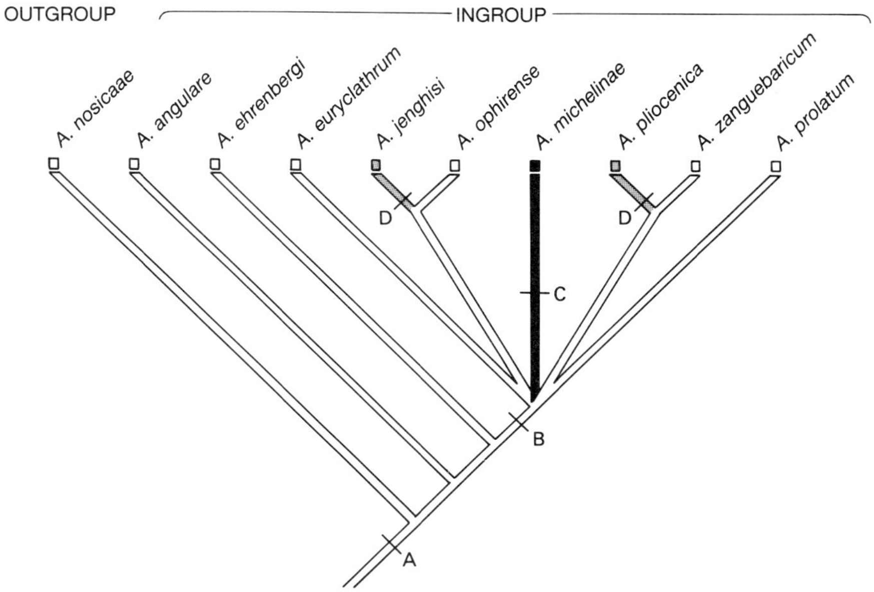

Figure 4.4 Consensus cladogram of 65 trees based on the data presented in Table 4.1. Character A is plesiomorphic for the entire group, character B is a synapomorphy for the unresolved group *A. euryclathrum–A. prolatum*, character C is an autapomorphy for *A. michelinae*, and character D is convergent in both *A. jenghisi* and *A. pliocenica*. The relationship between ingroup and outgroup is also marked.

These data might be encoded as a character-state tree (Fig. 4.5C) while the other characters might be either ordered or unordered (Fig. 4.5A, B).

A cladistic analysis using **parsimony** produced 65 trees all with 22 steps. A step is the number of states a character transforms through going between nodes on the tree. For example, for an ordered character a change between state 1 and 3 equals two steps, while for an unordered character the same change requires only a single step (Fig. 4.5). A **consensus** or the average tree representing the 65 most parsimonious trees (Fig. 4.4)

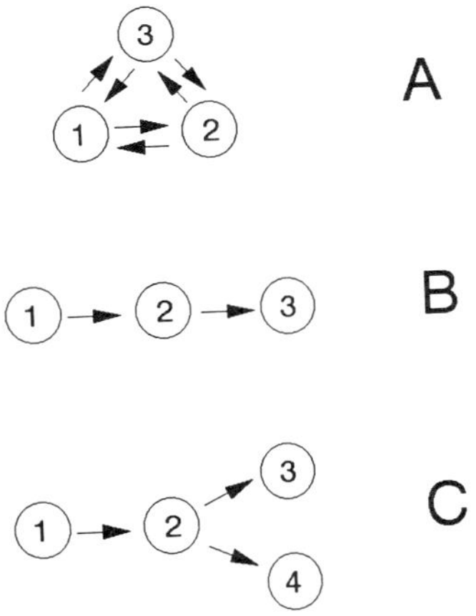

Figure 4.5 Possible character-state transformations. A, Unordered characters; all states can give rise to any other state. B, Ordered character transformation; state 1 gives rise to state 2, etc., reversals (e.g., 2 to 1 or 3 to 2) may or may not occur. C, Character-state tree; state 1 gives rise to state 2 which in turn gives rise to either state 3 or state 4.

reveals several monophyletic groupings (e.g., *A. nosicaae*; *A. pliocenica*, and *A. zanguebaricum*; *A. jenghisi*, and *A. ophirense*), as well as an unresolved branching pattern called a polytomy (*A. euryclathrum* through *A. prolatum*). Clearly, more characters are needed to resolve further the relationships within this group. This example analysis also demonstrates an autamorphic character and the discovery of convergent characters (Fig. 4.4). This latter character is said to be **homoplastic**, a character found in two or more taxa, but absent in their common ancestor.

Phylogenetic analyses of protist fossils are few (Fig. 4.6). The best deal with freshwater diatom groups while those that have been published on marine protist fossils are difficult to interpret because of the lack of outgroups or the presence of zero length branches in the cladograms. Also, a preponderance of cladistic terminology with no analyses is becoming increasingly common in the literature. For example, commonly, the origin or extinction of a specific microfossil "clade" is discussed, but the monophyly of the group under discussion is never demonstrated, only assumed based on past taxonomic conventions. However, if the terminology is becoming common, then perhaps the analyses are not far behind.

The three schools of taxonomy outlined above differ substantially in philosophy, terminology, assumptions, and expectations. Numerical taxonomic and cladistic studies require their respective practitioners to list the characters and their states in data matrices. In many cases phenetists and evolutionary taxonomists simply describe or illustrate their key characters and argue their evol-

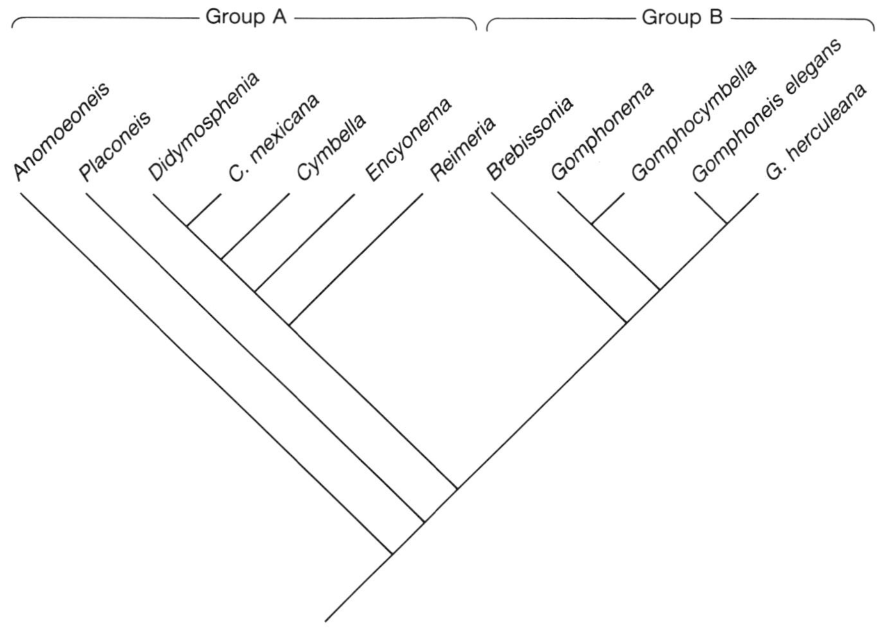

Figure 4.6 Cladogram of cymbelloid and gomphonemoid freshwater diatoms. This cladogram has been used to produce the unranked classification presented in Table 4.3. (Modified from Kociolek JP, Stoermer EF. *J Phycol* 1988; 24:377–85.)

utionary importance. Phenetic studies do not address the evolutionary histories of the study group. Evolutionary taxonomy purports to, but only cladistic methods require the worker to frame a taxonomic study in a testable hypothesis of relationships. Also, cladistic methodologies require the worker to state, before the formal analysis, his or her assumptions of character states, character transformations, outgroup comparisons, etc. All taxa, assumptions, and data are presented, and also become available for independent testing and reevaluation by others. And unlike phenetic and evolutionary taxonomic studies, the cladogram is not an endpoint, but a hypothesis to be falsified by additional data, restudy of characters and their transformations, or the addition of different outgroups, etc. Cladistic analysis is an iterative process.

Although not often implicitly stated, studies of biogeography, diversity, gradualistic, and punctuated evolutionary patterns all make inherent assumptions of relationships. Evolutionary scenarios for the presence of bipolar species are moot if the "species pair" do not share a common ancestor, and what is meant by a region having "low generic diversity?" Are all genera created equal and therefore a comparable form of organismal currency? Clearly, these studies require independent assessment and demonstration of the phyletic relationships of the organisms. It is neither sufficient nor robust to recognize a relationship between two morphologically similar taxa based on their sequence in the geologic column, describe the character transformations based on this sequence, and then construct an evolutionary scenario to explain the change. While the protist fossil record is the best available, its resolution does not abdicate the need for independent assessment of the relationships between protist taxa.

Classification

Classification is the arrangement of taxa into nested hierarchical sets. Typically, this has been done for convenience and to allow communication about these organisms. In most biologic disciplines (e.g., ecology, physiology, etc.), and especially in the case of taxa used in biostratigraphic or paleoceanographic studies, consistent identification of biologic species is important. However, nomenclatural stability should never take precedence over information content of a classification.

All three schools of taxonomy have sought to produce **natural classifications**. Pheneticists view natural groups as those taxa linked by the greatest similarity to one another. To evolutionary taxonomists natural groups are delimited by morphologic gaps between taxa and the distribution of characters for which an evolutionary scenario can be argued. Cladists consider natural groups to be monophyletic, and natural classifications to reflect organismal history.

The current Linnaean **ranked classification** scheme is reasonably compatible with the results of both phenetic and evolutionary taxonomic studies. However, it is inadequate in expressing the results of cladistic analyses. Here **unrank classifications** probably best reflect the nested sets of relationships that evolution has produced (Table 4.3).

Table 4.3 Unranked classification of the cymbelloid and gomphonemoid freshwater diatoms based on the cladogram presented in Figure 4.6

Group A
Reimeria
Encyonema
Cymbella
C. mexicana
Didymosphenia
Group B
Brebissonia
Gomphonema
Gomphocymbella
Gomphoneis elegans
G. herculeana

Supplementary reading

de Queiroz K. Systematics and the Darwinian revolution. *Phil Sci* 1988; 55:238–59. A philosophic and practical discussion of the reflection of phylogeny in taxonomy and classification.

Eldredge N, Novacek MJ. Systematics and paleobiology. *Paleobiology* 1985; 11:65–74. A review of the past, present, and future role of systematics in paleobiology.

Hennig W. *Phylogenetic Systematics.* Urbana: University of Illinois Press, 1966. The starting point and foundation of phylogenetic analysis.

Mayr E. *Principles of Systematic Zoology.* New York: McGraw-Hill, 1969. An introduction to evolutionary taxonomy.

Sneath PHA, Sokal RR. *Numerical Taxonomy.* San Francisco: WH Freeman, 1973. An introduction to numerical taxonomy.

Wiley EO. *Phylogenetics. The Theory and Practice of Phylogenetic Systematics.* New York: John Wiley, 1981. Also discusses evolutionary and numerical taxonomy, speciation, biogeography, nomenclature, etc.

Systematic software

COMPONENT. Ver. 1.5 for IBM-compatible computers. COMPONENT provides tools for constructing area cladograms, performing statistical tests of hypotheses of taxonomic and biogeographic congruence, etc. It interfaces with all tree-building software (Hennig86, MacClade, PAUP, PHYLIP). Contact: RDM Page, Department of Herpetology and Ichthyology, Central Park West at 79th St, New York, NY 10024–5192, USA.

DELTA. A general system for processing taxonomic descriptions. Ver. (19 August, 1990). IBM-compatible computers only. Organization and processing of taxonomic descriptions databases. Database produces natural language descriptions, keys, and PAUP data sets. Contact: MJ Dallwitz, CSIRO Division of Entomology, GPO Box 17000, Canberra, ACT 2601, Australia.

Hennig86. Ver. 1.5. For IBM-compatible computers. User unfriendly, but fast and powerful; a no-nonsense cladistic analysis package. Contact: JS Farris, 41 Admiral Street, Port Jefferson Station, New York, NY 11776, USA.

MacClade. Ver. 2.1. (Ver. 3.0 is scheduled for release in 1992.) Macintosh only. The most user-friendly program. It does not do analyses, but interfaces with PAUP and Hennig86. Use MacClade to maintain data sets, set options and assumptions, construct alternative trees, and manipulate branching patterns. Ver. 3.0 will have options for stratocladistic analyses. Contact: Sinauer Associates, Inc., Sunderland, MA 01375, USA.

NTSYS. Numerical Taxonomy System. Ver. 1.6 for IBM-compatible computers (Ver. 1.7, 1991). Provides basic programs for numerical taxonomic analyses, including similarity coefficients, cluster analysis, cophenetic values, etc. Modules may be linked to perform principal components analysis. Contact: Exeter Software, 100 North Country Road, Setauket, NY 11733, USA.

PAUP. Phylogenetic Analysis Using Parsimony. Ver. 2.4 for IBM-compatible computers (Ver. 3.0 is currently in β-test version) and Ver. 3.0 for Macintosh computers are extremely user friendly. Interfaces with MacClade (Mac) and Hennig86 (DOS). Contact: DL Swofford, Illinois Natural History Survey, 607 E Peabody Drive, Champaign, IL 61820, USA.

PHYLIP. Phylogeny Inference Package. (Ver. 3.0 for IBM-compatible computers. No Macintosh version.) Although slower and more cumbersome than either PAUP or Hennig86, PHYLIP provides the user with a large array of modules for discrete state as well as molecular sequence data, and allows the user to experiment with different assumptions and parsimony methods. Contact: J Felsenstein, Department of Genetics SK-50, University of Washington, Seattle, WA 98195, USA.

PALEOECOLOGY

Jere H. Lipps

The determination of the environment of deposition of ancient sedimentary rocks is of critical importance in deciphering the geologic history of a region and understanding the environmental framework for evolution. Paleontologists commonly refer to these kinds of studies as paleoecology, although only a few physical aspects of the environment are usually determined. Paleoenvironmental analysis is a more accurate term, especially as micropaleontologists have practiced it in the solution of geologic problems. Paleoceanography is a special kind of paleoenvironmental analysis that attempts to understand the oceanographic history of an ocean or area through analysis of microfossils.

Distributions

The environment in which an organism lives has innumerable components, including abiotic ones like temperature, salinity, light, substrate, and space, and biotic ones like nutrient supply, food availability, competition, and symbiosis. These affect organisms in different and, commonly, very complex ways. Although they are surely important in paleoenvironmental analysis, they can seldom be determined and are omitted. Paleoenvironmental analyses are based fundamentally on distributions and chemical composition of fossils.

Distributions of organisms

All of the conditions necessary for a species to live combine to produce three general distributions that are utilized in micropaleontology. Organisms are distributed by habitats, community assemblages, and provinces. Each is important in paleoenvironmental analysis, paleoceanography, and biostratigraphy of both benthic and planktic species.

At the smallest scale, different species of prokaryotes and protists preferentially live in different habitats. For example, in the Glass Mountains of Texas, fossil fusulinid foraminifera occur in different sedimentary habitats, characterized by different grain size and silt and clay content (Fig. 4.7). Each of these is a different habitat and each supports different species of protists because the conditions in each differ. The species in each may have special morphologic adaptations to cope with the different conditions. Any area of the sea floor or water column has many different habitats containing different assemblages. Within these habitats, organisms are not necessarily evenly distributed in time or space. Benthic and planktic species are commonly more abundant at some times of the year than at others (Fig. 4.8), and may even be more abundant in different years or decades than others. Paleontologists studying fossils derived from such seasonal or yearly situations interpret only conditions prevailing at particular times and not conditions representing the entire time sampled. For example, if the fossils were extremely abundant only in the summer, then estimates of ancient temperatures are not yearly averages but are warm summer temperatures only.

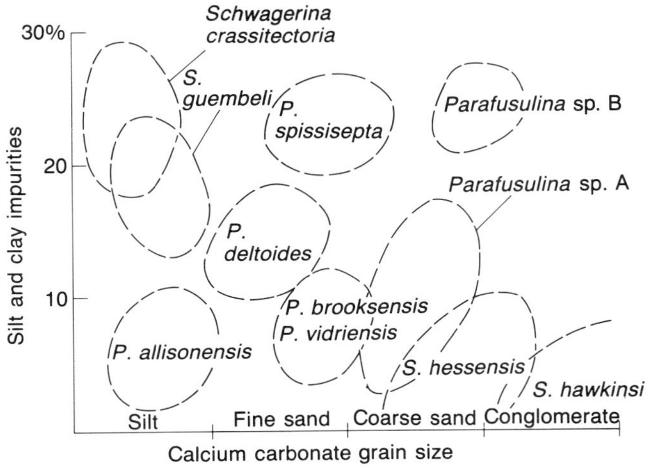

Figure 4.7 Sedimentary habitats of fusulinid species of the genera *Schwagerina* and *Parafusulina* in Permian rocks of the Glass Mountains, Texas. (Reproduced from Ross CA. *J Paleontol* 1961; 35:398–400.)

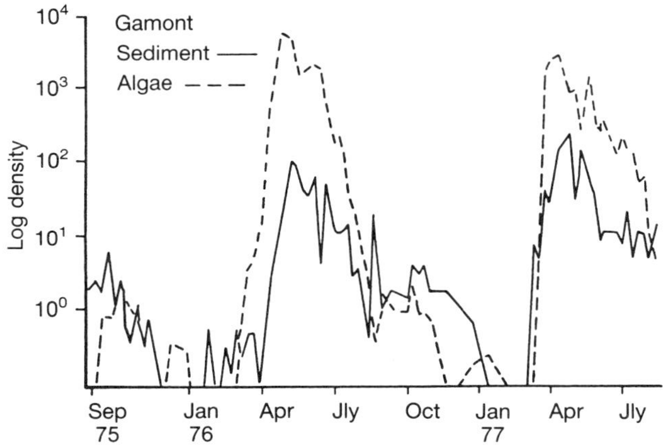

Figure 4.8 Seasonal distribution for 1975 through 1977 of the sexual stage of the foraminifera *Glabratella ornatissima* in algal (dashed line) and sediment (solid line) habitats in the subtidal off northern California (expressed as a semilog plot of the log of the density of individuals). This temporal variation in the fossil record would result in paleoenvironmental interpretation of only the spring and early summer conditions. (Reproduced from Erskian MG, Lipps JH. *J Foraminiferal Res* 1987; 17:240–56.)

Studies on living benthic shallow-water foraminifera and diatoms show that most species are statistically distributed randomly or are aggregated, and are seldom homogeneous (Fig. 4.9). Deep-sea benthic foraminifera are not homogeneously distributed within small (>1 m), similar areas either. However, most species of fossil benthic foraminifera are, in fact, distributed rather hom-

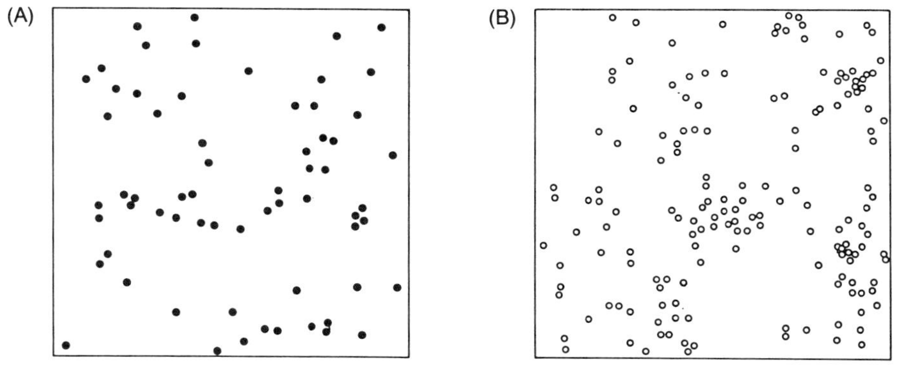

Figure 4.9 Within-habitat distribution of individuals in two species of foraminifera in 58 cm^2 areas in Rehoboth Bay, Delaware. A, *Haynesina germanica* shows a random distribution. B, *Ammonia beccarii* shows an aggregated distribution. (Reproduced from Buzas MA, Gibson TG. *Smithsonian Contrib Paleobiol* 1990; 68:1–35.)

ogeneously in the rocks, suggesting that life processes over generations (reproduction, dispersal) or taphonomic processes after death (transportation, diagenesis) smooth the live distributions over the times required to form the fossil record. Paleontologists probably do not have to be concerned about within-habitat variability in paleoenvironmental reconstructions.

Species are grouped by ecologists into communities, which are associations of species that generally co-occur. The species may co-occur for any number of reasons; for instance, because their habitats are similar or because they somehow are dependent on one another (predator and prey, for example). Within a particular geographic area or depositional basin, a number of different community associations or **biofacies** may exist. The differentiation of these biofacies is the principal activity involved in paleoenvironmental analysis.

At a still larger scale, communities co-occur with other communities over very large areas of the oceans or long segments of continental margins, and these assemblages of communities and species are referred to as provinces. Provinces are regions of similar biotic composition usually demarcated by significant community and species changes at particular places (see Figs 12.26–12.28). The changes are caused by some abrupt environmental change; commonly in the oceans to oceanographic changes, such as temperature, salinity, nutrient supplies, and associated currents. All groups of organisms respond to these changes, so that provincial boundaries affect not only protists and probably prokaryotes, but fishes, plants, and invertebrates as well. In planktic organisms, each province consists of single community assemblages (see Figs 11.17, 12.24).

Paleoenvironmental analyses

Three general approaches are used in paleoenvironmental analyses based on fossil prokaryotes and protists.

1 A direct phylogenetic method where the distributions of fossil species are compared with those of the most closely related modern species, and are assumed to have the same ecologic requirements.
2 An indirect approach in which various levels of organization, for example, similar morphology, community structure, or community distribution, are used to infer environments for similar levels in the fossil record.
3 An empirical chemical approach where the proportions of various elements or isotopes contained in a fossil are related directly to paleoenvironment by thermodynamic or chemical principles.

While the principles involved in each approach are relatively simple, many methods to acquire and analyze the data have been used.

The phylogenetic approach is based on the close relationship of living and fossil species or genera, with the assumption that the environmental tolerances of the living taxa are the same in the fossil taxa. The principle is simple: the species or genera of fossils are compared to the most similar taxa found in a modern environment, and the paleoenvironment is then equated with it (Fig. 4.10). Two major difficulties exist with this approach. First, it can be used only with taxa closely related to modern ones; thus, most fossil taxa older than Miocene are too distantly related to apply the method with reliability. Second, the method assumes that close phylogenetic relationship is equivalent to close environmental requirements, which is not always true. It is an application of a special kind of uniformitarianism, known as substantive uniformitarianism, which implies that everything—rates of processes, materials, and conditions—has remained constant through geologic time. The phylogenetic approach is best regarded as methodologic uniformitarianism which implies that the same laws of nature operated in the past but that the rates, conditions, or materials have not necessarily been constant. Every interpretation must be subject to verification by other

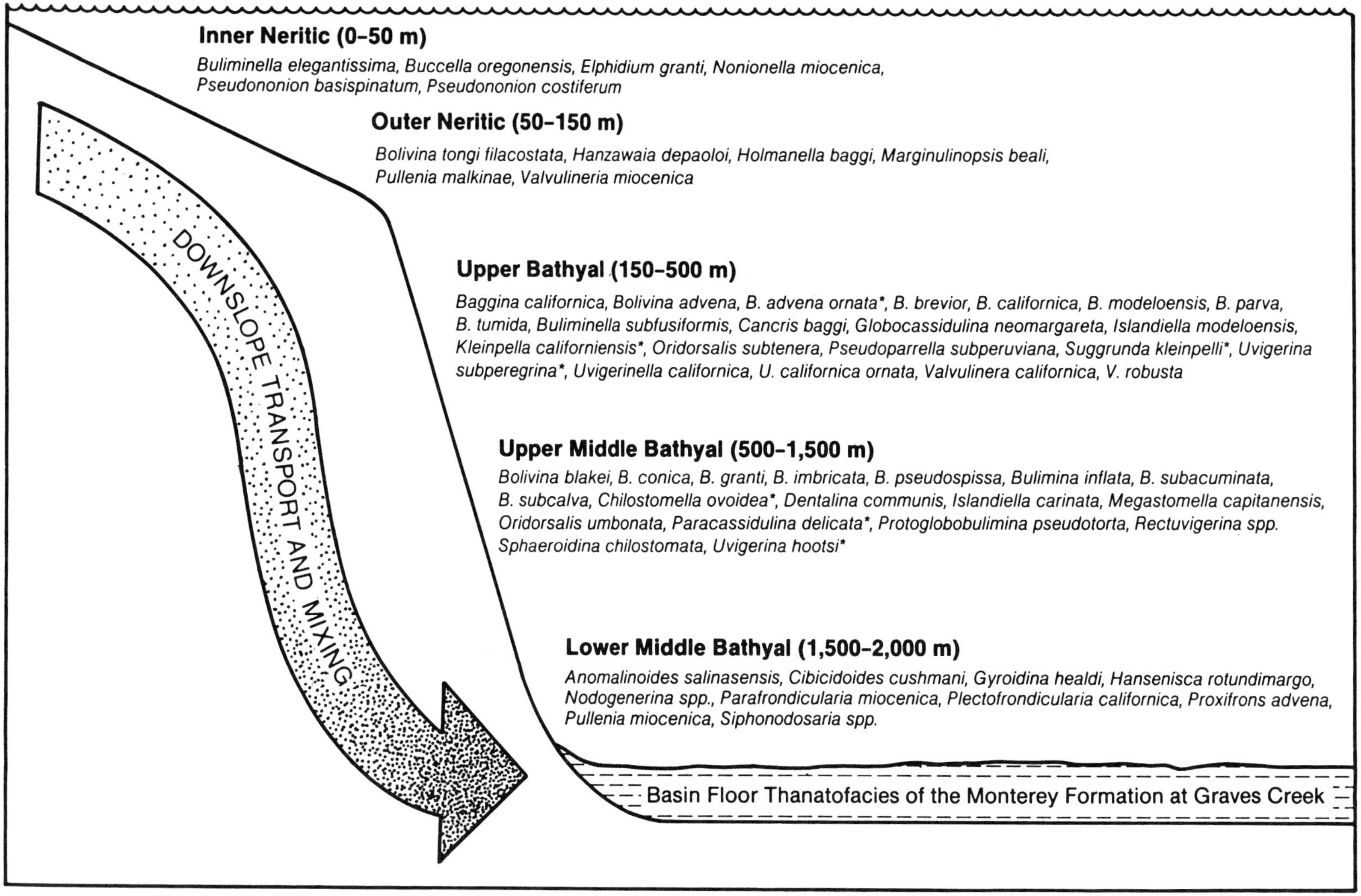

Figure 4.10 Paleoenvironmental reconstruction of the Miocene Monterey Formation in the Central Coast Ranges of California based on phylogenetic relationships of species of foraminifera. The rocks were deposited in a basin at lower middle bathyal (1500–2000 m) depths based on the occurrence of species of foraminifera whose closest living relatives live at those depths. In addition, species with living representatives from inner neritic (0–50 m) through upper middle bathyal (500–1500 m) occur with the deeper species indicating downslope transport by turbidity currents and slumping onto the basin floor. Asterisks indicate species tolerant of low oxygen conditions. (Reproduced from Finger KL, Lipps JH, Weaver JCB, Miller PL. *Micropaleontology* 1990; 36:1–55.)

means, either geologic or geochemical. In spite of this, the phylogenetic approach has yielded sound results most of the time, suggesting that organismal physiology and environmental tolerances have not usually changed without change in morphology.

Paleoenvironments may be inferred by examination of a number of relationships between living and extinct organisms and between fossils and their spatial distributions in rocks. Unrelated organisms with similar morphology may be assumed to have occupied similar environments. For example, Paleocene benthic foraminifera from the Gulf Coast of the USA have morphologies similar to modern species living in particular environments, although they are unrelated to each other. An analysis of the distribution of morphologic types reveals patterns similar to modern types, and the paleoenvironments are assumed to be similar as well (Fig. 4.11). The reason for the occurrence of similar morphologies in such patterns is unknown but may be related to adaptations to substrate, predation, sediment stability, or others. The ratio of agglutinated, porcelaneous, and hyaline tests in foraminifera has been characterized for a number of modern environments (see Fig. 12.30), and similar ratios have been used in a direct empiric comparison to infer paleoenvironments as old as Paleocene. Paleoenvironments may be reconstructed by knowing only the ratio of two groups from different habitats; for example, the ratio of planktic to benthic foraminifera is used to infer the depth of deposition—the more plankton the deeper the paleoenvironment, although the real relationship appears to be the degree of open ocean influence. In other cases, the paleoenvironment may be inferred by the spatial relationships of fossils to each other or to specific lithologies. In the Proterozoic rocks of Spitzbergen, prokaryotes occur in spatial relationships that allows a paleoenvironmental reconstruction from tidal flat to lagoon and an offshore shoal (Fig. 4.12). These and other kinds of inference have interpretive

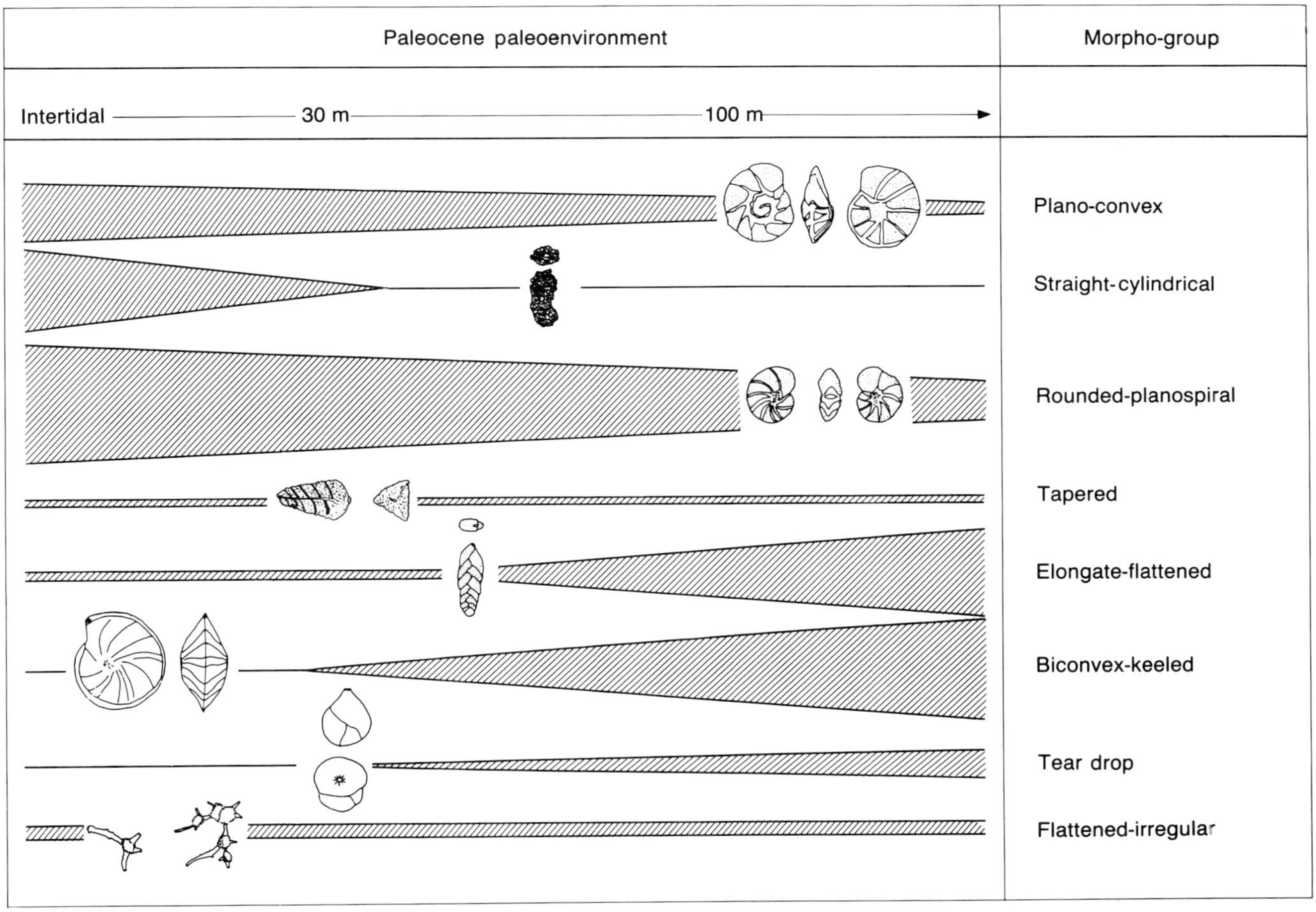

Figure 4.11 Morphology of Paleocene foraminifera and their paleoenvironments, as inferred by comparison of similar morphologies in modern environments. (Data and illustrations from KP Severin.)

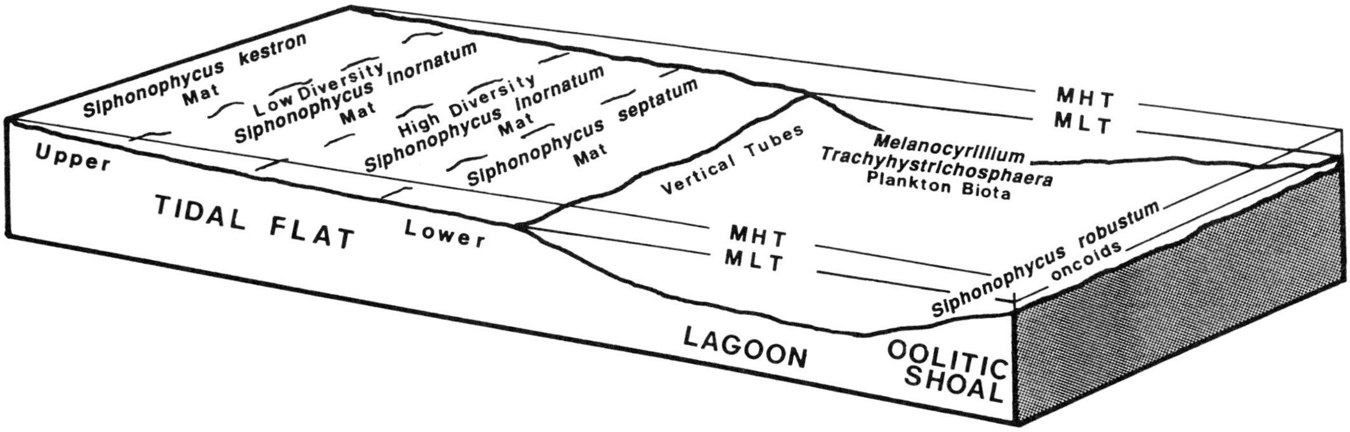

Figure 4.12 Seven hundred to eight hundred million-year-old paleoenvironments of the Proterozoic Draken Conglomerate Formation in Spitzbergen, inferred from the distributional relationships in the rocks of prokaryote assemblages. MHT, mean high tide; MLT, mean low tide. (Reproduced from Knoll AH, Swett K, Mark J. *J Paleontol* 1991; 65:531–70.)

power, but they must be used with care and confirmed by other means if possible.

The geochemistry of fossil protists yields environmental information. Although many geochemical measurements have been associated with paleoenvironmental analyses, the most useful is the ratio of the stable isotopes of carbon and of oxygen in carbonate skeletons. Stable isotopes do not undergo radioactive decay, hence are retained in the carbonate in the proportions incorporated when the skeleton was secreted, if they have not been subjected to diagenesis. Because the isotopes have fewer or greater neutrons in the nucleus of the atom, they have a different atomic weight. This difference causes slightly different chemical behavior, the results of which can be measured with a mass spectrometer. Changes in the ratios are compared by relating the change in the heavier isotope (^{18}O or ^{13}C) to a generally accepted standard. The changes are referred to as $\Delta^{18}O$ or $\Delta^{13}C$ and measured in parts per thousand (‰).

In the absence of postdepositional alteration, the isotopic composition of a skeleton reflects either the isotopic composition of the sea water, the composition of the metabolic products of the organism, or a mix of the two at the time of the secretion of the carbonate. Thermodynamically, colder water contains more ^{18}O than warmer water (the temperature effect). However, water masses have salinity differences as well as temperature differences. In water masses of higher salinity, ^{16}O decreases relative to ^{18}O because it evaporates so much more due to its lighter atomic weight (the water mass effect). Similarly, when ice caps form on land, they are enriched in ^{16}O relative to ^{18}O because the water evaporated from the sea to make the ice contains a little more ^{16}O (the glacial effect). Cold temperatures, increased salinity, or continental ice caps mean increased $\Delta^{18}O$, if metabolic effects are insignificant. The ratio of the isotopes of ^{13}C to ^{12}C is controlled largely by primary productivity and global changes in carbon cycling in and out of the oceans. Metabolism of the organism commonly affects both O and C isotopic ratios (the vital effect), in particular those fossils that may have photosynthesized or that contained photosynthesizing symbionts in their cells or tissues. Also, waters with high primary productivity, hence high photosynthetic rates, may have isotopic ratios that are displaced from those normally expected. Foraminifera that live in sediment may have $\Delta^{18}O$ and $\Delta^{13}C$ values that do not represent sea water present in the overlying water (the habitat effect). In addition, differential dissolution, reworking, and diagenesis may alter the original isotopic composition of the skeleton. Although all these effects may have operated on skeletons during secretion, deposition, and burial, through the careful selection of species for analysis, and comparison of results to others made on other materials, isotopic measurements have proved generally reliable. Now, paleontologists can closely estimate the temperature, salinity, glacial condition of the Earth, primary productivity, and aspects of carbon cycling. Estimates from isotope studies combined with inferences from the fossils themselves or their paleobiogeography are revealing the history of environmental change in the oceans in considerable detail.

Paleoceanography

Paleoceanography is based primarily on the paleobiogeography of both benthic and planktic protists and on the oxygen and carbon isotopic composition of carbonate fossils, protists in particular. The patterns that emerge for any particular time can best be understood in relation to modern oceanography.

The oceans are in constant motion at all scales, but paleoceanography is concerned with large-scale circulation, both at the surface and at depth. In the modern oceans, surface circulation is driven by the prevailing winds. The tradewinds, blowing out of the southeast and northeast, together with the Coreolis effect move water westward along the equator to be deflected to the north and south by the eastern coasts of continents. The currents, after passing along the continents, are blown eastward by westerly winds, and deflected south and north by the western coastlines of the opposite continents, thus completing a clockwise gyre in the northern hemisphere and a counterclockwise gyre in the southern hemisphere (Fig. 4.13A). This entire system ultimately is driven by the heat of the sun. The equatorial and western boundary currents are warm, while the high latitude and eastern boundary currents are cold; the central gyres have higher salinity because evaporation exceeds precipitation, and high latitude seas have lower salinity because of higher precipitation and runoff of fresh water from land

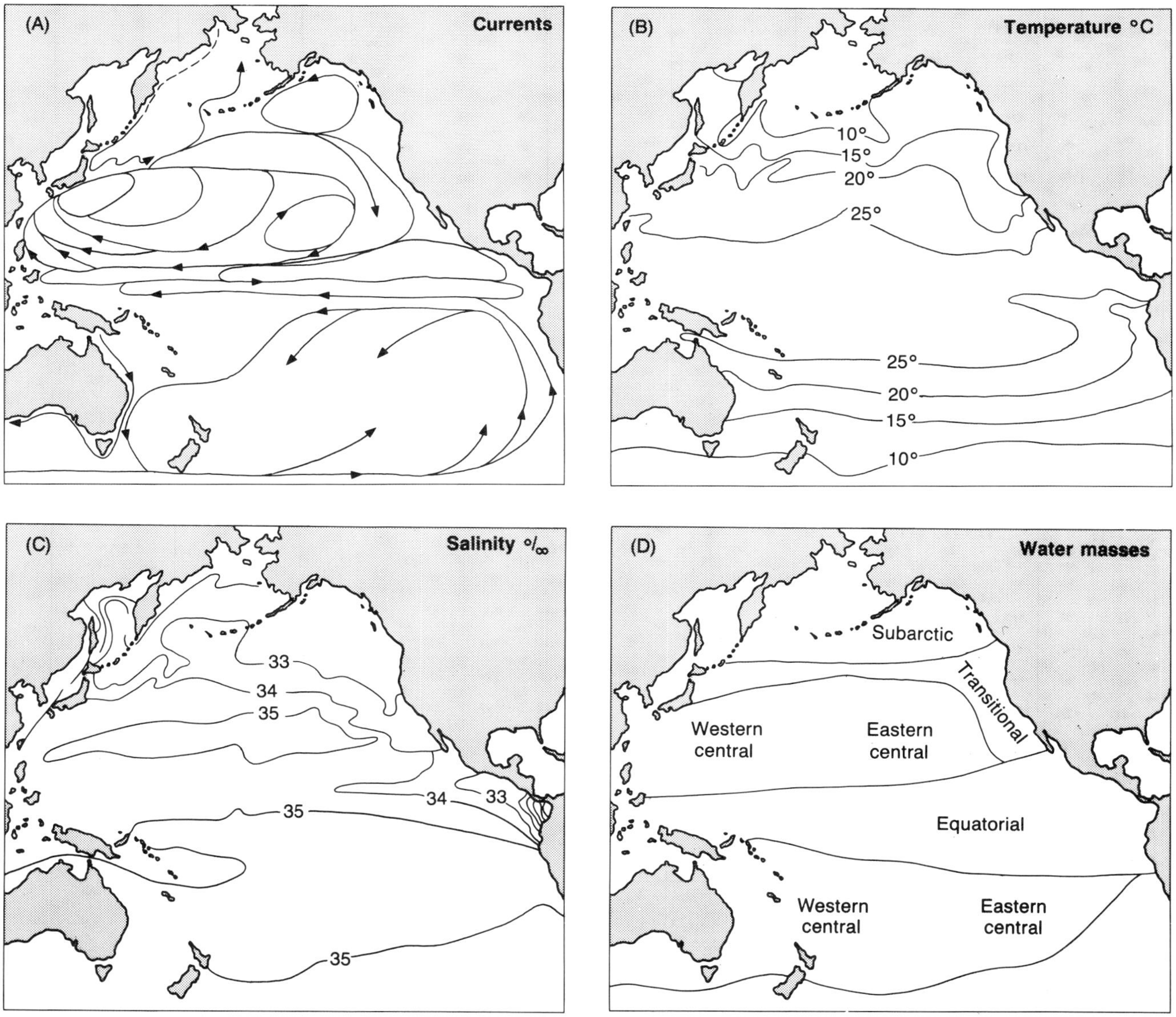

Figure 4.13 Modern oceanography of the Pacific Ocean. A, Currents, with clockwise gyre in the northern hemisphere and counterclockwise gyre in the southern hemisphere. B, Sea surface temperatures in summer. C, Sea surface salinities. D, Principal water masses of the Pacific Ocean. These can be subdivided into additional water masses based on finer recognition of temperature and salinity.

(Fig. 4.13B, C). Temperature, salinity, and the current patterns combine to produce distinctive water masses (Fig. 4.13D), each of which supports a distinctive biota.

These kinds of patterns can be expected for any time on the surface of the Earth because of the input of solar heat and Coreolis effect. Surface circulation may have varied in intensity and directions in the past, however, when ocean basins and continents had different configurations causing heat to be distributed differently and currents to be deflected in other directions. Such differences are usually detectable through the paleobiogeographic and oxygen isotopic measurements of planktic protists.

Deep-water currents are driven by density differences between water bodies, and since water density is dependent on the temperature and salinity of the water, these are known as **thermohaline** currents. Cold waters, because they are denser, tend to sink to levels of similar density, as do more saline waters. Today, most deep-water currents are driven chiefly by temperature differences whereas in the past, when temperature was more uniform throughout the oceans, deep circulation was chiefly salinity driven. Deep-water circulation is usually detectable through the paleobiogeography of benthic foraminifera (see Figs 12.35, 12.36).

Paleobiogeography reveals only paleoceanographic patterns, while isotopic measurements on fossil protists can give reasonable estimates of temperature and glacial conditions on Earth. These are important paleoceanographic measurements because the heat distribution of the oceans profoundly affects climates, water masses, intensity of circulation and upwelling (hence primary productivity), and atmospheric circulation, among others. Thus, special efforts have been made to determine for many geologic times the isotopic composition of benthic and planktic protists. For example, in the Tertiary, the general trend has been for surface waters in high latitudes to have cooled from perhaps a little more than 15°C to 0°C from the Paleocene to the Recent, most of this decrease having taken place since the beginning of the Miocene. Since cool high-latitude waters are dense and sink below warmer low-latitude waters, thermohaline circulation, derived chiefly around Antarctica, also originated and increased in intensity during this time. Isotopic measurements also reveal that during this time continental glaciers grew on the northern and southern continents, eventually producing the Pleistocene Ice Ages.

Supplementary reading

Buzas MA, Behrensmeyer AK. Ecology—paleoecology. In: Boardman RS, Cheetham AH, Rowell AJ, eds. *Fossil Invertebrates.* Oxford: Blackwell Scientific Publications, 1987: 19–28. Brief review of the basic concepts of ecology and paleoecology.

Hemleben CH, Spindler M, Anderson OR. *Modern Planktonic Foraminifera.* New York: Springer-Verlag, 1989. An excellent review of the biology of planktic foraminifera and its consequences for the geologic record. Especially good are the summaries of stable isotopic composition of skeletons.

Kennett JP. *Marine Geology.* Englewood Cliffs, NJ: Prentice-Hall, 1982. A complete treatment of the geology and paleoceanography of the oceans, and excellent discussions of how protists are used in deciphering ocean history.

Lipps JH, Berger WH, Buzas MA, Douglas RG, Ross CA. *Foraminiferal Ecology and Paleoecology.* Tulsa: Society of Economic Paleontologists and Mineralogists Short Course No. 6, 1979. A review containing much original data and interpretation of the ecology and paleoecology of foraminifera, including smaller and larger benthic foraminifera, planktic foraminifera, the use of stable isotopes, taphonomy of planktic foraminifera, and a clear explanation of the use of numerical techniques in ecology and paleoecology.

Valentine JW. *Evolutionary Paleoecology of the Marine Biosphere.* Englewood Cliffs, NJ. Prentice-Hall, 1973. A general review of ecologic concepts applied to the fossil record and evolutionary processes.

BIOSTRATIGRAPHY

Joseph E. Hazel

An attribute of most sedimentary rocks is the presence of fossils. The grouping of strata into units based on fossil content for the purposes of zonation and correlation of the rocks is referred to as **biostratigraphy** (**bio**, for life; **stratigraphy**, the science of the layered rocks). Human nature puts to practical use something as ubiquitous as fossils. Thus, in the last years of the eighteenth century and early in the nineteenth century, William Smith, an English civil engineer, recognized that fossils could be used to distinguish two similar lithologic units that were at different positions in the succession of strata because the fossil assemblages changed vertically in the sections of strata he studied. This is now known as the principle of faunal or biotic succession. Furthermore, Smith recognized that even a unit of uniform lithology could sometimes be divided by its fossil content. Smith's discovery that fossils occur in biologically consistent, repetitive sequences that are temporally restricted, but spatially widespread, is one of the most important contributions to geology. Without the ability to correlate geologic events historical geology would be impossible. Although other techniques are now available, fossils are and surely will remain foremost in the determination of the timing and correlation of geologic and biologic events. Smith's work led to the development of the relative geologic time-scale.

Biostratigraphy is concerned with the identification of taxa in strata, the tracing of their lateral and vertical extent, explanation of the observed distributional patterns, and the partitioning of the rock column into units based on the determined fossil distributions. Related paleontologic subdisciplines are paleobiogeography, biofacies analysis, and paleoecology.

Protists are among the best fossils for use in biostratigraphy. Most groups are widespread, abundant in rocks, and small enough to be easily recovered in relatively small samples, including rock chips generated during oil-well drilling. Because of their utility they are standard groups used by paleontologists in the oil and other industries and in marine geologic research, particularly where small samples or cores are most common. On land, rock outcrops can usually be closely inspected for larger fossils that have biostratigraphic utility, but even here protists are still very useful fossils because of their ubiquity and abundance.

For some parts of the geologic column, detailed biostratigraphies have been developed using each of the groups described in this book. In particular, Middle Paleozoic to Recent rocks are very well studied. For example, about 70 biostratigraphic zones based on calcareous nannofossils and on planktic foraminifers exist for the Cretaceous and Cenozoic for an average of about 2.0 My per zone. Similarly, there are about 30 zones based on radiolarians available for the Cenozoic also for an average of about 2.0 My per zone. These zones vary widely in duration, however, ranging from about 0.1 My or even less to as much as 5.0 My. The various protist zones are not equally recognizable everywhere for every time, but protists almost always provide a detailed and reliable biostratigraphy.

Evolution

The reason, of course, that Smith and his successors successfully used fossils to correlate rocks is because biologic species evolve through time. Smith and other early biostratigraphers knew nothing of evolution, and the explanation was later provided by Charles Darwin for the observed faunal succession through his theory of organic evolution by natural selection.

Three models seek to explain how evolution took place. The model that has come to be known as **phyletic gradualism** is based on Darwin's concept of evolution as a slow process. Morphologic changes in species populations, as a result of selection pressure, manifest themselves in the fossil record as a succession of subtly changing morphologies (Fig. 4.14). Such examples of phyletic gradualism are difficult to find in the fossil record. Proponents of this evolutionary process as the major way in which organisms evolve explain the lack of many clearcut examples by gaps in the rock record, incomplete sampling, and the lack of specimens at critical intervals because of preservation or other factors. The implication for biostratigraphy, if this is the dominant mode of evolution, is that where the rock record is complete species distinctions would be arbitrary with major intervals of time with transitional forms (Fig. 4.14). Therefore, biostratigraphic units based on such lineages would also not have sharp boundaries and would be difficult to recognize consistently. Many differences of opinion would exist between taxonomists as to where species start and end.

A second and newer model is termed **punctuated equilibrium** (Fig. 4.15) and holds that the reason new species seem to appear abruptly in the stratigraphic record is that randomly breeding large populations are genetically and, therefore, morphologically stable. Evolutionary change should then occur in small peripheral populations that are geographically isolated. Once a new species evolves, if it is ecologically tolerant and competitive, it expands its geographic range rapidly. Transitional forms

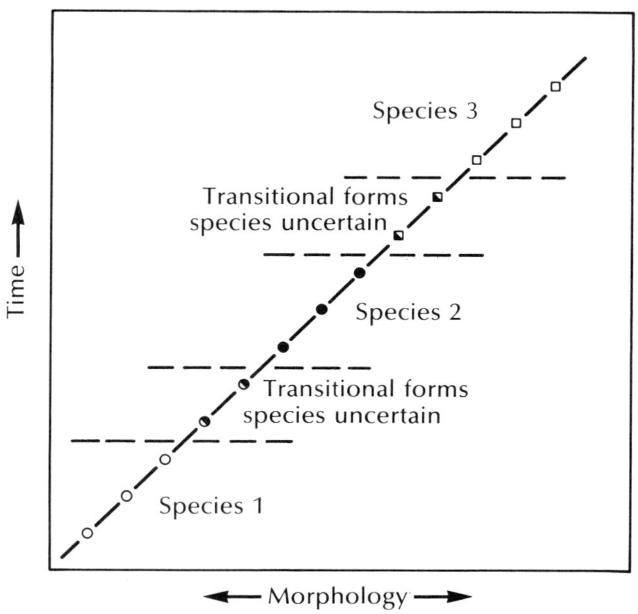

Figure 4.14 The phyletic gradualism model of evolution. There is slow morphologic change through time. Species boundaries are gradational and there are many transitional forms. Biozone boundaries based on such taxa would not be consistently recognizable.

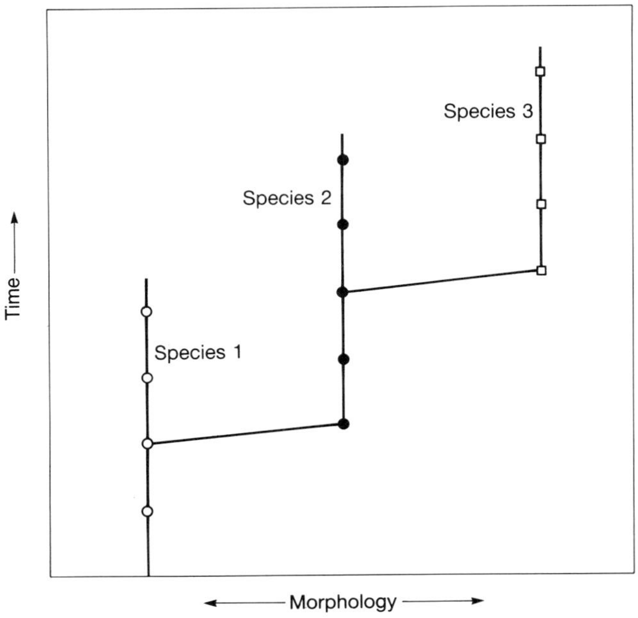

Figure 4.15 The punctuated equilibrium model of evolution. Species evolve in short time-periods in peripherally isolated populations and then quickly colonize broad areas. Transitional forms are few and difficult to find. Species boundaries are sharp and biozone boundaries, therefore, would also be easily recognizable. Parent species change little or not at all through time.

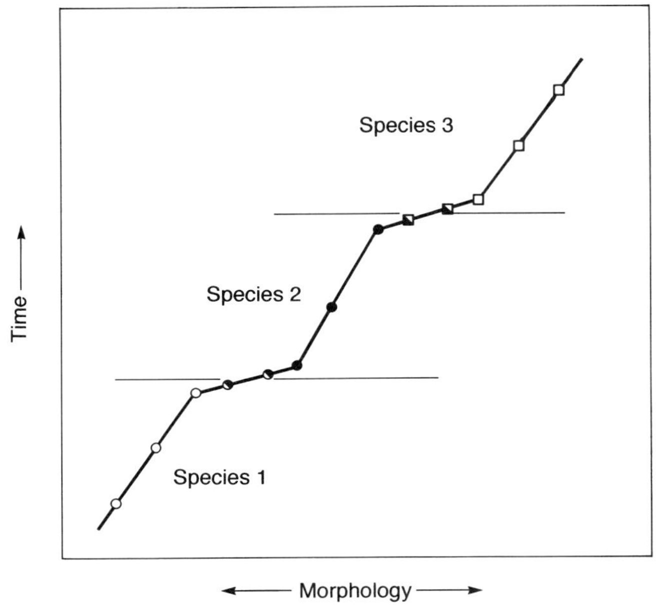

Figure 4.16 The punctuated gradualism model of evolution shows periodic acceleration of morphologic change in otherwise slowly changing lineages. Because of this acceleration, species boundaries are relatively sharp. Biozone boundaries based on such taxa are relatively easy to recognize.

between the parent and daughter species would be found only in the area of the original geographically isolated population, and the chances of collecting these forms would be slim. The fossil record itself suggests that most species' distributional patterns in time and space fit the punctuated equilibrium model and have evolutionary patterns similar to that shown in Figure 4.15. The major implication for biostratigraphy is that since species boundaries are clearly recognizable, the units of biostratigraphy that are based on such taxa would also be the subject of little debate.

Still more recently, another model for evolution has been suggested. Many protist groups are planktic, and were and are widely distributed in the world ocean. Their gene pools are large, and geographically isolated populations were probably few in number. Therefore, gradualism should be expected. What has been observed in some planktic foraminifers is a pattern of evolution termed **punctuated gradualism** (Fig. 4.16). It is a combination of the other two models and holds that daughter species evolve from parent species without branching, but instead of continuous gradual morphologic change, short periods of rapid morphologic (evolutionary) change occur in lineages. This could be in response to selection pressures generated, for example, by changing oceanic climate. Boundaries between species are relatively sharp, and, therefore, biostratigraphic units based on these lineages would also be easily recognized. Care must be taken, however, in interpreting observed patterns as shown in Figure 4.16 as representing punctuated gradualism. Gaps in the stratigraphic record of a lineage undergoing gradual change would yield a similar pattern.

Zones

The **biozone** is the basic tool of biostratigraphy. A zone is a body of strata defined by its fossil content. A variety of names and definitions have been proposed in the history of stratigraphic paleontology, but there are only three basic types of zones: **assemblage** zones, abundance zones, and **interval zones**. Assemblage zones are biozones characterized by the association of three or more taxa and recognized without regard to range limits of taxa. Such zones are usually of regional importance. Abundance zones are based on the numerical maxima of one or more taxa. This zone is strongly influenced by local ecologic conditions and therefore is usually of local biostratigraphic importance. Most zones based on protists and other microfossils are interval zones. This type of zone has boundaries defined by particular fossil events, specifically first appearance or last appearance datums (FADs and LADs) of taxa. An interval zone comprises those strata between the single evolutionary or extinction events that define the boundaries of the zone. There are five types of interval zones (Fig. 4.17).

The *taxon range zone* (Fig. 4.17A) is the interval between the FAD and LAD of a taxon. It is also referred to as the total range zone. The local range zone is a geographically restricted variation of the taxon range zone. The interval between the FAD of one taxon below and LAD of another above is the *concurrent range zone* (Fig. 4.17B). An interval zone defined as being between the LAD of a taxon below and the FAD of another above partitions the range of a third taxon that normally occurs commonly in the interval, but also above and below it. The zone is named after the third taxon. This zone is known as the *partial range zone* (Fig. 4.17C). The interval between the successive FADs of two taxa has been referred to as the *successive first appearance zone* (Fig. 4.17D). Conversely, the interval between two successive LADs can be referred to as the *successive last appearance zone* (Fig. 4.17E). This last type of interval zone is not as commonly used as the other zones in exposed or cored strata, but is the mainstay of industrial biostratigraphy because most of the samples studied are derived from well drilling cuttings and actual lowest occurrences are not determinable because of downhole contamination.

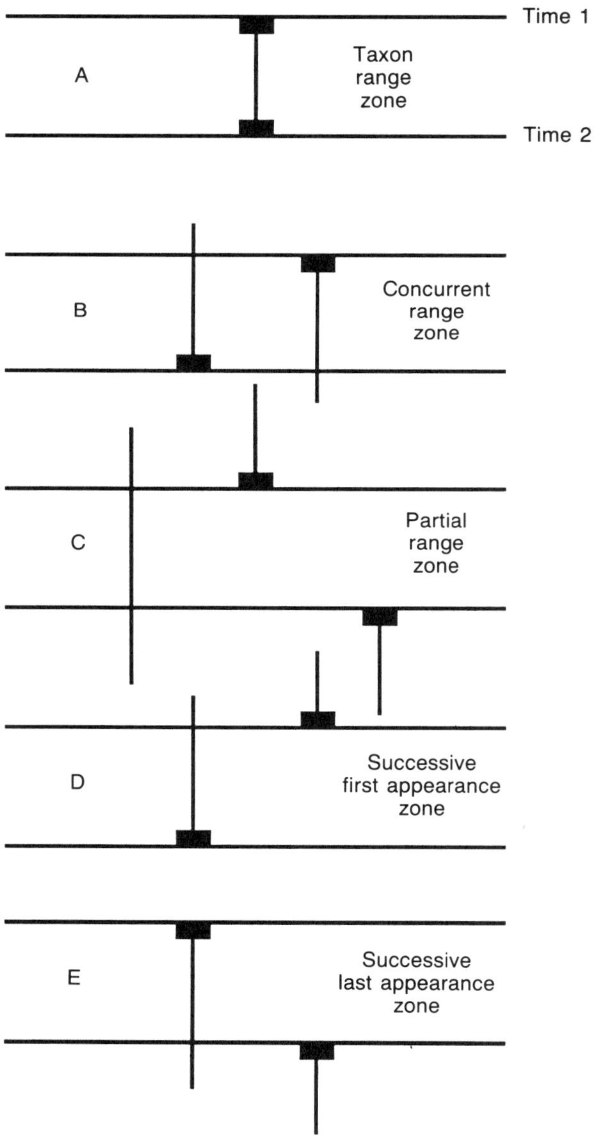

Figure 4.17 Types of biostratigraphic interval zones. Vertical lines are taxon ranges. Black bars indicate observed first and last occurrences.

Other events recorded by fossils may help recognize the zonal intervals. These fall into the category of *characterizing events*. Such events are evolutionary or extinction events that occur at the same or similar points in time as the events that define the zone boundaries. However, if the events that define the boundaries of an interval zone are not present in a stratigraphic section being studied, the zone is not identified. Characterizing events may lead the stratigrapher to conclude that a stratigraphic interval under study is correlative with intervals in other sections where the defining events are present, but the biozone term cannot be applied. In such cases, the *biochronozone*, a unit of *chronostratigraphy*, of the biozone is identified. A chronostratigraphic unit is a body of rock established to serve as the material reference for all rocks formed during the same interval of time. Therefore the biochronozone of a biozone encompasses all the rocks formed between the times of the events that define the biozone.

Biostratigraphy and the geographic distribution of species

Species vary from being very narrowly restricted in their geographic range to being very widely distributed. Taxa can have narrow longitudinal ranges and wide latitudinal ranges. An example is certain benthic forms living along north–south coasts. Conversely, some forms are restricted latitudinally and are widely distributed longitudinally, such as **thermophilic** planktic protists, certain calcareous nannofossils, diatoms, radiolarians, and foraminifers.

Figure 4.18 shows the hypothetical distribution of a species whose evolutionary first appearance and extinction might be used as zone-defining events. The illustrated envelope containing seven stratigraphic sections is the known distribution in space and time of the taxon. This is the zone. All the rocks formed between Time 1 and Time

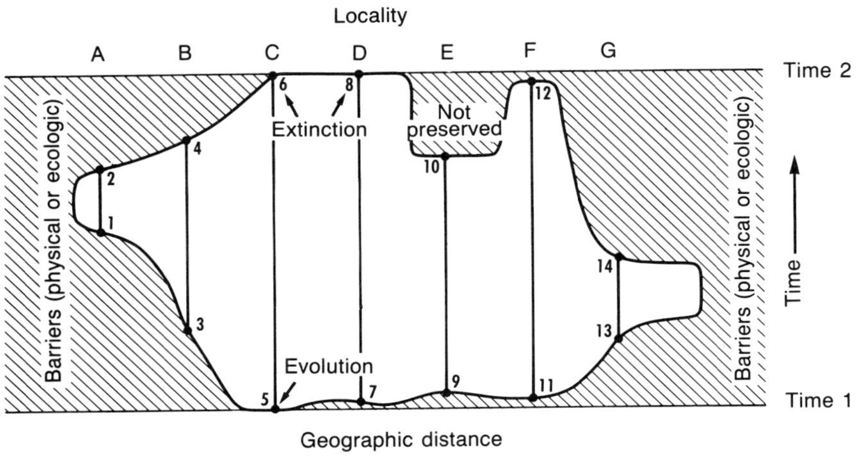

Figure 4.18 Hypothetical distribution of a single species in time (vertical) and space (horizontal). The total time range of the species from its evolution to its extinction is represented only at locality C. At all other localities, the species was either unable to occupy a particular site for its entire duration because of the time required for dispersal, ecologic or physical barriers, or it was not preserved for some reason.

2, both outside and inside of the envelope, make up the biochronozone of this taxon range zone. In this case only at section C do the zone and biochronozone coincide, that is, here the lower part of the range is filled. Only at sections C and D is the upper part of the range filled. At all the other sections the range of the taxon is unfilled to varying degrees. This, of course, has important implications for biostratigraphy. The occurrences where the range is unfilled are not the FAD or LAD of the species. If diagnostic characterizing information were not available, however, the biostratigrapher might not be able to conclude that the ranges were unfilled and varying degrees of inaccurate dating of the sections could occur. For example, the sections at A and at G are both in the taxon range zone of this species, but each represents only small and different parts of the biochronozone. Thus, although zonal biostratigraphy is a powerful stratigraphic tool and has served geology well for over 150 years, caveats are associated with its use.

Quantitative biostratigraphy

In the last 20 years geology in general has become a more quantitative discipline, and so has biostratigraphy. A large number of data manipulative techniques have been proposed to analyze different types of data sets. However, most microfossil biostratigraphic data are from measured sections or boreholes, and the important information conveyed is the stratigraphically lowest and highest occurrences of taxa in samples. These are event-stratigraphic data, and a family of techniques is available for the analysis of such data. Of these, those that are popular and powerful are based on graphing strategies. These use the geometry of graphs to determine the relative or absolute positions in time of fossil and other events. A major advantage of such techniques is that they allow the biostratigrapher to move from raw, event-type data to chronostratigraphy, and in doing so avoid the problems of zonal biostratigraphy discussed above. These techniques are excellent for determining the relationships of fossil events of various biotic groups to each other and to physical events such as reversals of the Earth's magnetic field and changes in the chemical isotopic composition of the ocean. The events are recorded on the same scale and provide a tool for high-resolution event stratigraphy.

The most widely used graphing technique is one proposed 28 years ago by A.B. Shaw and termed graphic correlation. It is used here to illustrate quantitative biostratigraphy. Sources of information on graphic correlation and other quantitative techniques can be found in the supplementary reading list at the end of this chapter.

Graphic correlation uses two-axis graphs to plot sedimentary rock thickness in any two stratigraphic sections against each other, or sediment thickness in a section against a composite (that is, a model) of two or more sections already analyzed using the technique. The locations of the first and last occurrences of fossils (magnetostratigraphic and other event-stratigraphic data may also be used) are plotted as points on the graph. As long as the section being plotted includes sedimentary rocks representing coincident or overlapping intervals of time with sections already graphed, the data in common may allow a correlation between the section and the model.

Graphic correlation may be used to compare any two measured stratigraphic sections with one another, but more importantly it may be used to combine data from many sections to produce a hypothesis concerning the position in time of all events. For multiple sections a correlation exists between each section and a composite of all sections being studied. A graphic correlation plot of data from a hypothetical measured section is plotted against a composite of previously studied sections in Figure 4.19. The model or composite is on the vertical axis. Units on this axis represent a composite of the feet or meters of the sections already compared and are referred to as composite units. The composite axis is a scale that measures time as expressed in terms of the rate of rock accumulation of the standard reference section. The standard reference section is the measured section chosen to represent the first model. In Figure 4.19 a solid black line is shown with three segments. This is the line of correlation (LOC) between the section and the model. The fossil first observed occurrences and last observed occurrences (open circles and plus signs, respectively) through which the LOC passes are therefore interpreted to be precisely in place in time in the section and the model. Those fossil events that are close to the LOC are nearly in place in time in the section and the model. These events provide the information needed to position the LOC. This placement is a subjective decision based on the evaluation of the importance of the individual plotted points. How the LOC is fitted through the data points is based on the skill and experience of the stratigrapher doing the graphing. Once it is positioned, an equation (or equations if there is more than one segment) can be calculated for the LOC using two points and simple linear regression.

First and last occurrences (Fig. 4.19) that plot considerably off the LOC represent distinctly unfilled ranges, either in the section or the model. Last occurrences above the LOC signify that the observed occurrence at this section is not as late in time as that event had been recorded at other sections used to build the model. First occurrences below the LOC represent the opposite case, that is, these events are known to occur earlier in time at at least one other section that is part of the model. Last occurrences below the LOC mean that this event occurred later in time in the new section and represents an unfilled range in the model and therefore needs range revision. This is done by projecting—either graphically or by use of the regression equation—to the LOC and then to the vertical axis to adjust the position of this event in the model. Conversely, first occurrences above the LOC signify that in the new section this event occurs earlier in time than in the model, and its position is adjusted in the model in the same manner. Any fossil events in the new section that are not as yet part of the model can now be added by projecting from the section axis to the LOC and then to the model axis. More sections are added, some of which only partially overlap the composite in time and therefore extend the coverage in time. New events are added to the database. Ranges are adjusted until the first evolutionary appearance and extinction events (FADs and LADs) for the taxa are discovered, and adjustments, therefore, are no longer needed.

In the context of graphic correlation, and providing there are data from other taxa that will allow LOCs

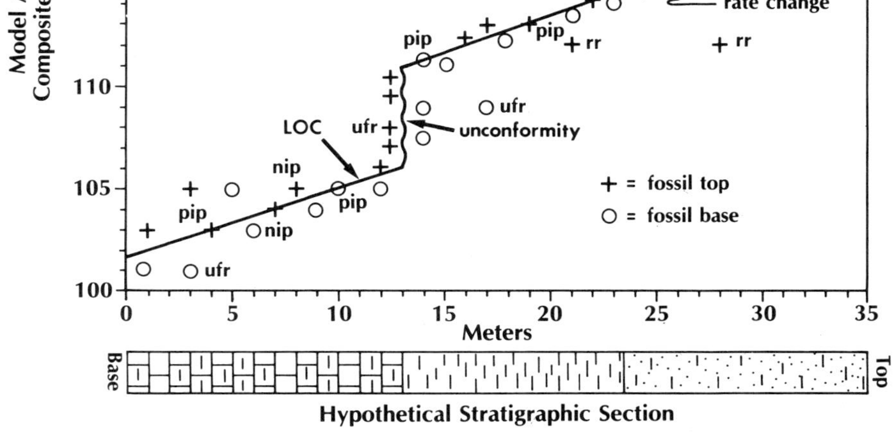

Figure 4.19 Graphic correlation plot of fossil highest and lowest occurrences at a hypothetical section against the position of these events in a model derived from previous plots. LOC, line of correlation; pip, precisely in place; nip, nearly in place; ufr, unfilled range; rr, range revision.

to be positioned for the sections, events 5, 6, and 8 of Figure 4.18 would be precisely in place. Events 7, 9, 11, and 12 would be nearly in place. Events 1, 3, and 13 will plot off the LOC below and 2, 4, 10, and 14 above; these are distinctly unfilled ranges. The sections of Figure 4.18 would be correlated without the use of zonal biostratigraphy.

The geometry of the graphic correlation plot (Fig. 4.19) also allows other geologic interpretations. In order to fit the LOC to the middle part of the stratigraphic section there must be a vertical offset, which indicates that in the section some time is not represented by rock; that is, there is an unconformity at 13.0 m. A normal fault or highly condensed section would also give this pattern. The duration of the hiatus at the unconformity can be determined by projecting the ends of the two LOC segments at the unconformity to the model axis. This gives a duration of 5.0 composite units (111.0−106.0=5.0). There is a bend in the LOC at 23.5 m. This indicates a change in the rate of rock accumulation from a relatively fast rate (more meters per composite unit; 3.0 m/cu) to a relatively slow rate (fewer meters per composite unit; 1.1 m/cu).

Supplementary reading

Boggs S Jr. *Principles of Sedimentology and Stratigraphy*. Columbus, OH: Merrill, 1987. Good chapter on biostratigraphy. Includes a complete reproduction of the North American Stratigraphic Code as an appendix.

Edwards LE. Insights on why graphic correlation (Shaw's Method) works. *J Geol* 1984; 92:583−97. An excellent summary of how and why graphic correlation works with references to important papers on graphing techniques.

Dowsett HJ. Application of the graphic correlation method to Pliocene Marine Sequences. *Marine Micropaleontol* 1989; 14:3−32. Discussion and application of graphic correlation using calcareous nannofossils and planktic foraminifera.

Gradstein FM, Agterberg FP, Brower JC, Schwarzacher WS. *Quantitative Stratigraphy*. Dordrecht: Reidel, 1985. Descriptions with examples of many quantitative techniques.

Hazel JE. Chronostratigraphy of upper Eocene microspherules. *Palaios* 1989; 4:318−29. Discussion of graphic correlation and creation of a high resolution model using three groups of protists.

Kauffman EG, Hazel JE, eds. *Concepts and Methods of Biostratigraphy*. Stroudsburg, PA: Dowden, Hutchinson & Ross, 1977. Still a major compilation of papers by many experts in biostratigraphy and closely related studies. Comprehensive list of references up to the mid-1970s.

Malmgren BA, Berggren WA, Lohmann GP. Evidence for punctuated gradualism in the Late Neogene *Globorotalia tumida* lineage of planktonic foraminifera. *Paleobiology* 1983; 9:377−89.

North American Commission on Stratigraphic Nomenclature. North American Stratigraphic Code. *Amer Assoc Petrol Geologists Bull* 1983; 67:841−75. A copy should be on every English-speaking geologist's bookshelf. Available from AAPG for only $1.00. Contains references to important papers and to other codes.

Taylor ME. Biostratigraphy and paleobiogeography. In: Boardman RS, Cheetham AH, Rowell AJ, eds. *Fossil Invertebrates*. Palo Alto, CA: Blackwell Scientific Publications, 1987: 52−66. Good summary of biostratigraphy and factors controlling animal distributions.

Tipper JC. Techniques for quantitative stratigraphic correlation: a review and annotated bibliography. *Geol Mag* 1988; 125:475−94. Useful compilation of references to quantitative techniques with a discussion of various methods and philosophies.

Prokaryotes

5

Stjepko Golubic, Andrew H. Knoll

Introduction

Prokaryotes, or bacteria, do not comprise a class, a phylum, or for that matter a **monophyletic clade** of any kind. The term prokaryote refers to microorganisms that exhibit a common grade of intracellular organization—organisms now commonly grouped in two separate kingdoms. Prokaryotes are small but ubiquitous. They occur in alpine snow fields and hot sulfur springs, in acid mine drainage and alkaline playa lakes, in marine plankton, and in the rumen of cows. Millions of prokaryotes reside on your skin and in every moist pore of your body. In fact, the number of bacterial cells within your digestive tract far exceeds the number of nucleated cells in your entire body. Prokaryotes are central to the cycling of materials through the biosphere and are of practical importance in disease, decay, and any number of industrial processes, including cheese making and the clean-up of oil spills.

Prokaryotes were originally defined in terms of eukaryotic features they lack—no membrane-bound nucleus, energy metabolism not localized in membrane-bound organelles, no mitosis or meiosis. Although clearly establishing prokaryotes as a group distinct from the nucleated cells of our common experience, this approach to differentiation unfortunately was accompanied by an interpretation of prokaryotic features as primitive and eukaryotic ones as advanced. In recent years, this negative definition of prokaryotes has been gradually replaced by one that stresses the features they do possess and how these differ from analogous characteristics of eukaryotes (Table 5.1).

In the geologic record, prokaryotes occur as **body fossils**, **chemical fossils**, and **trace fossils** (**stromatolites**, **oncolites**, and minute cavities bored by **endolithic** microorganisms—microborings). They are the oldest known fossils on Earth, providing a record of evolution 3500 My long. Although prokaryotes were probably diverse through most of this time, the majority of these fossils appear to belong to just one group of prokaryotes, the cyanobacteria.

History of study

Although prokaryotic organisms were among the animalcules observed by Antonie van Leeuwenhoek (1632–1723) using his newly invented microscope in the 1670s, knowledge of their biology and systematic relationships long lagged behind that of plants and animals. Carolus Linnaeus (1707–1778), the father of plant and animal systematics, explicitly recognized the existence of these organisms, but thought them too small to classify. He instead placed them with other microbes into a group he named *Chaos* (!). Bacteriology came of age during the second half of the nineteenth century as scientists such as Louis Pasteur (1822–1895), Ferdinand Cohn (1828–1923), Robert Koch (1843–1910), Martinus W. Beijerinck (1851–1931), and Sergei N. Winogradsky (1856–1953) determined the basic properties of prokaryotes and demonstrated their importance in pathology and environmental processes. Modern microbial ecology has been advanced by the research and teaching of the Dutch school of microbiology at Delft, especially by A.J. Kluyver (1888–1956) and C.B. van Niel (1897–1985).

Stromatolites were described from Proterozoic carbonates more than a century ago, and a number of early twentieth century paleontologists reported the presence of bacteria in Precambrian rocks. These reports notwithstanding, most scientists remained skeptical about the possibility of a Precambrian microfossil record until Stanley Tyler and Elso Barghoorn reported the discovery of well-preserved microorganisms in cherts of the 2000 My old Gunflint Iron Formation, Ontario. The abundant filaments and cocci found in Gunflint rocks were not easy to dismiss as either mineralogic artifacts or younger contaminants, and by the time that Barghoorn and Tyler published their seminal monograph on these fossils in 1965, the reality and antiquity of the prokaryotic fossil record had become accepted by the scientific community. In the relatively short time that has elapsed since Precambrian paleontology was born, this field has

Table 5.1 Major differences between prokaryotes and eukaryotes

Prokaryotes	Eukaryotes
Organismal groups	
Archaebacteria, eubacteria	Protists (protozoa and algae), fungi, plants, animals
Cell organization	
Simple, all cell functions take place within a single intracellular space bounded by a unit membrane. Cells (0.2–)0.5–2(–80) μm wide	Intracellular space is compartmentalized into membrane-bound organelles performing specialized functions. Cells (0.5–)10–50(–200 000) μm wide
Development	
Mostly unicellular and microscopic forms. Differentiation limited	Uni- and multicellular, micro- and macroscopic forms. Differentiation can be extensive
Cell wall	
Contains peptidoglycan only in eubacteria. Glycoproteins only in archaebacteria. Cell wall absent in mycoplasmas	Contains chitin or cellulose. Glycoproteins common. Cell wall absent in protozoa and animals
DNA	
A single molecule of DNA is in a closed-loop chromosome (nucleoid), attached to plasma membrane. Additional DNA in circular plasmids	DNA distributed in several linear chromosomes, complexed with proteins (histones), within a membrane-bound nucleus which also contains RNA
Sexual systems	
Absent or unidirectional (from donor to recipient). Genetic transfer and recombination by transformation, transduction, or conjugation	Regular, involving equal participation of both partners. Diploid and haploid forms alternate between fertilization and meiosis
Ribosomes	
All ribosomes with a sedimentation constant of 70S (Swedberg units, with subunits of 50S and 30S)	Ribosomes in cytoplasm with sedimentation constant of 80S (subunits 60S and 40S), those in mitochondria and plastids of 70S
Cell division	
Cell division by fission, following DNA duplication, and separation along plasma membrane	Cell division by various forms of mitosis, involving microtubules and mitotic spindle in chromosome separation
Motility	
Simple, rotating bacterial flagella composed of flagellin protein. No cytoskeleton, intracellular motility, or phagocytosis. Gliding motility common. Gas vesicles present in some forms	Complex, flexing 9 + 2 flagella and cilia, composed of tubulin and other proteins. Cytoskeleton, ameboid movement, and phagocytosis based on actin-like proteins. Gliding motility common. Gas vesicles absent
Spores	
Endospores containing dipicolinic acid, heat resistant. Actinospores, conidia, myxospores, akinetes	Endospores absent. Various reproductive and resting spores following mitosis, meiosis, or fertilization (zygospores)
Metabolism	
Extremely diverse. Obligately and facultatively anaerobic, microaerophilic, and aerobic forms	Almost all are aerobic; exceptions are few and mostly secondary
Glycolysis and respiration	
Several glucose metabolism pathways. Respiration enzymes bound to plasma membrane or mesosomes. Not packaged separately	Embden–Meyerhof glucose metabolism, followed by Krebs (CTA) cycle, and cytochrome-based electron transport. Respiration enzymes packaged within mitochondria
Photosynthesis	
Anoxygenic and oxygenic photosyntheses, with one or two photosystems. Various electron donors including H_2O. Enzymes bound to plasma membrane, chromatophores, thylakoids, or vesicles, not packaged separately	Only oxygenic photosynthesis involving two photosystems. H_2O as used as electron donor. Enzymes for photosynthesis in thylakoids, packaged within membrane-bound plastids
Lipids and secondary products	
Vaccinic and oleic acids, and hopanes common. Archaebacterial lipids ether-linked. Steroids rare. Various antibiotics common	Linoleic acid common. Steroids and alkaloids common

matured into an active and integral subfield of paleontology as a whole. Hundreds of microfossil assemblages have been discovered, stromatolites have received detailed study as both biologic and sedimentary objects, and geochemical insights on prokaryotic life have mushroomed.

Systematics and phylogeny

Prokaryotes are difficult to classify and describe in the same manner as eukaryotes. The commonly accepted definition of a biologic species—groups of interbreeding populations that are reproductively isolated from other such groups—cannot be applied to prokaryotes because they reproduce by fission and can take up genetic material from the environment, or from other distantly related bacteria. Among prokaryotes, species as defined for eukaryotes simply do not exist. Nevertheless, prokaryotes diversify and speciate. Their populations share phenotypic characters, perform common ecologic functions, and display common evolutionary roles and tendencies. Because of their morphologic simplicity, most prokaryotic species are described on the basis of physiologic and biochemical traits. Only in morphologically more complex taxa such as cyanobacteria and actinomycetes can size and shape be applied usefully to classification.

The principal feature of the Linnaean system of classification is the hierarchical ordering of taxa. When Linnaeus proposed his system, the grouping of species into genera, families, orders, and classes was explicitly artificial, but most modern biologists strive to achieve a classification that reflects phylogeny. For many years, the higher-level taxonomy of prokaryotes was a classification of convenience, grouping together species sharing common morphologic, ultrastructural, or physiologic characters. The use of computers permitted evaluation of a larger number of phenotypic traits, a practice often referred to as numerical taxonomy. However, few microbiologists had any confidence that systems of prokaryotic organisms could be interpreted in terms of evolutionary relatedness. Molecular biology, however, provides a means of understanding prokaryotic physiology, genetics, and phylogeny. The most successful application of these approaches in prokaryote systematics involves comparisons of nucleotide and amino acid sequences, determined from pure cultures of a large and representative number of microorganisms.

An approximation of the genetic relatedness between two microorganisms can be determined from their DNA base ratios. Because two of the four DNA bases are complementary, the fraction of guanine plus cytosine base pairs provides a gross formula of DNA base composition (% mole G + C), which is constant for each species. Organisms with widely differing %G + C values are unrelated; those with similar ratios, however, may or may not be related, since the sequential arrangement of the bases in this test remains unknown. DNA–DNA hybridization, which matches complementary sequences of DNA *in vitro*, has been successful in comparing closely related organisms. In the process of protein synthesis, the genetic code, written as a sequence of DNA and mRNA bases, is translated into a sequence of amino acids. Comparison of amino acid sequences in homologous enzymes in different organisms provides another tool to study taxonomic relatedness. In recent years, prokaryotic phylogeny has also been ascertained by comparing base sequences of 5S and 16S rRNA fragments of the smaller (30S) subunits of ribosomes. The phylogenetic distance between two organisms is assessed on the basis of the minimum number of changes that have occurred independently within particular conservative portions of the molecule since the two lineages departed from their common ancestry. The results of sequence analyses are cataloged and compared with the aid of a computer. This method has provided insights into the relations among higher prokaryotic and eukaryotic taxa.

Molecular genetics confirmed that prokaryotes differ significantly from eukaryotes in rRNA sequence. However, the comparisons revealed an additional deep phylogenetic split within the prokaryotes. Most common bacteria cluster together into a group recognized as the kingdom Eubacteria, while a number of unusual prokaryotes cluster separately into a kingdom-level group christened the Archaebacteria. Further research has confirmed the distinction between these two groups, demonstrating that they also differ in many important phenotypic features, justifying the division of prokaryotic microorganisms into two separate kingdoms, the Archaebacteria and Eubacteria.

The archaebacteria diversified into three functionally different groups (Fig. 5.1, left; Table 5.2), dominated by extreme thermophiles with sulfur-dependent metabolism, methanogenic bacteria and relatives, and extreme halophiles, respectively. The eubacteria diversified into about 10 distinct phyla and a few isolated basal branches that may represent additional, as yet poorly defined, phylum-level groups (Fig. 5.1, right; Table 5.3). As in archaebacteria, the basal, most isolated branches of the eubacterial tree are occupied by anaerobic thermophiles, suggesting that these traits may generally be considered ancient. The eubacteria have evolved a remarkable variety

Table 5.2 Major groups of archaebacteria. (Modified from Woese CR. *Microbiol Rev* 1987; 51:221–71)

Methanogenic bacteria Characterized by ability to generate methane by the reduction of CO_2 using H_2, or by the anaerobic metabolism of acetate
Extreme halophilic bacteria Distinctive phototrophy, using bacteriorhodopsin as a photoreactive pigment. High internal K^+ concentration
Extreme thermophilic bacteria Anaerobic, thermophilic organisms mostly requiring sulfur in metabolism (SH chemoautotrophy, SO_4^{2-} reduction, anaerobic S respiration of organics). Some have evolved aerotolerance and even aerobic metabolism

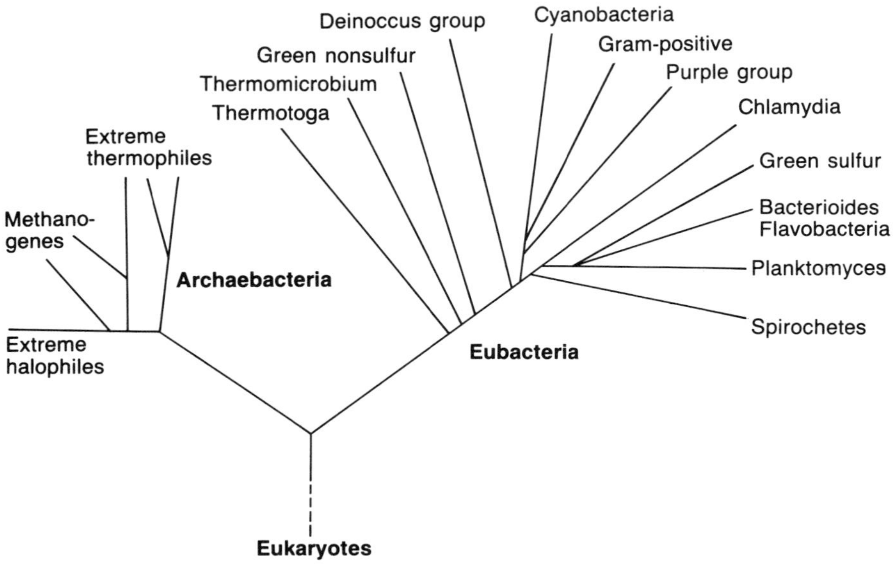

Figure 5.1 Prokaryotic phylogeny based on 16S rRNA sequence comparisons. Two kingdom-level clusters are distinguished: Archaebacteria and Eubacteria, both distant from the third cluster comprising the eukaryotic cytosol. (Modified from Woese CR. *Microbiol Rev* 1987; 51:221–71.)

Table 5.3 Major groups of eubacteria. (Modified from Woese CR. *Microbiol Rev* 1987; 51:221–71)

Purple bacteria
Metabolically the most diverse eubacteria. Thought to be fundamentally phototrophic, but later evolved other metabolic modes. Includes purple sulfur and purple nonsulfur phototrophic bacteria using photosystem I; chemolithotrophs, sulfate reducers, N_2 fixers (rhizobacteria); pathogens of both plants (agrobacteria) and animals (rickettsias, *Legionella*), and *Escherichia coli*. Genealogic ancestry of mitochondria

Gram-positive bacteria
Distinguishable on basis of cell wall that retains Gram stain (but exceptions exist). Includes the phototrophic heliobacteria (photosystem I); the predominantly branched filamentous, aerobic actinomycetes; aerobic, heterotrophic rods and cocci such as staphylococci; and anaerobic spore-forming bacteria such as the clostridia and streptococci

Cyanobacteria
The so-called blue–green "algae." Photoautotrophs using photosystems I and (usually) II. Contain chlorophyll *a*. Includes prochlorophytes (which contain chlorophyll *a* and *b*). Genealogic ancestry of the plastids found in algae, bryophytes, and vascular plants

Green sulfur bacteria
Phototrophic bacteria using photosystem I and distinctive architecture of photosynthetic membrane system

Spirochetes
Distinguishable by spiral shape and periplasmic, axially coiled fibrils. Anaerobic or (less commonly) aerobic heterotrophs

Bacterioides/flavobacterium group
Anaerobic and aerobic Gram-negative heterotrophs united largely on the basis of ribosomal RNA similarities. Includes gliding flexibacteria

Planctomyces group
Stalk-forming, aerobic heterotrophs. Cell walls lack peptidoglycan, contain proteins. Require low nutrient levels

Chlamydiae
Obligate intracellular parasites. Obtain ATP* for energy metabolism from hosts. Cell walls lack peptidoglycan. Cause of diseases in birds and mammals

Radiation-resistant deinococcus group
Atypical cell walls contain ornithine. Includes Gram-negative *Deinococcus* and Gram-positive *Thermus*

Green nonsulfur bacteria
Includes distinctive phototrophic (photosystem I) bacteria related to *Chloroflexus*, as well as other, as yet incompletely characterized, thermophilic and mesophilic bacteria

Other eubacteria
Although most eubacteria have 16S rRNA sequences that fit into the preceding 10 groups, some recently discovered bacteria of unusual phenotypes appear to represent additional major divisions of the kingdom

* ATP, adenosine triphosphate.

of metabolic pathways. However, the widespread distribution of phototrophy among eubacteria, known to occur in five phyla, suggests that this metabolic mode may have evolved early within the kingdom. Aerobic respiration appears to have evolved independently several times, possibly by modification of photosynthetic pathways.

While some eubacterial groups defined by molecular phylogeny correspond to traditionally recognized taxonomic groups—the cyanobacteria are a good example—many bacterial groups defined on the basis of conventional phenotypic characters cannot be reconciled with molecular phylogenetic schemes. Some phyla that cohere tightly using molecular techniques contain a wide variety of phenotypes; conversely, members of some traditionally defined taxonomic groups are dispersed in several different phyla. An explanation of this pattern necessitates widespread parallel and convergent evolution, widespread lateral transfer of genes coding for physiologic capabilities, or a combination of the two.

Two different opinions exist on relationships between the archaebacteria and eubacteria, and on the relationship between eukaryotes and the prokaryotic kingdoms. In one (Fig. 5.1), the three kingdoms are fundamentally separate, meaning that the last common ancestor of archaebacteria, eubacteria, and eukaryotic cytosol (cytoplasm and nucleus, excluding organelles bounded by a double membrane) was an extinct group of organisms more primitive than any found today. In this scheme, all kingdoms are monophyletic, although the eukaryotic cell is fundamentally chymeric (a composite cell made of components having different phylogenetic origins). In eukaryotes, the principal organelles responsible for energy metabolism, the mitochondrion and chloroplast, are derived from eubacteria, specifically the purple bacteria and cyanobacteria, respectively. An alternative phylogeny based on the same data (Fig. 5.2) holds that the eubacteria and the eukaryotic cytosol are descended from the archaebacterial methanogenic and extremely thermophilic branches, respectively. In this interpretation, the archaebacteria are considered as paraphyletic, having the two other monophyletic kingdoms nested within it. The eubacteria constitute a monophyletic clade with some archaebacteria, while the eukaryotic cytosol and other archaebacteria form a second primary, monophyletic group. Whichever phylogeny is preferred, the results show that prokaryotic organisms provide the keys to understanding the earliest diversification of life on Earth.

Biology

Morphology

Morphologically, the simplest bacteria are small, spherical, or rod-shaped cells about 0.5 μm in diameter. Most archaebacteria and many eubacteria (e.g., most pathogens) fall into this category. Simple prokaryotes have traditionally been named according to their shapes; "cocci" (spheres, e.g., *Streptococcus pyogenes*), "bacilli" (straight rods, e.g., *Bacillus anthracis*), "vibrios" (curved rods, e.g., *Vibrio cholerae*), and "spirilli" (helical rods, e.g., *Rhodospirillum rubrum*). While many prokaryotes are morphologically simple, others include complex and highly differentiated forms. **Budding** and prosthecate bacteria form cellular protrusions (prosthecae) and hyphae-like extensions. Sheathed and stalked bacteria produce extracellular polysaccharide envelopes (glycocalyx), sheaths, and stalks, often leading to ecologically important adaptations, by providing attachment and protection, or promoting pathogenicity. Some bacteria form special resistant resting stages, called **endospores**.

The most complex morphologies among prokaryotes are found in the myxobacteria, actinomycetes, and cyanobacteria. Myxobacteria exhibit the most elaborate behavior and life-cycle among prokaryotes. The rod-shaped, gliding cells of myxobacteria aggregate in a coordinated fashion to build large, 0.1–0.2 mm high, often strikingly colored fruiting bodies filled with millions of spores. Actinomycetes form mycelia, extensive networks and fungus-like filaments (hyphae). They reproduce by spores or conidia borne on specially differentiated structures. Cyanobacteria include many large and differentiated forms, with cells up to 60 μm in diameter. Some of them are considered to be truly multicellular.

Major differences between prokaryotic and eukaryotic cells can be seen in the ways their cellular constituents are organized. These differences are evident in their ultrastructural properties as revealed by electron mi-

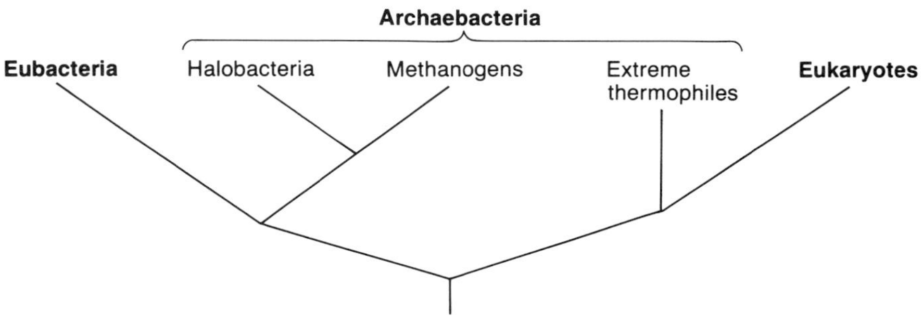

Figure 5.2 Phylogenetic tree showing possible alternative relationships between the three kingdoms based on the same data as Figure 5.1. (Modified from Lake JA. *Nature* 1988; 331:184–6.)

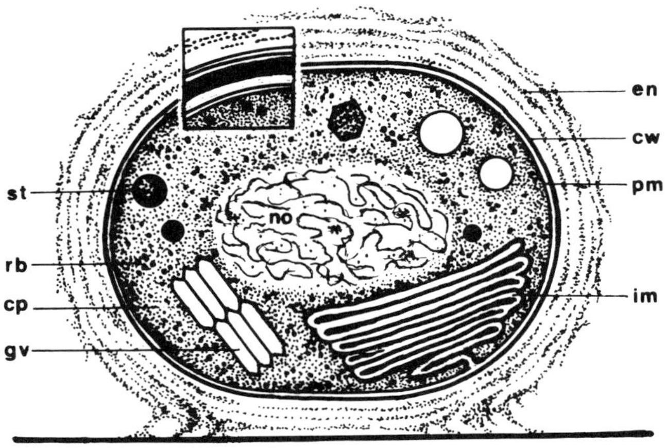

Figure 5.3 Ultrastructure of a prokaryotic cell: no, nucleoid (area with DNA); en, polysaccharide envelope (glycocalyx); cw, cell wall; pm, plasma membrane; im, intracytoplasmic membranes (e.g., thylakoids); gv, gas vesicles; cp, cytoplasm; rb, ribosomes; st, storage granules (e.g., lipids, polyphosphate, cyanophycine, sulfur, glycogen, PHB). (Modified from Schlegel HG. *Allgemeine Mikrobiologie*. Stuttgart: G. Thieme Verlag, 1985.)

croscopy. All constituents of a prokaryotic cell are located within a single noncompartmentalized intracellular space, that is surrounded by **plasma membrane** as shown schematically in Figure 5.3. The nucleoid is a centrally located area of fine fibrous texture containing the coiled DNA-loop (no). 70S Ribosomes (rb) are scattered within the **cytoplasm** (cp) as small, electron-dense dots, while the plasma membrane (pm) is resolved (Fig. 5.3, inset) as two electron-dense layers separated by an electrolucent one. No double membrane-bounded intracellular organelles exist as in eukaryotic cells; however, intracellular extensions of the plasma membrane in prokaryotes (im) may form tubular or lamellar involutions, stacks, and packages (chromatophores, thylakoids). These are usually sites of intensive enzymatic activities. In addition, there are inclusions bounded by a single membrane, such as gas vesicles (gv), chlorosomes, and carboxysomes. Cells contain various storage inclusions (st), such as lipid inclusions, glycogen, PHB (poly-β-hydroxybutyrate), polyphosphate, and sulfur granules. Cell-wall composition varies from one prokaryotic group to another, and some groups lack cell walls altogether (e.g., mycoplasmas). External features of prokaryotic cells include bacterial flagella, prosthecae, pili, and external envelopes (en; stalk, capsule, glycocalyx).

Genetics

Prokaryotes and eukaryotes share basic genetic properties: genetic information is encoded and replicated in DNA, and protein synthesis is mediated by RNA through processes of transcription and translation. Nevertheless, the ways in which genetic information is stored and transmitted are substantially different in prokaryotic and eukaryotic cells. DNA in prokaryotes is organized in a single closed-loop chromosome attached to the cell membrane. In addition, a small number (5–100) of bacterial genes are often located on separate smaller loops of DNA called plasmids, which can either replicate separately or become integrated into the main genome. Unlike eukaryotes, prokaryotic organisms do not have histone proteins associated with their DNA. More than 90% of DNA is engaged in coding for proteins, and transcription and translation are linked. Prokaryotic mRNA is usually polycistronic—it carries the code for more than one protein. Genetic information in prokaryotes can be transferred from cell to cell by one of the following means. Transformation involves incorporation of free DNA from the environment into cells. Transduction involves viruses as vectors that transfer DNA. Conjugation involves pairing and physical contact of bacterial cells during which DNA passes from the donor cell into the recipient cell. Genetic recombination in prokaryotes is limited, occurring by the occasional exchange of homologous or nonhomologous pieces of DNA between the genome and plasmids, genome and viral DNA, or genome and transformed DNA.

In contrast, genetic information in eukaryotic DNA is distributed on separate linear chromosomes. A large proportion of DNA (sometimes exceeding 90%) does not code for proteins. The noncoding sequences, called introns, are enzymically removed following transcription, so that only coding sections, or exons, enter the translation process. Transcription occurs within the nucleus, whereas translation takes place in the cytoplasm. Eukaryotic mRNA is monocistronic—it carries the code for a single protein. Genetic information in eukaryotes is transmitted sexually, and genetic recombination is regulated in the processes of mitosis and meiosis.

Metabolism

All organisms require energy to grow, reproduce, move, or simply to maintain their structure. Metabolism describes the totality of chemical transformations that take place in living organisms, closely linked with energy generation and expenditure. While eukaryotes are morphologically diverse, the prokaryotes are by far the more metabolically diverse organisms. All organisms are constructed of carbon polymers with an average C : H : O ratio of 1 : 2 : 1. The principal sources of carbon in the environment are CO_2 and CO_3^{-2}, compounds much more oxidized than organic carbon. In addition to a carbon source, all organisms require reducing power. Oxidations are usually accompanied by a release of energy, while most reductions require an input in energy. Energy is required for the synthesis of organic carbon compounds.

The metabolic diversity of microorganisms can be classified on the basis of combinations of three essential resources: energy, reducing power, and carbon source (Table 5.4). According to this scheme, the photo-litho-autotrophs (upper row) refer to photosynthesizing organisms: cyanobacteria, some other phototrophic eubacteria, eukaryotic algae, and green plants, which are the principal primary producers of organic matter. The term chemo-litho-autotrophs applies to chemosynthesizing primary producers, which play a minor role in terrestrial and shallow-water environments, but are prominent in the type of communities found at deep-sea vents. The combination chemo-organo-heterotrophs (lower row) describes anaerobic and aerobic heterotrophic bacteria, as well as fungi and animals. They constitute the consumers and decomposers in various ecosystems. Those microorganisms which fall into the remaining combinations, or those which can opt for more than one combination, are called mixotrophs. These combinations are not equally represented among organisms; some are metabolically favored; others are linked, or interconvertible. Hence, these terms are often used loosely. The term **heterotroph**, for example, designates dependency on organic carbon sources; a dependency on organic electron donors is implicit.

Central to all processes of biologic energy generation is the capture of chemical energy released in oxidation processes. During this process, electrons pass from one compound that serves as an electron donor to another, the electron acceptor. In living systems, electrons are transferred through a series of oxidation–reduction (redox) reactions carried out by specialized enzymes and coenzymes. This transfer may involve electrons alone (usually two at a time), or electrons in conjunction with protons (dehydrogenation–hydrogenation reactions). The energy released in oxidation processes can be converted into a pH (proton) gradient across membranes, which creates an electric potential that can subsequently be employed for concentrating nutrients, for cell motility, or for the synthesis of high-energy phosphate bonds (mostly ATP). Phototrophic organisms are able to convert light energy into chemical energy by the photoexcitation of electrons within light-harvesting pigments such as chlorophyll.

Steep redox gradients often develop across sediment–water interfaces in aquatic systems, and this is reflected in distinct horizons dominated by particular, metabolically different microbial populations (Fig. 5.4). Various types of respirers constitute the principal heterotrophs near the sediment surface. They use organic matter as the electron donor and oxidize it to CO_2, while using external electron acceptors via a sequence of coordinated, membrane-resident, cytochrome-carrying enzymes. Aerobic respirers use molecular oxygen as the electron acceptor. A change in redox potential in the course of aerobic respiration of about 1.2 volts results in considerable energy yield. Intensive microbial respiration can deplete ambient

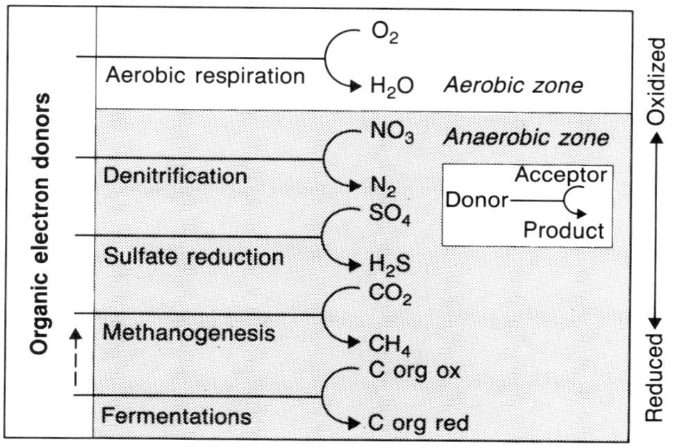

Figure 5.4 Aerobic and anaerobic processes near the sediment–water interface. The upper portion (mostly above the sediment–water interface) is the oxygenated zone promoting aerobic processes (i.e., aerobic respiration). Below the oxycline (line of O_2 depletion) are increasingly reduced sediment layers, supporting a series of anaerobic processes, that employ various electron acceptors. Anaerobic respiration includes denitrification closer to the sediment surface and sulfate reduction deeper in the reduced zone. Methanogenesis and various types of fermentation operate under extremely reduced conditions. They may be functionally linked: fermentations that release molecular hydrogen provide an electron donor for methanogenesis (dashed line). In well-oxygenated, oligotrophic environments the oxidized zone moves below the sediment–water interface; in stagnant, eutrophic environments the oxycline moves above it, with anoxic conditions spreading into the water column. (Modified from Schlegel HG. *Allgemeine Mikrobiologie*. Stuttgart: G. Thieme Verlag, 1985.)

Table 5.4 Energy, reducing power, and carbon source in prokaryotic metabolism

Energy	Reducing power	Source of carbon
Light: phototrophs	Inorganic: lithotrophs	Inorganic: autotrophs
Chemical: chemotrophs	Organic: organotrophs	Organic: heterotrophs

oxygen causing anoxic conditions. **Facultative** anaerobic respirers are then able to switch to nitrate (NO_3^-) as the electron acceptor. These bacteria are also known as denitrifiers, because they release molecular nitrogen (N_2). A few types of bacteria can use ferric iron (Fe^{3+}) as an electron acceptor. Deeper in the anaerobic sediment **obligate** anaerobic respirers occur which use sulfate (SO_4^{-2}) as the electron acceptor. They are also called sulfate reducers. One product of this energy-generating (dissimilatory) sulfate reduction is hydrogen sulfide (H_2S), which is released into the environment. Energetically, the least efficient processes are carried out by the fermenters. Fermentable compounds such as carbohydrates and amino acids contain considerable chemical energy; however, fermenters are able to extract only a small fraction of the potential energy because they use internally generated, only slightly less reduced organic substrates as electron acceptors. Fermentation products, such as alcohols and organic acids, remain reduced, thus retaining high caloric value. They accumulate in the environment and are good energy sources for other microorganisms. Fermentation reactions are often accompanied by the release of gases, such as CO_2 and H_2. Fermenters often co-occur with another specialized group of bacteria—the methanogens. These anaerobic archaebacteria are able to extract energy using H_2 as the electron donor and CO_2 as the acceptor, releasing methane (CH_4) as their product.

Some specialized microorganisms are able to use reduced inorganic compounds as electron donors, and molecular oxygen as the electron acceptor. Their energy supply is independent of organic sources. These organisms are also able to use inorganic compounds as a source of reducing power in their biosynthesis, hence they are called *chemolithotrophs*. If they also use CO_2 as the sole source of carbon, they qualify to be called *chemolithoautotrophs*. Chemolithotrophs occur in interface environments, along the contact between the anaerobic and aerobic zones. Their energy yield depends on the redox potential of their electron donors, which range from molecular hydrogen at the negative (reduced) end of the redox scale (hydrogen-oxidizing bacteria), to ferrous iron at the positive (oxidized) end of the scale (iron-oxidizing bacteria). Together with sulfide-(H_2S) oxidizing bacteria, methane-(CH_4) and ammonia-(NH_3) oxidizing bacteria are the principal primary producers in deep-sea vent and seep communities. Colorless sulfur-oxidizing bacteria can be autotrophic and mixotrophic. Their product, elemental sulfur (S°), can serve either as an electron donor or as an electron acceptor for other microbial reactions. Sulfur-oxidizing bacteria are found in both major groups, the archaebacteria and the eubacteria. They live in a wide variety of marine and freshwater environments. Chemolithotrophs that oxidize Fe^{2+} and Mn^{2+} are thought to be active in formation of iron ores and deep-sea manganese nodules.

An unusual group of aerobic archaebacteria are the extreme halophiles, which require high concentrations of environmental Na^+ (ca. 3 × sea water up to saturation) internally balanced by equivalent concentrations of K^+. Extreme halophiles evolved a light-driven proton-pump system, using the pigment bacteriorhodopsin, which is similar to the visual pigment of the eye.

The most effective energy generation is carried out by phototrophic bacteria. Anoxygenic phototrophs include purple (sulfur and nonsulfur) bacteria, green (sulfur and nonsulfur) bacteria, and heliobacteria. Their light-harvesting systems are based on different types of bacteriochlorophyll. Energy is generated by a cyclic photophosphorylation process in which electrons are excited by long wavelength photons and recycled. Reducing power is obtained either from organic compounds (nonsulfur purple bacteria and heliobacteria), or from H_2S (sulfur purple bacteria and green bacteria). Anoxygenic phototrophic bacteria are either obligate or facultative anaerobes. However, they can use light only under anaerobic conditions; in the presence of oxygen, their metabolism is heterotrophic. The pigments of purple bacteria are embedded in various tubular and lamellar internal membrane systems; in green bacteria they are located along the plasma membrane within special bodies called chlorosomes; in heliobacteria, they are embedded in the plasma membrane.

Oxygenic phototrophs include cyanobacteria (blue—green algae), eukaryotic algae, and plants. Their light-harvesting system is based on chlorophyll a. Energy generation is by noncyclic photophosphorylation involving two linked photosystems. Water molecules are split by photon energy and used as external electron source. Electrons pass through both photosystems, and are used in reduction reactions. In some cyanobacteria, facultative anoxygenic photosynthesis is possible under anoxic conditions.

This brief review of metabolic diversity among prokaryotes shows that microorganisms are able to make use of each other's metabolic wastes and, in proper combination, to recycle resources. Examples of complementary metabolic combinations include sulfur-oxidizing and sulfur-reducing organisms, anaerobic respirers and chemolithotrophs, and aerobic respirers and oxygenic phototrophs. Some bacteria that participate in such mutually beneficial metabolic combinations have become integrated into close symbiotic relationships.

Giant worms in deep-sea rift environments have no mouths and guts, but harbor millions of symbiotic chemolithotrophs within a special organ. Indeed the eukaryotic cell can be considered a highly integrated symbiosis between an ancient eukaryotic host, that was probably anaerobic and heterotrophic, an aerobic respiring eubacterium (mitochondrion), and, in some instances, an oxygenic phototrophic eubacterium (the chloroplast).

Ecology

Microbial ecology deals with the interrelationships between microorganisms and their environments. These interactions are bidirectional. Microorganisms respond to environmental conditions and are distributed according to their growth requirements and competitiveness. Their activity, on the other hand, modifies the environments in which they grow. Commensurate with the small size of microorganisms, these interactions take place within microenvironments. Because of their numbers, growth rates, and the intensity of microbial metabolic activities, however, their impact extends from the microscopic to the global scale. Over geologic time, microbial activities have significantly modified the lithosphere, hydrosphere, and atmosphere of our planet.

Although prokaryotes are nearly ubiquitous in Earth surface habitats, individual clones, strains, or species do not tolerate the entire range of environments occupied by the entire group. Microbial populations may *survive* under a broad range of conditions, but the range under which they will *grow and reproduce* is generally much more restricted. The range of habitats within which a population grows optimally is still narrower, as is the distribution of populations in nature. Ranges of tolerance to growth of prokaryotes as related to some important parameters are summarized in Table 5.5. As used here, tolerance range refers to the continued ability to grow and reproduce under a given set of environmental conditions (ranges of survival in a latent, i.e., inactive, state are much broader). The ranges listed here represent reliably reported extreme values for prokaryotic growth in general.

Table 5.5 Environmental ranges of prokaryotic growth. (After Knoll AH, Bould J. *Trans R Soc Edinburgh: Earth Sci* 1989; 80:209–23)

	Minimum	Maximum
Temperature	−10°C	110°C
Matric water activity	0.75	1.0
Salinity	0	Halite saturation
Pressure	<1 MPa*	120 MPa*
pH	0.5	11.0
Po_2	0	>1 atm

* MegaPascals.

Natural populations thrive only where they *compete* successfully. This range may correspond to growth optima, but it need not do so. For example, prokaryotes that live in the deep ocean usually exhibit optimal growth at temperatures well above the 0–5°C found in bathypelagic environments; these bacteria live in frigid waters because they compete best under these conditions rather than at the conditions specified by optimal growth. Similarly, halophilic cyanobacteria and other prokaryotes that occur in extreme environments have their growth optima in less extreme ranges. In general, suboptimal growth can confer competitive success if potential competitors grow even more poorly under the same conditions.

Microbial community composition varies greatly as a function of environment. On a microenvironmental scale, species distributions may be strongly demarcated along environmental gradients. Thus, in Shark Bay, Australia, populations of mat-building cyanobacteria closely track gradients of average water supply within the intertidal zone; contacts between communities are often sharp, probably in response to competitive thresholds.

In broader perspective, particular metabolic groups of prokaryotes may show limited environmental tolerances. For example, distinct limitations characterize photosynthetic bacteria. No phototrophic prokaryote is known to be capable of growth in environments more acidic than about pH 4–4.5; in fact, relatively few species occur below pH 6. Similarly, phototrophs appear to have an absolute temperature limit of 70–74°C, well below the extreme tolerances of ca. 110°C exhibited by some chemolithotrophic archaebacteria. Other archaebacteria, such as extreme halophiles, are restricted to **hypersaline** habitats. The environmental ranges exhibited by modern prokaryotes can help in the interpretation of microfossils collected from sedimentary rocks of well-characterized depositional paleoenvironments. Conversely, the general tolerances exhibited by major prokaryotic taxa may provide important insights on the nature of early Earth surface environments.

A significant part of microbial metabolism utilizes energy generated by the exploitation of existing differences in redox potential in the environments prokaryotes occupy. Microbial activities also depend on the distribution of compounds that can be used as resources for their metabolic functions. Solubility and, hence, availability of these nutrient compounds, in turn, depend on redox conditions (Eh) and hydrogen-ion concentrations (pH). Environmental Eh and pH, however, are also strongly affected and modified by microbial metabolic activities. For example, the release of organic acids and CO_2 by fermenting bacteria decreases ambient pH, while the release of ammonia by proteolytic bacteria increases it. The relationship between the release of molecular oxygen by oxygenic phototrophs and its removal by aerobic respirers can have a most profound effect on environmental redox conditions. Other Eh-dependent microbial processes that enrich or deplete the pool of biologically usable nitrogen include, for example, N_2 fixation and denitrification, respectively.

The synthesis and complete microbial degradation of organic matter is possible under both anaerobic and aerobic conditions. A complete recycling under anaerobic conditions, however, requires a succession of several specialized metabolic partners. Under aerobic conditions, the same task is readily accomplished individually by most aerobic respirers (Fig. 5.5). The diminishing chances for the several appropriate partners to be recruited at the right time and place account for the fact that under anoxic conditions organic matter often remains incompletely degraded, favoring its preservation, burial, and fossilization.

Cyanobacteria

Most fossil prokaryotes are thought to be cyanobacteria (blue–green algae). The cyanobacteria constitute a coherent and distinctive phylogenetic group within eubacteria, characterized by unique metabolic capabilities and rRNA-nucleotide sequences. Functionally, cyanobacteria are distinguished from all other prokaryotes, including photosynthetic bacteria, by their ability to carry out plant-like oxygenic photosynthesis using H_2O as electron donor. The name cyanobacteria (the Greek word *cyanos* = blue) indicates their possession of phycocyanin, a blue accessory photosynthetic pigment, and their bacterial structure and origin. The older name blue–green algae referred to the pigment combination of phycocyanin and chlorophyll a, and to their functional similarity with eukaryotic algae.

Morphology and systematics of cyanobacteria

Cyanobacteria are the morphologically most diverse prokaryotes. The simplest cyanobacteria are unicells, only a fraction of a micrometer in size, and similar to ordinary rod-shaped and coccoid bacteria; the most complex ones are macroscopic, multicellular forms, similar in size and appearance to mosses. The smallest individual cells are about 0.6 μm wide. More commonly, cells are between 5 and 15 μm, although some coccoid and filamentous taxa exceed most unicellular eukaryotes in size (Fig. 5.6), and range up to 60 μm in diameter. Structural and functional differentiation into cells that perform specialized functions occurs both in unicellular and multicellular cyanobacteria. These include various **reproductive cells**, **apical cells**, **heterocysts**, and **akinetes**.

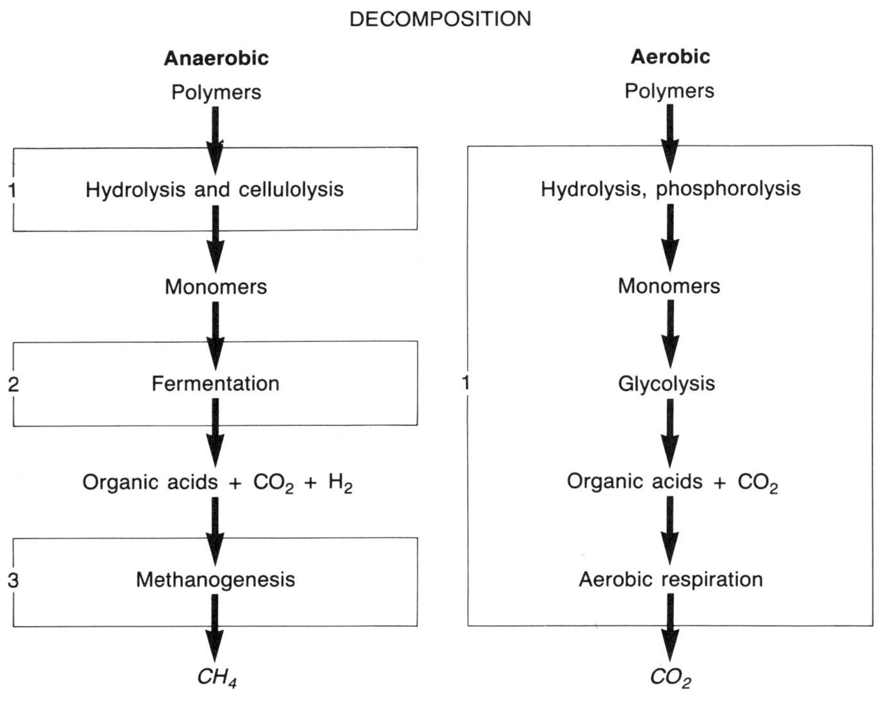

Figure 5.5 Comparison between typical **anaerobic** (left) and **aerobic** (right) decomposition of organic matter. Processes occurring within a single microorganism are placed in a box. At least three microorganismal types are required for a complete anaerobic decomposition, as compared with one under aerobic conditions.

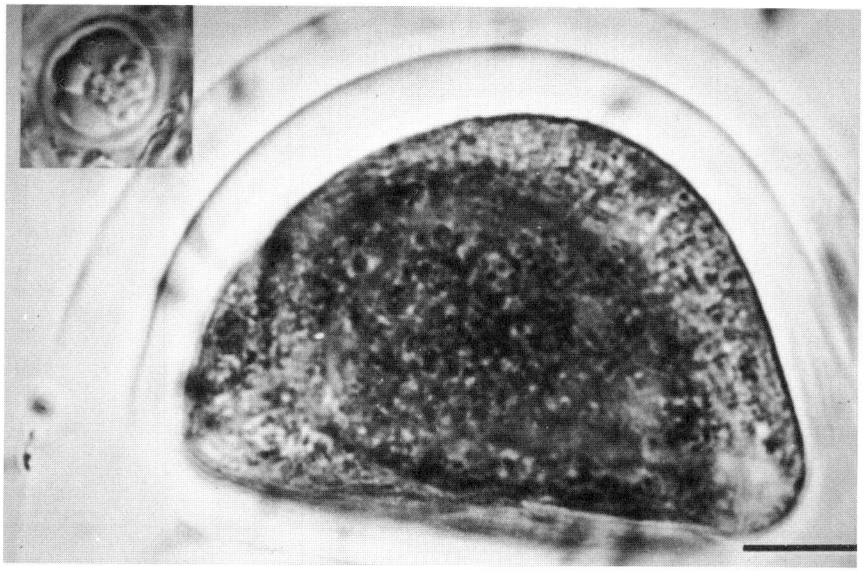

Figure 5.6 Cell of a large unicellular prokaryote, *Chroococcus turgidus* (Cyanobacteria), as compared with a common unicellular eukaryote, *Chlorella* sp. (Chlorophyta, inset upper left). Scale bar, 10 μm.

Because of the morphologic complexity and functional uniformity of cyanobacteria, their classification is based largely on morphologic rather than physiologic or biochemical properties. This circumstance facilitates comparisons between modern and fossil cyanobacteria. Traditionally, cyanobacterial systematics has been based on studies of natural populations and documented according to botanical usage, with preserved holotypes. This system has been complemented by bacteriologic practice in which collections of type cultures are used as taxonomic references. Since the number of cyanobacteria available in *axenic* culture is still relatively small and far from representative of the diversity of the group, both approaches continue to be practiced. Recent advances in molecular phylogenetic research have demonstrated that the cyanobacteria are a coherent group, with several morphologically and phylogenetically justified taxonomic subdivisions. Only the morphologically simplest forms are phylogenetically heterogeneous. Cyanobacterial classification that employs complementary data from traditional field-based studies, bacteriologic cultures, and molecular phylogeny is currently under development.

Morphologic properties that are used to characterize and classify cyanobacteria fall into two categories.

1 Properties of cells (shape, size, pigmentation, division planes and timing, degrees of cell separation and differentiation).

2 Properties of extracellular envelopes (shape, size, pigmentation, and layering).

In combination, these features express a considerable morphologic diversity in cell arrangements, colony formation, developmental cycles, and modes of reproduction. Some of these properties are relatively constant and genetically stable, while others are subject to environmental modifications within genetically determined limits. Understanding the degree and causes of this variability is essential in the taxonomic characterization and determination of cyanobacteria.

Both morphologic and phylogenetic systematics are in agreement that cyanobacteria constitute a distinctive phylum of microorganisms. Unicellular and multicellular cyanobacteria are placed for convenience in separate classes (Fig. 5.7). The unicellular or coccoidal forms, class *Coccogoneae*, are characterized by a complete separation of cells following division. These cells may disperse thereafter or form spherical, prostrate, or serial colonial aggregates held together by extracellular polysaccharide envelopes. Multicellular cyanobacteria, class Hormogoneae, maintain close contact between adjacent cells after division.

In the class Coccogoneae, the simplest forms are classified in the order Chroococcales. These do not differentiate into cells with specialized functions, and generate no specialized reproductive cells. Cell division occurs as **binary fission**, in which divided cells grow to their full size and separate before dividing again. In this respect, chroococcalean cyanobacteria are not different from most heterotrophic bacteria. Subsequent cell divisions may proceed in one, two, or three planes. Unless disturbed, the cells originating from a division in one plane assume a linear arrangement; those resulting from division in two alternating planes are arranged in a two-dimensional plate; and those following division in three perpendicular planes form a three-dimensional cuboidal arrangement.

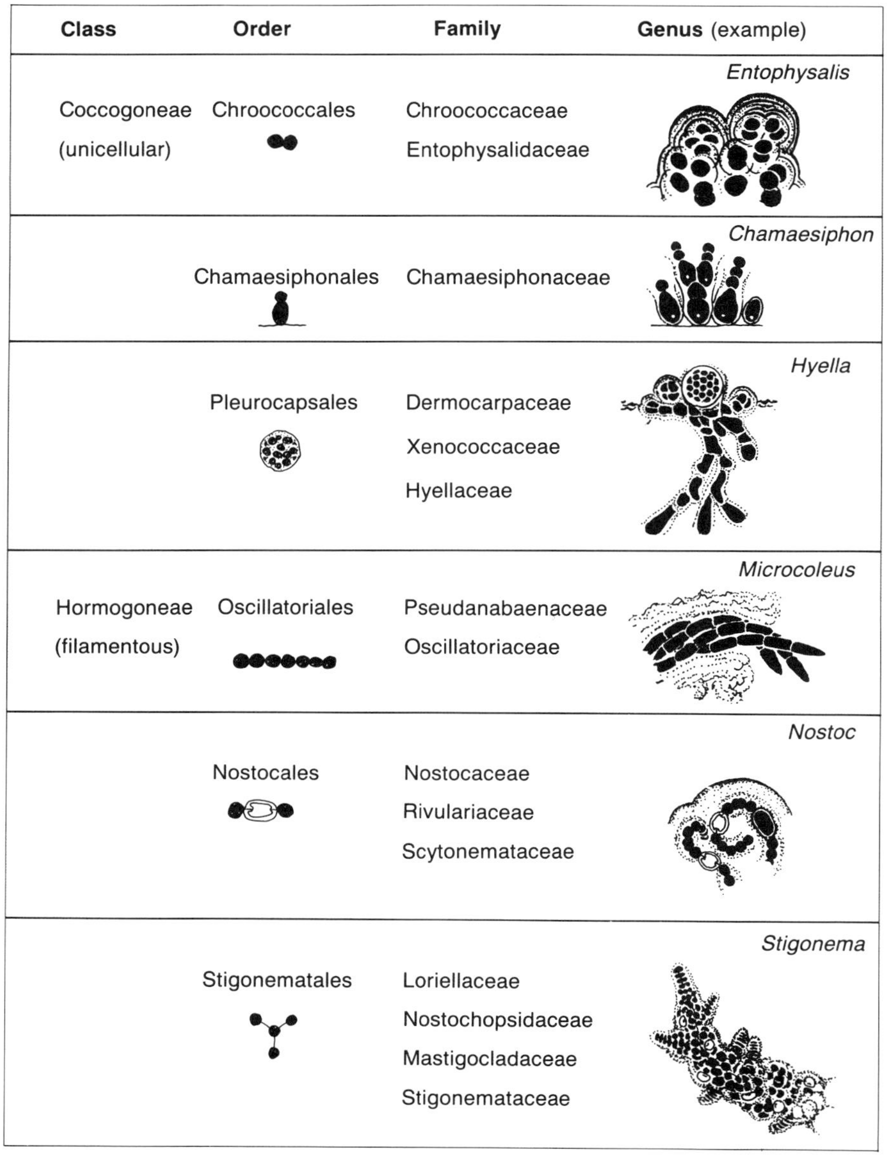

Figure 5.7 Systematic grouping of modern cyanobacteria. Orders are accompanied by symbols showing principal distinguishing features. Binary fission is common to all cyanobacteria. Budding by asymmetric binary fission and baeocyte formation by multiple fission are exclusive modes of reproduction in Chamaesiphonales and Pleurocapsales, respectively. All filamentous cyanobacteria form trichomes and hormogones. Heterocysts are common to members of Nostocales and Stigonematales. True branching is the exclusive property of Stigonematales.

These basic patterns are usually modified by crowding during intensive cell division, under constraints imposed by the intermittent production of extracellular gelatinous envelopes. As the envelopes are produced by the cell and excreted through the cell wall, the youngest layers are deposited adjacent to the cell, while older ones are pushed outward (Fig. 5.8). The excretion of extracellular polysaccharides is not always uniform and may occur in pulses, apparently synchronized with the rhythm of cell division. This results in a distinct layering of envelopes, discernible under the light as well as the electron microscope. Sequential cell divisions are often marked by a distinct encapsulation in which the outer, parental envelopes enclose the internal, filial ones. The elasticity of these envelopes varies from taxon to taxon. In some taxa, the outer envelopes burst after a few cell divisions, keeping the colonies small; in others they expand, holding large cell clones together. Reproduction in chroococcalean cyanobacteria is accomplished either by the dispersal of single cells or by the fragmentation of colonies.

All naked unicellular taxa of this order, as well as those that form spherical, nonpolar colonies, are classified within the family Chroococcaceae. Genera are distinguished on the basis of the cell division planes and the formation of extracellular gel. Cells divide in one plane (*Synechococcus, Aphanothece, Gloeothece, Johannesbaptistia*), two alternating planes (*Synechocystis, Merismopedia, Gomphosphaeria*), or three planes (*Aphanocapsa, Gloeocapsa, Chroococcus*). Fossil examples include *Eosynechococcus* (Fig. 5.9), *Sphaerophycus*, and *Gyalosphaera*.

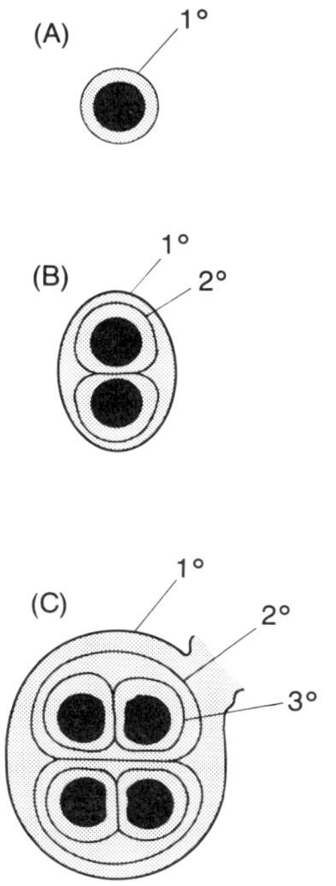

Figure 5.8 Cell division and polysaccharide envelope formation in coccoidal cyanobacteria. A, Primary envelope (1°) is excreted by the cell. B, Cell divides and each daughter cell produces a new envelope (2°), while the primary envelope is stretched outward, enveloping both daughter cells. C, Following division of the two daughter cells, their progeny form new envelopes (3°), stretching the old ones, until the oldest (1°) breaks open.

Chroococcalean taxa with polarized growth that form prostrate and hemispherical aggregates or produce unidirectional gelatinous stalks are placed in the family Entophysalidaceae. Genera are distinguished on the basis of the orientation and symmetry of colonies that results from the relationship between cell division and the production of extracellular gel. Entophysalidacean cyanobacteria form billowing prostrate colonies (*Entophysalis*; Fig. 5.10A), or grow branched, gelatinous stalks (*Cyanostylon*). Endolithic forms grow **geopetally**, penetrating calcareous substrates (*Hormathonema*). Fossil *Eoentophysalis* (Fig. 5.10B) is found frequently in Proterozoic cherts, often forming stromatolitic structures. A *Cyanostylon*-like microfossil has been found in Late Proterozoic strata and named *Polybessurus* (Fig. 5.11).

Coccoid cyanobacteria that have specialized reproductive structures are placed in separate orders. The order Chamaesiphonales, with a single family and only a few genera, includes attached forms with polar, asymmetrical cell division that produces a larger, sessile basal cell, and a smaller, distal reproductive cell (bud or exospore). The sessile cell continues to divide asymmetrically, delivering small reproductive cells that are released to settle elsewhere and repeat the cycle.

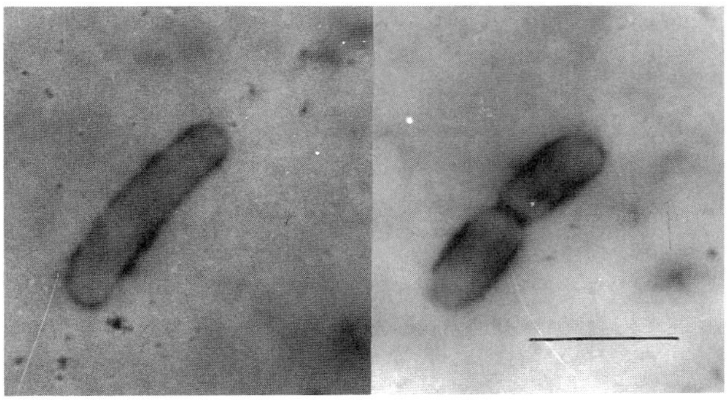

Figure 5.9 Cells of the microfossil *Eosynechococcus thuleensis* in states of elongation and division, from 1100–1200 My old Narssarssak Formation, northwest Greenland. Scale bar, 10 μm.

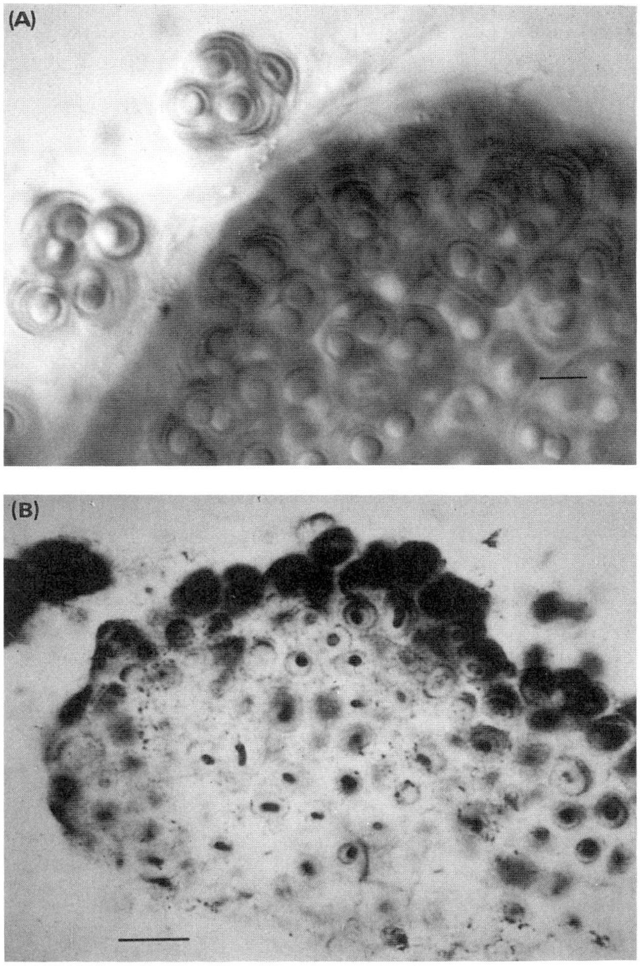

Figure 5.10 Stromatolite-building coccoid cyanobacteria. A, Modern *Entophysalis major* from Shark Bay, Australia. B, Silicified *Eoentophysalis belcherensis* from ca. 1900 My old Belcher Supergroup, Canada. Scale bars, 10 μm.

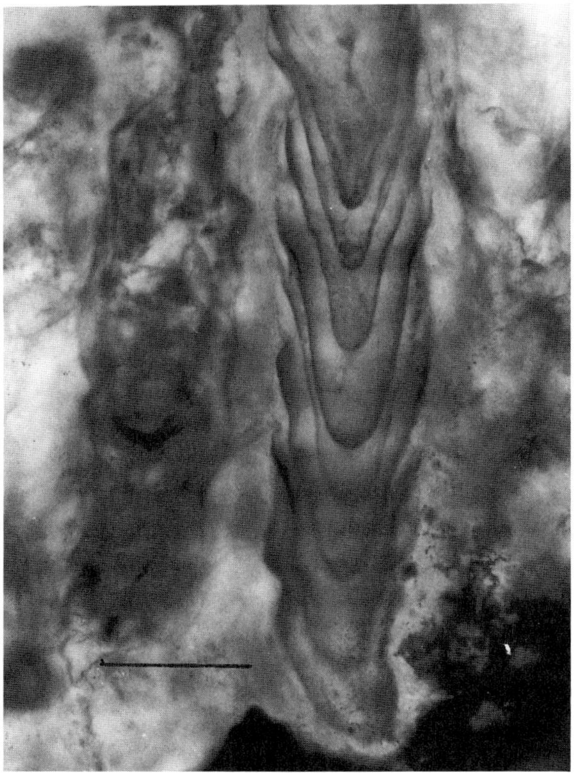

Figure 5.11 *Polybessurus bipartitus*, a *Cyanostylon*-like fossil cyanobacterium from the 700–800 My old Eleonore Bay Group, east Greenland. Scale bar, 100 μm.

Within the order Pleurocapsales, reproduction occurs when a cell enlarges and then undergoes multiple fission to produce a large number of small reproductive cells that are released by the disintegration of the original cell envelope. These reproductive cells, or baeocytes, are either motile or nonmotile after the release. They settle, grow to several times their initial volume, and then repeat the cycle. Distinctions among pleurocapsalean genera are difficult and still a subject of dispute. The simplest members of this group, classified in the family Dermocarpaceae, show only multiple fission. All released baeocytes grow until they reach several tens or hundreds times their initial volume before undergoing multiple fission and release of baeocytes (Chroococcidiopsis, Dermocarpa). Taxa of the family Xenococcaceae show various combinations of binary and multiple fission in the course of the life-cycle. The resulting cell clusters range from simple polarity (Dermocarpella) to compact three-dimensional cell packages (Myxosarcina, Xenococcus). The most complex coccoid cyanobacteria belong to the family Hyellaceae. Their cell divisions appear to be regulated. They occur at particular locations and follow an internal rhythm, resulting in apical growth and formation of linear cell series. Some genera are **epilithic** (*Pleurocapsa*); others are **endolithic**, penetrating carbonate substrates (*Hyella*; Fig. 5.12A, *Solentia*). A dividing apical cell extends the series, and performs the boring. Fossil cyanobacteria belonging to this group include Palaeopleurocapsa and Eohyella (Fig. 5.12B).

Within the multicellular class Hormogoneae, cells are integrated in tight linear arrangements called **trichomes**. Polysaccharides excreted along trichomes can have a fluid consistency that promotes gliding motility, or they may form tubular gelatinous **sheaths**. Trichomes together with their sheaths are referred to as **filaments**. Some taxa are characterized by filaments that contain only single trichomes, while others contain multiple bundled trichomes surrounded by a common sheath, which is often compartmentalized. True branches are caused by occasional changes in the plane of cell division. Alternatively, false branches can be caused by trichome fragmentation followed by the lateral protrusion of fragments, or by the subdivision of trichome bundles, each false branch

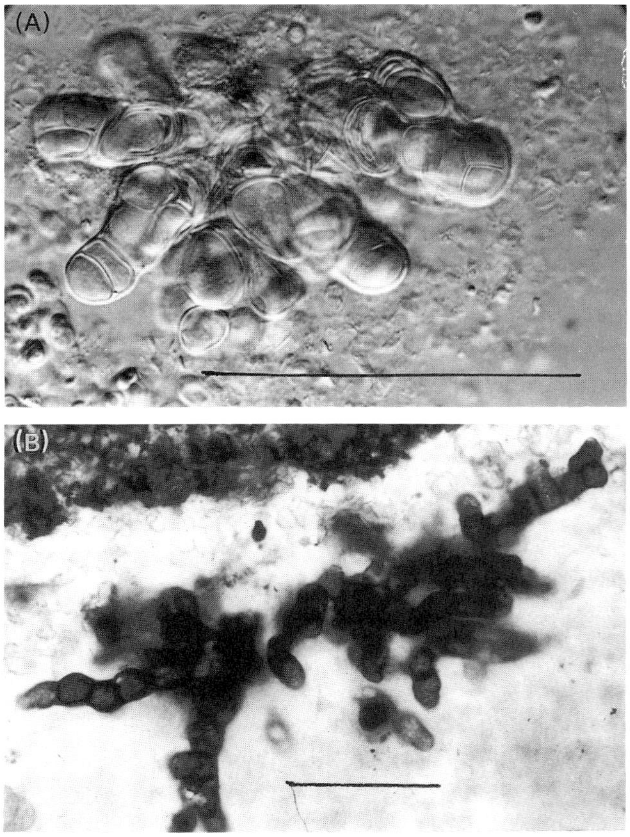

Figure 5.12 Modern and fossil endoliths penetrating ooid grains. A, Young colony of modern *Hyella* sp. extracted from an ooid grain, Andros Island, Bahamas. B, *Eohyella rectoclada* in oolites of 700–800 My old Eleonore Bay Group, east Greenland. Scale bars, 100 μm.

forming a separate sheath. Filamentous cyanobacteria disperse by unsheathed, actively gliding trichome fragments called **hormogones**.

The simplest filamentous forms without any cell differentiation are placed in the order Oscillatoriales. They retain gliding motility throughout their life-cycle. The family Pseudanabaenaceae includes small gliding forms with light-refracting cross walls that frequently fragment. Several species of *Pseudanabaena* are characterized by terminal gas vesicles. The larger family Oscillatoriaceae includes naked gliders (*Oscillatoria*; Fig. 5.13A, *Arthrospira*, *Spirulina*), sheathed forms containing single trichomes (*Phormidium*; Fig. 5.14A, *Lyngbya*), and forms with bundled trichomes within a common sheath (*Schizothrix*, *Microcoleus*, *Hydrocoleum*).

Filamentous cyanobacteria, classified in the orders Nostocales and Stigonematales, are characterized by a higher degree of morphologic and functional differentiation of their cells. Most have the ability to differentiate heterocysts, and some also form akinetes. The order Nostocales includes forms with linear, unbranched trichomes. As in the Oscillatoriales, cells divide exclusively in one plane. False branching occurs by trichome breakage and divergence with new sheaths forming in the process. The family Nostocaceae includes nonpolarized forms. Differentiation of heterocysts and akinetes is common, but the rest of the trichome cells remain undifferentiated. Cell division is evenly distributed along the trichome. Some planktic genera form water blooms of free-floating trichomes (*Anabaena*), or bundled, floating colonies (*Aphanizomenon*). Others are mostly benthic, surrounded by copious amounts of gel (*Nostoc*). Genera are distinguished on the basis of colony formation and the distribution of heterocysts and akinetes. More complex nostocaleans, in addition to forming heterocysts, differ-

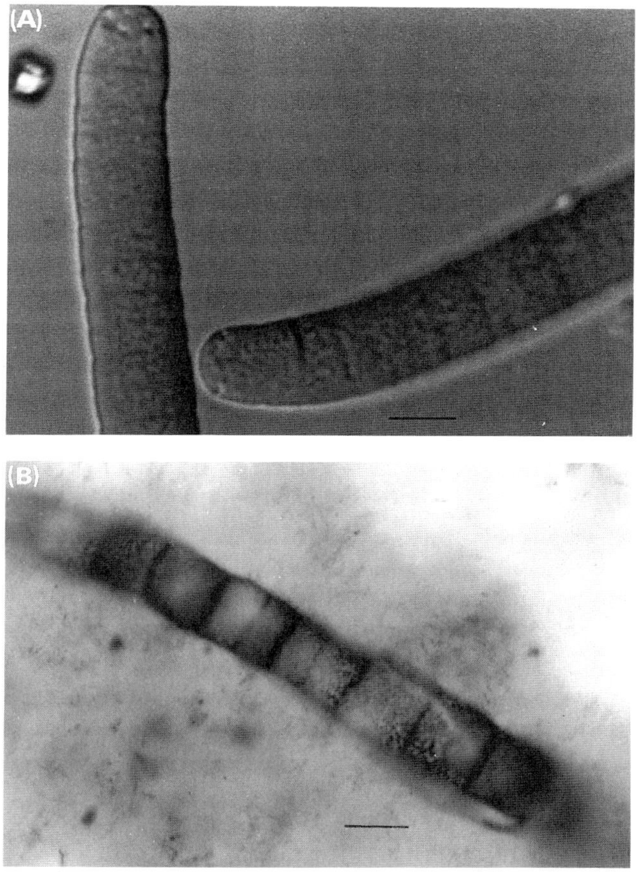

Figure 5.13 Modern and fossil filamentous cyanobacteria. A, Gliding trichomes of *Oscillatoria* sp. from a salt marsh. Scale bar, 10 μm. B, *Oscillatoriopsis obtusa*, a problematic oscillatoriacean cyanobacterium. An identity with chemolithotrophic bacteria is also possible. From 2000 My old Duck Creek Formation, Australia. Scale bar, 10 μm.

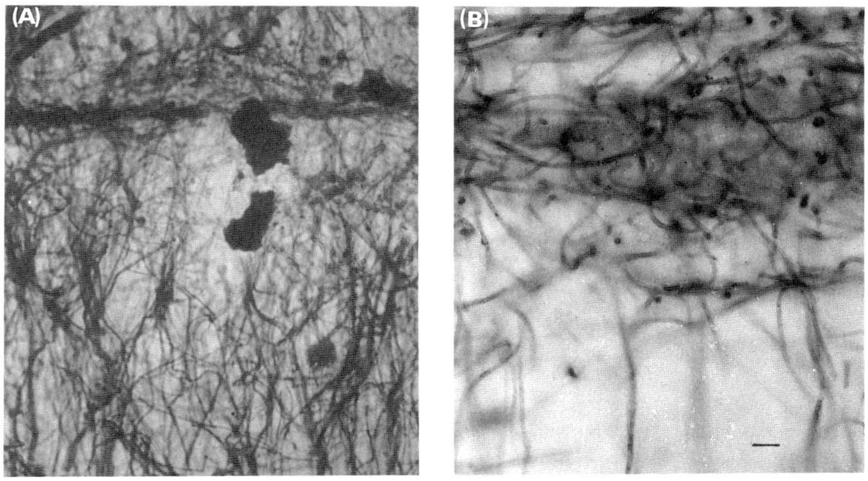

Figure 5.14 Vertical section through a stromatolitic lamina, showing sheaths oriented according to phototactic movement of cyanobacterial trichomes: upright during the day (lower part of A and B) and prostrate during the night (upper part of A and B). A, Modern stromatolite from Bonair Island, built by *Phormidium hendersonii*. B, Fossil cyanobacterial mat with *Eomycetopsis* filaments in 700 My old Backlundtoppen Formation, Spitsbergen. Sheaths in both cases are about 2 μm wide. Scale bar, 10 μm.

entiate trichomes into regions of active cell division (**meristems**), and regions in which rates of cell division are slow, usually producing cells of different shape and structure. The family Rivulariaceae includes taxa with pronounced polarized growth, showing cell differentiation along trichomes from a basal or intercalary meristem upward, to vacuolized narrow cellular whips at the distal end of the trichome. Heterocysts are basal, rarely intercalary, and are connected with the rest of the trichome by a single pore. Filaments are falsely branched, with single trichomes within individual sheaths (*Calothrix, Dichothrix*), bundled (*Gardnerula*; Fig. 5.15), or surrounded by common gel (*Rivularia, Gloeotrichia*). The members of the family Scytonemataceae have a subapical meristematic region and differentiate basipetally into longer and narrower cells. Some members have a distinct directional, upright growth. Some are unbranched (*Leptochaete, Fremyella*). Predominantly single false branches are associated with basal single-pore heterocysts and upright growth (*Tolypothrix*). Prostrate filaments that form mostly paired upright false branches are associated with intercalary double-pored heterocysts (*Scytonema*).

The order Stigonematales includes the most complex multicellular cyanobacteria. Trichomes exhibit true branching by occasional changes in the plane of cell division from transversal to longitudinal or oblique. A true branching point is defined by a cell connecting with neighboring cells on three sides. In many representatives, cell division leaves a central cytoplasmic connection between all cells, resembling the pore of a heterocyst. Heterocysts occur intercalated along the trichome or terminally on side-branches with limited growth. In the family Loriellaceae the branches are dichotomous. Members of other families branch laterally. In the family Nostochopsidaceae, branches are perpendicular to the main filament, forming a T-shape. Filaments are embedded in massive gel resembling that produced by *Nostoc* (*Nostochopsis*). Endolithic forms differentiate branches with terminal whips, short branches with a terminal heterocyst, as well as branches with unlimited growth originating from an apical, substrate-penetrating cell (*Mastigocoleus*). The family Mastigocladaceae is characterized by intercalary meristematic zones positioned at the branching points. The secondary growth that ensues produces Y-shaped branching. The members include interwoven filamentous forms with cylindrical and *torulose* (beaded, with constricted cross-walls) trichomes in thermal springs (*Mastigocladus*), marine taxa with filaments embedded in *Nostoc*-like masses of common gel (*Brachytrichia*), and endoliths (*Herpyzonema*). Large, elaborately branched cyanobacteria belong to the family Stigonemataceae. They are characterized by transversally dividing apical meristems that differentiate into filaments with torulose cells and lateral branching. Simple members of the family have uniserial filaments (*Hapalosiphon*), others form prostrate filaments of multiserial appearance that grow simple upright branches (*Fischerella, Westiella*). The more complex members of this family have two types of branches. Main branches with unlimited growth differentiate from linear, short-celled, apical meristems. Lateral branches with limited growth are embedded in a common sheath providing the main branches with a multiserial appearance, ultimately crowding into a series of cell clusters that resemble coccoid cyanobacteria (*Stigonema*).

Fossil remains of *Oscillatoria*-like trichomes have been named *Oscillatoriopsis* (Fig. 5.13B). They are relatively rare and usually deformed. Remains of shriveled

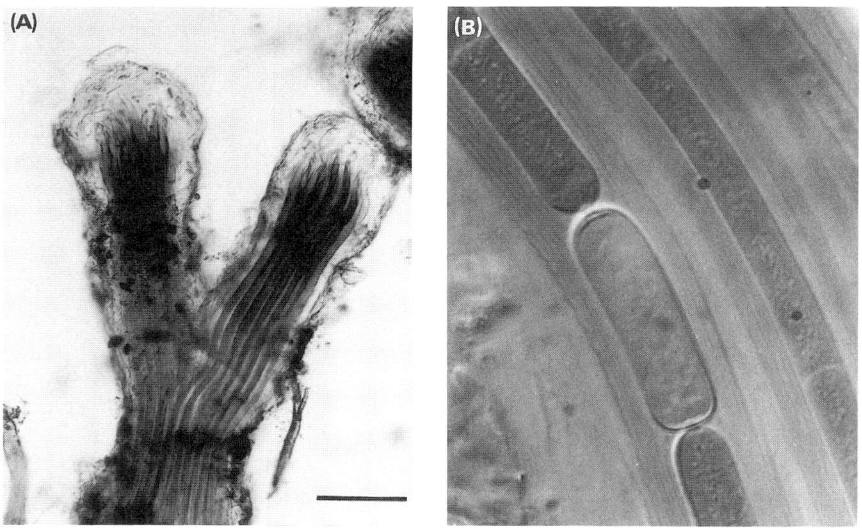

Figure 5.15 Stromatolite building filamentous heterocystous cyanobacterium *Gardnerula corymbosa* from Shark Bay, Australia. A, Dichotomously branched bundles of trichomes embedded in common gel. B, Detail view of the bundle, showing a heterocyst. Scale bar shown in A: 100 μm for A, 10 μm for B.

trichomes are known under the name *Biocatenoides*. More commonly, only sheaths preserve. Narrow sheaths are named *Eomycetopsis* (Fig. 5.14B), and the wider ones *Siphonophycus* (Fig. 5.16). Trichomes of oscillatorialean cyanobacteria can glide within their sheaths or abandon the sheaths entirely, commonly in response to environmental stress. Such reaction has been observed in modern and fossil stromatolites (Fig. 5.14). To date no fossil heterocystous cyanobacteria have been found. Despite several reports in the literature, a convincing heterocyst has yet to be demonstrated in the fossil record.

Cell structure and function

The cellular organization of cyanobacteria is similar to that of other Gram-negative eubacteria. The structural integrity of the living cell is maintained by cell walls which contain peptidoglycan, a cross-linked polymer of sugar derivatives, and amino acids. Strong folding of this layer, however, also seems responsible for the tendency of prokaryotic cells to shrivel after death, a fact which limits their recognition and interpretation in the fossil record. Inside the peptidoglycan layer is a **plasma membrane** involved in the transport of ions, and the exchange of nutrients and metabolic products between cell and environment. The plasma membrane does not provide the cell with any rigidity or structural integrity. Outside the peptidoglycan layer is a thin, relatively permeable, membrane-like wall layer. Cyanobacterial cell walls are commonly surrounded by polysaccharide envelopes or sheaths of variable consistency and thickness. These envelopes are involved in various external functions, including movement, flotation, attachment, colony

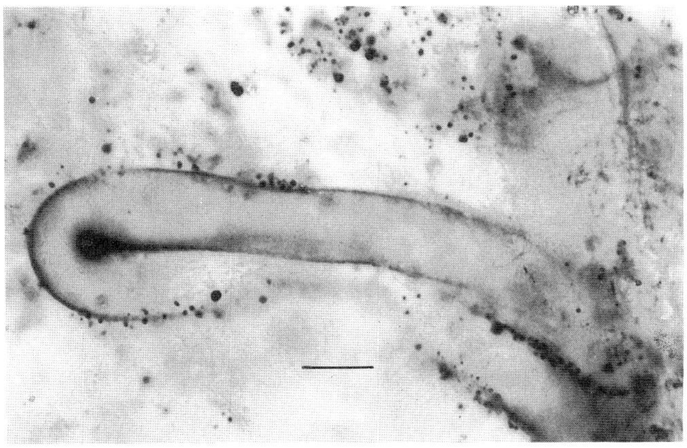

Figure 5.16 *Siphonophycus kestron*, a probable cyanobacterial sheath, from 700–800 My old Rysso Formation, Nordeustlanded, Svalbarg. Scale bar, 10 µm.

formation, water storage, light screening, and general protection.

The interior of a cyanobacterial cell is similar to those eubacteria that have intensive membrane-bound metabolic activities. These activities require functional surfaces provided by an inward folding of the plasma membrane, resulting in various intracellular membranous structures. In cyanobacteria, these membranes, or **thylakoids**, are the site of photosynthetic activity.

The photosynthetic reaction centers equipped with chlorophyll *a* are thought to be embedded deeply within thylakoid membranes, covered by stacks of additional light-channeling chlorophyll *a* molecules. On the outside, the thylakoid system is complemented by specialized light-harvesting bodies, or *phycobilisomes*, that act as light-collecting lenses. Phycobilisomes contain phycobilin pigments that widen the photosynthetically useful light spectrum by harvesting photons not well absorbed by chlorophyll directly. Phycobilisomes are constructed so efficiently that they conduct light energy almost without loss to chlorophyll *a*. Dark reactions of photosynthesis are believed to occur outside thylakoid membranes.

Cyanobacterial cells contain various inclusions, some of which represent nutrient storage. Unlike other prokaryotes, cyanobacteria commonly contain polyunsaturated fatty acids. Some cyanobacteria contain gas vesicles, an exclusively prokaryotic intracellular structure which provides the cells with buoyancy.

Functionally, cyanobacteria are a uniform group, for the most part obligately phototrophic. Although some species can use simple organic compounds as a source of carbon, they remain light-dependent for their energy supply. Cyanobacterial photosynthesis is oxygenic, but under anoxic conditions some species of cyanobacteria are able to switch facultatively to anoxygenic photosynthesis. As in some green and purple photobacteria, this process uses H_2S as an electron donor and employs only photosystem I. Sulfur granules eliminated in the process are deposited outside cells.

A number of cyanobacteria have the ability to fix molecular N_2. This energy-requiring process is carried out by the oxygen-sensitive enzyme dinitrogenase. In many cyanobacteria, N_2 fixation occurs under spatially or temporally localized low oxygen conditions; however, some cyanobacteria have evolved specialized cells called heterocysts, which are functional sites of N_2 fixation under aerobic conditions. Heterocysts differentiate from ordinary cells by decreasing the permeability of their cell walls. These cell walls increase in thickness, leaving only one or two small openings or pores for controlled exchange of gases and metabolic products with neighboring cells. **Photosystem II** with its oxygen-producing activity is discontinued; only **photosystem I** functions to provide energy. The pigment composition of heterocysts is also modified, so that they often differ in coloration from vegetative cells. Because they are morphologically distinct, heterocysts have the potential to be recognized in the fossil record.

Cell division in cyanobacteria occurs by binary fission, as in other prokaryotes. The simplest unicellular cyanobacteria are dispersed following each cell division. More complex coccoidal forms reproduce by specialized reproductive cells. Buds form by asymmetric fission and baeocytes (nannocytes) by multiple fission—repeated binary fission without intervening growth. Filamentous cyanobacteria reproduce by short motile trichome fragments called **hormogones**. In some taxa, hormogones may be encased in multiple envelopes and released as specialized nonmotile structures called hormocysts (e.g., *Westiella*). In this case they perform the dual functions of reproduction and survival under unfavorable conditions. Survival under unfavorable conditions is also promoted by akinetes, nonmotile, enlarged, and granule-loaded cells that are common in filamentous heterocyst-bearing taxa.

Motility, behavior, and ecology

Cyanobacteria move mostly by gliding, usually in conjunction with the production of extracellular slime and in contact with solid substrates. Unicellular (coccoid) cyanobacteria sometimes move in a jerky fashion, exhibiting a sort of limited gliding. Reproductive cells in some coccoid taxa are nonmotile when released; in others they retain gliding motility for a limited time after release, before their extracellular envelopes form. Some recently discovered, very small, marine planktic species of the genus *Synechococcus* swim by an unknown mechanism. Unlike other swimming bacteria, the cyanobacteria are not flagellated. Filamentous gliding cyanobacteria such as *Oscillatoria* and *Spirulina* remain motile throughout their life-cycle; others alternate between stationary and motile phases. Several theories have been advanced to explain the mechanism of the gliding movement, but none is completely satisfactory.

The responses of cyanobacteria to environmental stimuli can be viewed as behavior. Being phototrophic, they exhibit both phototactic and **phototropic** behavior, the former expressed as movement and the latter as growth in response to light. Motile taxa and baeocytes are positively phototactic. In addition, motile cyanobacteria show **chemotactic** responses. Sheathed motile cyanobacteria regularly abandon their sheaths in response to stress or the onset of unfavorable conditions. The sheaths left behind in sediments contribute to the binding and stabilization of sediment grains. Under favorable conditions these sheaths may become fossilized. The orientation of these sheaths in sediments reveals the behavior of the organisms at the time the sheaths were made (Fig. 5.14). Heterocystous forms, which are normally not motile, often grow in an oriented fashion. A number of cyanobacteria are able to penetrate calcareous substrates and live inside carbonate rocks or skeletal particles. Boring behavior is associated with a change in growth orientation toward the interior of the substrate, and has been demonstrated in the fossil record.

Cyanobacteria are ecologically diverse, with a large number of species occurring in both marine and freshwater environments. They include **stenotopic**, sensitive species that require near-optimal and constant conditions, as well as **eurytopic** ones that can withstand extreme and sudden changes in living conditions. Freshwater and open ocean planktic species are commonly stenotopic, whereas eurytopic species inhabit extreme environments and commonly suffer osmotic (hypersaline) or matric (dryland) water shortages.

Marine cyanobacteria have adjusted their nutrient requirements to the chemistry of the sea water. Small unicellular cyanobacteria belonging to the genus *Synechococcus* are the most numerous organisms in the marine plankton, and are significant contributors to oceanic primary production. They are the main component of the **picoplankton** (defined by a small size range of organisms: 0.2–2.0 μm). The large filamentous cyanobacterium *Trichodesmium* is common in the plankton of tropical oceans. Its trichomes are packed in tight bundles within which oxygen is depleted and nitrogen fixed. Coastal waters and estuaries are inhabited by cyanobacterial species that tolerate wider ranges in salinity and higher nutrient levels.

Marine benthic cyanobacteria occur as **periphyton** on rocks and **epiphyton** on algae, but they are also well adapted to settle on loose sediments and stabilize sand particles, often constructing extensive microbial mats. They are particularly common in tropical shallow waters of relatively high salinity, including intertidal flats. The most common genera dominating these environments are *Entophysalis* and *Aphanothece* among coccoid cyanobacteria, and *Microcoleus*, *Lyngbya*, and *Schizothrix* among filamentous ones.

Freshwater cyanobacteria are well known as causes of water blooms, consisting of particularly dense planktic populations that accompany the **eutrophication** and pollution of lakes and ponds. *Microcystis flos-aquae* is a coccoid cyanobacterium that blooms, as its name implies. Some *Microcystis* strains are toxic to cattle that drink pond water. Filamentous cyanobacteria that occur in water blooms include species of the non-heterocystous genus *Oscillatoria* (e.g., *O. rubescens* and *O. prolifica*), as well as species of several heterocystous genera: *Anabaena, Cylindrospermum,* and *Aphanizomenon*. Some of these forms differentiate akinetes which sink as the bloom subsides, and hibernate in the sediment. They are thought to return to the water column during the spring mixing of lake waters. Picoplanktic cyanobacteria are also numerous in lakes.

Thermophilic cyanobacteria, particularly species of *Synechococcus, Oscillatoria, Spirulina, Phormidium,* and *Mastigocladus,* commonly dominate moderately hot springs of different chemical compositions. Species of *Gloeocapsa, Gloeothece, Nostoc, Scytonema,* and *Calothrix* colonize bare cliffs periodically wetted by water seepage or sea spray. The trapping and binding of soil particles, soil consolidation in deserts, and formation of desert crust are pioneered by the cyanobacteria *Microcoleus vaginatus, Scytonema tenellum,* and *Schizothrix* spp. *Chroococcidiopsis* spp. occupy pore spaces in sandstones of hot and cold deserts. Such aridity-tolerant cyanobacteria may have occupied the land surface as primary producers during Precambrian times, long before the colonization of continents by plants.

Other cyanobacteria, including the coccoid species of *Hormathonema, Hyella,* and *Solentia,* and the filamentous species *Plectonema terebrans, Mastigocoleus testarum,* and *Kyrtuthrix dalmatica* can live endolithically by penetrating carbonate substrates such as limestone, coral skeletons, shells, shell fragments, and ooids in shallow-water marine environments. **Euendoliths** in freshwater environments are less diversified and less well known.

Cyanobacteria are commonly found in symbiotic relationships with fungi, forming lichens and less commonly with protists, plants (in the fern *Azolla* and cycad roots), and animals.

Paleobiology

Bacterial remains are relatively common in Phanerozoic rocks, but little remarked upon. Empty bacterial walls and envelopes can be recognized in the particulate organic matter in petroleum reservoir rocks and other sediments, and casts of probable sulfate-reducing bacteria have been observed in electron micrographs of Paleozoic pyrite grains. Perhaps most interesting, analyses of the organic films that preserve a record of the soft parts of Eocene mammals from the Messel Shale, Germany, show these films to be carpets of bacterial cells. The Proterozoic record of common bacteria is spotty and unreliable, because bacterial remains cannot be easily distinguished from detrital **kerogen** or particles of mineral matter.

Despite their small size and seemingly ephemeral nature, cyanobacteria and cyanobacteria-like prokaryotes have an excellent fossil record. Phanerozoic cyanobacteria are most commonly preserved by micritic rinds that coated organisms either during life or early in diagenesis; examples include *Girvanella*, a filamentous form common in Paleozoic carbonates, *Renalcis*, a globular fossil found in Early Paleozoic reef assemblages, and *Aeolisaccus*, a possible scytonematacean cyanobacterium found in carbonate deposits of Mesozoic carbonate platforms (Fig. 5.17). Organically preserved cyanobacteria are much less common, but do occur in silicified peats such as the Lower Devonian Rhynie Chert of Scotland. In Precambrian rocks, the relative abundance of these two preservational types is reversed. Micrite-lined filaments have been observed in stromatolitic carbonates of Middle to Late Proterozoic age, but the overwhelming majority of cyanobacteria-like fossils are preserved as organic remains in chert or fine-grained siliciclastic sediments. Evidence of cyanobacterial activity is preserved as an extensive record of stromatolites, which is particularly diverse in the Precambrian strata. A different kind of trace fossil especially

Figure 5.17 Cross-section of a possible cyanobacterial filament preserved in carbonate. The sheath is preserved as a micritic rind around the filament lumen, filled with sparry calcite. SEM of polished and slightly etched rock section. Scale bar, 10 μm.

common in Phanerozoic strata are microborings of endolithic cyanobacteria. Chemical indications of past biologic activity are also widespread in sedimentary rocks. These include both isotopic and molecular signals of the physiology or taxonomy of ancient prokaryotes.

Although the direct and indirect records of microorganisms are abundant in at least Phanerozoic and Proterozoic rocks, their paleobiologic interpretation is not necessarily simple. The confidence with which fossil remains can be interpreted depends largely on the quality of preservation encountered and the means by which this evidence is evaluated.

Preservation potential and early diagenesis

The widespread preservation of cyanobacteria may seem surprising in that these microorganisms do not form robust mineralized skeletons of the type conventionally associated with high preservation probability. Yet many cyanobacteria do produce preservable structures of a different type—extracellular sheaths and envelopes that are resistant to decay. Studies of the postmortem degradation of cyanobacteria in microbial mats of hypersaline, **sabkha**-type environments show that their cells tend to collapse following death and loss of turgor. Cell walls undergo strong folding, so that the cell remains shrivel down to a minute fraction of their former volume. In contrast, layers of extracellular envelopes retain their generally convex outlines, and undergo a lesser reduction in volume (Fig. 5.18).

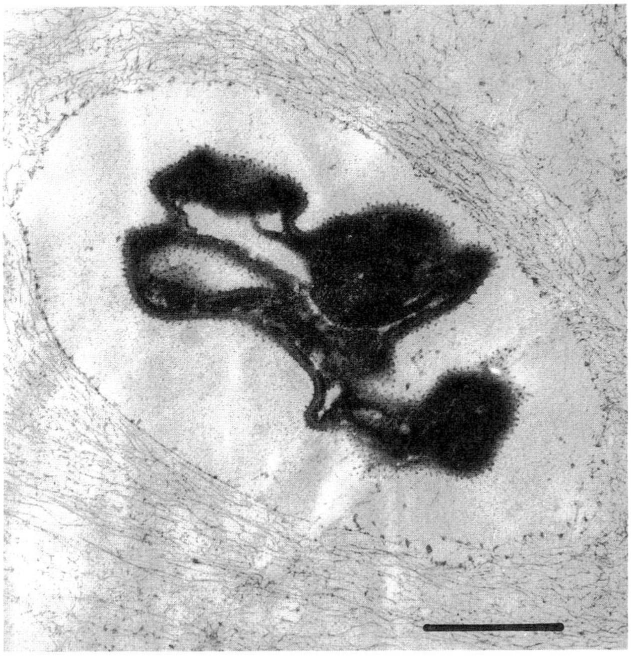

Figure 5.18 Electron micrograph of the cell of *Entophysalis major* in early stages of degradation. Note the dark strongly convoluted peptidoglycan cell wall in contrast to polysaccharide envelopes which maintain their rounded outlines. Scale bar, 1 μm.

These observations are consistent with the fossil record, where the remains of cyanobacterial cells often appear as small deformed granules surrounded by intact envelopes (Fig. 5.10B). The envelopes provide the closest available approximation of original cell size and morphology. The arrangements of cell envelopes often reveal sequences of cell division, permitting the reconstruction of life-cycles for ancient cyanobacteria.

Fossilization is a selective process in which taphonomic deformation and information loss afflict different cell structures differently. Accordingly, the fossil record will be biased toward more resistant cellular structures. Similarly, the chances of preservation and fossilization are not equal in all of the different environments in which cyanobacteria occur. Therefore, the fossil record is biased in favor of environments that promote preservation and fossilization. Modern environments that are characterized by relatively stable conditions and maintain high species diversities of cyanobacteria and algae correlate with a low preservation potential because of the fragile cellular structures of most cyanobacteria that live there and the efficient degradation of dead organic matter by a correspondingly diverse community of heterotrophic microorganisms. Chances for preservation are much better in environments where conditions, such as water and oxygen supply, salinity, sediment flux, pH, and Eh, fluctuate. In such environments the species diversity is lower; however, the cellular structures of resident organisms are tougher, and their degradation is often interrupted and incomplete.

Morphology, behavior, and paleoecology

The small size of prokaryotes and protists requires special preparation techniques associated with the use of the light and electron microscope. The small size, however, has its advantages. Fossil populations of cyanobacteria, for example, can be preserved in rock fragments no larger than a grain of sand. Rocks that have suffered from erosional destruction to an extent that their macrofossils have become obliterated beyond recognition may still harbor recognizable microfossils. While macrofossils are frequently incomplete and fragmentary, microfossils are often preserved intact and even in large numbers representing entire populations (Fig. 5.19A) or microbial communities (Fig. 5.19B). Large populations and entire assemblages of fossil cyanobacteria permit close statistical comparison of ancient and modern taxa.

Populations of living microorganisms change in time. Those changes that contribute to the maintenance of a microbial population are cycles of cell division and reproduction (e.g., Fig. 5.9). Complex microorganisms such as cyanobacteria exhibit morphologic and functional differentiation in the course of these cycles. Such processes can be observed in living forms, but must be reconstructed for fossils. The method is similar to that of demographic census analyses in which dynamic properties such as growth rates can be reconstructed from age distributions, assuming that the timing in the life-cycle of each individual in a large population is randomly distributed among individuals rather than synchronized. Consequently, the death and fossilization of an entire population will preserve individuals caught in different stages of their life-cycles. The frequency distribution of these stages at that moment will contain information on the average rates of change and duration of these stages within the average life-cycle for the population.

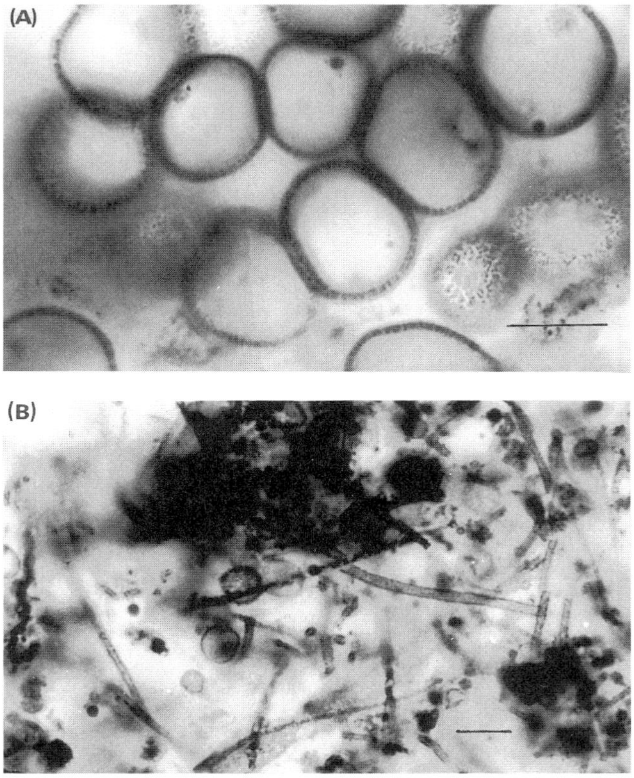

Figure 5.19 Problematic microfossils occurring in dense assemblages. A, Problematic spheroidal microfossils with heavy sculptured outlines and small inclusions, from Draken Formation, east Greenland. B, Problematic prokaryotes, from Gunflint Iron Formation, ca. 2000 My. Scale bars, 10 μm.

The ability of microorganisms to detect a change in the environment and respond to it according to their requirements is behavior. Fossil microbial behavior can be detected by assessing the consequences of this behavior, if they should preserve. In fossil coccoid cyanobacteria (e.g., *Eoentophysalis*) a pigment deposited in the envelopes in response to excessive light marks the colony surface at the time of this response (Fig. 5.10). In fossil filamentous cyanobacteria prostrate filaments (or sheaths) may be the consequence of light optimization in a stable environment or simply the physically most likely position along the sediment—water interface. A change of this position from prostrate to perpendicular to that interface is physically unlikely, and can be interpreted as a response of motile cyanobacteria to changes in light (phototaxis) or chemistry (chemotaxis) associated with burial. Alternation of filament orientation may indicate cyclic recurrence of the conditions that caused these behavioral changes, as has been demonstrated in the formation of stromatolitic laminae (Fig. 5.14). Penetration into carbonate substrates is a behavior of endolithic cyanobacteria that can be recognized in the fossil record by the orientation of organismal remains or borings into the substrate (Fig. 5.12), rather than away from it as would be expected for epilithic forms.

Paleoecologic inferences can be drawn from distributions of fossil populations within or between paleoenvironments which, in turn, can be recognized on the basis of independent sedimentologic criteria. This is of particular importance in determining whether a fossil population is *autochthonous* (buried in place) or *allochthonous* (transported to its burial ground from elsewhere). A significant proportion of cyanobacteria are benthic organisms, commonly associated with loose and shifting sediments. By virtue of their habitat, these taxa are frequently buried and have a good chance to be preserved and fossilized in place. However, populations of benthic cyanobacteria also may become dislocated during storms. In addition, fossil assemblages may consist of indigenous benthic microorganisms and planktic microorganisms that sank to the bottom following death and became incorporated in the sediment.

In summary, most of the information used in classic, morphology-based cyanobacterial taxonomy, in conjunction with information derived from population ecology, can be brought to bear on fossil populations. Given this fact, how precise is the systematic interpretation of cyanobacteria-like fossils?

Fossil cyanobacteria: precision of identification

Confidence in systematic interpretation depends critically on two factors: the amount of morphologic information extractable from the fossil population, and the degree to which we can distinguish between cyanobacteria and other prokaryotes or algae with convergent morphology. For isolated individual fossils, poorly preserved specimens, or populations that consist solely of simple spheroidal vesicles, systematic affinities may be impossible to infer; such fossils are best considered as *incertae sedis* (of uncertain relationship). Populations that have distinctively cyanobacterial morphologies and/or behavioral orientations, and that have close morphologic counterparts among extant cyanobacteria found in comparable physical environmental settings, can be compared closely with modern forms. The best cases include *Eoentophysalis*, colonies of intertidal, mat-building cocci that are indistinguishable from modern mat-forming species of *Entophysalis* (Fig. 5.10); the unicellular, stalk-forming, tidal-flat dweller *Polybessurus* (Fig. 5.11) that closely resembles living *Cyanostylon*-type populations in the Bahama Banks; and endolithic fossil *Eohyella* species in Proterozoic ooids that are clear counterparts of *Hyella* species that bore into present-day ooids in shallow carbonate platforms (Fig. 5.12).

The interpretation as cyanobacterial of the filamentous sheaths that constitute a major component of most benthic Proterozoic assemblages is not as well founded. Commonly, traces of trichomes are preserved in these sheaths, thus they are most likely prokaryotic. Some of these sheaths are large (>10 μm in cross-sectional diameter; *Siphonophycus* sp.; Fig. 5.16), and clearly belonged to phototactic mat builders in tidal-flat environments; these are probably cyanobacteria. Some other filamentous sheaths are minute (<2 μm in diameter), for example *Tenuofilum* and *Eomycetopsis*. These could be the remains of cyanobacteria, or anoxygenic photosynthetic bacteria comparable, for example, to modern *Chloroflexus* species that occur in **peritidal** microbial mats.

Oscillatoriopsis is a microbial fossil genus consisting of preserved trichomes comparable in morphology to the modern cyanobacterium *Oscillatoria* (Fig. 5.13). Such fossils have been interpreted as fossil cyanobacteria, but similar shapes are also found among modern sulfur-oxidizing bacteria (*Beggiatoa*), phototrophic bacteria, and even sulfate reducers. The uncertainty increases in older rocks, where paleoenvironmental interpretations are less reliable.

Thus, although cyanobacteria-like fossils are distributed abundantly through Proterozoic rocks and even occur sporadically in Archean metasediments, the confidence with which these remains are interpreted taxonomically varies widely. The problem is not at the specific or generic levels, but principally in relating fossils to higher taxa.

Most prokaryotic fossils resemble cyanobacteria, although at least some of them may well represent other eubacteria. For example, rod-shaped microfossils found in the Middle Proterozoic Thule Group, northwestern Greenland, and elsewhere could be cyanobacteria comparable to species of *Synechococcus*, or they could be, among other possibilities, the remains of purple photosynthetic bacteria. Morphologically distinctive iron-metabolizing bacteria occur in the Gunflint and other Proterozoic iron formations, and heterotrophic bacteria of various types have been recognized on the basis of distinctive morphology in several other Proterozoic formations. However, the record is spotty and few strong evolutionary conclusions can be drawn on the basis of these remains.

Prokaryote fossil record

Body fossils

Most cyanobacterial and cyanobacteria-like fossils compare closely with living populations. Many fossils can be related to present-day taxa and assemblages that have specific and strictly delineated environmental ranges. The group, therefore, affords little biostratigraphic utility, but it is quite useful in paleoenvironmental investigations.

These fossils testify that life originated extremely early in Earth history, that morphologic diversification occurred rapidly, and that many lineages show extreme morphologic (and in some cases, by inference, physiologic) conservatism. Coccoid and filamentous microfossils occur in weakly metamorphosed sedimentary rocks of the Swaziland Supergroup, South Africa, and the Warrawoona Group, Western Australia, both dated at 3500 My. These fossils have morphologies much like those of younger cyanobacteria, but they are not diagnostic in this regard. The fossils indicate that several major morphologic classes were present in Early Archean communities. The true morphologic diversity in these deposits is likely underrepresented for taphonomic reasons. As diagenetic and metamorphic alteration increase, apparent microfossil diversity decreases as a result of selective obliteration and loss of morphologic detail.

Fossils comparable to most extant cyanobacterial groups (save for the Chamaesiphonales and Stigonematales) occur in Proterozoic rocks, and, taken at face value, cyanobacterial diversity appears to have increased throughout the Proterozoic. However, there are reasons to believe that this apparent diversification stems, at least in part, from differential sampling—younger Proterozoic epochs are represented by more assemblages from a greater number of environments than the older ones. Indeed, the existing record is compatible with the hypothesis that the major outlines of cyanobacterial diversity were well established by the Early Proterozoic. The stratigraphic distribution of selected microfossil assemblages is shown in Figure 5.20.

Trace fossils

Prokaryotes produce two kinds of trace fossils—stromatolites and microborings—in carbonate substrates. **Stromatolites** are the trace fossil record of microbial communities. They are initiated by the establishment of a thin mat of microbes on a sediment surface. As sediment particles accumulate on top of mats, they are trapped and bound into a coherent layer by the microorganisms. Microbially mediated precipitation of calcium carbonate can also contribute to sediment accumulation and stabilization. Most mat-building microorganisms require light, and they cannot function when buried beneath the accumulating sediment. They respond by moving or growing upward through the thin sediment covering, or, as in the case of coccoid Entophysalidaceae, by flooding the sediment surface with cells released from polysaccharide envelopes. The microorganisms establish a new mat surface, which will, in turn, trap, bind, and/or precipitate new sediments. A stromatolite is a product of a continuing interaction between microbial activities and sedimentation (Figs 5.21, 5.22). Through time, commonly, a laminated structure accretes, each lamina marking a former position of the living mat community. Despite their apparent simplicity, microbial mats are complex microbial communities, containing a physiologically diverse assortment of prokaryotes. On the present-day Earth, filamentous cyanobacteria are the predominant mat-builders, but coccoid cyanobacteria, other types of bacteria, and a variety of eukaryotic algae produce well-defined mats.

The dual biologic and sedimentologic nature of stromatolites makes them particularly useful in studies of Archean and Proterozoic life. Although the mat organisms may themselves be obliterated during diagenesis, the

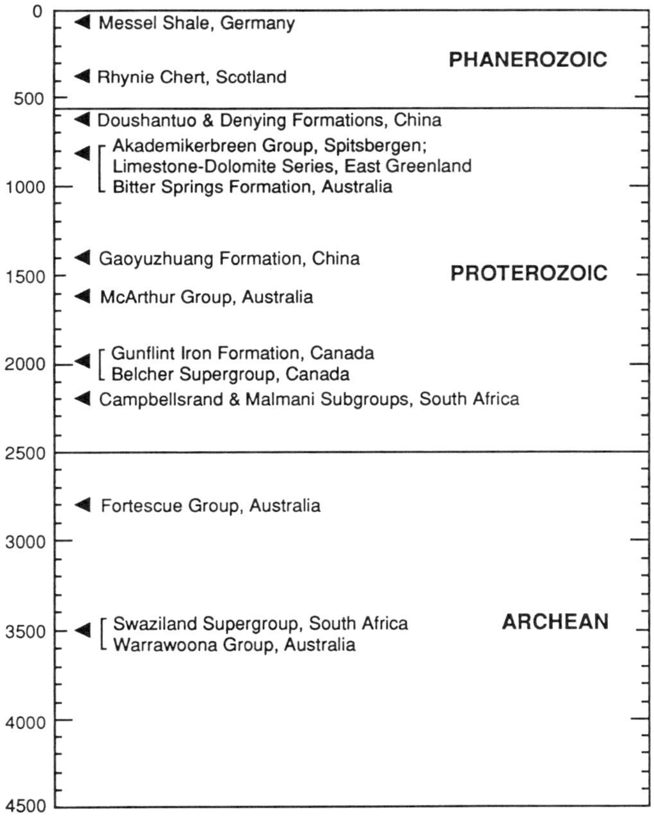

Figure 5.20 Time-table in millions of years (left) of major prokaryotic microfossil assemblages.

Figure 5.21 Elongated modern stromatolites in the intertidal range of Hamelin Pool, Shark Bay, Australia. The rims of stromatolites are actively accreting, colonized by *Entophysalis major*. The central areas are lithified.

Figure 5.22 Laminated columnar stromatolites from the upper part of the Onverwacht Group, South Africa. (Photograph courtesy of DI Lowe, GR Byerly.)

unique sedimentary structures they form are as permanent as the rocks in which they occur. While organically preserved fossils of microbial mat populations are known from only a limited range of Proterozoic paleoenvironments, stromatolites are abundant and widely distributed in Proterozoic carbonates and, to a lesser extent, iron formations.

Stromatolites vary widely in morphology and microstructure. Some are stratiform, or flat laminar; others are hemispheric or elongated (Fig. 5.21), while others may be columnar (Fig. 5.22), or cone shaped. **Oncolites** are essentially spheroidal stromatolites. Stromatolites may occur as small isolated structures, or they may be found in **bioherms** and **biostromes** whose dimensions rival those of Phanerozoic reefs. Clearly, the overall morphology of stromatolites correlates closely with sedimentologic circumstance, but microstructural features, or fabrics, are under at least partial control of mat-building organisms. Stromatolites represent distinctive fossil microbial communities.

Given their sedimentologic attributes, stromatolites can be useful in paleoenvironmental studies of Proterozoic sedimentary basins. They have also proven useful in stratigraphic correlation within basins where particular biostromes form extensive and distinctive marker beds. The hypothesis that stromatolite morphologies can be used in intercontinental biostratigraphic correlation of Proterozoic successions is more controversial. Lower, Middle, and Upper Proterozoic carbonates do contain distinctive assemblages of stromatolites, but until systems for the description and naming of stromatolites have been standardized, and until more biostratigraphic studies are completed in conjunction with sedimentologic research, the stratigraphic promise of stromatolites will remain incompletely fulfilled.

The change in stromatolites through time does not necessarily indicate directional evolution in mat-building cyanobacteria. Changes could equally well result from directional shifts in environments, evolution in eukaryotic components of mats, and/or evolution in algae or animals that modified the distribution of mats through competition or predation.

Endolithic (euendolithic) cyanobacteria leave morphologically distinctive microborings within carbonate substrates. Together with boring traces of other prokaryotic and eukaryotic euendoliths, cyanobacterial microborings are often preserved empty, contributing to the porosity of the rock. They can be studied by infiltrating them with polymerizing resins (epoxy, araldite) to form casts that are then exposed by dissolving the rock, and observed by SEM. Similarly prepared modern microborings provide materials for comparison and interpretation of fossil ones. Microborings represent protected microenvironments where the endoliths lived, but they are also conducive to preservation and fossilization following the death of the resident endoliths.

Microborings and well-preserved endolith body fossils have been found in silicified stromatolites as old as 1500 My and in silicified Late Proterozoic 700–800 My old ooids (Fig. 5.12B). Endoliths with distinguishable biologic ultrastructure are preserved in carbonate rocks of Silurian age (425 Ma). Fossil microborings are often filled by sediments and/or secondary precipitates. The filling materials may differ from the bored matrix in color, texture, or solubility, which facilitates their study. In petrographic thin sections borehole outlines can be recognized by the color contrast between the fill and substrate, or by differences in grain size and patterns; less soluble fills are sometimes weathered out from the rock as natural casts. In many cases fossil microborings have remained empty. Microborings are commonly observed in Phanerozoic rocks; however, their detailed study requires special preparation techniques. Of all trace fossils, microborings probably most closely reveal the morphology of organisms that caused them.

Chemical evidence

In addition to body fossils and trace fossils the rock record preserves evidence of prokaryotic metabolic activities and their biochemical composition, although in most cases this biogeochemical record cannot be related to particular fossil populations. Much of the organic matter originally present in an organism is destroyed soon after the death of the cell in which it was synthesized. Fortunately, some biomolecules, especially lipids and other hydrocarbons, can be preserved in ancient sedimentary rocks, and some of these biomolecules provide evidence for specific groups of prokaryotes. For example, archaebacterial cell membranes are unique in their possession of isoprenoid-ethers; the presence of geochemical derivatives of these molecules in ancient sediments provides diagnostic evidence for the presence of archaebacteria at the time of deposition. Hopanoids, membrane components widespread in eubacteria, were actually identified first as significant components of sedimentary organic matter and only later connected to their microbial source.

Molecular organic geochemistry is a useful tool in hydrocarbon exploration and provides valuable insights into the bacterial components of Phanerozoic ecosystems. Analysis of Proterozoic biomarkers is more difficult, both because of the thermal history of many (but not all) Proterozoic basins, and because early workers experienced problems with contamination. In recent years, however, the molecular geochemistry of Proterozoic organic matter has been reactivated. Improved technology and experimental controls are beginning to provide new insights into the diversity and activities of early prokaryotes.

Among the most useful geochemical tools in the paleobiologic study of prokaryotes is the isotopic geochemistry of carbon and sulfur. During the fixation of CO_2, photoautotrophic organisms incorporate the lighter carbon isotope, ^{12}C, in preference to the heavier stable isotope, ^{13}C. Because of this fractionation, organic matter derived from photosynthetic sources can be identified in the geologic record. Certain secondary processes can further fractionate the carbon isotopes in organic matter, so that metabolisms such as methanogenesis may leave an isotopic signature. During the process of dissimilatory sulfate reduction, sulfate-reducing bacteria fractionate the isotopes ^{32}S and ^{34}S in a manner analogous to that of carbon, leaving another biogeochemical imprint in sedimentary rocks.

The isotopic record indicates that autotrophic CO_2 fixation (and, most likely, photosynthesis) existed at least 3500 Ma. Surprisingly, the earliest unequivocal isotopic evidence of bacterial sulfate reduction is found only in Late Archean sedimentary rocks. This would mean that the origin of bacterial sulfate reduction postdates the evolution of photosynthesis by 1000 My, a conclusion that is at variance with theoretical expectations. No consensus has yet been reached on the interpretation of little-fractionated sulfur isotopic signatures in Early Archean pyrites and sulfates.

In a few cases, prokaryotic microorganisms precipitate distinctive minerals that can be recognized in the geologic record. A notable example are the morphologically distinctive, single-domain crystals of magnetite formed by **magnetotactic** bacteria. Bacterial magnetite is widely distributed in modern sediments, and has been recognized in Proterozoic rocks.

Chemical fossils provide a valuable complement to the morphologic record of prokaryotic evolution. They help fill in the sketchy history provided by body fossils and additionally provide useful stratigraphic and paleoenvironmental information. Molecular fossils in particular have proven valuable in the evaluation of petroleum source rocks.

Summary of the evolutionary record of prokaryotes

Molecular phylogeny indicates that the archaebacteria and eubacteria diverged early in biologic history, each group diversifying rapidly into a number of metabolically distinct, but essentially anaerobic, phyla. Paleobiologic studies indicate that this initial diversification took place at least 3500 Ma. By that time, and possibly much earlier, the Earth supported complex ecosystems fueled by photosynthesis. The Late Archean formation and stabilization of large continents resulted in a much richer paleobiologic record by the Early Proterozoic. There is a problem in determining which, if any, features of this expanded record actually document evolutionary diversification, as opposed to improved sampling. Nevertheless, during the Early Proterozoic atmospheric oxygen levels clearly increased, leading to a new radiation of aerobic prokaryotes. Microfossils, stromatolites, and geochemical markers document the richness of the Proterozoic microbial world. Later Proterozoic microfossil assemblages contain more species than earlier ones, but to what extent evolution contributed to this apparent growth in diversity is not clear. Later Proterozoic environments are much more completely sampled than earlier ones, and there is at present no compelling reason to reject the hypothesis that apparent changes in Proterozoic prokaryotic diversity are largely a product of paleontologic sampling. The latest Proterozoic radiation of animals and, later, plants provided a new and continually changing set of biologic environments that have made possible continuing bacterial evolution during the Phanerozoic.

Supplementary reading

Barghoorn ES, Tyler SM. Microorganisms from the Gunflint chert. *Science* 1965; 147:563–77. The classic reference in Precambrian paleontology.

Brock TD, Madigan MT. *Biology of Microorganisms*. 5th edn. Englewood Cliffs, NJ: Prentice Hall, 1988. General discussion of prokaryote biology.

Carr NG, Whitton BA, eds. *The Biology of Cyanobacteria*. Berkeley, CA: University of California Press, 1982. A detailed account of the biology of this most important prokaryotic group.

Cohen Y, Rosenberg E, eds. *Microbial Mats: Physiological Ecology of Benthic Microbial Mats*. Washington, DC: American Society for Microbiology, 1989. Up-to-date essays on modern microbial mats, the biology of mat microorganisms, mat biogeochemistry, and implications for the study of the fossil record.

Golubic S, Barghoorn ES. Interpretation of microbial fossils with special reference to the Precambrian. In: Flügel E, ed. *Fossil Algae*. Berlin: Springer-Verlag, 1977: 1–14. A primer on the taphonomy of microorganisms.

Green JW, Knoll AH, Golubic S, Swett K. Paleobiology of distinctive benthic microfossils from the Upper Proterozoic Limestone-Dolomite "Series," central east Greenland. *Amer J Bot* 1987; 74:928–40. An illuminating case study of fossil prokaryotes and their living counterparts.

Knoll AH. The distribution and evolution of microbial life in the Late Proterozoic Era. *Annu Rev Microbiol* 1985; 39:391–417. A review and interpretation of the paleobiologic evidence for microbial evolution during the Proterozoic Eon.

Knoll AH, Golubic S. Anatomy and taphonomy of a Precambrian algal stromatolite. *Precambrian Res* 1979; 10:115–51. A study of microfossils within their sedimentary context.

Leadbetter ER, Poindexter JS, eds. *Bacteria in Nature. Bacterial Activities in Perspective*. vol. 1. New York: Plenum Press, 1985. The science of microbiology seen in historical perspective.

Schopf JW, ed. *Earth's Earliest Biosphere: Its Origin and Evolution*. Princeton, NJ: Princeton University Press, 1983. A detailed review and interpretation of the geologic and biologic evidence for Archean evolution.

Walter MR, ed. *Stromatolites. Developments in Sedimentology*. vol. 20. Amsterdam: Elsevier Scientific, 1976. A comprehensive compendium on stromatolites.

Woese CR. Bacterial evolution. *Microbiol Rev* 1987; 51:221–71. Classic account of the molecular phylogeny of prokaryotes and its implications for archaebacterial and eubacterial evolution.

Woese CR, Wolfe RS, eds. *The Bacteria. Archaebacteria*. vol. 8. London: Academic Press, 1985. A comprehensive treatise on archaebacterial biology.

Acritarchs and prasinophytes

6

Carl V. Mendelson

Introduction

Acritarchs are organic-walled microfossils that cannot be placed confidently into any existing classification of organisms. In 1963 W.R. Evitt coined the word for "small microfossils of unknown and probably varied biological affinities consisting of a central cavity enclosed by a wall of single or multiple layers and of chiefly organic composition..." Thus, acritarchs are what is left over—the residue—after all other organic-walled microfossils have been assigned to a systematic category. The name is derived from the Greek *akritos* (uncertain or confused) + *arche* (origin).

Acritarchs embrace a heterogeneous and polyphyletic collection of organic-walled microfossils of uncertain affinities. Most of them are probably the remains of a thick-walled resting stage, or **cyst**, in the life-cycle of planktic eukaryotic algae. Some acritarchs are believed to be dinoflagellate cysts (see Chapter 7). Many others can be assigned confidently to the green algae, the chlorophytes. In particular, some acritarchs bear a striking resemblance to the nonmotile stage (**phycoma**) in the life-cycle of modern **prasinophytes**, which are phylogenetically primitive and primarily marine green algae that are sometimes assigned to their own class. Even though fossil prasinophytes are now well known, they commonly are referred to the acritarchs for convenience. For example, the titles of many major monographs include only the word "acritarch" even though the authors explicitly acknowledge the green algal, and especially prasinophyte, affinities of many of the studied microfossils. Therefore, this chapter includes microfossils of the green algae known informally as prasinophytes because micropaleontologists traditionally study them with acritarchs. Some workers avoid the word "acritarch" altogether, using instead terms such as microplankton, microphytoplankton, or organic-walled microfossils.

Micropaleontologists who study organic-walled microfossils are called **palynologists**. Their objects of investigation include spores and pollen grains of vascular plants, dinoflagellate cysts (see Chapter 7), fungal spores, and acritarchs, among many others. Collectively, such microfossils are called **palynomorphs**.

The study of acritarchs and prasinophytes is important because these microfossils provide clues to paleoenvironments, and they can be used for biostratigraphic correlation. However, perhaps more importantly, such microfossils most probably represent the remains of the primary producers of the Proterozoic and early Paleozoic, a role later assumed by other groups of phytoplankton. Such a fundamental ecologic role for acritarchs and prasinophytes has been recognized only recently.

History of study

The history of acritarch research blends with that of fossil dinoflagellates because of similarities in morphology and techniques of study. Microfossils that are now referred to the acritarchs were probably first reported by M.C. White in 1862 from Ordovician, Silurian, and Devonian strata of western New York. White suggested they were diatoms, sponge gemmules (special types of reproductive bodies), or **sporangia** of the distinctive types of green algae known as desmids. In fact, many workers had identified similar microfossils as **xanthidia** ever since Ehrenberg, in 1836, reported a find of such objects in Cretaceous flints and had referred some of them to *Xanthidium* (a living desmid); but Ehrenberg's fossils later turned out to be dinoflagellate cysts (see Chapter 7). The next 70 years witnessed few advances in the study of acritarchs. Workers continued to report finds of xanthidia, some suggesting that they were eggs of trilobites or graptolites, or bryozoan statoblasts (asexual reproductive bodies).

The first significant breakthrough came with the work in the early 1930s of the late Alfred Eisenack, who described many species that now are placed in the acritarchs from lower Paleozoic pebbles in Baltic glacial deposits. Initially, Eisenack thought that these microfossils were the eggs of planktic invertebrates, but by the late 1930s and especially after World War II, he came to regard them as remains of fossil phytoplankton. A few other European workers were also active at this time. In a 1933 monograph on microfossils from Upper Cretaceous and Danian (lower Paleocene) Baltic flints, O. Wetzel introduced the genus *Hystrichosphaera*. (The nomenclatural history of this generic name, which in full is *Hystrichosphaera* O. Wetzel, 1933 *ex* Deflandre, 1937, is most colorful; the name no longer enjoys currency, because it was determined to be a subjective junior synonym of *Spiniferites* Mantell, 1850.) Subsequent studies frequently referred to spiny cyst-like objects as **hystrichospheres**. Then, in 1961, Evitt showed that many hystrichospheres, including *Hystrichosphaera*, were dinoflagellate cysts, and 2 years later he allocated the nondinoflagellate, organic-walled microfossils to the group Acritarcha. A few years prior to Evitt's important work, the description of acritarch genera entered the current log phase of growth.

The publication history of fossil prasinophytes extends even farther back than that of acritarchs. The first **tasmanitids**, which are distinctive thick-walled and pore-bearing fossil prasinophytes, were described by Hooker in 1852 as possible lycopod seed cases. The next significant report was in 1875, when Newton named the genus *Tasmanites* after a peculiar deposit of white coal (composed largely of tasmanitids) in the Permian of Tasmania. By this time a few investigators considered such fossils algal in origin. Further advances were made, as in the acritarch case, through the efforts of Eisenack in the 1930s. Subsequent work by Eisenack and other workers helped clear up some of the confusion in nomenclature that had so far accumulated. In the course of resolving the hystrichosphere controversy in 1963, Evitt assigned tasmanitids and similar forms to the Acritarcha; later the same year Downie, Evitt, and Sarjeant referred the tasmanitids to the green algae, specifically the class Chlorophyceae. Over the course of the next decade, other studies confirmed that tasmanitids and related forms are very similar morphologically to the nonmotile stage of modern prasinophyte algae.

Morphology

Acritarchs have not been rigorously analyzed in terms of theoretical morphology; if they were, they would probably occupy a relatively large volume of theoretical morphospace. This is not surprising, inasmuch as the acritarchs are an unnatural grouping. The following three sections address morphology and its application: in this section the morphologic attributes of acritarchs are reviewed; in "Systematics," these attributes are evaluated in terms of taxonomic utility; and in "Paleobiology," the phylogenetic placement and paleoecologic significance of acritarchs are considered.

Acritarch morphology may be explored through study of the **vesicle**, **processes**, and **excystment structures** (Fig. 6.1).

Vesicle

The vesicle, or hollow body of an acritarch, is described in terms of size, shape, wall structure, surface sculpture, and composition. It ranges from a few micrometers to a few millimeters in diameter, although individual taxa commonly exhibit a narrow size distribution. The vesicle of most acritarchs ranges between 50 and 100 μm; however, the diversity of smaller acritarchs (<20 μm) is underestimated because fine organic debris in preparations obscures them. Most acritarchs were probably spherical, spheroidal, or at least inflated originally, but they are typically preserved as flat, two-dimensional compressions in shales. This preservational bias has complicated comparisons between two- and three-dimensionally preserved assemblages, and has resulted in the establishment of superfluous taxa.

Acritarchs have an astounding variety of shapes (summarized diagrammatically in Fig. 6.2 and illustrated by Upper Ordovician microfossils in Figs 6.8–6.15): spheroidal (Figs 6.2A, 6.8B_1, 6.9B, 6.11A_1–A_2); ellipsoidal (Figs 6.2B, 6.11D); fusiform (Figs 6.2C, 6.9E, 6.14B_1); crescentic (Figs 6.2D, 6.9F); pad shaped (with a cross-section that looks like an inflated triangle) (Figs 6.2E, 6.9C); bean shaped (Figs 6.2F, 6.9G); egg shaped (Fig. 6.2H); and pillow shaped (Fig. 6.9D). Other vesicles can be cylindrical, flask-like, pear shaped, or

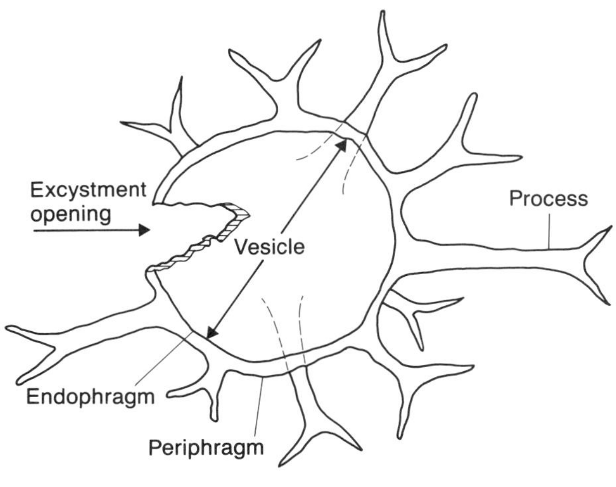

Figure 6.1 Major morphologic features of an acritarch. (Modified from Williams GL. In: Haq BU, Boersma A, eds. *Introduction to Marine Micropaleontology*. New York: Elsevier North-Holland, 1978: 316.)

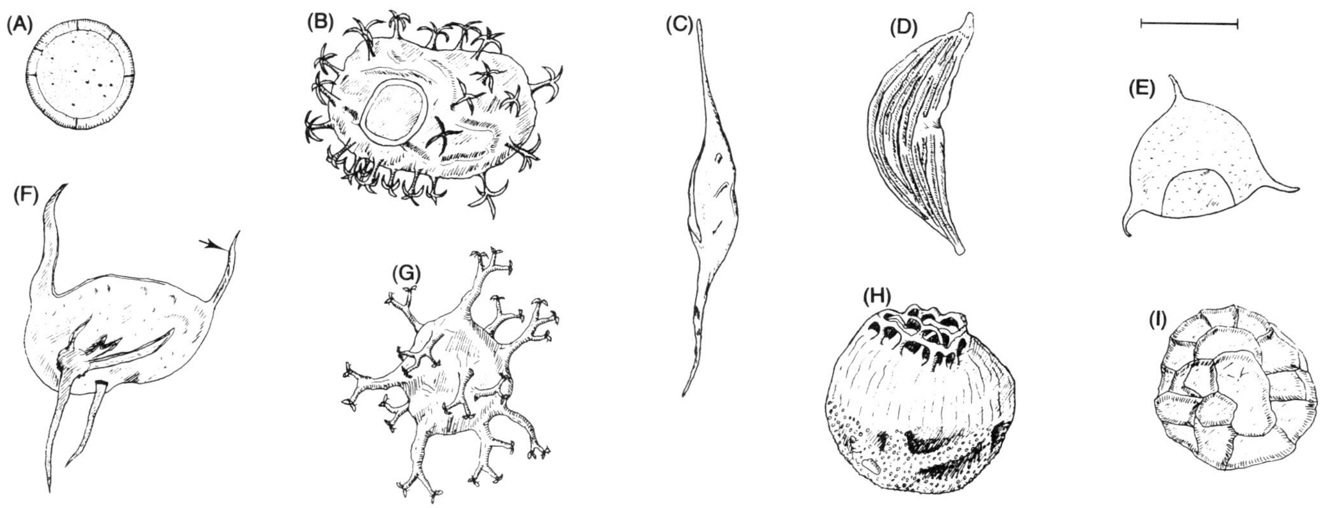

Figure 6.2 Prasinophytes (A, I) and acritarchs (B–H) showing range of morphologic features. Consult Figure 6.4 for surface sculpture terms. Specimens from the Lower Jurassic (Lias) of Dorset, England (A); the Upper Ordovician Sylvan Shale of south-central Oklahoma (B–G); the Lower Ordovician *Obolus* beds of the St Petersburg region, Russia (H); and the Lower Devonian of Oklahoma (I). A, *Tasmanites newtonii* Wall, 1965. Note circular outline of vesicle (surface view) and thick wall (optical section), which is enlarged in Figure 6.3A. Scale bar, 60 µm. B, *Peteinosphaeridium* sp. Note elliptical outline of vesicle, distinctive processes (simplified; see Fig. 6.5E), and prominent subcircular pylome surrounded by thickened rim and plugged by laevigate operculum. Scale bar, 25 µm. C, *Leiofusa litotes* Loeblich & Tappan, 1978. Elongate fusiform vesicle is produced into gradually tapering, hollow processes; longitudinal slit may be excystment opening. Scale bar, 65 µm. D, *Baiomeniscus eumecus* Loeblich & Tappan, 1978. Vesicle has lunate–crescentic outline and an ornament of longitudinally directed folds. Scale bar, 55 µm. E, *Villosacapsula setosapellicula* (Loeblich, 1970) Loeblich & Tappan, 1976. Note typical inflated-triangular outline, short processes in plane of vesicle, and incipient epityche. Scale bar, 20 µm. F, *Sylvanidium paucibrachium* Loeblich, 1970. Typical bean-shaped vesicle; arrow indicates process shown in Figure 6.3D. Scale bar, 35 µm. G, *Multiplicisphaeridium* sp. Subcircular vesicle has laevigate surface and broad-based processes that exhibit up to three orders of bifurcation distally. Scale bar, 15 µm. H, *Ooidium rossicum* Timofeev, 1957. Egg-shaped vesicle is granulate in lower half; upper half characterized by discontinuous striae that merge into bases of anastomosing trabeculae. Scale bar, 20 µm. I, *Cymatiosphaera* sp. Note reticulocristate surface divided into polygonal fields. Scale bar, 30 µm. (A Redrawn from Wall D. *Geol Mag* 1962; 99:353–62. H Redrawn from Loeblich AR Jr. *Proc North Amer Paleontological Convention*. 1970: G705–88. I Redrawn from Tappan H. *The Paleobiology of Plant Protists*. San Francisco: WH Freeman, 1980: 821.)

stellate. Compression and other diagenetic factors (e.g., pyrite growth, impressions of detrital grains) typically modify shape, so the inflated form should be reconstructed in descriptions. Shape has figured prominently in some artificial classifications.

The wall layers of acritarchs can be important diagnostic characters. Many acritarchs have a single layer; others have two layers that may vary in thickness (Fig. 6.1). Some acritarchs that appear to have one wall layer may actually have two, because the process wall (which is considered to be continuous with the outer layer) is commonly much thinner than the vesicle wall. Thin-walled taxa include *Leiosphaeridia* (Fig. 6.8A) and *Leiofusa litotes* (Figs 6.2C, 6.9E). In these forms the wall is commonly less than 0.5 µm thick, although the loose generic concept of *Leiosphaeridia* includes forms having walls much thicker. Thick-walled forms illustrated include *Tasmanites newtonii* (Figs 6.2A, 6.3A), which has a wall 2–7 µm thick; *Tasmanites* sp. (Fig. 6.8B_1–B_2), which has a wall 4–5 µm thick; *Baltisphaeridium* sp. (Figs 6.3B, 6.12A_1–A_2), with a wall 2–3 µm thick; and *Orthosphaeridium insculptum* (Figs 6.3C, 6.12B), with a wall about 2 µm thick. The thickness of the wall probably controls, to some degree, the nature and extent of folding and radial cracking in compressed acritarchs; other factors might be wall composition, wall structure, the ratio of wall thickness to vesicle diameter, and the original shape of the vesicle.

Wall ultrastructure has been investigated in only a few acritarchs. Those acritarchs studied seem to have two general types of ultrastructure. The first type may be summarized as a single-layered wall penetrated by narrow radial canals (Figs 6.3A, 6.8B_2), which resembles modern and fossil prasinophytes (e.g., *Tasmanites*), and includes members of such acritarch genera as *Baltisphaeridium*, *Peteinosphaeridium*, and *Veryhachium*; examples of these three genera are illustrated here, but the wall ultrastructure cannot be discerned at the scale shown. The second type of ultrastructure consists of a very thin (less than 0.5 µm thick) but apparently two-layered wall; a zone of pores, 0.5–2.0 µm in diameter, separates the inner from the outer layer. This wall structure has been studied in

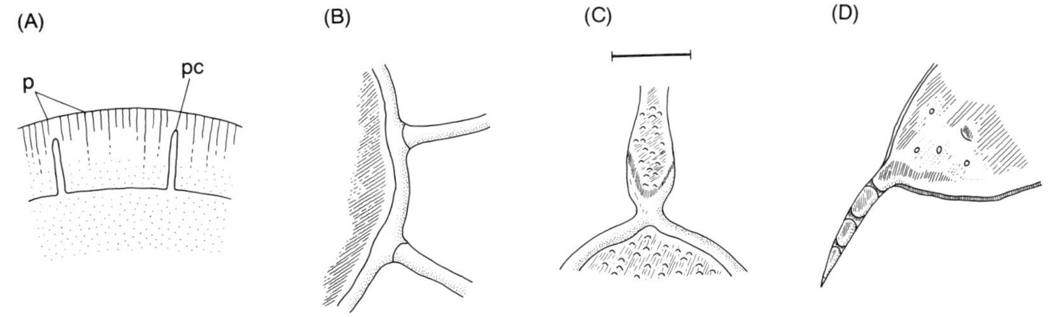

Figure 6.3 Optical sections of wall structure and processes. All except A are from the Upper Ordovician Sylvan Shale of south-central Oklahoma. A, *Tasmanites newtonii* Wall, 1965 (enlargement of Fig. 6.2A). Thick wall is characterized by radially arranged fine punctae (p) and coarse pore canals (pc). Scale bar, 6 μm. B, *Baltisphaeridium* sp. Note relatively thick wall, which thickens slightly opposite processes, and solid proximal part of processes. Scale bar, 15 μm. C, *Orthosphaeridium insculptum* Loeblich, 1970. Note typical constriction and prominent plug at base of process, and coarsely microgranulate ornament. Scale bar, 10 μm. D, *Sylvanidium paucibrachium* Loeblich, 1970 (enlargement of process in Fig. 6.2F). Shows partial fill (secondary?) resulting in internal ladder-like structure. Scale bar, 15 μm. (A Redrawn from Wall D. *Geol Mag* 1962; 99:353–62, and from Williams GL. In: Haq BU, Boersma A, eds. *Introduction to Marine Micropaleontology*. New York: Elsevier North Holland, 1978: 322.)

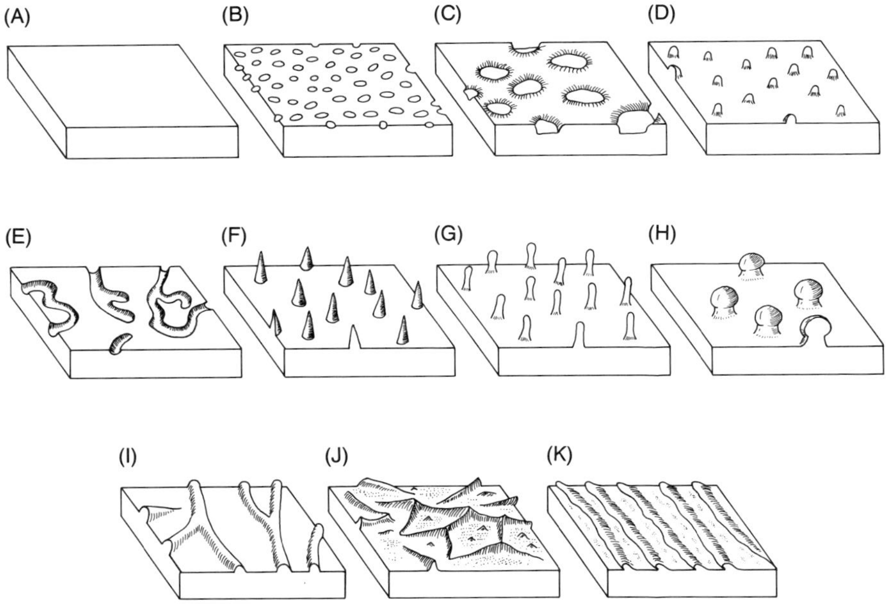

Figure 6.4 Important surface sculpture types; sculptural elements are typically smaller than 5 μm. A, Laevigate. B, Punctate. C, Foveolate. D, Granulate. E, Fossulate. F, Echinate. G, Baculate. H, Gemmate. I, Rugulate. J, Cristate. K, Costate. (A, C–I Redrawn from Williams GL, Sarjeant WAS, Kidson EJ. *Amer Assoc Stratigraphic Palynologists Contrib Ser* 1978; 2A:1–21. B, J, K Redrawn from Tappan H, Loeblich AR Jr. *Micropaleontology* 1971; 17:385–410.)

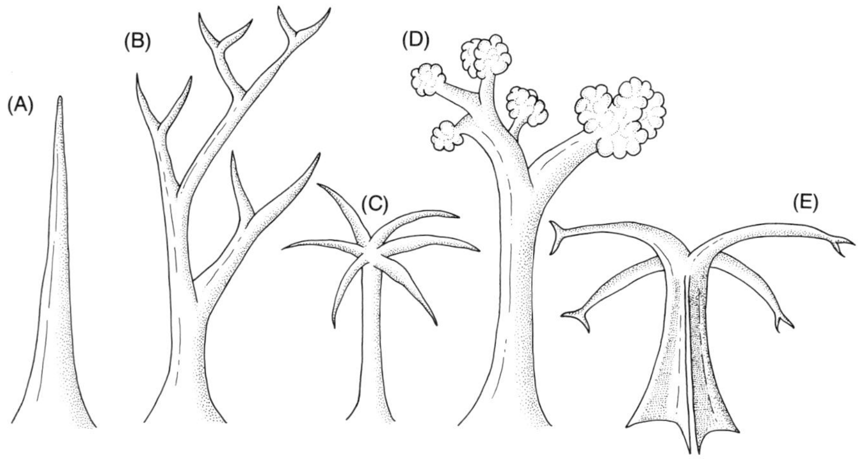

Figure 6.5 Selected types of process terminations in acanthomorph acritarchs; processes are typically larger than 5 μm. A, Simple. B, Dichotomously branched (common in *Multiplicisphaeridium*). C, Palmate. D, Lobulate. E, Complex process of *Peteinosphaeridium* sp. (enlargement of Fig. 6.2A). Each process is supported by stiff buttresses that expand at base; process divides distally into four solid, flexible branches that bifurcate at tip. (A–D Redrawn from Williams GL, Sarjeant WAS, Kidson EJ. *Amer Assoc Stratigraphic Palynologists Contrib Ser* 1978; 2A:1–21.)

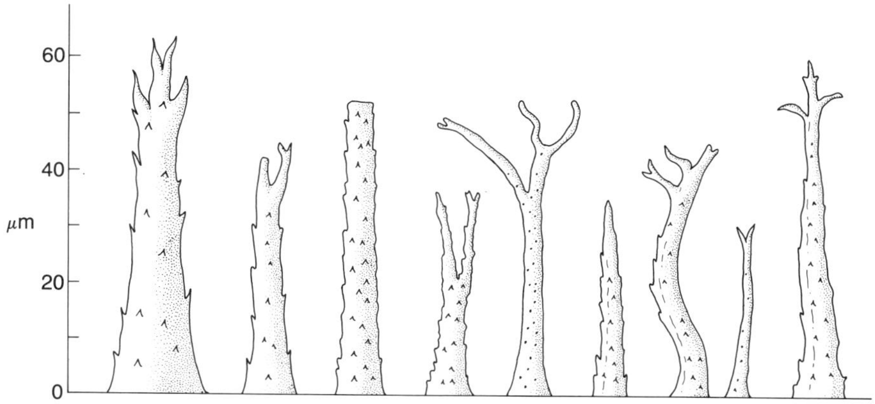

Figure 6.6 Diagrammatic representation of process variability, in lateral view, of *Diexallophasis remota* (Deunff, 1955) Playford, 1977. (Redrawn from Playford G. *Geol Surv Can Bull* 1977; 279:1–87.)

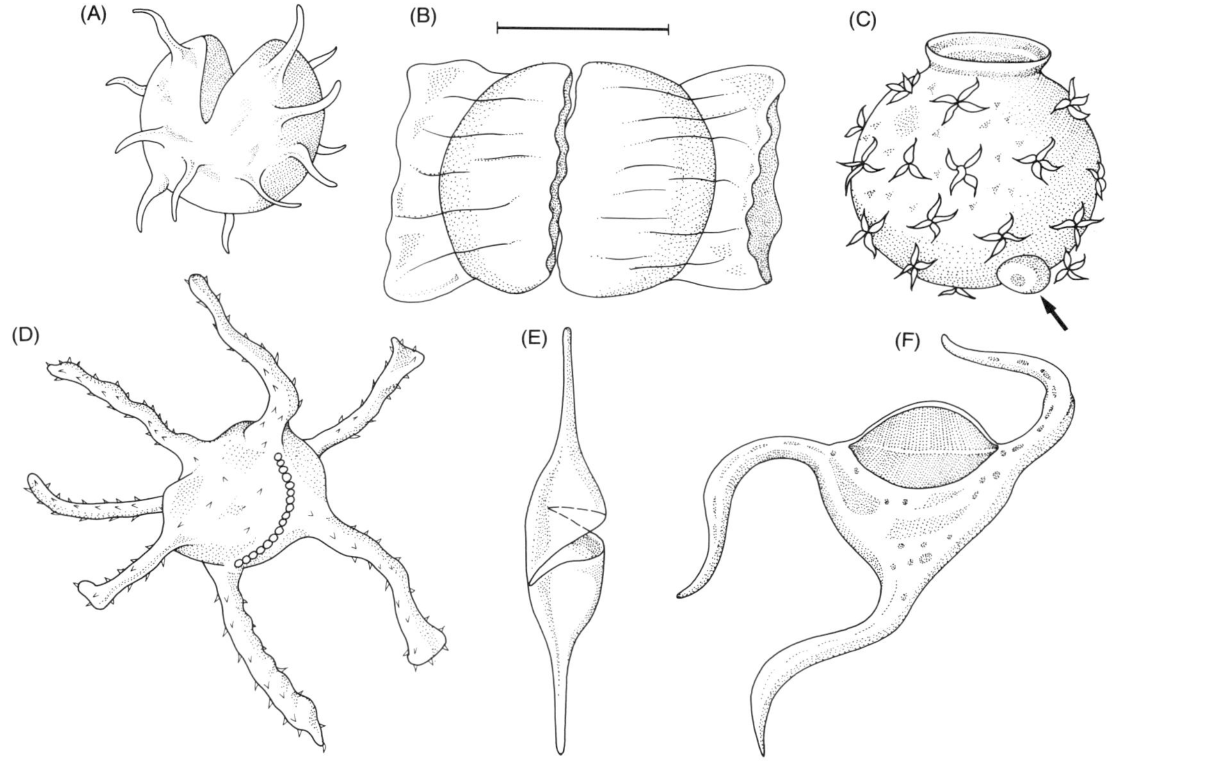

Figure 6.7 Excystment openings in acritarchs (scales approximate). A, Partial rupture in *Micrhystridium*. Scale bar, 15 μm. B, Median split in *Riculasphaera*. Scale bar, 30 μm. C, Pylome with bordering rim and plug-like operculum in *Polyancistrodorus*; this genus also characterized by antapical pseudopylome (arrow). Scale bar, 40 μm. D, Ornamented border delimiting site of partial rupture in *Diexallophasis gotlandica* (Cramer, 1970) Le Herissé, 1984. Scale bar, 25 μm. E, Trochospiral suture, idealized. Scale bar, 50 μm. F, Epityche in *Veryhachium*. Scale bar, 40 μm. (A,B,C, F Redrawn from Loeblich AR Jr, Tappan H. *Rev Española Micropaleontol* 1969; 1:45–7. D Based on Le Herissé A. *Rev Palaeobot Palynol* 1984; 43:217–36. E Redrawn from Miller MA, Wood GD. *Neues Jahrbuch Geol Paläontol Monatshefte* 1982: 547–52.)

the acritarch *Acanthodiacrodium divisum*, and is similar to that in some dinoflagellates. These results are based on only a few of the more than 400 validly proposed genera of acritarchs and fossil prasinophytes, and therefore generalizations are not yet possible. Wall ultrastructure is difficult to study by scanning electron microscopy; light microscopy and transmission electron microscopy are the best approaches.

Acritarchs vary greatly in the complexity of their surface ornamentation or surface sculpture (Fig. 6.4). This phenomenon has been appreciated best since the advent of scanning electron microscopy methods because subtle variations are difficult to discern by light microscopy alone. As work progressed, the number of terms used to describe surface sculpture proliferated; many terms have accumulated multiple meanings, a situation that has

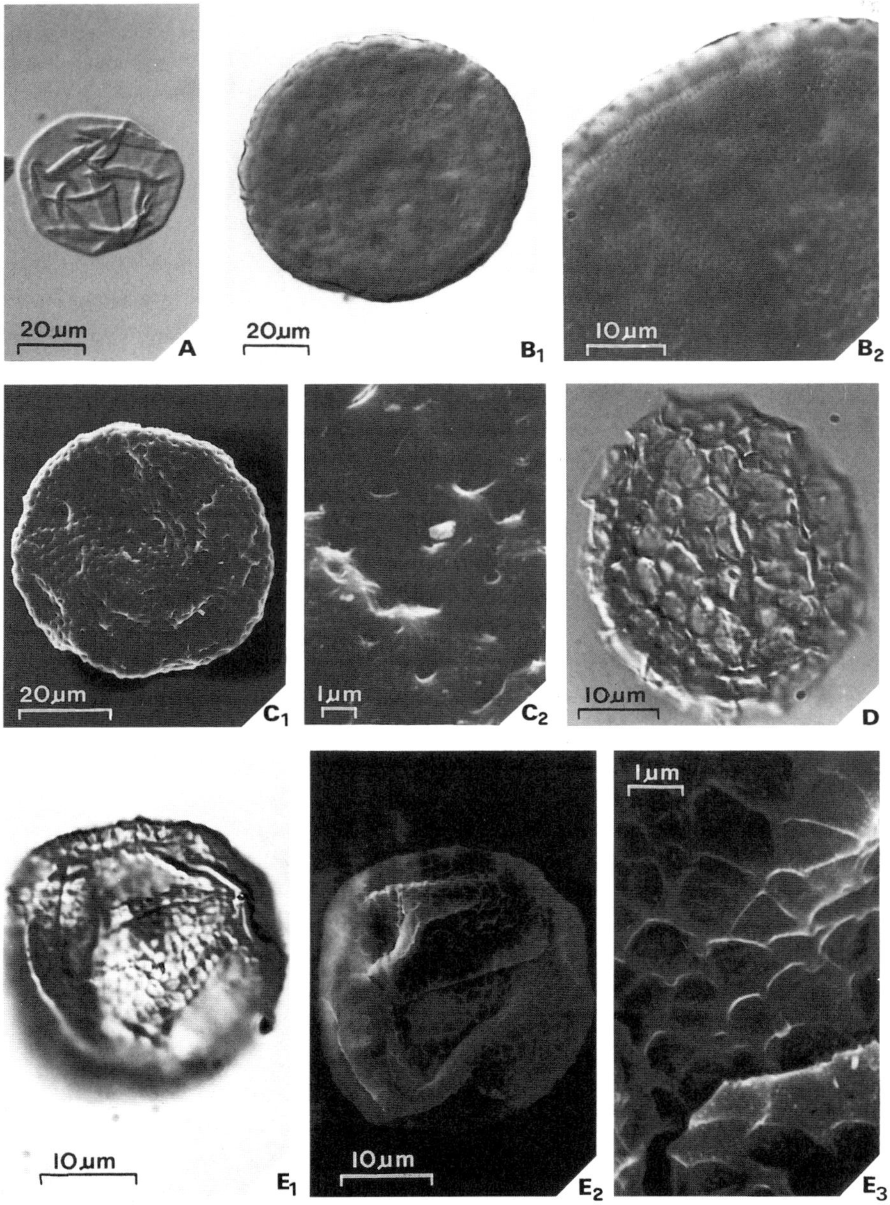

Figure 6.8 Acritarch (A) and prasinophytes (A?, B–E) from the Upper Ordovician Sylvan Shale of south-central Oklahoma. (A, B_1–B_2, D, E light micrographs; C_1–C_2, E_2–E_3 scanning electron micrographs.) A, *Leiosphaeridia* sp.; note readily folded, relatively thin wall exhibiting laevigate surface. B, *Tasmanites* sp. B_1, Typical circular outline of vesicle. B_2, Enlargement of B_1 at lower focus, showing thick wall in optical section and punctate inner surface of vesicle. C, *Tasmanites* sp. C_1, Degraded specimen. C_2, Enlargement of central part of C_1, showing pore openings. D, *Cymatiosphaera* sp.; shows surface divided into polygonal fields that are surrounded by diaphanous crests about 4 µm high (visible at upper left). E, *Cymatiosphaera* sp. E_1, Subcircular outline of vesicle. E_2, Same specimen as E_1 viewed from opposite side, showing polygonal fields that are smaller than those of species in D. E_3, Enlargement of right-central part of E_2 and rotated 180°, showing microgranulate surface within polygonal fields.

hampered communication among palynologists.

Many types of surface sculpture are now clearly defined and well illustrated. The terms are designed for double duty: when used alone, they apply to features seen with the optical microscope (e.g., **granulate**); addition of the prefix "micro-" allows description of features less than 1 µm in diameter or width, and thus applies to features resolvable only by electron microscopy (e.g., **microgranulate**). Some acritarchs display more than one type of ornament, which may be localized on particular parts of the vesicle (e.g., Fig. 6.2H). Certain process-bearing ("spiny") acritarchs have a process ornament that differs from the vesicle ornament. Figure 6.4 depicts some of the more common sculptural elements encountered in acritarch studies; compare these with the surface sculpture of actual specimens in Figures 6.8–6.15.

Few analyses have been made to determine the chemical composition of the vesicle. Condensed fatty acid derivatives are known in some Paleozoic leiospheres (simple spheroidal acritarchs that lack processes and have a relatively smooth wall), a chemistry similar to **sporopollenin**, the compound that makes up the resistant outer coating common to spores and pollen grains of vascular plants. The exact structure and composition of sporopollenin is still being investigated; it seems to be a high-molecular weight and acid-resistant polymer of carbon, hydrogen, and oxygen. Because of their acid-resistant nature, acritarchs are thought to be composed of

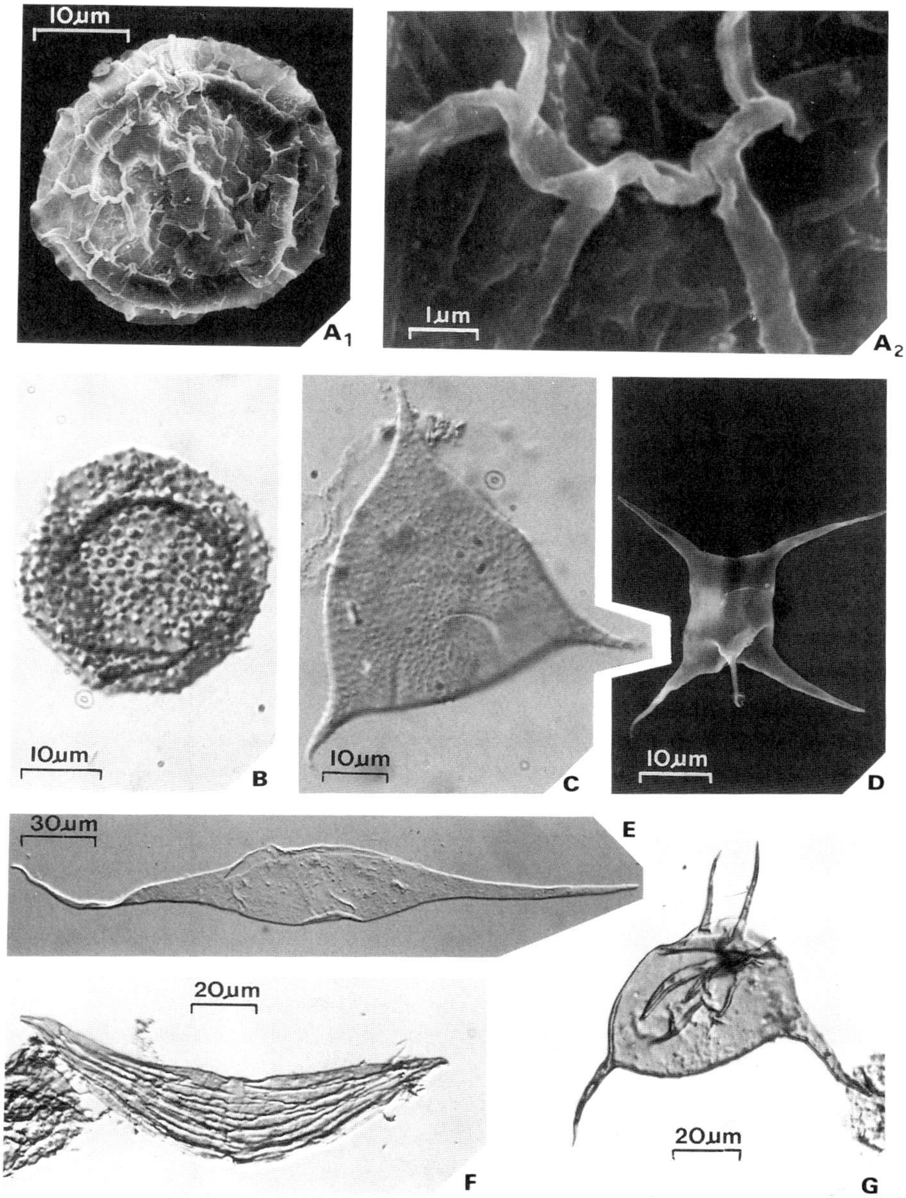

Figure 6.9 Acritarchs (B–G) and prasinophyte (A) from the Upper Ordovician Sylvan Shale of south-central Oklahoma. (B, C, E–G light micrographs; A_1–A_2, D scanning electron micrographs.) A, *Cymatiosphaera* sp. A_1, Surface divided into polygonal fields. A_2, Enlargement of left-central part of A_1, showing smaller polygonal fields within the larger ones. B, *Lophosphaeridium* sp., showing typical folds and subcircular outline of vesicle, and ornament of low, variably spaced, blunt-tipped, conical elements. C, *Villosacapsula setosapellicula* (Loeblich, 1970) Loeblich & Tappan, 1976, showing typical inflated-triangular outline of vesicle with short processes in plane of vesicle, and prominent incipient epityche at bottom center. D, *Veryhachium oklahomense* Loeblich, 1970, showing rectangular outline of vesicle, laevigate surface, and rare fifth process arising from face of vesicle. E, *Leiofusa litotes* Loeblich & Tappan, 1978, showing elongate fusiform vesicle produced into gradually tapering, hollow processes; longitudinal slit along upper surface may be excystment opening. F, *Baiomeniscus eumecus* Loeblich & Tappan, 1978, showing vesicle with lunate-crescentic outline, and an ornament of prominent longitudinally directed folds. G, *Sylvanidium paucibrachium* Loeblich, 1970, showing typical bean-shaped vesicle; note partially filled process at lower left, resulting in internal ladder-like structure.

sporopollenin. However, the definition requires only that they be organic walled; thus, microfossils composed of less resistant organic matter, such as cellulose, still qualify as acritarchs.

Processes

The shape of an acritarch may be further modified by one or more **processes**. Processes are relatively major morphologic features that project from the surface of the vesicle and are typically more than 5 µm long (Fig. 6.1); sculptural elements (Fig. 6.4) are typically smaller, rarely projecting more than 5 µm from the vesicle surface. Most processes appear to arise from the outer wall layer (**periphragm**; Figs 6.1, 6.3B) in bilayered vesicles. Processes vary greatly in size, shape (e.g., conical, hair-like), nature of junction with vesicle (e.g., constricted, angular), and surface sculpture. Some processes are simple: they do not branch and have a more or less circular cross-section (Fig. 6.5A). Simple processes are highly varied, and may be long and flexible (Fig. 6.13C_1), short and ornamented by micrograna (Fig. 6.13B_2), and thin and filmy (Fig. 6.10A). Complex processes may branch dichotomously (Figs 6.5B, 6.10D_1–D_2, 6.11A_1–A_3) or irregularly (Fig. 6.6). The distal ends of processes may be extremely elaborate (Figs 6.5C, D). The very complex

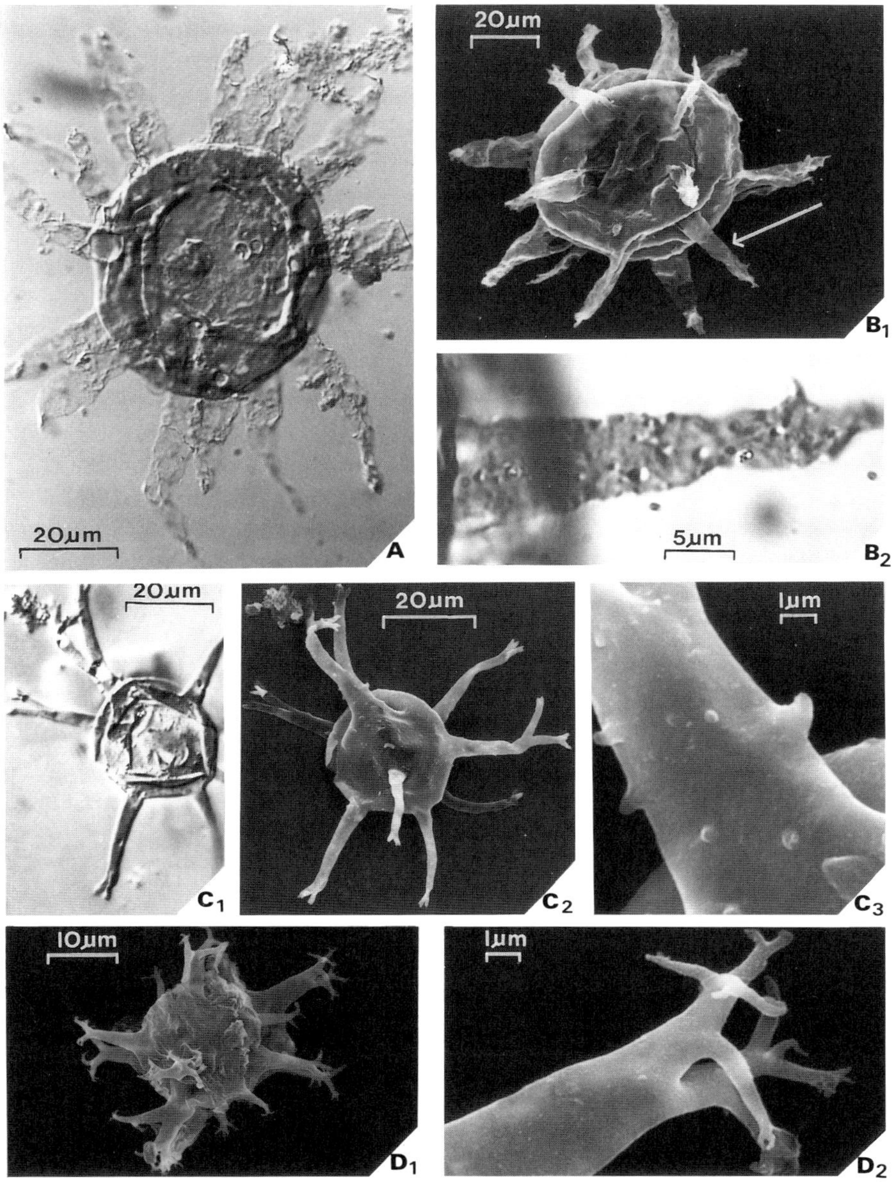

Figure 6.10 Acritarchs from the Upper Ordovician Sylvan Shale of south-central Oklahoma. (A, B_2, C_1 light micrographs; B_1, C_2–C_3, D_1–D_2 scanning electron micrographs.) A, *Actipilion druggii* Loeblich, 1970, showing numerous, thin-walled, transparent processes. B, *A. druggii*, different specimen. B_1, Arrow indicates process enlarged in B_2. B_2, Enlargement of process in B_1 viewed from opposite side, showing filmy, delicate nature. C, *Diexallophasis* sp. C_1, Vesicle nearly circular in outline and produced into processes that freely communicate with vesicle interior. C_2, Same specimen as C_1 viewed from opposite side; some processes bifurcate, others simple with digitate tips; rupture at left accidental(?). C_3, Enlargement of process at upper left in C_2, showing ornament of scattered grana and spinules (small spines). D, *Multiplicisphaeridium* sp. D_1, Vesicle subcircular in outline with laevigate surface (diagenetically altered in places) and with broad-based processes. D_2, Enlargement of process at upper right of D_1, showing nearly laevigate surface and up to four orders of bifurcation.

processes of *Peteinosphaeridium* (Figs 6.5E, 6.11D, 6.15B) are quadrate in cross-section, and supported by stiff buttresses that expand at the base; distally they become diaphanous, and each process divides into four solid, flexible branches that bifurcate at the tips. Similarly complicated processes occur in other acritarchs and can be diagnostic at the genus or species level. Process variability in a single species is shown in Figure 6.6.

Processes on an individual vesicle are termed **homomorphic** if all of them are similar; if more than one type of process projects from the vesicle, the processes are termed **heteromorphic**. Some processes are apparently solid (e.g., Figs 6.5E, 6.14C), whereas others are hollow (e.g., Figs 6.5B, 6.6, 6.9E, 6.10C_1). This is an important taxonomic distinction, because individual species do not typically have both hollow and solid processes.

Another taxonomic character that may have systematic importance, but which is controversial, is whether hollow processes consistently communicate freely with the vesicle interior. Direct communication is characteristic of such genera as *Villosacapsula* (Fig. 6.9C), *Diexallophasis* (Figs 6.6, 6.10C_1), and *Veryhachium* (Fig. 6.15A_2). On the contrary, in some species of *Baltisphaeridium*, and in *Orthosphaeridium* (Figs 6.3C, 6.12B), the wall is modified into a meniscus-like plug in the constricted proximal part of the process. In other species assigned to *Baltisphaeridium*

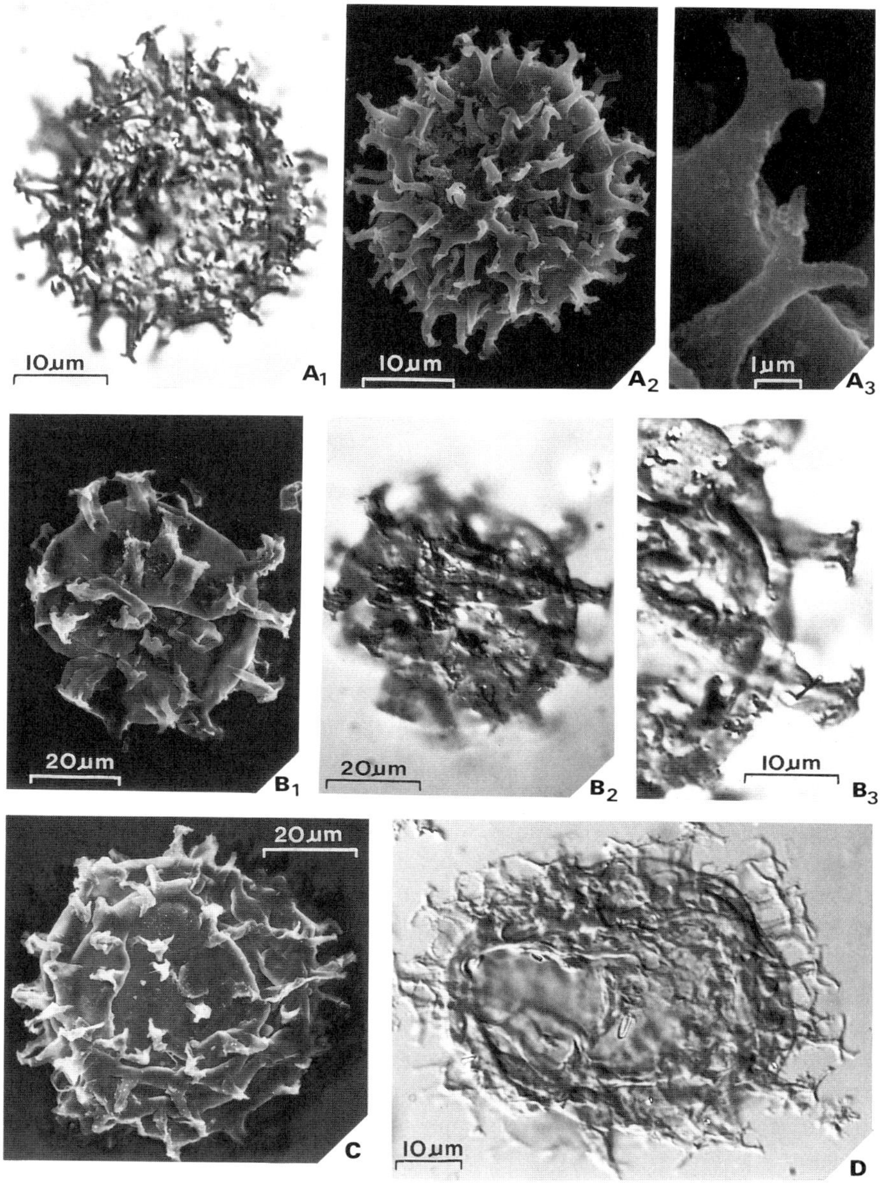

Figure 6.11 Acritarchs from the Upper Ordovician Sylvan Shale of south-central Oklahoma. (A_1, B_2-B_3, D light micrographs; A_2-A_3, B_1, C scanning electron micrographs.) A, Unidentified form. A_1, Vesicle relatively small and circular in outline. A_2, Same specimen as A_1 viewed from opposite side, showing short, distally branched processes. A_3, Enlargement of upper right part of A_2, showing process detail. B, *Peteinosphaeridium* sp. B_1, Vesicle with prominent, finned processes that appear to be triangular or quadrate in cross-section; rupture probably accidental. B_2, Same specimen as B_1 viewed from opposite side. B_3, Enlargement of right side of B_2, showing ribbed processes. C, Unidentified form, showing possibly ribbed processes that are somewhat degraded and thus difficult to characterize. D, *Peteinosphaeridium* sp., showing processes supported by stiff buttresses that expand at base; processes divide distally into four solid, flexible branches that bifurcate at tips.

(Figs 6.3B, 6.12A_2), the wall may slightly thicken opposite the attachment site of the process. Variations on these themes are common. For instance, although the processes freely communicate with the vesicle in *Sylvanidium* (Figs 6.3D, 6.9G), distally, they may be partly infilled resulting in internal ladder-like structures. The superiority of light microscopy over scanning electron microscopy is apparent if the internal nature of processes is of systematic importance.

The number and distribution of processes are important features, and they may modify acritarch symmetry. Many species have numerous processes that are evenly distributed over the vesicle (Figs 6.11A_2, 6.13B_1), and thus the original symmetry of the vesicle is maintained. Other acritarchs display one process at each pole; such processes may be fairly distinct (Fig. 6.9E), or they may merge imperceptibly with the vesicle (Fig. 6.14B_1). The processes of *Villosacapsula* (Fig. 6.9C) and *Veryhachium* (Fig. 6.15A_1-A_2) are concentrated in one plane, thus accentuating the bilateral symmetry of the vesicle.

Excystment structures

If some acritarchs are cysts comparable to those produced by dinoflagellates (see Chapter 7), then perhaps they had similar life-cycles in which a motile vegetative cell alter-

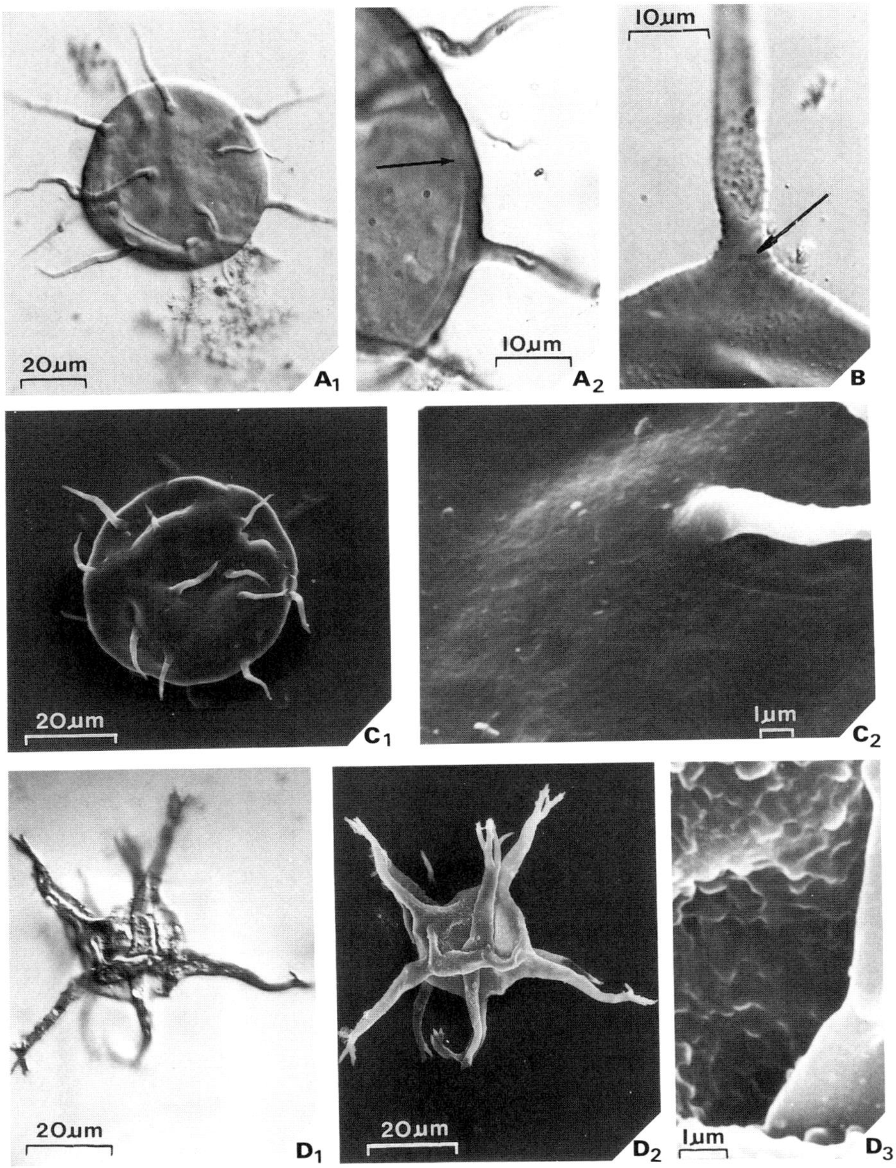

Figure 6.12 Acritarchs from the Upper Ordovician Sylvan Shale of south-central Oklahoma. (A_1–A_2, B, D_1 light micrographs; C_1–C_2, D_2–D_3 scanning electron micrographs.) A, *Baltisphaeridium* sp. A_1, Circular vesicle outline and uncrowded, flexible processes. A_2, Enlargement of right side of A_1, showing in optical section the relatively thick wall and solid proximal part of process; arrow indicates inner boundary of wall. B, *Orthosphaeridium insculptum* Loeblich, 1970, enlargement of prominent process in upper half of specimen illustrated in Figure 6.15D, showing typical constriction and prominent plug at base of process; plug cup-shaped distally and straight proximally (arrow); note coarsely microgranulate ornament throughout. C, *Baltisphaeridium* sp., probably same species as in A. C_1, Circular outline of vesicle and uncrowded processes. C_2, Enlargement of right-central part of C_1, showing laevigate surface. D, *Diexallophasis* sp. D_1, Vesicle ornamented with scattered spinules. D_2, Same specimen as D_1 viewed from opposite side, showing digitate process terminations. D_3, Enlargement of center of D_2, showing complex surface ornament (diagenetic?).

nated with a cyst, or resting stage. If so, the vegetative contents of the acritarch must have escaped through an opening upon resumption of the reproductive cycle or perhaps upon restoration of favorable environmental conditions. Evidence of such escape openings, or **excystment structures**, is not found on all acritarchs, but sufficient examples are known to indicate that different types of structures characterize particular taxa. Excystment modes may be classified into five major types:

1 **partial rupture**,
2 **epityche**,
3 **median split**,
4 **pylome**, and
5 **cryptopylome**.

The partial rupture (Figs 6.7A, 6.13A_1, 6.14C, D) is the simplest and probably most common type of excystment structure among acritarchs. It involves the partial splitting of the vesicle, leaving a gap that varies in its degree of development. An accidental break may be indistinguishable from the partial rupture, in which case ruptures of many specimens should be analyzed for consistency of occurrence, geometry, and position on the vesicle. This type of excystment characterizes forms assignable to *Micrhystridium* and *Baltisphaeridium*, among many others. The epityche, which might be considered a specialized variant of the partial rupture, is characterized by a curving slit that allows a flap of the vesicle to fold out. It is especially common in *Villosacapsula* (Fig. 6.9C)

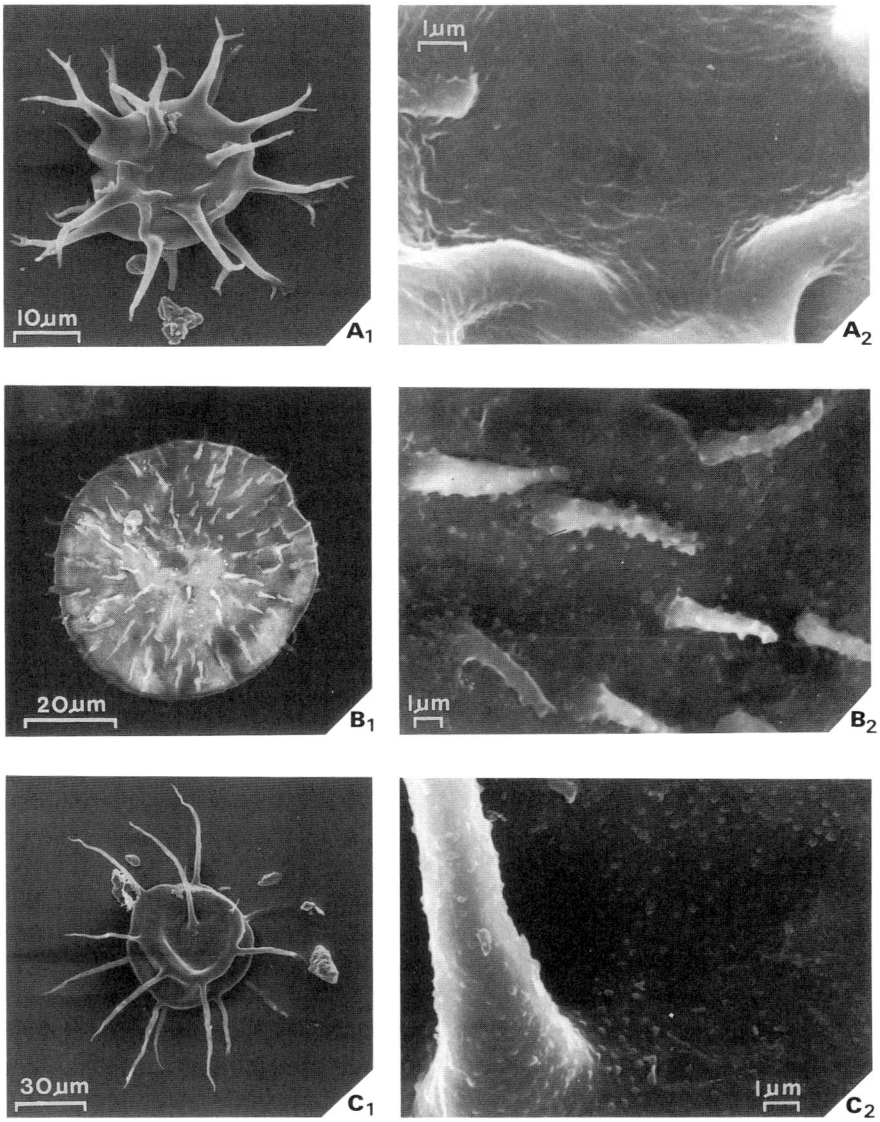

Figure 6.13 Acritarchs from the Upper Ordovician Sylvan Shale of south-central Oklahoma. (All scanning electron micrographs.) A, Unidentified form; resembles *Multiplicisphaeridium* in terms of process morphology and *Cheleutochroa* Loeblich & Tappan, 1978, in terms of surface ornament. A_1, Subcircular vesicle outline, heteromorphic processes, and partial rupture (excystment structure). A_2, Enlargement of A_1, showing reticulate surface ornament that grades into ridges at base of process. B, *Gorgonisphaeridium* sp. B_1, Circular outline of vesicle and numerous short processes. B_2, Enlargement of right side of B_1, showing microgranulate ornament on both processes and vesicle. C, *Baltisphaeridium* sp. C_1, Folded vesicle with long, flexible processes. C_2, Enlargement of central process in C_1, showing microgranulate ornament throughout.

and *Veryhachium* (Figs 6.7F, 6.15A_1–A_3).

The median split results in the separation of the vesicle into two nearly equal parts. It is particularly well developed in *Riculasphaera* (Figs 6.7B, 6.27J) and *Orthosphaeridium* (Fig. 6.15D).

An opening having a rounded or subangular outline, and which is closed by a plug-like **operculum** prior to excystment, is commonly termed a pylome. Because the pylome is preformed, its position can be detected on the vesicle surface before excystment; in many cases it is bordered by a thickened rim or collar (Figs 6.7C, 6.15B, C_1–C_2), or even by a tubular neck.

In some specimens a particular part of the vesicle wall is discarded, leaving a rounded to subangular opening. No evidence of the position of the opening on the vesicle exists before excystment, hence this type of structure is termed a cryptopylome.

Some acritarchs possess a **pseudopylome**. It is a thickening of the vesicle wall, resembling the pylome rim or collar, but which typically occurs opposite the true pylome (e.g., in *Polyancistrodorus*; Fig. 6.7C). A true *second* pylome has been reported in a few species and is especially well documented for *Asketopalla formosula*. Perhaps both of these phenomena might be attributed to mutations in genes that regulate the cyst-construction program. Such mutations apparently never became fixed in any acritarch population, perhaps because they conferred no adaptive advantage; for whatever reason, no acritarch species consistently exhibits double pylomes.

Acritarchs employ additional excystment modes

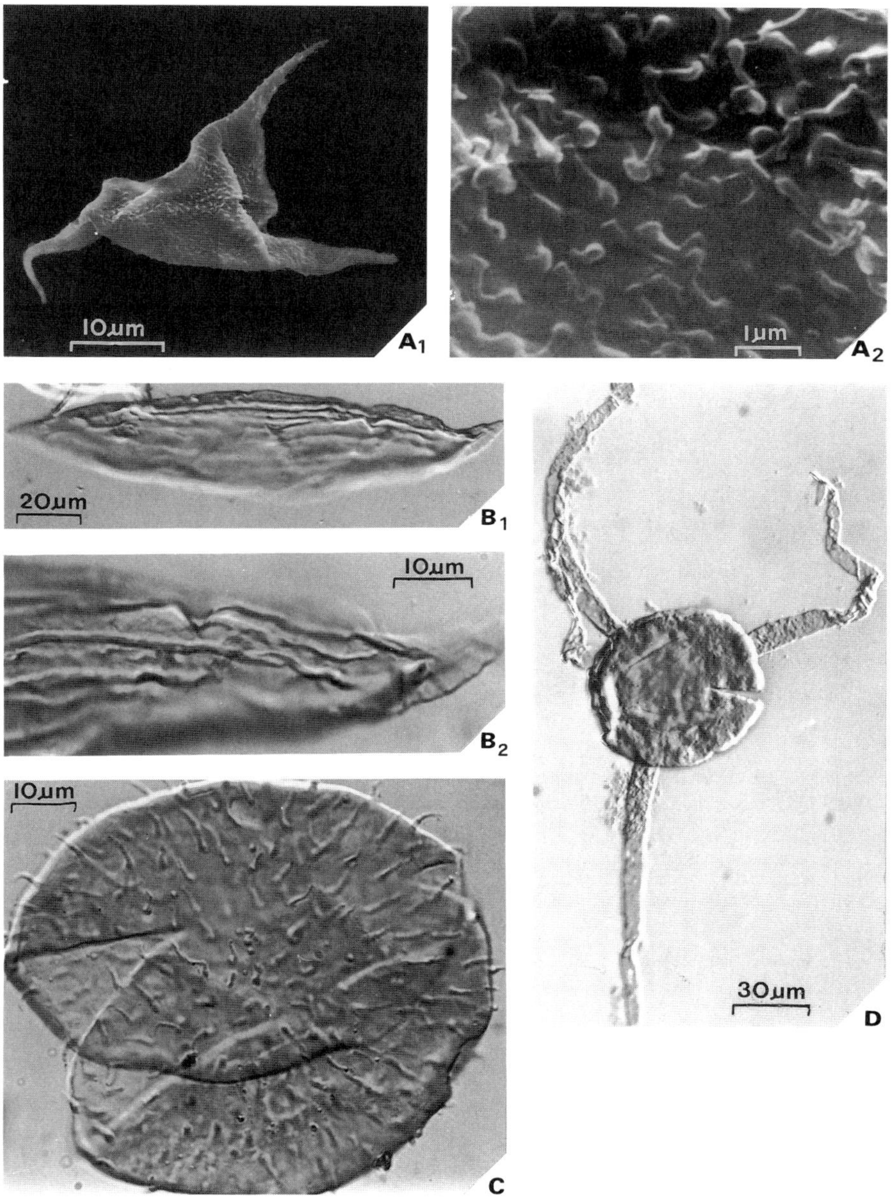

Figure 6.14 Acritarchs from the Upper Ordovician Sylvan Shale of south-central Oklahoma. (B_1–B_2, C, D light micrographs; A_1–A_2 scanning electron micrographs.) A, *Villosacapsula setosapellicula*. A_1, Triangular outline of vesicle, with processes in plane of vesicle. A_2, Enlargement of central part of A_1, showing typical surface ornament of variably spaced spinules. B, *Eupoikilofusa* sp. B_1, Nearly perfectly symmetrical elongate fusiform vesicle. B_2, Enlargement of right side of B_1, showing prominent, longitudinally directed ridges up to 2 µm in width. C, *Gorgonisphaeridium* sp., showing partial rupture mode of excystment. D, *Baltisphaeridium aliquigranulum* Loeblich & Tappan, 1978, showing circular outline of vesicle, few long tapering processes, and excystment by partial rupture.

beyond these five main types. For example, a spiral-shaped excystment structure, called a **trochospiral suture** (Fig. 6.7E), occurs in *Leiofusa tumida* and *Eupoikilofusa cantabrica*. The partial rupture of some species may be bordered by complex ornamentation; such borders may indicate the position of the rupture on the vesicle prior to excystment (Fig. 6.7D). Considering the variability now known, the taxonomy of excystment structures most likely will be modified significantly in the next few years.

Systematics

Acritarchs and prasinophytes display a wide range of characteristics that might prove useful in classificatory schemes. Both artificial and natural approaches to classification have been proposed, but a host of taxonomic problems continues to impede their development. Such problems plague the systematics of acritarchs more than most other groups.

Artificial classifications

The most popular artificial classification is not a phylogenetic statement, but a classification of convenience. It lies totally outside the jurisdiction of any codes of nomenclature. The group Acritarcha is divided into sub-

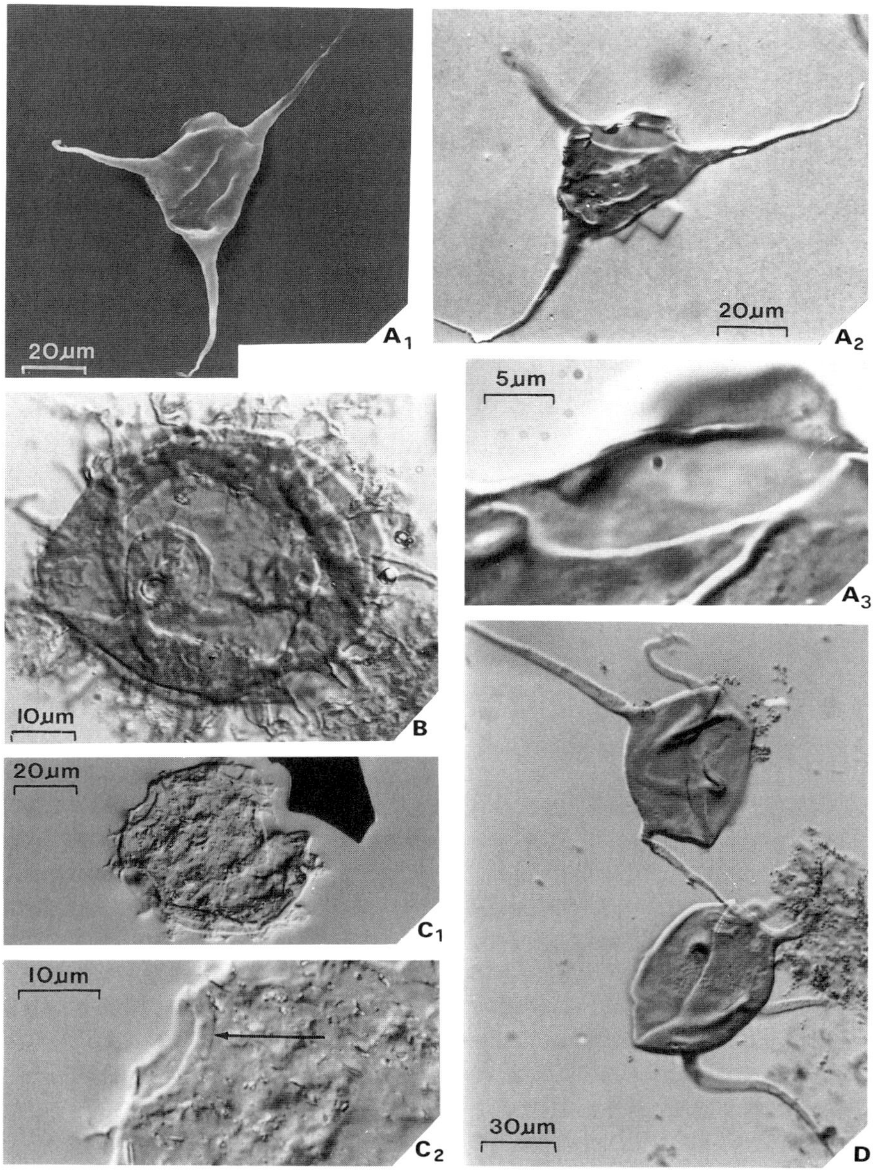

Figure 6.15 Acritarchs from the Upper Ordovician Sylvan Shale of south-central Oklahoma. (A_2–A_3, B, C_1–C_2, D light micrographs; A_1 scanning electron micrograph.) A, *Veryhachium* sp. A_1, Typical inflated-triangular outline of vesicle, which is produced into long, flexible processes that are hair-like distally; prominent flap of epityche visible along upper margin of vesicle. A_2, Same specimen as A_1 viewed from opposite side, showing well-developed epityche; rectangular features along lower margin due to damage from electron beam. A_3, Enlargement of A_2, showing epityche. B, *Peteinosphaeridium* sp., showing subcircular to ovate outline of vesicle, and prominent subcircular pylome plugged by laevigate operculum, and surrounded by elevated rim. C, *Peteinosphaeridium* sp., different specimen. C_1, Note that operculum has been lost. C_2, Enlargement of pylome area in C_1, showing prominent thickened border (arrow). D, *Orthosphaeridium insculptum*, showing typical subquadrate vesicle that has separated into two halves (median split mode of excystment); vesicle probably split when cover slip was pressed down during mounting. (C_1–C_2, Courtesy of Merrell A. Miller.)

groups on the basis of a characteristic morphologic feature such as general shape (Table 6.1). A subgroup is not based on a type genus, so transfer of genera from one subgroup to another, or from a subgroup to a Linnaean family, presents no problems in taxonomic nomenclature. Subgroups are identified by the termination "-morphitae."

The descriptive nouns derived from the subgroup names in this classification are very useful, even if the system is not used formally (see Table 6.1; subgroup definitions below are abbreviated). **Acanthomorphs** (from subgroup Acanthomorphitae) are spherical forms that lack an inner body and crests, and have prominent simple or branching processes (Figs 6.10, 6.12, 6.13). **Polygonomorphs** (Polygonomorphitae) encompass vesicles with a polygonal outline, and which typically have simple processes (Figs 6.9C, D, 6.14A_1, 6.15A_1–A_2). **Sphaeromorphs** (Sphaeromorphitae) include forms that have a spherical to ellipsoidal vesicle and which lack processes, although the wall may be variously ornamented (Figs 6.8A, 6.9B). **Netromorphs** (Netromorphitae) have elongate to fusiform vesicles that may be produced into processes or spines at one or both poles (Figs 6.9E, F, 6.14B_1). **Diacromorphs** (Diacromorphitae) have a spherical to ellipsoidal vesicle; they are characterized by an equatorial zone free of ornamentation, whereas polar areas display a variety of surface sculptures. **Herkomorphs** (Herkomorphitae) have a spherical to subpoly-

Table 6.1 Important subgroups of the Acritarcha

Subgroup	Probable biologic affinity (if well documented)
Acanthomorphitae	
Polygonomorphitae	
Sphaeromorphitae	prasinophytes (some)
Netromorphitae	
Herkomorphitae	prasinophytes (most)
Diacromorphitae	
Pteromorphitae	prasinophytes (most)
Prismatomorphitae	prasinophytes (most)
Oömorphitae	

Subgroups having proven dinoflagellate affinities are excluded. See text for references to illustrations of the subgroups.

gonal vesicle that is divided by crests into polygonal fields (Figs 6.8D, E_1-E_3, 6.9A_1-A_2). An equatorial flange or **ala** typifies the **pteromorph** genera (Pteromorphitae; Figs 6.23D, 6.27F, H). **Prismatomorphs** (Prismatomorphitae) have a prismatic to polygonal vesicle, the edges of which may be produced into a flange (Figs 6.20C, 6.27E). **Oömorphs** (Oömorphitae) have an egg-shaped, polarized vesicle, which is relatively smooth on one end and heavily ornamented on the other (Fig. 6.2H). Most herkomorphs, pteromorphs, and prismatomorphs are now considered to be prasinophytes or other green algae.

Originally, 13 subgroups were proposed to encompass the range of acritarch morphologies. Since 1963 over 20 additional subgroups have been proposed; some of these proposals duplicated earlier efforts, whereas others helped distinguish distinctive morphologies. The constituent genera of many of these subgroups are now considered to be dinoflagellates.

Certain organic-walled microfossils are excluded, some explicitly, by the original definition of acritarch. The informal group **Cryptarcha** was created to address this situation. The Cryptarcha comprises organic-walled hollow vesicles or cellular bodies (which may be associated in pairs, clusters, or colonies) and aseptate filamentous microfossils; cryptarchs do not have processes. Sphaeromorphs are considered cryptarchs, and are thereby excluded from the Acritarcha, under this classification. Two other cryptarch subgroups (comparable to acritarch subgroups) were added in order to house the filamentous and colonial or aggregated microfossils. The cryptarchs include most Precambrian microfossils, which are dominated by nondescript simple spheres and clusters and by filaments.

An additional way to organize acritarchs is to omit all suprageneric categories, and simply arrange genera alphabetically. An advantage is the ease in finding genera in a long systematic paper; a disadvantage is the inconvenience in comparing major subgroups of acritarchs, such as acanthomorphs and netromorphs, with specimens of similar shape.

Natural classifications

Acritarchs

To date, few natural classifications of acritarchs have been attempted, and none of the schemes has gained widespread acceptance. For example, the class Hystrichophyta was proposed for the orders Tasmanales, Leiosphaeridiales, and Hystrichosphaerales. Most of the Tasmanales and some of the Leiosphaeridiales are now considered to be prasinophytes, and almost all of the remaining members of the Leiosphaeridiales would be called acritarchs today. The Hystrichosphaerales contained only two genera, which were later discovered to be dinoflagellates. The Hystrichophyta was later divided into three families: Baltisphaeridaceae, Leiosphaeridaceae, and Leiofusidaceae. The two classifications are similar in emphasizing the general shape of the vesicle and the character of its periphery, but neither is widely accepted today.

Natural classifications based on general form may be inherently faulty. Similar vesicle shapes may be evolved in unrelated lineages through convergent evolution and maintained through parallel evolution, as a common solution to the problem of planktic existence. An analogous situation obtains for the order Foraminiferida, especially among benthic foraminifera, in which distinct test morphologies (e.g., trochospiral, planispiral, biserial) occur in more than one suborder (see Chapter 12). Thus, general shape might be influenced greatly by environmental conditions. Those acritarchs considered to be cysts probably formed internally because nearly all protists make their cyst or test internally (e.g., dinoflagellates; see Chapter 7). The general morphology of an internal cyst might be coupled genetically with that of the vegetative parent, and therefore its morphology might change as a function of environment. On the other hand, details of the wall ultrastructure, composition, and surface sculpture would tend to be independent of environmental influences; such details might form the basis for a truly natural classification, from which phylogeny might be inferred. Distinctive surface sculpture, in particular, might be a key to tracing evolutionary lineages of acritarchs.

Another clue to natural groupings might be mode of excystment. Variants of particular excystment modes may characterize certain taxa. For example, median splits may be bordered by a row of granules in some species. The orientation of the median split with respect to vesicle symmetry may also prove to be important. Pylomate taxa may differ in:

1 the diameter of the pylome relative to the vesicle;
2 the morphology of the operculum;
3 the ornamentation of the operculum;
4 modifications of the pylome rim.

Increasing importance will be attached to excystment structures in future attempts to classify the acritarchs.

Prasinophytes

The prasinophytes are a group of green flagellates characterized by a scaly quadriflagellate (or biflagellate) motile stage; the flagella also typically bear distinctive scales. Some oceanic prasinophytes encyst during their life-cycle (Fig. 6.16), producing a highly resistant, and apparently sporopollenin-containing, nonmotile stage called the **phycoma** (plural, phycomata). Because the nonmotile stage actually grows considerably during its existence, the term "phycoma" is preferred over "cyst." The term "cyst" connotes a physiologically quiescent state. Palynologists believe that fossil prasinophytes are preserved in the phycoma stage of development.

Experts disagree on the phylogenetic placement of the prasinophytes. Many prasinophyte genera are considered to exhibit primitive green algal characters, suggesting that the group might represent the ancient stock from which other types of green algae evolved. Many workers emphasize this key role for the prasinophytes by placing them in their own botanical division, the Prasinophyta, whereas others have argued for a class Prasinophyceae within the division Chlorophyta. Still other microbiologists have, on the basis of detailed studies of cellular ultrastructure, allocated prasinophyte genera to other classes of green algae, thereby rejecting the concept of a natural group of prasinophytes. Finally, a recent cladistic analysis of green algae failed to separate the prasinophytes as a natural grouping.

In spite of the somewhat chaotic state of prasinophyte systematics, reasonable and useful natural classifications are available. An example is shown in Table 6.2.

Impediments to progress in acritarch systematics

A variety of problems obstructs the path, if there is one, to a well-accepted natural classification of acritarchs and of fossil prasinophytes. These obstacles include failure by workers to adhere to rules of nomenclature, poor taxonomy, and difficulties in comparing microfossils from different lithologies.

Acritarch studies, almost from their inception, have been plagued by problems of nomenclature. Some early

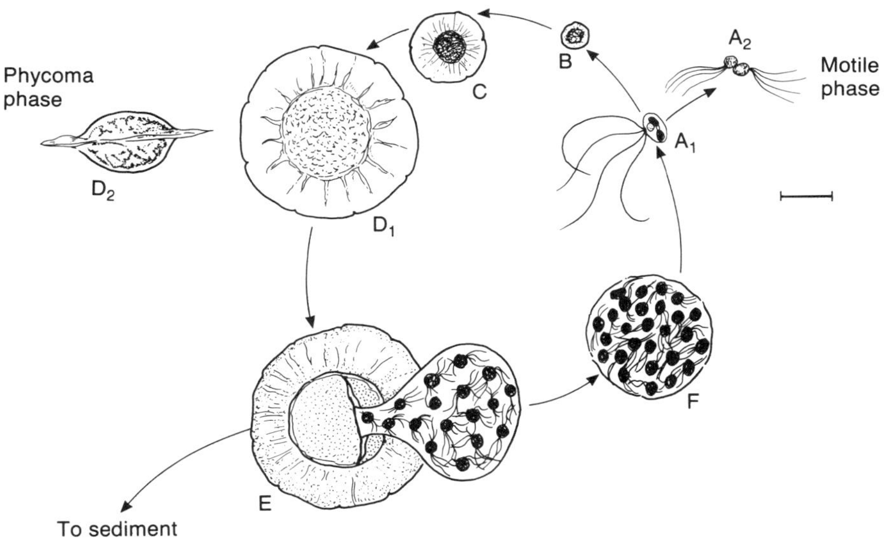

Figure 6.16 Idealized life-cycle of the modern prasinophyte *Pterosperma*, common in temperate seas. A_1, Motile quadriflagellate cell bearing scale- and hair-covered flagella. A_2, Reproduction by fission, producing two motile cells; such fission may occur repeatedly. B, Beginning of nonmotile phycoma phase; cell in A_1 has lost flagella and developed a thick wall and equatorial flange or ala. C, D_1, Phycoma at successive stages of development, which requires 2 weeks to 3.5 months (in culture). D_2, Side view of D_1, showing equatorial nature of ala. E, Escape of inner sac (surrounded by thickened inner wall of phycoma) through release suture (excystment opening). The inner sac contains vegetative cells that are developing flagella. The resistant phycoma, which may descend to the sea floor and be preserved, strongly resembles pteromorph acritarchs. F, Inner sac containing flagellated motile forms. Upon release from phycoma, inner sac may grow up to 16 times its initial diameter. The wall thins, and the motile cells burst out of the sac. Scale bar, 10 μm for A_1; and 50 μm for A_2, B–F. (Modified from Tappan H. *The Paleobiology of Plant Protists*. San Francisco: WH Freeman, 1980: 810.)

Table 6.2 A classification of the prasinophytes, including some common modern and fossil genera. (Compiled and modified from Tappan H. *The Paleobiology of Plant Protists*. San Francisco: WH Freeman, 1980:812–22)

	Modern genera	Fossil genera	Stratigraphic range of order
Division Chlorophyta			
Class Prasinophyceae			
Order Pedinomonadales	*Pedinomonas*		Holocene
Order Pyramimonadales	*Pyramimonas* *Halosphaera*	*Chuaria* *Leiosphaeridia* *Trachysphaeridium*	Proterozoic to Holocene
Order Prasinocladales	*Platymonas* *Prasinocladus* *Tetraselmis*		Holocene
Order Pterospermatales	*Micromonas* *Nephroselmis* *Pachysphaera* *Pterosperma*	*Crassosphaera* *Maranhites* *Tasmanites* *Pterospermopsis* *Pterospermella* *Quisquilites* *Cymatiosphaera* *Melikeriopalla* *Muraticavea* *Retialetes* *Duvernaysphaera* *Polyedryxium*	?Proterozoic, Cambrian to Holocene

For the order Pyramimonadales the listing of a fossil genus implies only that *some* of its constituent species were probably prasinophytes.

workers treated acritarchs under the **International Code of Zoological Nomenclature**; later workers opted to use the **International Code of Botanical Nomenclature**. As the literature multiplied certain genera became acutely overused, thus becoming so-called waste-basket genera. Unfortunately, waste-basket genera are still common, and include *Baltisphaeridium* (including over 300 species), *Cymatiosphaera*, *Leiosphaeridia*, *Micrhystridium*, *Multiplicisphaeridium*, and *Veryhachium*. Recent attempts to define these genera more narrowly have helped to address this situation.

Another problem is insufficient description. Some genera have been based on only a very brief description, which results in an extremely broad concept of the genus. For example, the original description of the important genus *Lophosphaeridium* Timofeev *ex* Downie, 1963 (see Fig. 6.9B) stated only that the vesicle was "thick and tuberculose." This example illustrates a further problem: failure to adhere to the rules of the *International Code of Botanical Nomenclature*. Timofeev, in 1959, neglected to cite the **type species** of *Lophosphaeridium*, so the genus remained invalid until Downie designated one in 1963. In the past, some genera were defined in such a way as to include a great diversity of forms. In certain cases, the included species might have been allocated to at least six different genera. In order to rectify such situations, taxonomists must study carefully the type species (if one has been designated) to establish the true nature of the genus; many of the other species may then need to be transferred to more appropriate genera.

A severe problem, which particularly afflicts Precambrian studies, concerns difficulties in comparing microfossils three-dimensionally preserved in chert with compressions obtained from shales. Chert microfossils are commonly studied in thin section; shale microfossils are typically studied in acid macerates, as are most other palynomorphs. This dichotomy in technique and preservation has led to different taxonomic approaches, and has resulted in superfluous names because the names are based on diagenetic variants. Fortunately, guidelines are currently being developed for the comparative study of microfossils preserved in chert and those preserved in shales.

Other problems frustrate attempts to arrange the acritarchs naturally. The major obstacles include the following:

1 some type specimens (especially those in the USSR) are not easily obtained for comparison;

2 many taxonomic treatments are in obscure publications of limited availability;

3 many illustrations are of low quality;
4 invalid names are common; and
5 generic and specific characters, especially those of sphaeromorphs, may be modified during degradation and diagenesis.

Paleobiology

What are acritarchs?

The short answer to this question is, "Nobody knows, by definition." Informed inference is possible on the basis of supporting evidence, if available.

According to the original and a slightly more comprehensive subsequent definition, various types of microfossils are explicitly excluded from the acritarchs: chitinozoans (problematic Paleozoic urn-shaped microfossils); "microforaminifera" (organic chamber linings of foraminifera, common in some acid-resistant concentrates); spores and pollen grains of higher plants; and intimately connected clusters of cells, although loose chains or aggregates are allowed. Acritarchs probably represent various stages in the life-cycle of unicellular algae, including cysts and vegetative cells; some might also be nondistinctive higher plant spores or eggs of various metazoans. By definition, therefore, acritarchs comprise a polyphyletic assemblage of microfossils. For convenience, and because most of them are probably algal in origin, acritarchs are named according to the rules of the *International Code of Botanical Nomenclature*.

Many experts argue that most acritarchs are resting stages, or cysts, of eukaryotic, planktic algae. Some workers favor the cyst affinity because most acritarchs have a capsule-like morphology, with an imperforate wall made of highly resistant material. Putative excystment structures, such as operculum-plugged pylomes and consistently oriented vesicle ruptures, support this view. However, the definition of acritarch does not rule out vegetative cells. In fact, the walls of some algal vegetative cells contain chemically resistant biopolymers, such as sporopollenin. For example, the outer wall layer of the common chlorococcalean green alga *Chlorella fusca* contains sporopollenin, and is indistinguishable from many Proterozoic and Paleozoic sphaeromorphs. Thus, discrimination between vegetative cells and cysts in fossil material very likely will be impossible. The eukaryotic affinities of acritarchs are indicated by their generally large diameter (typically >20 μm), robust walls, and relatively complex wall structure and morphology. Yet, none of these criteria is diagnostic.

Many acritarchs have been transferred to the dinoflagellates. Before a transfer can be effected legitimately, the acritarch should demonstrate features that clearly evidence affinity. Some acritarchs resemble dinoflagellate cysts in having processes, multiple wall layers, a degree of symmetry, and distinctive excystment structures (other than archeopyles); but most acritarchs do not have an ordered, easily interpreted arrangement of surface features that corresponds to paratabulation in dinoflagellates (see Chapter 7). Some modern unarmored dinoflagellates, and at least one armored type, produce cysts that would be considered acritarchs if found as fossils.

Evidence that selected acritarchs are the remains of prasinophyte phycomata is very compelling. For example, the morphologic and ultrastructural similarities between *Tasmanites* and *Pachysphaera*, and between *Leiosphaeridia* and *Halosphaera*, are striking. The latter genus of each pair is a modern prasinophyte that has a phycoma stage in its life-cycle. Other microfossil species, such as those assigned to *Pterospermopsis* and related genera (Figs 6.23D, 6.27F), and to *Fimbriaglomerella* (Fig. 6.27H), resemble closely the phycoma stage of the modern prasinophyte *Pterosperma* (Fig. 6.16). Despite the number of transfers to the Prasinophyceae, and although information about these algae is increasing rapidly, prasinophytes may not constitute a natural association of green algae and probably represent a variety of unicellular green phytoflagellates grouped together for convenience.

Some acritarchs may eventually be allocated to other extant algal divisions, but first **phycologists** need to investigate more thoroughly the cyst stages of a variety of modern forms, and establish criteria for differentiating one cyst stage from another. For example, the biologic affinities of some of the simple fossil sphaeromorphs (e.g., *Leiosphaeridia*; see Fig. 6.8A) include the following possibilities:

1 sheath of a cyanobacterial unicell;
2 outer common sheath of a colonial cyanobacterium;
3 outer wall of a cyanobacterial spore or sporangium;
4 vegetative eukaryotic alga, and
5 cyst of a eukaryotic alga!

Paleoecology

Almost all acritarchs have been recovered from marine strata and/or were associated with marine fossils. Many lines of evidence suggest that acritarchs were phytoplankton. First, they occur in a variety of sediment types: shale, siltstone, sandstone, carbonate rocks, and chert; mudrocks are probably the best source because most acritarchs behave hydrodynamically as mud-sized particles. Second, many forms exhibit a cosmopolitan distribution in coeval sediments. Third, the morphology of many of them suggests adaptations for a floating existence. Fourth, their (largely putative) sporopollenin composition argues for plant rather than animal affinities. Fifth, acritarchs are small and occur in great numbers; by analogy with today's ecosystems, they were probably primary producers rather than consumers. Finally, they are commonly distributed randomly in thin sections, which seems more consistent with a planktic rather than benthic lifestyle.

Some acritarchs do seem to have certain preferred habitats. For example, Middle and Upper Ordovician rocks from Kentucky, Ohio, and New York contain three acritarch associations that correspond to the following environments: nearshore shallow water; offshore shoal; and offshore open marine. Similar acritarch associations and diversity gradients occur in the subsurface Ordovician of northeastern Kansas.

In the Silurian of Wales and western England, trends in acritarch abundance and diversity suggest the following conclusions:

1 nearshore areas are characterized by low-diversity assemblages, primarily composed of thin-walled sphaeromorphs and the genera *Veryhachium*, *Micrhystridium*, and *Diexallophasis*,

2 offshore shelf environments are characterized by high-diversity assemblages, which consist of a great variety of acritarch types and which are not dominated by any particular taxon, and

3 deep-water environments are characterized by low-diversity assemblages, dominated by thick-walled sphaeromorphs.

In Late Devonian reefs in western Canada, sphaeromorphs predominated in near-reef (inshore) facies; about 1.6 km away from reefs, diversity rose as thin-spined acanthomorphs appeared; farther from the reefs, typically more than 6.5 km away, thick-spined acanthomorphs and polygonomorphs raised the diversity (Fig. 6.17). The abundance (number of specimens recovered) of all species increased away from the reef. The sedimentary rocks representing the quiet, deeper-water environment preserved the most acritarchs, and may reflect the abundance of acritarchs in the water column.

Acritarchs, together with dinoflagellate cysts, spores, and pollen grains, track Early Jurassic transgressive and regressive events in England and Wales. Many acanthomorphs preferred a relatively inshore, but basinal, environment, whereas certain polygonomorphs and netromorphs favored an open-sea environment. Assemblages closer to shore tended to be dominated by one species; the offshore assemblages were more heterogeneous and species-rich.

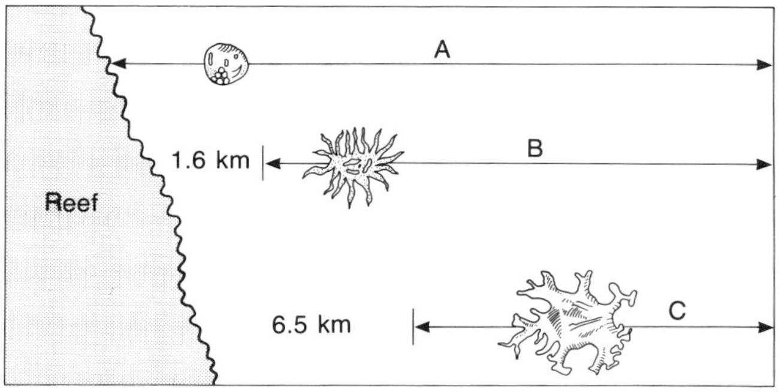

Figure 6.17 Diagrammatic representation of the distribution of acritarch associations near an Upper Devonian reef, central Alberta. Association A is characterized by simple, spherical forms; association B is dominated by thin-spined forms; association C includes polyhedral and thick-spined forms. (Redrawn from Staplin FL. *Palaeontology* 1961; 4:392–424, and from Williams GL. *Introduction to Marine Micropaleontology*. New York: Elsevier North Holland, 1978: 321.)

These few paleoecologic studies do not agree in detail, partly because they are not completely comparable. They are based on rocks that differ widely in age (i.e., some of the differences may be due to evolutionary change) and in degree of sedimentologic characterization. Still, they suggest that acritarch assemblages may be sensitive to distance from shore or perhaps to water depth, and that such assemblages may be used to define paleoenvironments. The generally low-diversity nearshore assemblages and higher diversity offshore assemblages parallel the situation in the ecology of modern phytoplankton. Sample extraction biases, however, may alter diversity values in fossil assemblages; such biases need to be addressed if environmental information is to be inferred confidently from diversity gradients.

Excystment structures, in addition to being good taxonomic characters, might hold paleoecologic information. If a species displays a high percentage of germinated cysts (i.e., microfossils having open excystment structures), this might suggest favorable conditions at the time of fossilization. Pylomate species probably should not be used for this analysis, because the operculum may fall off diagenetically rather than being lost as a natural consequence of excystment.

Acritarchs display a wide variety of shapes. Are such shapes functional? Do they have any adaptive value? Little work has been done specifically on acritarchs, but the literature on phytoplankton provides some general ideas.

Commonly, the acritarch vesicle is interpreted as an adaptation for floating. This interpretation may be too simplified. The diverse morphologic features in plankton may not be adaptations for floating; rather, they may permit more effective use of the euphotic zone by promoting twisting and tumbling of cells. In addition, even though members of the freshwater plankton live in a less viscous medium, they typically do not display more elaborate form-resistance features, such as branching spines, than do marine varieties. But the freshwater types are, in most cases, smaller, thus increasing their surface area to volume ratio and helping to retard sinking. Some also produce extensive mucilage, which may aid in suspension. Shape is less important in determining sinking rates than other factors; for instance, in diatoms cell density (which can be adjusted by metabolic products) might be the most important factor.

If spines and other form-resistance features are not characters that increase the surface area to volume ratio, how else might they function? They could be used to increase the effective diameter of the cell or cyst, thereby discouraging grazing by zooplankton. Their processes might cause them to get entangled in a layer of so-called marine snow, as many modern diatom resting spores do, and thus never settle to the sea bottom. Marine snow, amorphous organic particulates partly derived from mucus structures of planktic organisms, occurs at many depths in the ocean, including the deep sea. It is particularly well developed in near-surface waters, where layers of it typically occur to a depth of about 15 m. A common component of this material is a population of phytoplankton and zooplankton. In effect, marine snow acts as a firm substrate in an open-water environment. Diatom resting spores may reside in the marine snow until more favorable nutrient conditions trigger their germination. Perhaps some acritarchs were similarly adapted to this material: their processes, with all manner of branching and grapnel-like hooks and spines, might be adaptations for lodgment in marine snow (e.g., Figs 6.10D_1–D_2, 6.26C, G, 6.27B, D). Then, when their reproductive cycle resumed or perhaps when favorable environmental conditions returned, excystment took place and the vegetative cell was positioned in the water column where it could photosynthesize and assimilate nutrients immediately. Thus, some acritarchs may have been able to colonize surface waters of the ocean wherever marine snow was abundant; today, marine snow is most common in neritic environments.

Biostratigraphy

Acritarchs are widespread throughout the geologic column, both laterally and vertically. Their lateral variability allows inferences about acritarch paleoecology; vertical occurrences, which extend from the Lower Proterozoic (in strata up to 2 By in age) to the Holocene, provide a basis for biostratigraphy, and they also permit a general understanding of major evolutionary episodes (Fig. 6.18).

Acritarchs are currently being used as biostratigraphic markers in Upper Proterozoic rocks, where few other types of fossils even occur. They are used extensively in

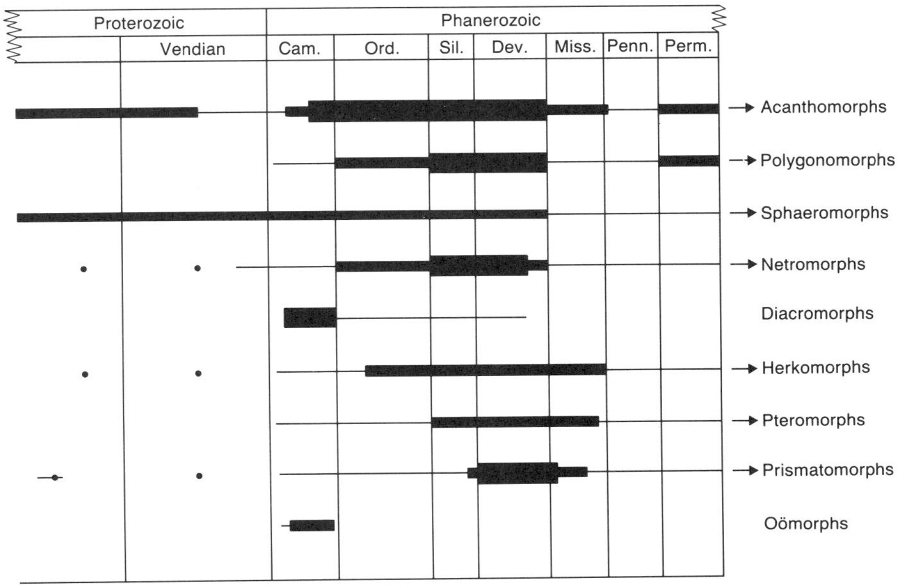

Figure 6.18 Generalized stratigraphic ranges of the major subgroups of acritarchs from about 700 Ma (within the Late Proterozoic) to the end of the Permian (about 245 Ma). Arrows indicate those subgroups that have a post-Permian record; this record is poorly known and thus is not depicted. Width of lines roughly corresponds to number of species recorded for the particular interval. Well-documented Proterozoic occurrences are indicated by dots. The bold line within the Proterozoic in the acanthomorph column represents the remarkably large acanthomorphs discussed in the text.

correlating lower Paleozoic rocks, where traditionally trilobites, brachiopods, graptolites, and conodonts have been used. Post-Devonian strata, especially Mesozoic and Cenozoic rocks, have not been characterized sufficiently in terms of acritarch composition (except locally); accordingly, the biostratigraphic application of acritarchs in these rocks is very limited.

The utility of acritarchs in zoning the upper part of the Proterozoic has been amply demonstrated. Acritarchs have been used to zone Upper Proterozoic strata of the Varanger Peninsula (East Finnmark), northern Norway, and to correlate them with strata in southern Sweden, Svalbard, East Greenland, and arctic regions of the USSR. The Vendian (uppermost Proterozoic) of the East European Platform has also been zoned biostratigraphically using acritarchs. Five acritarch complexes succeed each other in time; this relatively fine level of zonation in the Vendian, however, has not been extended confidently to other areas.

Acritarchs might be used to zone the important Proterozoic–Phanerozoic transition strata. The possible Late Proterozoic extinction of many acritarchs might mark an important datum. Biostratigraphic schemes based on acritarchs have been attempted for transition strata in many areas, such as the sequence on the Varanger Peninsula and the East European Platform; the compiled range charts suggest that stages of the Lower Cambrian, and perhaps the lower boundary itself, can be identified by acritarchs. Specifically, acritarch zonations have been proposed for otherwise sparsely fossiliferous Upper Proterozoic and Lower Cambrian strata in southern Scandinavia and east-central Poland (Fig. 6.19). Here a radiation of diagnostic Early Cambrian acritarchs replaced an assemblage of morphologically simple latest Proterozoic forms. The contrasting assemblages help to define the Proterozoic–Phanerozoic boundary in that part of the world.

Acritarchs have been used extensively in lower Paleozoic biostratigraphy, including work on defining boundary stratotypes. For example, five distinctive floras of acritarchs are found across the Cambrian–Ordovician boundary in Dayangcha, China; and eight acritarch zones are recognized for the Tremadocian (lowermost Ordovician) of Shropshire, England. In one interesting study in the Caradocian (Upper Ordovician) of Shropshire, the vertical sequence of acritarchs was inconsistent with those studied elsewhere. The strata appeared to be inverted because the acritarchs had been reworked (eroded and redeposited) from older deposits. The important Upper Ordovician section on Anticosti Island, Quebec, has been zoned in detail by acritarchs. Finally, acritarchs have been used successfully to zone and correlate many Devonian rock units in Europe, North America, and elsewhere.

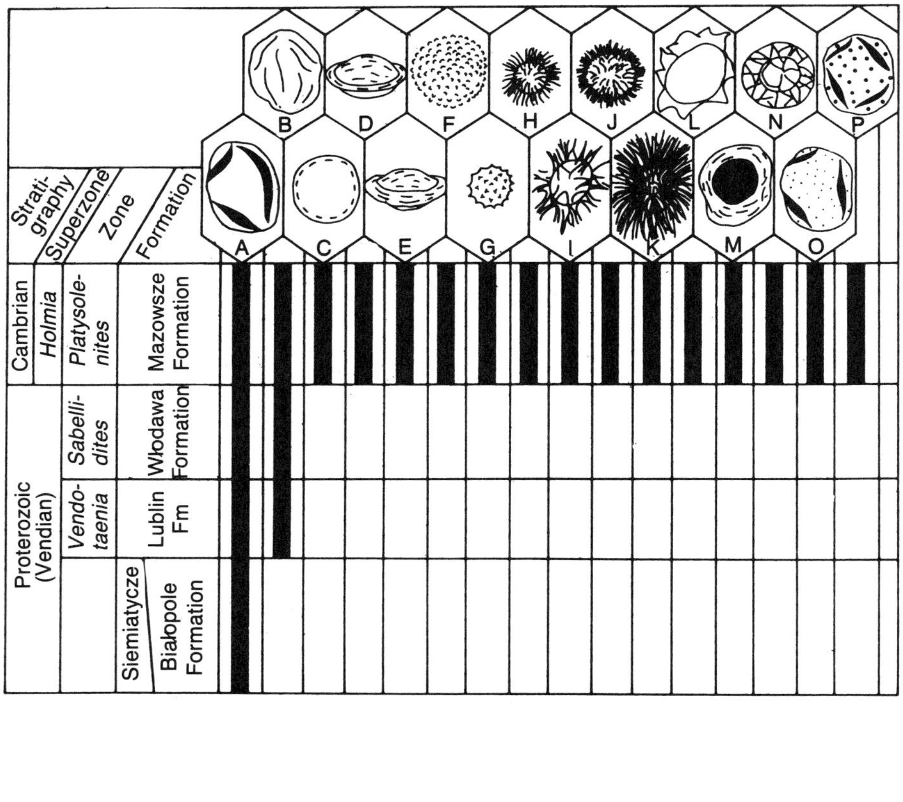

Figure 6.19 Distribution of acritarchs and prasinophytes in the Upper Proterozoic and Lower Cambrian of the Lublin Slope, East European Platform, in east-central Poland. Distribution bars correspond to the taxa sketched at top: A, *Leiosphaeridia* sp. 1. B, *Leiosphaeridia* sp. 2. C, *Leiomarginata simplex*. D, *Granomarginata prima*. E, *G. squamacea*. F, *Lophosphaeridium tentativum*. G, *Micrhystridium tornatum*. H, *M. lanatum*. I, *Comasphaeridium* sp. 1. J, *Comasphaeridium* sp. 2. K, *C. strigosum*. L, *Pterospermella* sp. 1. M, *Pterospermella* sp. 2. N, *Cymatiosphaera minuta*. O, *Tasmanites tenellus*. P, *T. bobrowskae*. (Modified from Moczydłowska M, Vidal G. *Geolog Förenin Stockh Förhandlin* 1986; 108:201–23.

Evolutionary history

Archean and Proterozoic

About 50 categories of microfossils and microfossil-like objects have been described from the Archean, but only two or three of these categories include unequivocal microfossils. A similar evaluation of the Early Proterozoic record (from 1.6 to 2.5 Ga) revealed 86 categories, 23 of which contain true microfossils. Are some of these Archean and Early Proterozoic microfossils acritarchs? Their simple morphology and relatively poor preservation permit many interpretations. Filamentous types have been assigned bacterial or cyanobacterial affinities; the objective approach would label them cryptarchs. The simple spheroids might also be bacteria or cyanobacteria; most are probably not eukaryotic algae, considering their small size. Again, in a purely descriptive mode they would be called sphaeromorphs, and hence either acritarchs or cryptarchs.

Most Precambrian acritarchs are sphaeromorphs having a surface sculpture varying from smooth to reticulate. Some of these surface sculptures are rather intricate (Fig. 6.20A) and therefore provide a means for recognizing distinct types of sphaeromorphs that are otherwise nondescript. A few forms from Lower Proterozoic rocks, including the Gunflint Iron Formation (about 2 By old), very questionably resemble acanthomorphs and polygonomorphs. Occurrences of nonsphaeromorphs (primarily acanthomorphs) increase in the Upper Proterozoic, and are confined primarily to strata younger than about 950 My. Although the classification in Table 6.1 is artificial, the first appearance and early evolution of each subgroup might have phylogenetic significance. The nonsphaeromorphs include the earliest prismatomorphs (e.g., Fig. 6.20C; see Fig. 6.18); pylome-bearing types with echinate opercula (Fig. 6.20B); and curious vase-shaped forms called **melanocyrillids**, which superficially resemble problematic lower Paleozoic urn-shaped microfossils called chitinozoans.

The melanocyrillids are perhaps the most complex Proterozoic microfossils known (Fig. 6.21). They have been recovered from at least 15 localities around the world, and seem to be limited to rocks about 950–700 My old. Most of them are between 40 and 100 μm in length, although much larger specimens (up to 1 mm) are known. At one end, melanocyrillids exhibit a distinctive triangular opening, which is fringed by a hexagonal or triangular lip. Many workers interpret this opening as an excystment structure and consider the melanocyrillids to be cysts or perhaps sporangia of eukaryotic algae. Other micropaleontologists emphasize the similarity of melanocyrillids to protozoans such as tintinnids (see Chapter 14), thus arguing that they represent the earliest remains of heterotrophic protists. Until this controversy is resolved, melanocyrillids will remain organic-walled microfossils of uncertain affinities and thus be placed among the acritarchs.

Among the other nonsphaeromorphs, early acanthomorphs are remarkable (see Fig. 6.18). They include relatively complex double-walled forms with short cylindrical processes (Fig. 6.20E_1–E_2) and forms bearing distinctive processes that are sturdily built and flare distally (Figs 6.20D, F). In a few cases, an acanthomorph has been discovered *inside* what appears to be a large, unornamented sphaeromorph (Fig. 6.20G). Such exceptionally preserved microfossils strengthen the argument that certain acritarchs were internally formed cysts, similar to the **chorate cysts** of many fossil and living dinoflagellates (see Chapter 7).

The significant disparity in size between Proterozoic and Early Cambrian acanthomorphs is equally remarkable (Fig. 6.22). The Proterozoic contains some relatively giant acanthomorphs, a few over 400 μm in diameter. Only 53 occurrences of unequivocal Proterozoic acanthomorphs have been reported, but of these 15 have a maximum vesicle diameter exceeding 200 μm. In contrast, *all* Early Cambrian acanthomorphs (321 reported occurrences) are less than or equal to 73 μm in diameter, and only 20 have a maximum diameter exceeding 50 μm (Figs 6.22, 6.23E, G–J).

As recent knowledge indicates, the fabric of acritarch evolution in the Proterozoic is complicated and not well understood. The relatively complex acritarchs of the Late Proterozoic may represent a radiation of cyst-forming eukaryotic plankton. By about 600–650 Ma (during the Vendian) many of these taxa disappeared, leaving only a few species, primarily sphaeromorphs; perhaps 70% of the species became extinct, which would be considered a major Late Proterozoic extinction event. Rediversification was rather slow: a second radiation that restored high planktic diversity did not begin until well into the Early Cambrian. This hypothesis needs better stratigraphic control and is supported by relatively limited data, but it is intriguing because the proposed extinction occurs near the time of the apparent extinction of the Ediacaran metazoan fauna of medusae and other cnidarian-like animals, and the appearance of invertebrates and eukaryotic algae with mineralized skeletons.

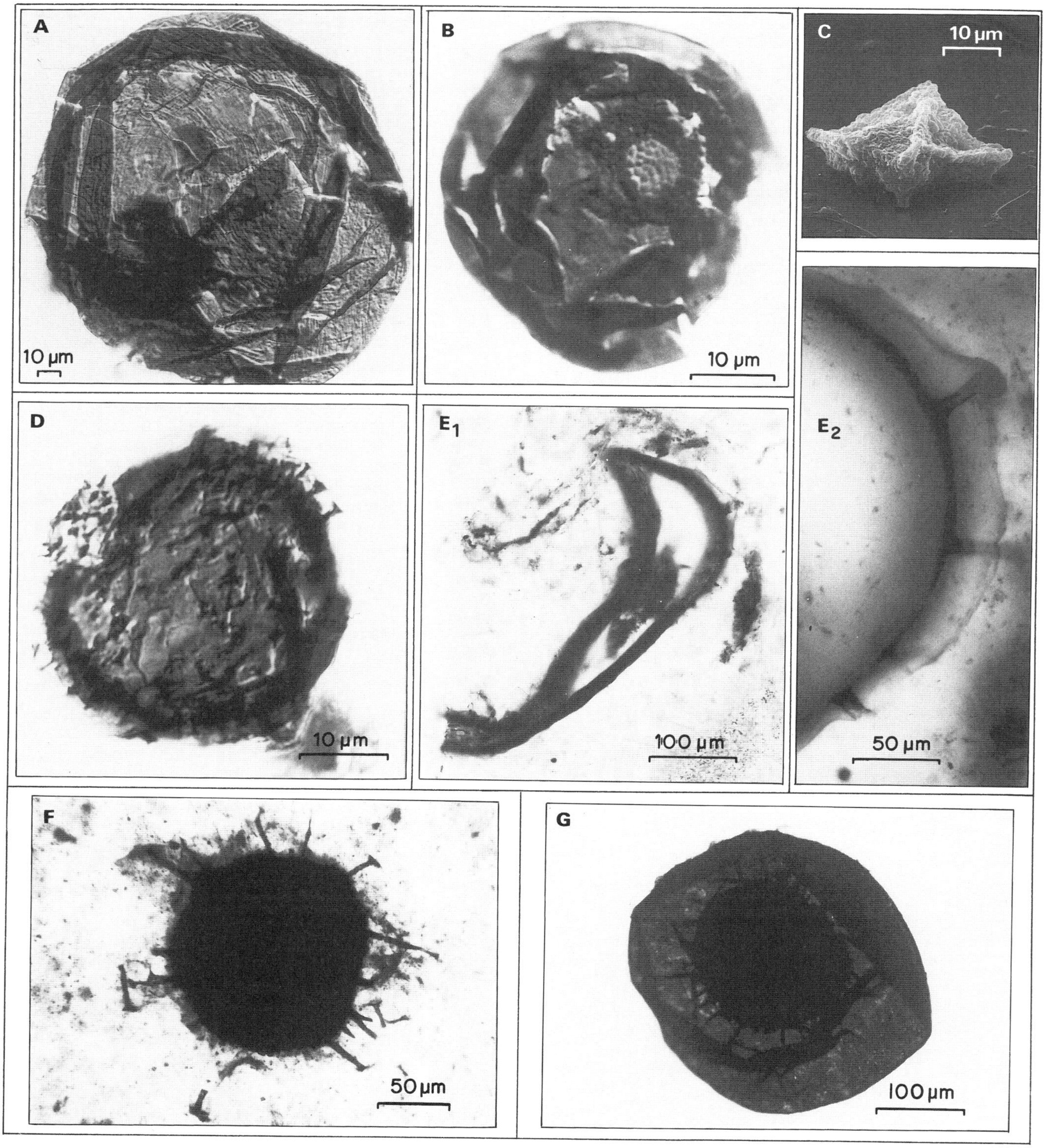

Figure 6.20 Late Proterozoic acritarchs from the Kwagunt Formation, Chuar Group of northern Arizona (A, B, D); Visingsö Group of southern Sweden (C); Hunnberg Formation, Roaldtoppen Group of Nordaustlandet, Svalbard (E_1-E_2); and Svanbergfjellet Formation, Akademikerbreen Group of northeastern Spitsbergen, Svalbard (F, G). (A, B, D, E_1-E_2, F, G light micrographs; C scanning electron micrograph.) A, *Kildinosphaera lophostriata* (Jankauskas, 1979) Vidal, 1983, note distinctive surface sculpture of concentric striae. B, *Leiosphaeridia* sp. A, small hemispherical granulate structure may be operculum covering excystment opening. C, *Octoedryxium truncatum* Rudavskaja, 1973, prismatomorph characterized by octahedral vesicle. D, *Vandalosphaeridium walcottii* Vidal & Ford, 1985, note distinctive processes. E, *Trachyhystrichosphaera vidalii* Knoll, 1984. E_1, Collapsed double-walled vesicle. E_2, Hollow cylindrical processes arising from inner wall and supporting outer wall or membrane. F, Large acanthomorph; note long (up to 60 μm) processes that flare at the tips. G, Unornamented vesicle containing acanthomorph; microfossil similar to that in F is preserved within a larger, unornamented vesicle. (A, B, D Reproduced from Vidal G, Ford TD. *Precambrian Res* 1985; 28:349–89. C Reproduced from Vidal G. *Fossils Strata* 1976; 9:1–57. E_1-E_2 Reproduced from Knoll AH. *J Paleontol* 1984; 58:131–62. F, G Reproduced from Butterfield NJ, Knoll AH, Swett K. *Nature* 1988; 334:424–27.

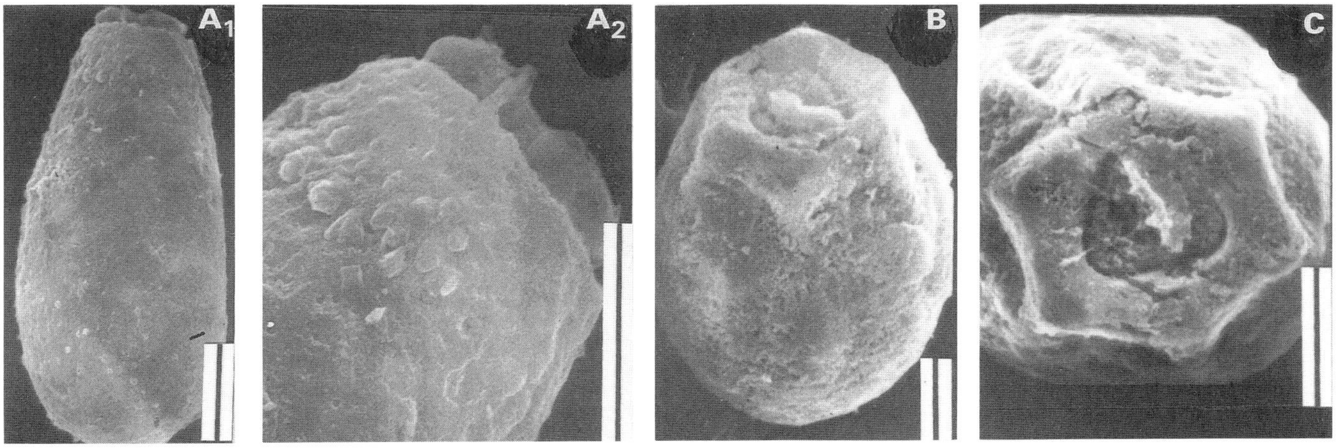

Figure 6.21 Scanning electron micrographs of melanocyrillids in longitudinal (A), oblique (B), and polar or oral view (C); specimens from the Kwagunt Formation (ca. 850 My old) of the Chuar Group, Grand Canyon, northern Arizona. A, *Melanocyrillium fimbriatum* Bloeser, 1985. A_1, Note tear-shaped vesicle. A_2, Enlargement of A_1, showing ornamentation along oral fringe. B, *M. hexodiadema* Bloeser, 1985, note hexagonal oral margin. C, *M. hexodiadema*, note subtriangular excystment(?) opening. All scale bars, 25 μm. (Reproduced from Bloeser B. *J Paleontol* 1985; 59:741–65.)

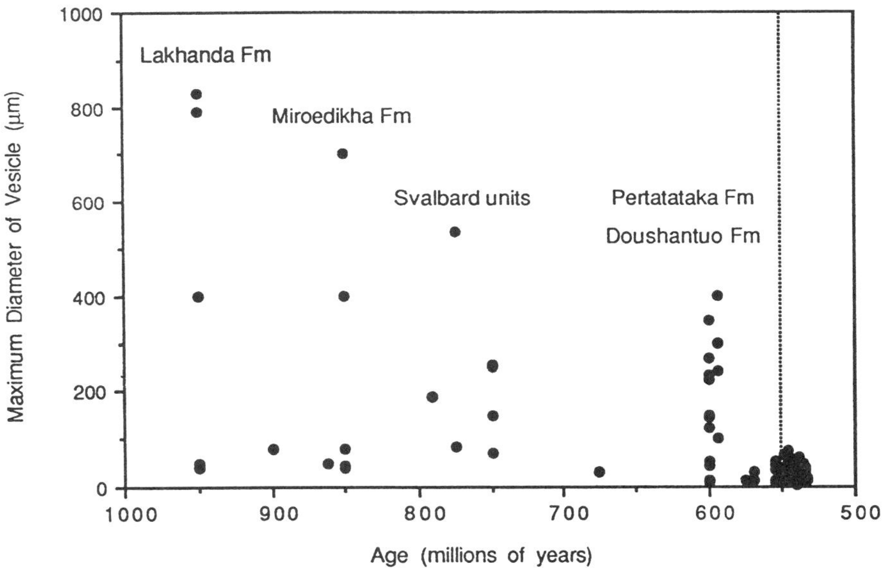

Figure 6.22 Temporal distribution of Proterozoic and Early Cambrian acanthomorph acritarch occurrences as a function of maximum vesicle diameter (n = 373). Note that data do not extend beyond the Early–Middle Cambrian boundary (about 530 Ma). Important formations yielding large acanthomorphs appear at their approximate ages. The Lakhanda and Miroedikha Formations are in eastern Siberia; the Doushantuo Formation is in east-central China; and the Pertatataka Formation is in north-central Australia. The vertical line at 550 Ma represents the Proterozoic–Cambrian boundary.

Phanerozoic

The Early Cambrian contains well-documented representatives of many of the informal subgroups of acritarchs (e.g., herkomorphs, Fig. 6.23B; polygonomorphs, Fig. 6.23C; and pteromorphs, Fig. 6.23D), as well as the best documented early forms exhibiting the characteristic tasmanitid wall structure of radial pores (Figs 6.23F_1–F_2). After the Early Cambrian rebound, acritarchs continued to increase in abundance and diversity. The diacromorphs and oömorphs appeared and became fairly common components of the Cambrian microflora, but by Ordovician time oömorphs and most diacromorphs disappeared, while other groups were involved in a great radiation (Fig. 6.18). This radiation is particularly characterized by the origination of numerous species of *Baltisphaeridium* and *Veryhachium*; these and other genera display a broad range of surface sculptures at this time. One of the few well-documented evolutionary lineages in acritarchs has been proposed for the genus *Orthosphaeridium* during the Middle and Late Ordovician. Apparent phylogenetic trends include decreased vesicle

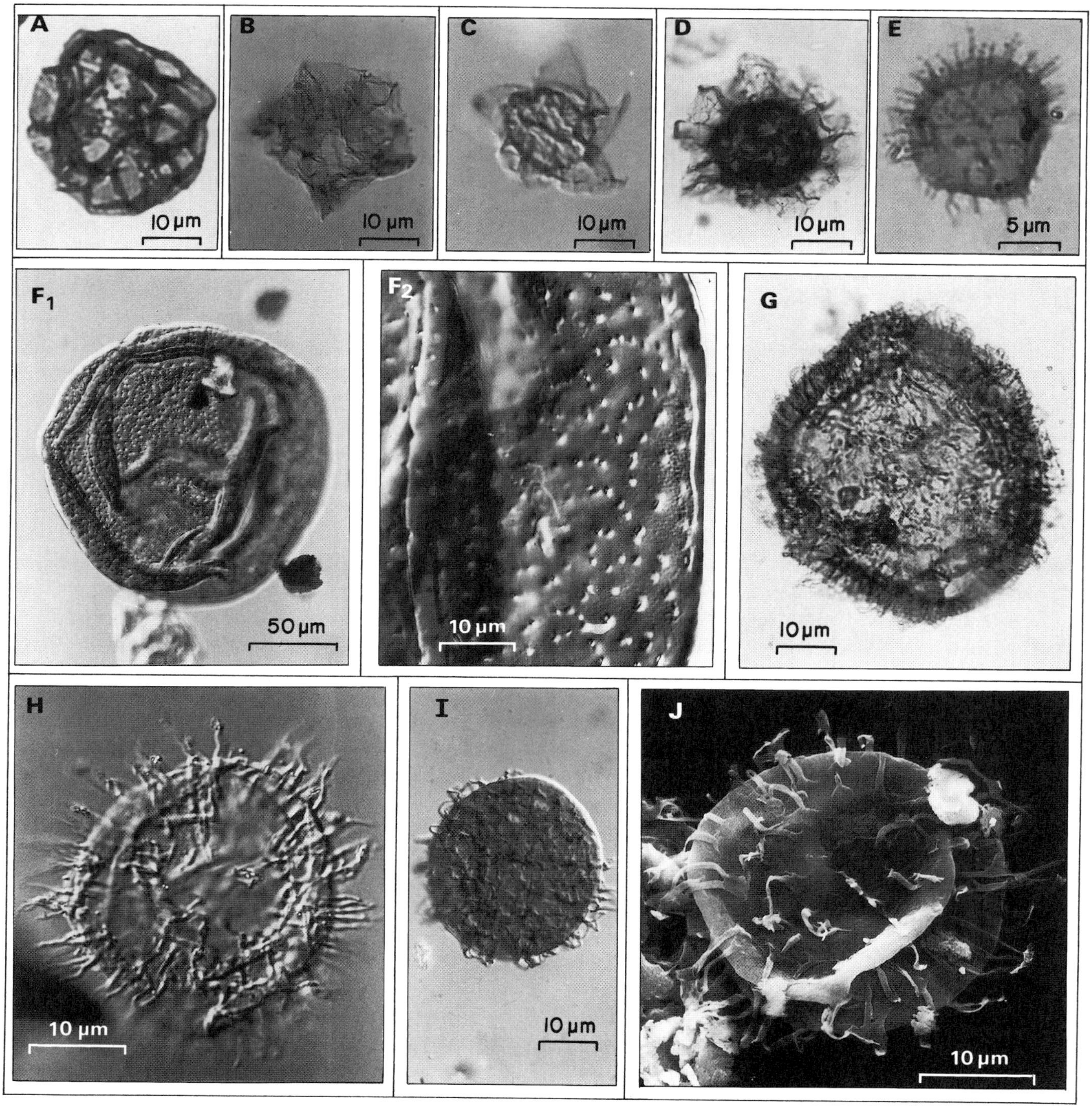

Figure 6.23 Early Cambrian acritarchs and prasinophytes from the Fucoid Beds of northwestern Scotland (A, E); Tokammane Formation of northeastern Spitsbergen, Svalbard (B, C, D, I, J); *Mickwitzia* Sandstone of Västergötland, Sweden (F_1-F_2, H); and the Hardeberga Sandstone of Skåne, southern Sweden (G). (A–I light micrographs; J scanning electron micrograph.) A, *Retisphaeridium dichamerum* Staplin, Jansonius & Pocock, 1965, perhaps a herkomorph, or a sphaeromorph: the reticulate pattern of folds may be due to compression. B, *Cymatiosphaera* sp. C, *Estiastra minima* Volkova, 1969, distinctive polygonomorph displaying five broad deltoid processes. D, *Pterospermella* cf. *P. solida* (Volkova, 1968) Volkova, 1979, prasinophyte bearing distinctive equatorial flange (i.e., pteromorph acritarch). E, *Micrhystridium ordense* Downie, 1982, note rather blunt terminations of processes. F, *Tasmanites volkovae* Kirjanov, 1974. F_1, Note thick wall of this prasinophyte. F_2, Enlargement of right part of F_1; note well-preserved pores. G, *Comasphaeridium brachyspinosum* (Kirjanov, 1974) Moczydłowska & Vidal, 1988, note dense covering of thin, hair-like processes. H, *Baltisphaeridium cerinum* Volkova, 1968. I, J, *Skiagia ciliosa* (Volkova, 1969) Downie, 1982, note flaring at tips of processes. (A, E Reproduced from Downie C. *Trans R Soc Edinb Earth Sci* 1982; 72:257–85. B, C, D, I, J Reproduced from Knoll AH, Swett K. *J Paleontol* 1987; 61:898–926. F_1-F_2, G, H Reproduced from Moczydłowska M, Vidal G. *Geolog Förenin Stockh Förhandlin* 1986; 108:201–23.)

ornament, an increase in number of major processes, and a decrease in vesicle inflation (Fig. 6.24). An abrupt turnover in the acritarch microflora occurs across the Ordovician–Silurian boundary in the southern Appalachians. Evidence from other parts of the world support the view that, within the limits of resolution, the terminal Ordovician extinction event in benthic invertebrates and an acritarch extinction event were synchronous. The great variety of Late Ordovician acritarchs may be appreciated in Figures 6.8–6.15.

The acritarchs then rebounded strongly and may have reached their highest diversity in the Silurian (Fig. 6.25). Silurian forms are larger than their Ordovician ancestors, and the acanthomorphs are characterized by complex processes (Fig. 6.26). Diversity remained high through the Devonian, with contributions from a great variety of acritarch subgroups and from prasinophytes (Fig. 6.27). Toward the end of the Devonian acritarch diversity decreased sharply. This fall reflects the great Late Devonian extinction at the end of the Frasnian Age, from which the acritarchs and many macroscopic invertebrates were never fully to recover (Fig. 6.18). Acritarchs remain relatively sparse during the rest of the Paleozoic, except for the appearance of a few specialized forms in the Permian. Post-Paleozoic types are overshadowed by other organic-walled microfossils, although a minor diversity peak in Late Triassic and Jurassic acanthomorphs, polygonomorphs, and herkomorphs has been documented. Mesozoic and Cenozoic palynologists concentrate on dinoflagellate cysts, spores, and pollen grains; they commonly place unfamiliar forms into waste-basket genera such as *Micrhystridium* and *Veryhachium*. More concentrated work on post-Paleozoic acritarchs will probably alter the diversity curve significantly.

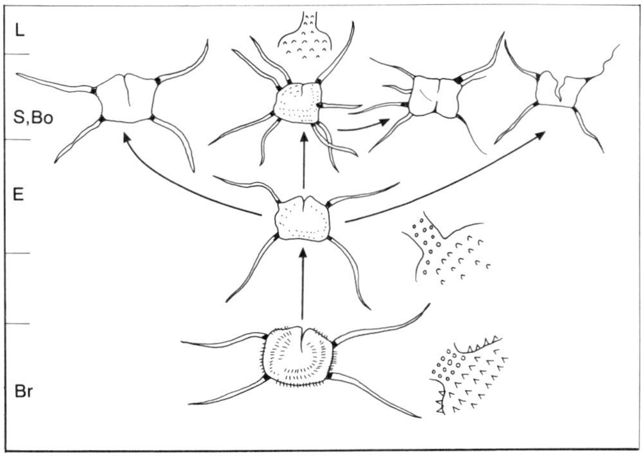

Figure 6.24 Possible phylogenetic trends in *Orthosphaeridium* (not to scale) in the Middle to Late Ordovician; see text for explanation. Vertical scale is a composite stratigraphic column indicating relative age of species depicted. (Middle Ordovician: Bromide Formation (Br); Upper Ordovician: Eden Formation (E), Sylvan Shale (S), Borkholmer Stufe (Bo), Lyckholmer Stufe (L).) (Redrawn from Loeblich AR Jr, Tappan H. *Amer Microscop Soc Trans* 1971; 90: 182–88.)

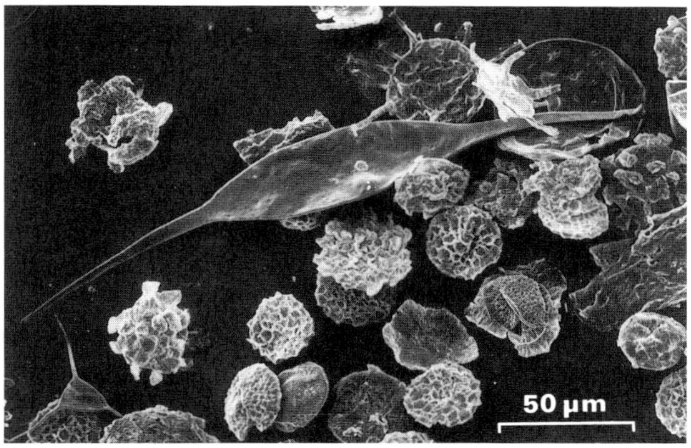

Figure 6.25 Surface of stub used in scanning electron microscopy showing a very rich preparation of acritarchs from the Upper Silurian Dadas Formation of Turkey. Note the netromorph, polygonomorph, and abundant acanthomorphs and herkomorphs. (Photomicrograph courtesy of Merrell A. Miller.)

Conclusion

The organic-walled microproblematica called acritarchs are beginning to attract the attention they deserve. If, indeed, they are cysts of eukaryotic phytoplankton, acritarchs were probably the most important primary producers for at least a billion years of Earth history, during much of the Proterozoic and the early Paleozoic.

Those paleontologists who study problematica may dwell too much on questions of affinity. Acritarchs may simply have been participants in the general evolutionary phenomenon of phytoplankton differentiation on this planet. That differentiation may have led to the evolution of many bizarre types of organic-walled microfossils, analogous to the situation in early radiations of invertebrate groups, such as echinoderms and arthropods. Thus, the acritarchs may evidence an evolutionary phase of early experimentation for eukaryotic phytoplankton during the Late Proterozoic and early Paleozoic. Late standardization occurred as many lines of acritarchs died out, leaving the more familiar organic-walled phytoplankton of the Mesozoic and Cenozoic.

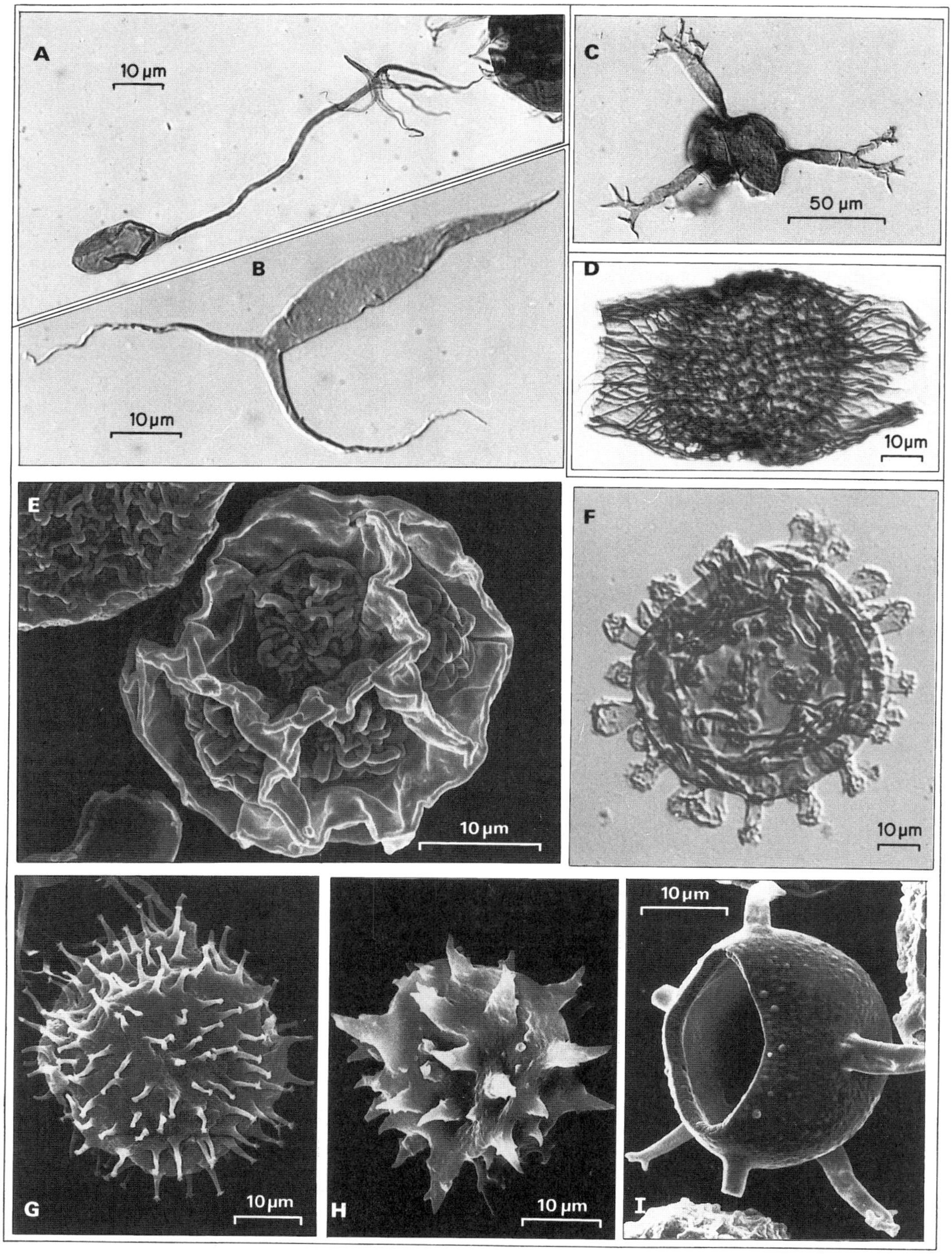

Figure 6.26 Acritarchs and prasinophytes from the Lower Silurian Clinton Group of New York (A–D, G, H), Upper Silurian Dadas Formation of Turkey (E, I), and the Upper Silurian Wenlock Shales of England (F). (A–D, F light micrographs; E, G–I scanning electron micrographs.) A, *Deunffia* cf. *D. ramusculosa* Downie, 1960, a very distinctive netromorph. B, *Domasia* sp., another very distinctive netromorph. C, *Oppilatala cara* (Cramer & Díez) Loeblich & Wicander, 1976, note prominent processes and the nature of their branching. D, *Carminella maplewoodensis* Cramer, 1968, note tubular, skirt-like extension of outer wall (vellum), a feature that characterized some Silurian and Devonian acritarchs. E, *Cymatiosphaera* cf. *C. mariae* Cramer *et al.*, 1976, note intricate sculpture within polygonal fields. F, *Visbysphaera dilatispinosa* (Downie, 1963) Lister, 1970, this form is a good biostratigraphic marker for the Wenlockian (lower Upper Silurian) in the British Isles. G, *Ammonidium waldronense* (Tappan & Loeblich, 1971) Wicander, 1974, abundant processes terminate in a rosette of branches. H, *Tylotopalla digitifera* Loeblich, 1970, note broad-based processes that terminate in a point or in short bifurcations. I, *Cymbosphaeridium pilar* (Cramer) Lister, 1970, characterized by a distinctive subpolygonal excystment opening, which strongly suggests the dinoflagellate affinities of this form. (A–E, G–I, Courtesy of Merrell A. Miller. F Reproduced from Downie C. *Palaeontology* 1963; 6:625–52.)

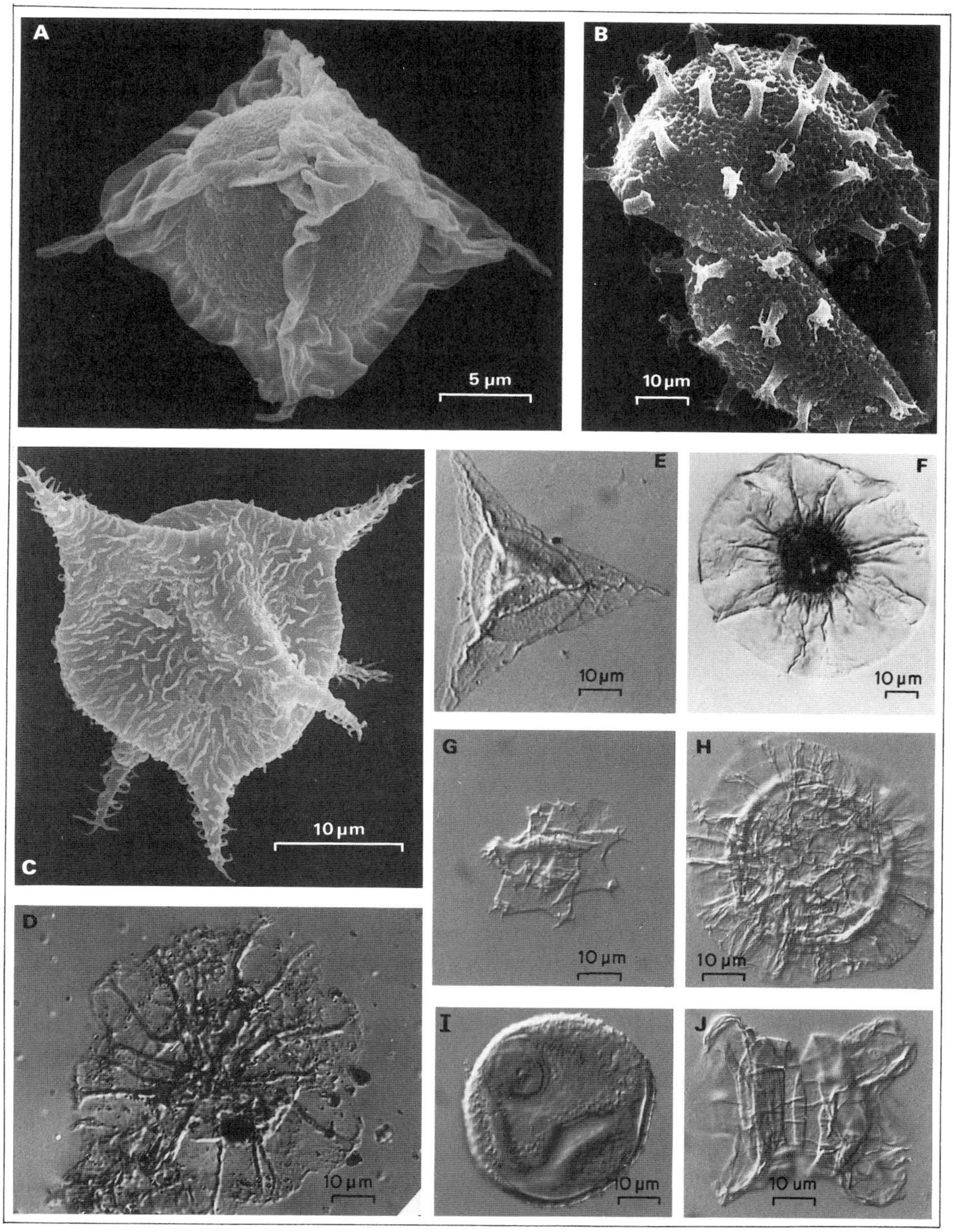

Figure 6.27 Acritarchs and prasinophytes from the Upper Devonian Lime Creek Formation of Iowa (A, C) and Chagrin Shale of Ohio (B); Middle Devonian Silica Formation of Ohio (D–F); and the Lower Devonian Haragan Formation of Oklahoma (G–J). (D–J light micrographs; A–C scanning electron micrographs.) A, *Daillydium pentaster* (Staplin, 1961) Playford, 1981, a characteristic Upper Devonian form. B, *Acriora petala* Wicander, 1974, note distinctive surface sculpture and processes that divide into many small branches at the tips. C, *Villosacapsula ceratioides* (Stockmans & Willière) Loeblich & Tappan, 1976, compare this Upper Devonian form with the Upper Ordovician species illustrated in Figure 6.14 A_1–A_2. D, *Tunisphaeridium tentaculaferum* (Martin) Cramer, 1971, this acanthomorph has distinctive hair-like processes that branch distally to form a dense, membrane-like mat that surrounds the vesicle. E, *Polyedryxium fragosulum* Playford, 1977, a probable prasinophyte. F, *Pterospermella reticulata* Loeblich & Wicander, 1976, a characteristic Middle Devonian prasinophyte. G, *Thysanoprobolus polykion* Loeblich & Tappan, 1970, characteristic Lower Devonian form. H, *Fimbriaglomerella divisa* Loeblich & Drugg, 1968, distinctive Lower Devonian prasinophyte. I, *Nanocyclopia aspratilis* Loeblich & Wicander, 1976, note well-developed laevigate operculum. J, *Riculasphaera fissa* Loeblich & Drugg, 1968, this form exhibits the median split mode of excystment. (A, C Reproduced from Wicander R, Playford G. *Micropaleontology* 1985; 31:97–138. B Reproduced from Wicander ER. *Palaeontographica* 1974; B148:1–137. D–F Reproduced from Wicander R, Wood GD. *Amer Assoc Stratigraphic Palynologists Contrib Ser* 1981, 8:9–43. G–J Reproduced from Wicander R. *Rev Palaeobot Palynol* 1986; 47:327–65.

Supplementary reading

Downie C. Observations on the nature of the acritarchs. *Palaeontology* 1973; 16:239–59. Review of acritarch morphology and evolution.

Downie C. Acritarchs in British stratigraphy. *Geol Soc Lond Spec Rep* 1984; 17: 26 p. Compilation of detailed range charts, and summary statements on the biostratigraphic utility of acritarchs and prasinophytes (primarily Paleozoic).

Downie C, Evitt WR, Sarjeant WAS. Dinoflagellates, hystrichospheres, and the classification of the acritarchs. *Stanford Univ Public Geol Sci* 1963; 7:1–16. Original subgroups of acritarchs and their definitions and included genera; see Tappan (1980) for update, including post-1963 subgroups.

Evitt WR. A discussion and proposals concerning fossil dinoflagellates, hystrichospheres, and acritarchs, I, II. *Natl Acad Sci (USA) Proc* 1963; 49:158–64, 298–302. Original paper in which acritarch was defined; also includes good discussion of the hystrichosphere controversy.

Loeblich AR Jr, Tappan H. Acritarch excystment and surface ultrastructure with descriptions of some Ordovician taxa. *Rev Española Micropaleontol* 1969; 1:45–57. Clear definitions of terms useful in the systematic description of acritarchs.

Muir MD, Sarjeant WAS. An annotated bibliography of the Tasmanaceae and of related living forms (Algae: Prasinophyceae). In: Jardiné S, ed. Microfossiles organiques du Paléozoïque 3. *Les Acritarches*. Paris: CIMP and CNRS, 1971: 51–117. Valuable resource for studying fossil and living prasinophytes.

Playford G, Dring RS. Late Devonian acritarchs from the Carnarvon Basin, Western Australia. *Palaeontological Assoc Spec Papers Palaeontol* 1981; 27:1–78. A model for illustration and description of acritarchs and prasinophytes.

Tappan H. *The Paleobiology of Plant Protists*. San Francisco: WH Freeman 1980: 1028 p. Start here because this book has the most comprehensive and best-illustrated treatment of acritarchs and fossil prasinophytes; see especially Chapters 3 (acritarchs) and 10 (prasinophytes).

Tappan H, Loeblich AR Jr. Surface sculpture of the wall in Lower Paleozoic acritarchs. *Micropaleontology* 1971; 17:385–410. Definitions of various types of surface sculpture, documented by scanning electron micrographs.

Traverse A. *Paleopalynology*. Boston: Unwin Hyman, 1988: 600 p. A comprehensive treatment of all types of palynomorphs, arranged stratigraphically rather than systematically.

Vidal G. The oldest eukaryotic cells. *Sci Am* 1984; 250:48–57. Account of Late Proterozoic evolution and extinction in acritarchs.

Williams GL, Sarjeant WAS, Kidson EJ. A glossary of the terminology applied to dinoflagellate amphiesmae and cysts and acritarchs: 1978 edition. *Amer Assoc Stratigraphic Palynologists Contrib Ser* 1978; 2A:121 p. Handy guide to the many terms used in descriptions.

Dinoflagellates

7

Lucy E. Edwards

Introduction

Dinoflagellates are protists that during some part of their life-cycle have a free-swimming stage with two dissimilar **flagella**. Some are plant-like, containing chloroplasts and producing organic compounds by photosynthesis, and some are animal-like, ingesting other organisms for food. Here the dinoflagellates are considered to be algae in the division Pyrrhophyta (flame algae).

Many people, although not familiar with dinoflagellates, are familiar with their effects. **Bioluminescent** dinoflagellates cause a sparkling of the sea at night as the waves break, and some dinoflagellates may produce blooms called red tides, which poison marine life or cause toxins to accumulate in shellfish, poisoning those who eat them. The Old Testament (Exodus 7: 20–21) and the writings of the ancient Greeks refer to red water and the killing of fish. The Red Sea may have gotten its name from dinoflagellate blooms.

Dinoflagellates take their name from the Greek word *dinos* meaning "whirling" (not to be confused with *deinos* meaning "terrible") and from their characteristic flagella, which propel them in a spiral motion. They live in a variety of habitats from the open ocean to freshwater lakes and ponds. Their means of nutrition is also varied. Many are photosynthetic; others may be grazers, predators, parasites, or symbionts. Dinoflagellates are second only to diatoms as **primary producers** in the oceans of the world.

The dinoflagellates are among the most primitive of the eukaryotes, and have been regarded as intermediate between prokaryotes and eukaryotes. The cells have the organelles typical of the eukaryotes such as **endoplasmic reticulum**, Golgi apparatus, and mitochondria, but the dinoflagellate nucleus is distinctive. The chromosomes remain condensed throughout life, not just during cell division, and the spindles which are present during cell division are outside of the nuclear membrane.

Some dinoflagellates include an encysted stage in their life-cycles. Although originally thought to be a means of surviving harsh conditions, the cyst stage seems to be an integral part of the sexual reproductive cycle. Cysts may be composed of cellulose, which probably does not fossilize, or of an extremely resistant, readily fossilized, complex organic wall. A few cysts contain calcite or silica. Cysts are the primary interest of paleontologists. Dinoflagellate cysts, or **dinocysts**, are abundant fossils in Mesozoic and Cenozoic strata. A single Silurian species is known, and many of the fossils now called acritarchs may actually be dinoflagellate cysts.

History of study

In 1753 Henry Baker first described living dinoflagellates as "Animalcules which cause the Sparkling Light in Sea Water." The first species was formally named by Müller in 1773.

In the 1830s, the German microscopist C.G. Ehrenberg examined many water and plankton samples and proposed several dinoflagellate genera that are well known today, including *Peridinium* (Fig. 7.7), *Prorocentrum* (Fig. 7.12D), and *Dinophysis* (Fig. 7.1B). More important to paleontologists, he first described fossil dinoflagellates from Cretaceous flint chips. He recognized as dinoflagellates forms that closely resemble living *Peridinium*. He also described spherical fossils with numerous spines that he believed to represent spores of the freshwater desmid alga *Xanthidium*. These spiny forms were named *Spiniferites* (Fig. 7.9B) by Mantell in 1850, but this work was largely ignored for over a century.

Although most early studies of modern dinoflagellates emphasized the motile stage, the cysts of *Ceratium* were recognized by Stein in 1883. The formation of a cyst inside the cellulosic outer covering, the **theca**, of living dinoflagellates was noted by Delage and Hérouard in 1896.

Classification of living thecate dinoflagellates depends heavily on the number and arrangement of the thecal plates. From the 1880s onward, workers recognized that the thecal plates could be grouped in various ways into plate series for descriptive purposes. The plate-series terminology presented by Kofoid in the early 1900s did much to standardize descriptions of dinoflagellate tabulation and remains in wide use today.

In 1933, Wetzel established the genus *Hystrichosphaera* for the spiny spherical forms such as Ehrenberg had seen, giving rise to the general term "hystrichosphere." Deflandre and Cookson in 1955 and Cookson and Eisenack in 1958 studied the morphology and biostratigraphy of marine plankton, including both the forms recognized as dinoflagellates and those termed hystrichospheres.

The 1960s witnessed both an increased understanding of the nature of dinoflagellate fossils and an increased application of fossil dinoflagellates to biostratigraphic problems. In 1961, Evitt reported that fossils recognized to be dinoflagellates have openings that follow definite, preformed lines to allow the emergence (excystment) of the organism. Evitt coined the term **archeopyle** for such openings and noted that dinoflagellate fossils with archeopyles must represent fossil cysts, not motile, thecate organisms. He also observed that many hystrichospheres have polygonal openings that must also be archeopyles and that the arrangement of the spines (processes) on many hystrichospheres reflects the morphology of the theca of the dinoflagellate that surrounds the cyst as it forms. In 1968, Wall and Dale documented the excystment of the living dinoflagellate *Gonyaulax* from a *Spiniferites*-type cyst.

Although Ehrenberg studied both living and fossil dinoflagellates, the biology and the paleontology of the group remained quite separate fields of study until the 1970s. Sarjeant's book in 1974 summarized what was then known about the morphology, reproduction, and encystment of living dinoflagellates in the context of fossil dinoflagellates. The Penrose Conference on living and fossil dinoflagellates, held in Colorado Springs in 1978, brought about integrated studies and increased understanding between the biologists working with dinoflagellates and the paleontologists studying their fossils. Subsequent joint meetings have been held every 3 or 4 years since the 1978 Penrose Conference.

Morphology

Living dinoflagellates show great morphologic variety (Fig. 7.1). The characteristic spiral movement of dinoflagellates results from the motion of two flagella. In a typical thecate dinoflagellate (Fig. 7.2), the **longitudinal flagellum** extends in the direction opposite to movement; that is, posteriorly. The **transverse flagellum**

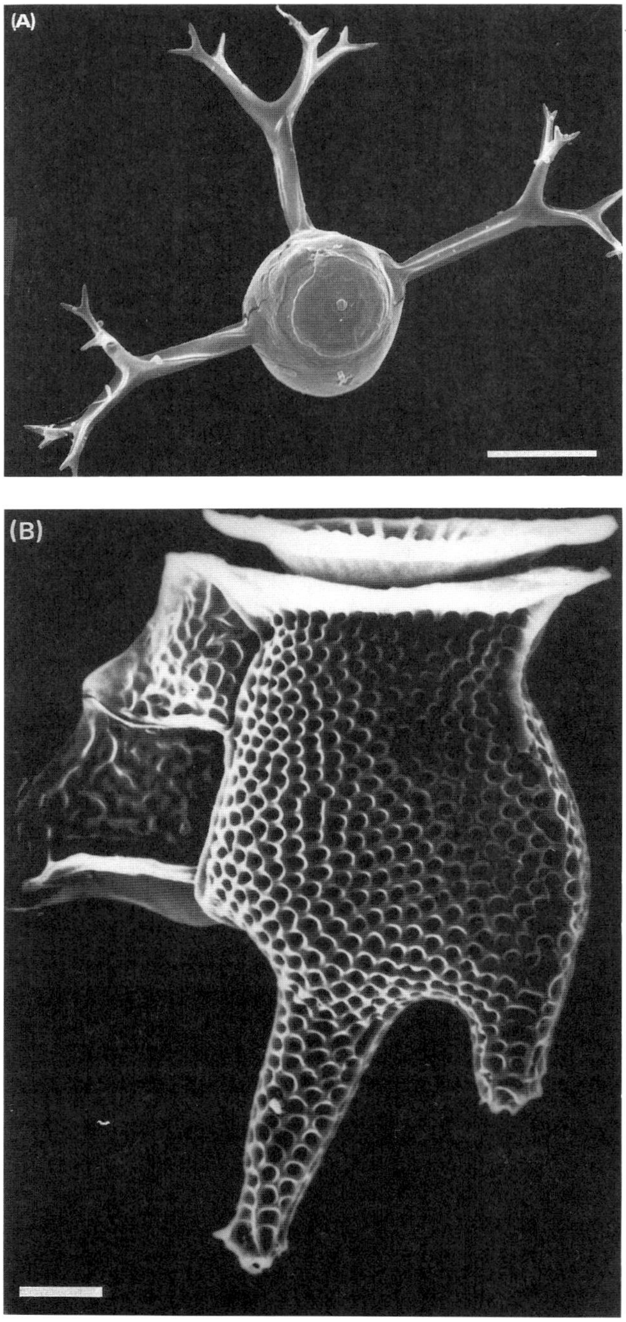

Figure 7.1 Scanning electron micrographs of dinoflagellates. A, *Cladopyxis brachiolata* Stein, apical view, an oceanic species with long bifid projections. Scale bar, 10 µm. B, *Dinophysis tripos*, left-lateral view. Scale bar, 10 µm. C, *Ceratium compressum* Gran, dorsal view. Scale bar, 20 µm. D, *Protoperidinium pentagonum* (Gran) Balech, dorsal view with thecal plates beginning to break up to expose cyst inside. Scale bar, 10 µm. E, *Gonyaulax grindleyi* Reinecke, ventral view, theca. Scale bar, 10 µm. F, *G. grindleyi* Reinecke, dorsal view, cyst showing archeopyle; this form also goes by the cyst name *Operculodinium centrocarpum* (Deflandre & Cookson) Wall. Scale bar, 10 µm. (Scanning electron micrographs courtesy of John Dodge (A–E) and Jane Lewis (F). A, B, C, E Reproduced from Dodge JD. *Atlas of Dinoflagellates*. London: Farrard Press, 1985.)

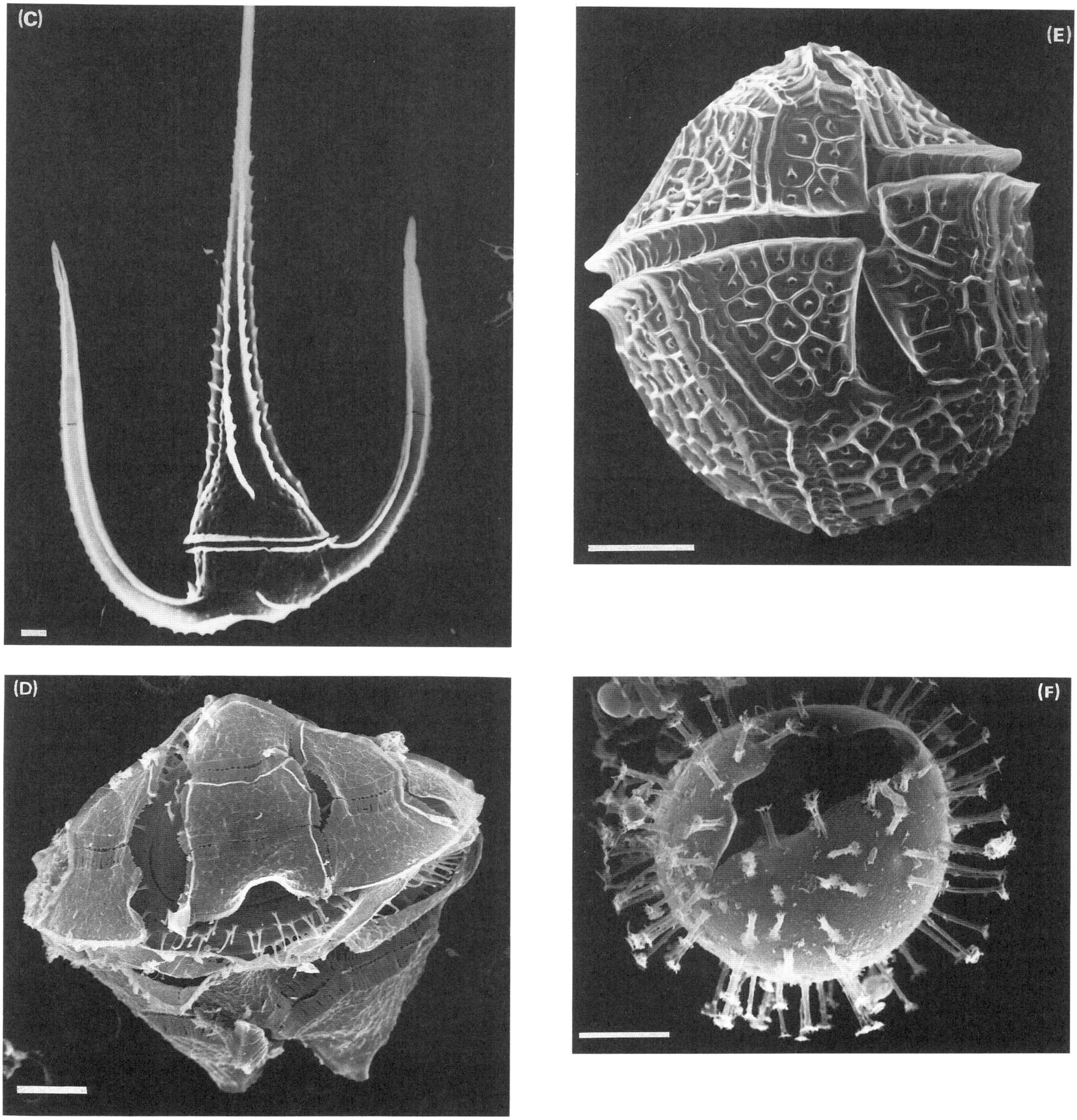

Figure 7.1 *continued.*

encircles the cell occupying a depression called the **cingulum**. The cingulum divides the cell into the anterior **epitheca** and the posterior **hypotheca**. The flagella emerge from flagellar pores, which define the **ventral surface**. A second depression, roughly perpendicular to the cingulum and containing the flagellar pores, is the **sulcus**. It divides the cell into left and right halves. The anterior end of the cell is also called the **apex**; the posterior end, the **antapex**. Many dinoflagellates have a theca that is constructed of thin cellulosic pieces called **plates** that fit together along linear **sutures**. The arrangement of the plates is called their **tabulation**.

Dinoflagellate fossils represent the more durable cyst, not the cellulosic theca. Most of the dinoflagellate cysts

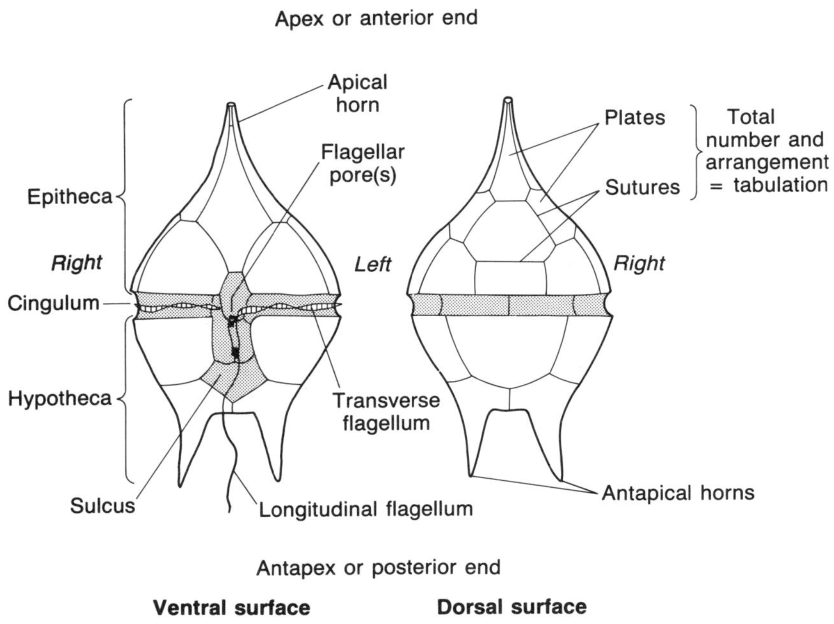

Figure 7.2 Features of the theca of a peridinialean dinoflagellate. (Adapted from Evitt WR. *Sporopollenin Dinoflagellate Cysts. Their Morphology and Interpretation*. Austin: American Association of Stratigraphic Palynologists Foundation, 1985:8.)

studied by paleontologists are composed of **sporopollenin**, the inclusive name for a high molecular weight polymer of carbon, hydrogen, oxygen, and nitrogen such as that which forms the outer wall of spores and pollen.

Terminology for dinoflagellate cysts parallels that for the theca (Fig. 7.3). Many paleontologists use the prefix "para-" to distinguish features of cysts from the presumed thecal counterparts of these features.

If the dinocyst closely resembles the theca in which it formed in both size and shape, it is termed **proximate** (Fig. 7.4). The cyst presumably formed in close contact with, or only slightly separated from, the parent theca. Proximate cysts closely resemble living dinoflagellates, and until the early 1960s these fossils were widely thought to represent thecae, not cysts.

Other dinocysts consist of a main body with spine-like projections (processes) or other prominent ornamentation. The term **chorate** is used for dinocysts in which the diameter of the inner body is considerably less than the overall diameter (Figs 7.3C, 7.5). The tips of the processes were presumably in contact with the theca as the cyst formed. Until recently, chorate cysts were not recognized as fossil dinoflagellates. The term "hystrichosphere" is sometimes found in older literature. Some hystrichospheres are now known to be chorate dinocysts. Spiny spheres that do not have obvious dinoflagellate-like characteristics are termed **acritarchs**, meaning, literally, of uncertain origin; they are dealt with in Chapter 6.

Dinoflagellate cysts may consist of one or several wall layers (Fig. 7.6). A cyst with a single layer is termed an **autocyst**; its wall is an **autophragm** (Fig. 7.6A). The terminology of multiple layers becomes more complex. If the cyst consists of two wall layers separated by numerous processes or other extensively distributed structures, the inner layer is called the **autophragm** and the outer layer is the **ectophragm** (Fig. 7.6B). If the two wall layers are not connected by such supporting structures, whether these layers are widely separated (**cavate**) or locally touching, the inner layer is called the **endophragm** and the outer layer is the **periphragm** (Fig. 7.6C). In the cases of three wall layers, these are **endophragm**,

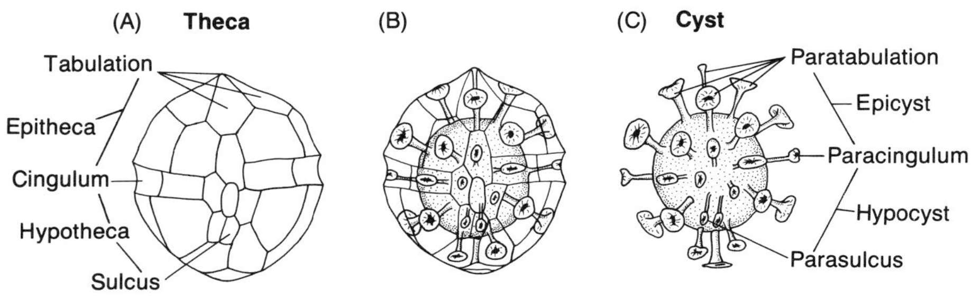

Figure 7.3 Theca and cyst terminology. A, Ventral view of theca showing tabulation. B, Chorate cyst forming inside theca. C, Cyst paratabulation reflects thecal tabulation. (Adapted from Evitt WR. *Sporopollenin Dinoflagellate Cysts. Their Morphology and Interpretation*. Austin: American Association of Stratigraphic Palynologists Foundation, 1985:56.)

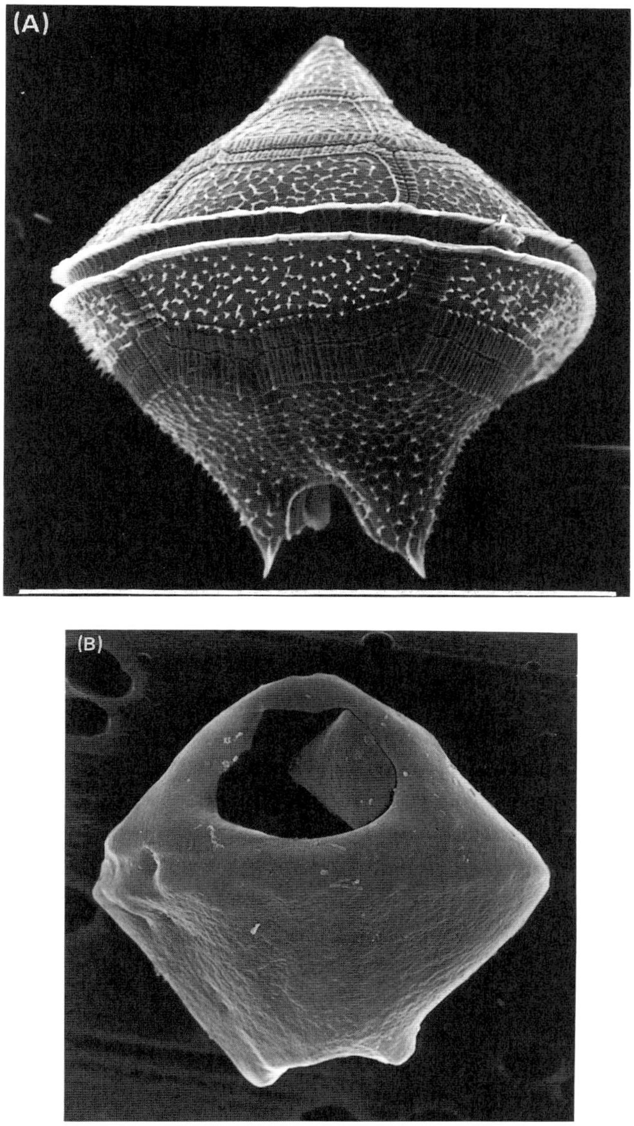

Figure 7.4 *Protoperidinium leonis* (Pavillard) Balech. A, Theca. B, Cyst. The term proximate is applied to cysts such as this that resemble the theca in size and shape. Scale bar, 10 μm. (Scanning electron micrographs courtesy of John Dodge and Jane Lewis.)

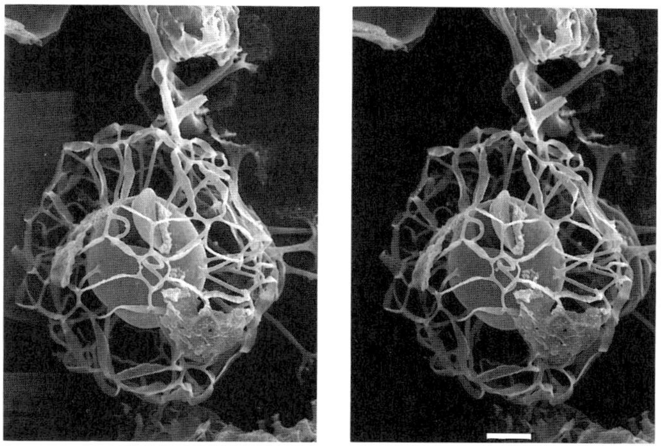

Figure 7.5 *Nematosphaeropsis pusulosa* (Morgenroth) Stover & Evitt, a chorate cyst. Scale bar, 10 μm. Scanning electron micrographs, stereo-pair.

mesophragm, and **periphragm** if the wall layers are not connected by supporting structures (Fig. 7.6D), and they are **endophragm**, **periphragm**, and **ectophragm** if the outer two layers are connected by supporting structures (Fig. 7.6E). Cavities beneath wall layers are named using the appropriate prefixes (auto-, endo-, meso-, peri-, and ecto-) for the wall layer which surrounds them, coupled with the suffix -cele.

Tabulation (theca) and paratabulation (cyst)

Individual plates on dinoflagellate thecae are grouped into series and numbered systematically. In a typical theca, plates are considered to be arranged in latitudinal series from apex to antapex: **apical** (touching the apex), **precingular** (touching the cingulum anteriorly), **cingular** (forming the cingulum), **postcingular** (touching the cingulum posteriorly), and **antapical** (touching the antapex). Plates lying between the apical and precingular series are termed **anterior intercalary**, and those lying between the postcingular and antapical series are **posterior intercalary**. Plates in the area of the sulcus are, logically, **sulcal**. Very small plates within the apical series have had a variety of names applied to them, such as preapicals, apical closing plates, and canal plates. Plates are numbered from the organism's left to right; that is, counterclockwise when viewed from the apex. A convenient shorthand, Kofoidian notation (Fig. 7.7), is used both to indicate the number of plates in each latitudinal series and to give a unique label to each specific plate.

Paratabulation is used to divide fossil dinoflagellate cysts into two major groups: the *peridiniaceans* and the *gonyaulacaceans*. Peridiniacean dinocysts have paratabulation similar to that of the modern genus *Peridinium*, typically seven precingular and five postcingular paraplates. Gonyaulacacean dinocysts have paratabulation similar to that of the modern genus *Gonyaulax*, typically six precingular and six postcingular paraplates. Figure 7.7 shows the complete formulas and plate designations for these two important genera.

The Kofoidian system is a relatively straightforward means of describing patterns of plates or paraplates. When applied to the paratabulation of fossil cysts, the Kofoidian system should be objective and descriptive. Paleontologists and neontologists may also want to make compari-

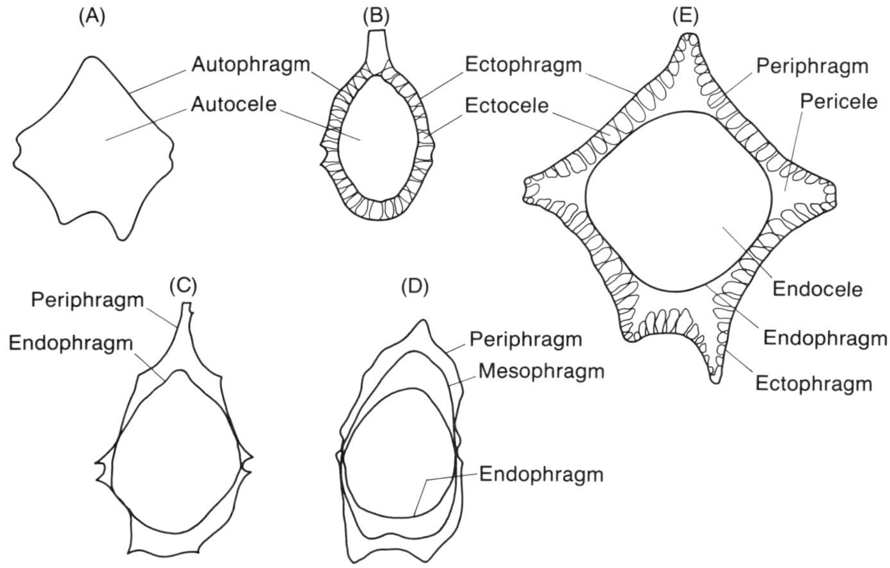

Figure 7.6 Wall-layer terminology. (Reproduced from Evitt WR. *Sporopollenin Dinoflagellate Cysts. Their Morphology and Interpretation*. Austin: American Association of Stratigraphic Palynologists Foundation, 1985:63.)

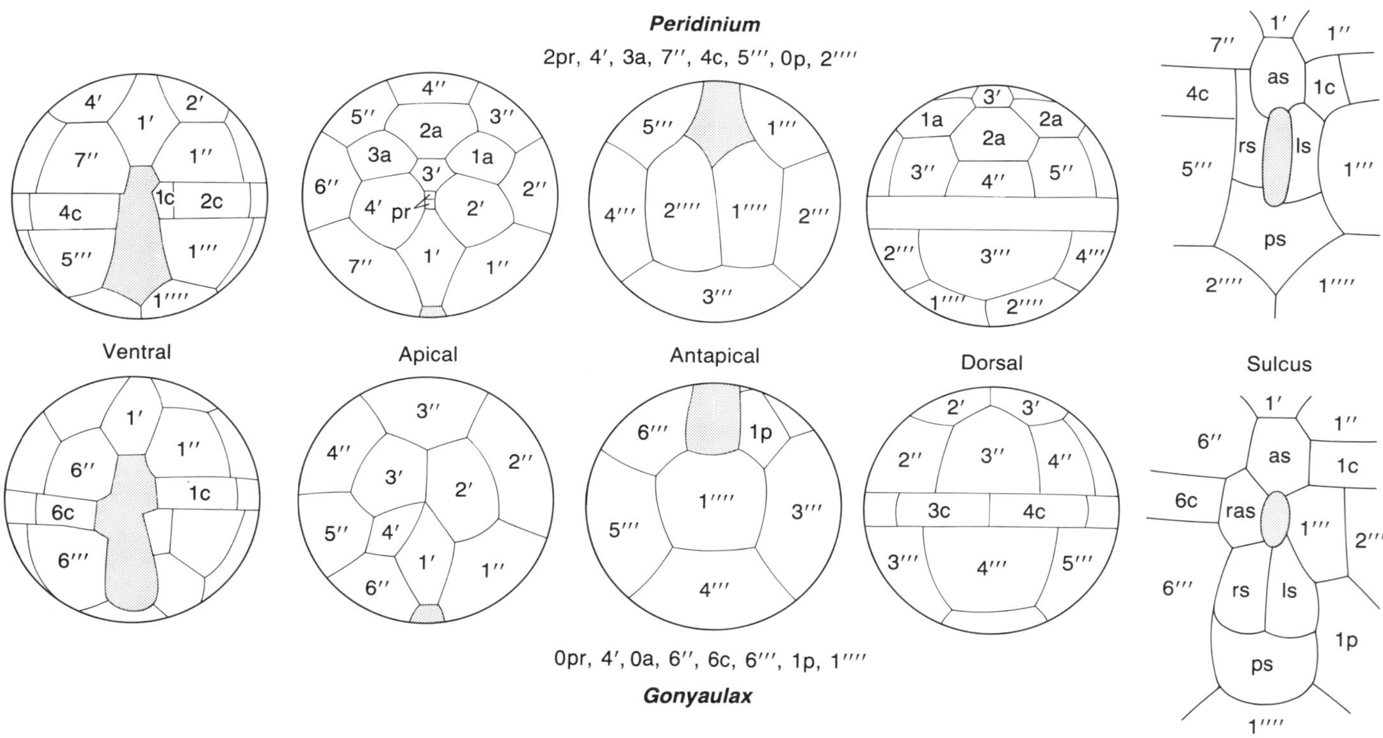

Figure 7.7 Schematic tabulation in *Peridinium* and *Gonyaulax*. Individual plates are labeled according to the Kofoidian system. In the formulas given, the numbers represent the number of plates in each plate series. A zero indicates that no plates are present in the specified series. (Modified from Evitt WR. *Sporopollenin Dinoflagellate Cysts. Their Morphology and Interpretation*. Austin: American Association of Stratigraphic Palynologists Foundation, 1985:90.)

sons between patterns and to make inferences about possible homologies involving specific paraplates or plates. In contrast to the Kofoidian system, other systems use idealized models to draw attention to homologous plates or paraplates (Fig. 7.8).

Archeopyle

Another distinctive feature in fossil dinoflagellates is the **archeopyle**, the opening in the cyst wall through which the cytoplasm emerges during excystment (Fig. 7.9). The

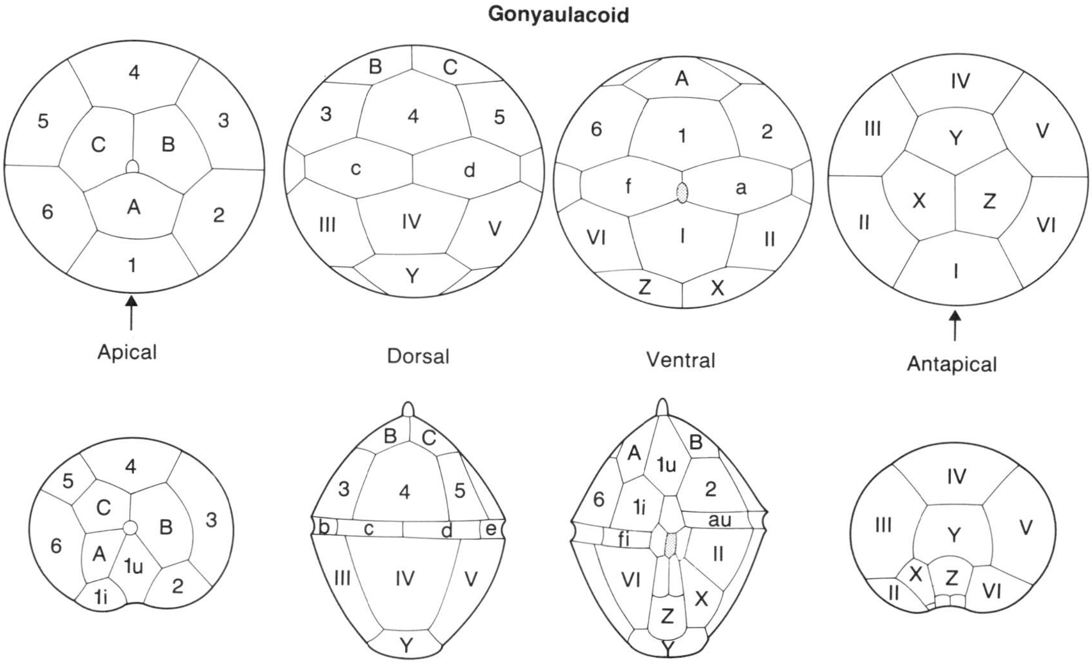

Figure 7.8 Taylor–Evitt notation for tabulation in gonyaulacaceans. Upper row shows the primary model; lower row shows the pattern as expressed in many species of *Gonyaulax*. (Reproduced from Evitt WR. *Sporopollenin Dinoflagellate Cysts. Their Morphology and Interpretation*. Austin: American Association of Stratigraphic Palynologists Foundation, 1985:296.)

opening forms along definite preformed lines that correspond to the boundaries of paraplates. The position and nature of the archeopyle is genetically determined and is of primary importance in the classification of dinoflagellate cysts. The archeopyle is the hole; its cover is the **operculum**.

Archeopyle terminology is based on the paraplate series in which the opening occurs. The archeopyle may consist of one or several plates within a series or it may consist of plates from more than one series. Common types are: apical archeopyle (Fig. 7.9A), intercalary archeopyle (Fig. 7.9C), and precingular archeopyle (Fig. 7.9B). In cysts with more than one wall layer, each layer may form a distinctive archeopyle (Fig. 7.9C). Intercalary and precingular archeopyles are most commonly formed from the middorsal paraplate or paraplates.

The operculum may be **simple**, consisting of a single piece made up of one or several paraplates (Fig. 7.9C, operculum in periphragm; Fig. 7.9D), or **compound**, consisting of several pieces (Fig. 7.9C, operculum in endophragm). A **free operculum** (Fig. 7.9D) is completely separated from the rest of the cyst by sutures. An **adnate operculum** is incompletely surrounded by sutures and thus permanently attached (Fig. 7.9C, operculum in periphragm). A free operculum that remains touching the cyst is termed **adherent** (Fig. 7.9C, operculum in endophragm).

Surface features on the cyst

Surface features on the cyst (Fig. 7.10) include granules, spines, and the prominent projections called processes. An assemblage of processes representing a single paraplate forms a process group. Linear elements on the cyst include ridges, folds, and septa. If the surface ornament does not give some indication of paratabulation, it is **nontabular** (Fig. 7.10B). **Tabular** distribution is **parasutural** if it emphasizes boundaries between paraplates (Fig. 7.10C, F) and **intratabular** if it emphasizes the central parts of paraplates (Fig. 7.10E). **Penitabular** features are those that lie approximately parallel to paraplate boundaries within individual paraplates (Fig. 7.10D). Processes that occur at the intersection of three (or more) paraplates are **gonal**; those that occur along paraplate boundaries away from such triple junctions are **intergonal** (Fig. 7.10G).

Classification

The dinoflagellates have been claimed by both botanists and zoologists and therefore have been classified under

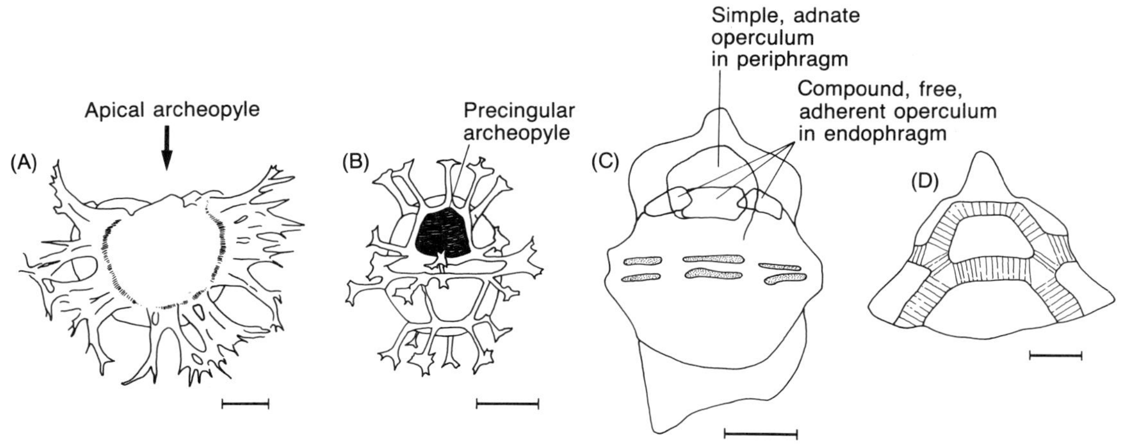

Figure 7.9 Archeopyle and operculum variety. A, Apical archeopyle in *Areoligera*, shown in ventral view. B, Precingular archeopyle in *Spiniferites*, shown in dorsal view; archeopyle is formed by the loss of a single, middorsal precingular paraplate. C, Intercalary archeopyle in *Chatangiella*, shown in dorsal view; periphragm has a single-plate intercalary archeopyle and a simple, adnate operculum; endophragm has a three-plate archeopyle and a compound, free, adherent operculum. D, Operculum of *Palaeoperidinium*, exterior view; the free, simple operculum consists of one apical, three intercalary, and three dorsal precingular paraplates.

In a shorthand notation for archeopyle types, uppercase letters identify the series. Parentheses are used to denote a single opercular piece consisting of several paraplates, lower case "a" indicates adnate opercular pieces, and a slash is used to separate the archeopyle notations for different wall layers. Thus the notations for the figures above are: A, (4A); B, P; C, Ia/3I; D, (A3I3P). Scale bars, 20 μm.

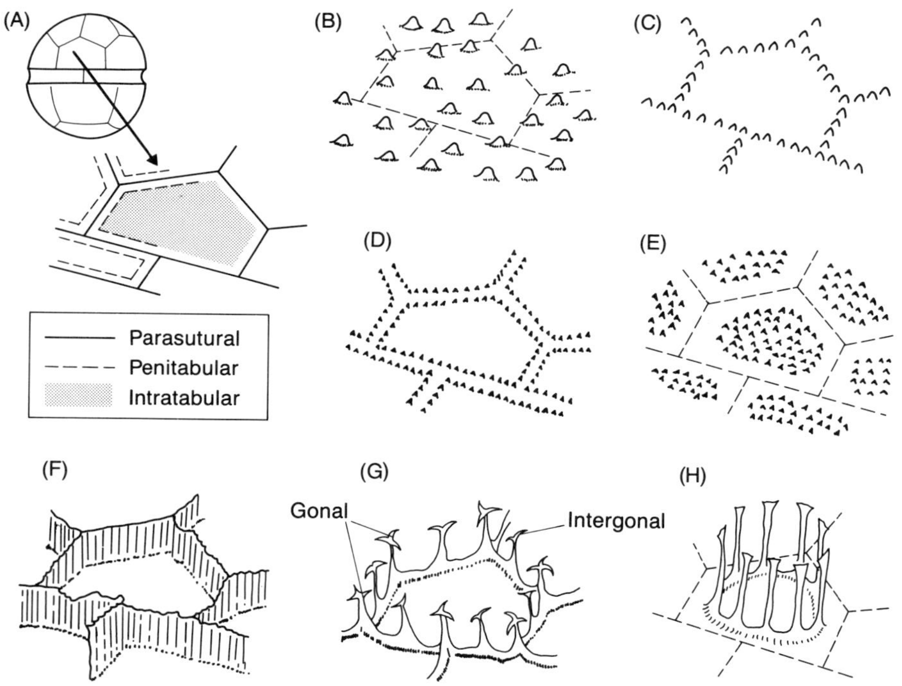

Figure 7.10 Surface ornament. A, Orientation diagram. B, Nontabular features. C, Parasutural features. D, Penitabular features. E, Intratabular features. F, Parasutural septa. G, Gonal and intergonal processes. H, Process group. (Modified from Evitt WR. *Sporopollenin Dinoflagellate Cysts. Their Morphology and Interpretation*. Austin: American Association of Stratigraphic Palynologists Foundation, 1985:70.)

both the **International Code of Botanical Nomenclature** and the **International Code of Zoological Nomenclature**. Although the dinoflagellates form a relatively coherent group, roughly half of the living species possess chloroplasts and are photosynthetic and half are nonphotosynthetic. Classification is further complicated by the fact that biologists rely on the theca and internal cell characteristics, whereas paleontologists must rely almost exclusively on preserved cysts.

In recent years, paleontologists (but not neontologists) have reached a consensus and use the botanical code. The dinoflagellates are considered to be among the most primitive eukaryotes. One possible phylogenetic position of the dinoflagellates is shown in Figure 7.11.

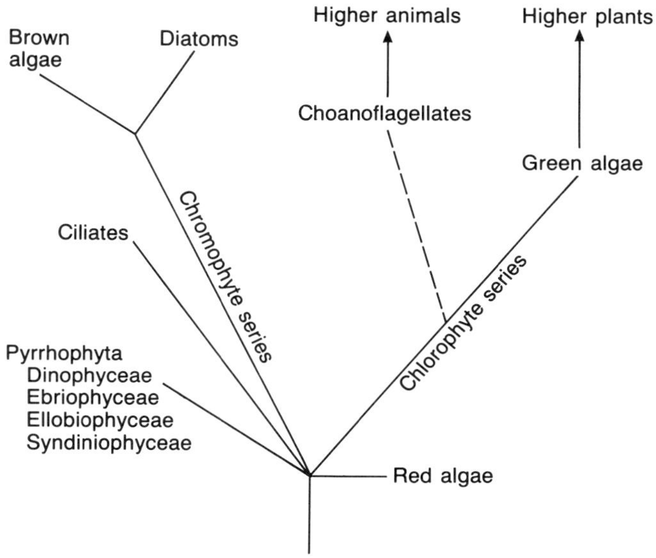

Figure 7.11 Possible phylogenetic relations of the dinoflagellates. (Reproduced from Evitt WR. *Sporopollenin Dinoflagellate Cysts. Their Morphology and Interpretation*. Austin: American Association of Stratigraphic Palynologists Foundation, 1985:7.)

Within the Pyrrhophyta, the dinoflagellates form the class Dinophyceae. The closely related class Ebriophyceae includes the ebridians, which have internal skeletons, are extant, and have a fossil record beginning in the Paleocene. Classes Ellobiophyceae and Syndiniophyceae are parasitic and have no fossil record.

No consensus has been reached on classification within the Dinophyceae. The abbreviated classification shown in Table 7.1 emphasizes groups that are important to paleontologists. The Prorocentrales (Fig. 7.12D) are characterized by two flagella that are located anteriorly (relative to the swimming direction). The cell is armored, and the theca consists mostly of two large plates. There are no certain fossil representatives of this order, but some acritarchs may belong to the Prorocentrales.

The remaining dinoflagellate orders are characterized by two flagella that are located ventrally, with one flagellum located in a transverse flagellar groove (the cingulum). The Dinophysiales (Fig. 7.1B) are armored and consist of two large thecal plates and about a dozen smaller plates. A single Jurassic species may be a fossil representative of this order. The Gymnodiniales lack obvious thecal plates and appear naked or unarmored. The Gymnodiniales include diverse genera such as *Polykrikos* (Fig. 7.12C), a pseudocolonial dinoflagellate with a fossilizable organic-walled cyst, the Cretaceous *Dinogymnium* (Fig. 7.18 no.17), which probably represents the covering of the motile stage, and *Actiniscus* (Fig. 7.12A), which has a fossilizable internal siliceous

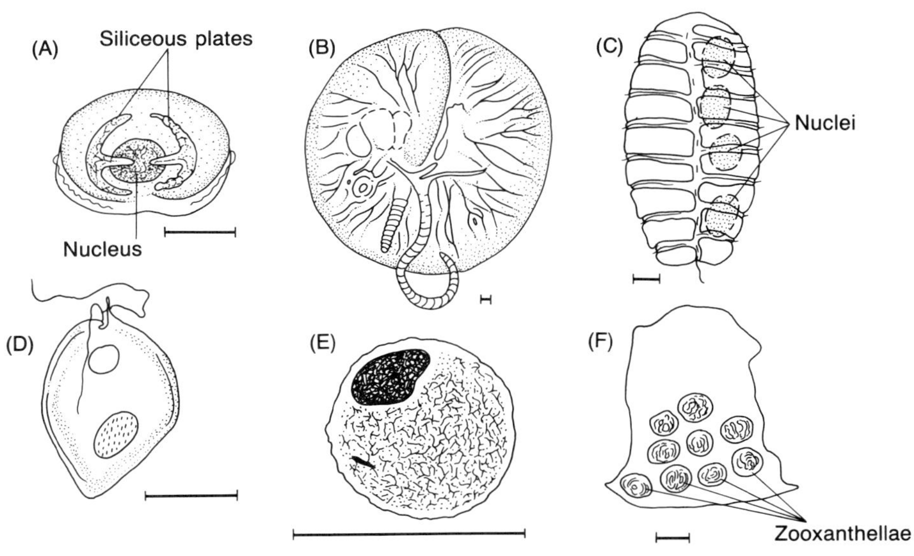

Figure 7.12 Examples of living dinoflagellates. A, *Actiniscus*, an unarmored marine dinoflagellate with star-shaped siliceous plates surrounding the nucleus. B, *Noctiluca*, a large, bioluminescent dinoflagellate, having an elongate tentacle. C, *Polykrikos*, marine, a pseudocolony, here with four nuclei. D, *Prorocentrum*, showing anterior flagella. E, *Thoracosphaera*, with a calcareous cell wall covering the coccoid vegetative stage. F, A single cell from a hydroid and its nine symbiotic zooxanthellae. All scale bars, 20 μm. (E Redrawn from photo in Tangen K, Brand LE, Blackwelder PL, Guillard RRL. *Marine Micropaleont* 1982; 7:196. F Redrawn from photo in Trench RK. In Taylor FJR, ed. *The Biology of Dinoflagellates. Botanical Monographs, 29*. Oxford: Blackwell Scientific Publications, 1987:559.)

Table 7.1 Dinoflagellate classification, selected orders

	Fossil record
Division Pyrrhophyta Pascher 1914	?Silurian–Holocene
Class Dinophyceae Fritsch 1929	?Silurian–Holocene
Order Prorocentrales Lemmermann 1910 Figure 7.12D (anterior flagella, armored, theca consists mostly of two large plates)	???
Order Dinophysiales Lindermann 1928 Figure 7.1B (ventral flagella, armored, theca consists of two large and several smaller plates)	?Jurassic
Order Gymnodiniales Lemmermann 1910 Figures 7.12A, C, 7.18 no.17 (ventral flagella, unarmored)	Cretaceous–Holocene
Order Peridiniales Haeckel 1894 Figures 7.7, 7.9 (ventral flagella, armored, about two dozen thecal plates)	?Silurian, Triassic–Holocene
Order Nannoceratopsiales Piel & Evitt 1980 Figure 7.18 no.4 (probably ventral flagella, cyst with epicyst-like Peridiniales, hypocyst with two large plates like Dinophysiales)	Jurassic
Order Thoracosphaerales Tangen 1982 Figure 7.12E (ventral flagella during brief, unarmored motile stage, calcareous cell covering on vegetative stage)	Triassic–Holocene
Order Zooxanthellales Loeblich 1970 Figure 7.12F (ventral flagella during brief, unarmored motile stage, symbiotic coccoid stage)	indirect evidence ?Silurian–Holocene

skeleton. The Peridiniales (Figs 7.7, 7.9) are armored and generally have about two dozen thecal plates but may have more than 40. Most of the dinoflagellates in the fossil record belong to this order. The Nannoceratopsiales (Fig. 7.18 no.4) are known only from fossil cysts. The epicyst resembles that of the Peridiniales (numerous plates) and the hypocyst resembles that of the Dinophysiales (two large plates, two small plates).

Both the Thoracosphaerales and the Zooxanthellales have coccoid vegetative stages and brief, free-swimming stages that are like the Gymnodiniales. The Thoracosphaerales have a calcareous cell covering on the coccoid stage (Fig. 7.12E). The Thoracosphaerales do not produce sporopollenin cysts, but rather calcareous-walled, generally spherical fossils that are common in Cretaceous and younger sediments. The coccoid forms of the Zooxanthellales (Fig. 7.12F) are symbionts in corals, protozoans, and mollusks. The evidence for the presence of the Zooxanthellales in the fossil record is indirect and speculative, but this order may have played an important role in evolution and diversification within major invertebrate groups.

Classification of living dinoflagellates is based on the morphology of the motile stage (location of the flagella, presence or absence of rigid thecal plates, number and arrangement of plates, general shape), the cell contents, and aspects of the life history of the organism. Unfortunately, the complete life-cycles of relatively few dinoflagellate species have been determined. By necessity, fossil dinoflagellates are classified on the basis of features of the cysts: paratabulation, archeopyle, wall relationships, and ornamentation. Studies of dinoflagellates in culture have shown that thecae that have been assigned to the same modern species or species-complex can produce cysts that are so different that paleontologists place them in different genera. Are modern taxa too broad, or are fossil taxa too finely split? Conversely, a number of species that can readily be distinguished on thecal morphology all produce similar nondescript, round, brown cysts. Clearly, further research is needed.

Paleontologists working with dinocysts take a variety of stances on higher-level classification of dinoflagellates. Some authors recognize families, suborders, and orders based on cyst characteristics. Other authors have grouped morphologically similar cyst-based genera into informal lineages. Still others do not attempt to group cyst-based genera into higher taxa and rely on alphabetical listings. Work is currently in progress on unifying the taxonomy of fossil and living dinoflagellates.

The number of described genera of living dinoflagellates is approximately 130, and these genera include more than 2000 marine and 200 freshwater species. A recent compilation of fossil dinoflagellates includes 525 valid genera and 2958 valid species; these numbers have more than doubled in the last two decades.

Biology

Although the dinoflagellates are considered to be primitive eukaryotes, they are amazingly complex organisms (Fig. 7.13). The group shows wide diversity in morphology, nutrition, and habitat. They range in size from 7 to 2000 μm, although most forms are less than 200 μm.

Cell organization

Historically, the dinoflagellates were grouped into thecate

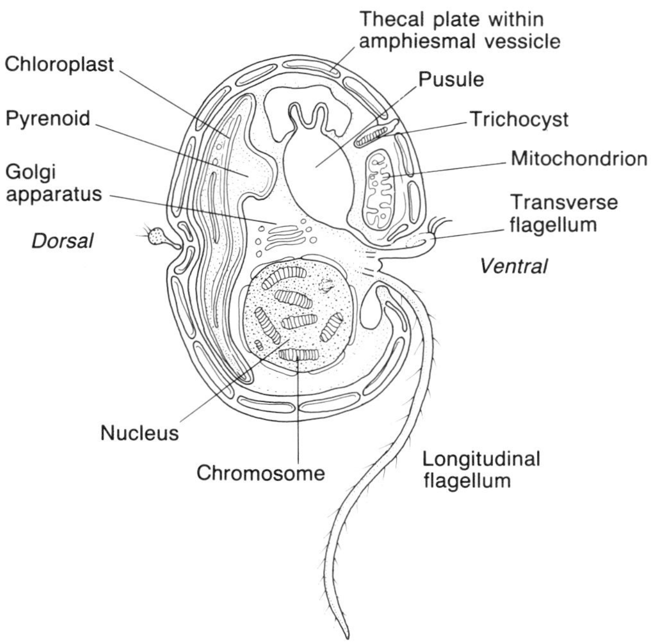

Figure 7.13 Diagrammatic section through a generalized thecate, photosynthetic dinoflagellate. (Modified from Taylor FJR. *The Biology of Dinoflagellates. Botanical Monographs, 29.* Oxford: Blackwell Scientific Publications, 1987:2.)

(armored) and naked (unarmored) forms, based on the presence or absence of a rigid, tabulate layer below the flexible outer membrane. Electron-microscope studies show an intergradation between these types. All free-living dinoflagellates have a cell covering called the amphiesma which consists of an outer membrane, a thin layer of cytoplasm, and a layer of flattened vesicles. The cell cover may include additional membranes and layers below the vesicle layer. In unarmored forms, the vesicles are fluid-filled or contain only granules; other seemingly unarmored forms have very thin plates. In armored forms the plates are prominent. The growth patterns of these plates and the direction of overlap of adjacent plates may provide important clues to plate homologies.

Dinoflagellates contain typical eukaryotic organelles in the cytoplasm: smooth and rough endoplasmic reticulum, Golgi apparatus, mitochondria, lipid and starch grains, and various accumulation bodies such as food vacuoles. Dinoflagellates may also have any of several distinctive cytoplasmic inclusions or organelles. Trichocysts and nematocysts are ejectile organelles of unknown function. The peduncle is a retractable protoplasmic extension which may play a part in attachment to substrate or host. The pusule consists of a series of membrane-bound tubules of uncertain function—nutrition, waste, flotation, and osmoregulation have all been suggested. Light-sensitive organelles include the eyespot, made of pigment-containing globules, and the more elaborate ocellus, which has retinoid and lens.

The cytoplasm of photosynthetic dinoflagellates contains chloroplasts. Both chlorophyll *a* and *c* occur, with chlorophyll *a* usually dominant. Carotenoids common in dinoflagellates include xanthophylls such as peridinin and pyrrhoxanthin. Unlike those in most other algae, dinoflagellate chloroplasts are surrounded by three, not two, membranes. Possibly, the chloroplasts in dinoflagellates were originally symbiotic eukaryotic algae. Pyrenoids are specially differentiated regions of the chloroplasts that are composed of proteinaceous material and have a granular appearance. Their specific function is not known; they may be for storage or production of proteins, starch, or enzymes.

As in the rest of the eukaryotes, the dinoflagellate nucleus is bounded by a nuclear membrane. The dinoflagellate nucleus is relatively large in proportion to the cytoplasm and contains a prominent nucleolus. Chromosomes number from four to 500. The dinoflagellate chromosomes remain in a permanently condensed state, in contrast to typical eukaryote chromosomes, which are condensed only during cell division.

Dinoflagellate mitosis is distinctive. With one known

exception, dinoflagellates develop extranuclear spindles. The spindle apparatus forms within the cytoplasm and then sinks into the nucleus as a series of microtubule-containing invaginations in the nuclear envelope. Chromosomes remain associated with the persistent nuclear envelope, but with some sort of microtubule–chromosome interaction. The envelope eventually constricts and divides to form two daughter nuclei.

During at least a brief part of their life-cycle, most dinoflagellates have two dissimilar flagella. In the order Peridiniales (Fig. 7.2) and most other dinoflagellates, the two flagella arise from the ventral surface, one encircling it, one trailing it. In the order Prorocentrales (Fig. 7.12D), the flagella arise from what is considered the anterior end, but both structural and functional differences in the two flagella suggest that they are homologous to the longitudinal and transverse flagella of the Peridiniales. A few species of dinoflagellate have only a single flagellum.

Nutrition

About half of the known species of dinoflagellates are autotrophic, producing organic compounds by photosynthesis. Together with the diatoms, these dinoflagellates form the base of the food web in both marine and freshwater communities. Diatoms have a faster generation time than the dinoflagellates and therefore are more likely to predominate in initial seasonal algal blooms. Dinoflagellate blooms often succeed diatom blooms. As nutrients such as phosphates and nitrates are depleted in the photic zone, the motile dinoflagellates move down in the water column at night to seek more nutrients and up in the day to seek sunlight. Most photosynthetic dinoflagellates are, strictly speaking, auxotrophs; they require externally supplied vitamins. Some dinoflagellates, such as freshwater *Ceratium*, rely on both photosynthesis and digestion of the remains of bacteria and algae.

In general, the nonphotosynthetic dinoflagellates rely on the ingestion of other organisms for nutrition. Some species engulf their food with a membranous pseudopod and then liquefy the cell contents leaving an intact but empty cell wall. Dinoflagellates feed on a variety of algae (including other dinoflagellates), on the eggs of copepods or other marine plankton, on nematodes, and on plant spores. Dinoflagellates also may be saprophytes feeding on decaying organisms, external parasites on fish and invertebrates, or intracellular parasites. Some dinoflagellates rely on symbiotic cyanobacteria for nutrition.

Dinoflagellates are found as symbionts with a variety of marine organisms, such as foraminifers, radiolarians, other dinoflagellates, mollusks, and cnidarians. Tropical reef-building corals rely on symbiotic zooxanthellae (so named from the genus *Zooxanthella*). The dinoflagellates benefit from the protection offered by their hosts, as well as the nitrogen, phosphorus, and carbon dioxide they produce. The hosts get the benefit of waste removal and use the oxygen and carbon compounds that the dinoflagellates produce. The presence of symbiotic zooxanthellae enhances growth and leads to increased calcification. For unknown reasons, when stressed, corals may expel their symbionts.

Locomotion

The actions of the two flagella are responsible for the motility of the dinoflagellates (Fig. 7.14). The organism moves in a spiral path, rotating about an axis roughly aligned with its direction of movement. In photosynthetic dinoflagellates, the rotation of the cell allows all plastids equal exposure to sunlight. Vertical movement of dinoflagellates in the water column results from a combination of flagellar movement, gravity, and physiologic adjustment of buoyancy. Diurnal movement is especially important to the dinoflagellates and allows the cells to migrate downward at night, away from nutrient-depleted surface waters, and to return to the photic zone in anticipation of daylight.

In dinoflagellates with ventrally inserted flagella (Fig. 7.14), the transverse flagellum supplies both a forward and a rotational component of movement. The longitudinal flagellum trails behind and produces some forward movement but may serve primarily as a rudder. In dinoflagellates with terminal flagella (Fig. 7.12D), the

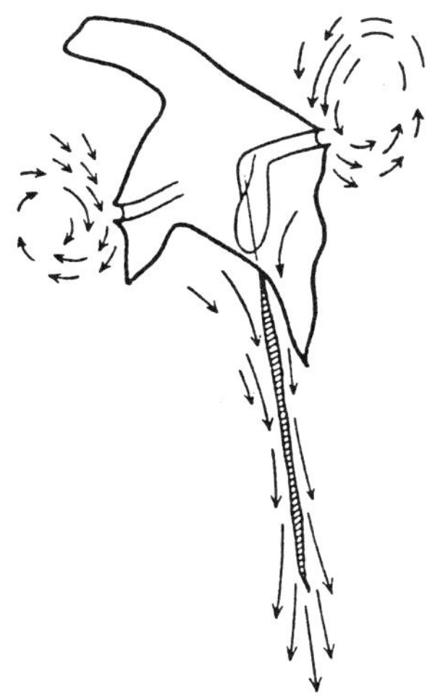

Figure 7.14 Flagellar currents of *Ceratium*. (Reproduced from Metzner P. *Z Bot* 1929; 22:225–65.)

equivalent of the transverse flagellum projects forward and beats in a circular fashion, while the longitudinal flagellum trails behind, again acting as a rudder.

Life-cycle and reproduction

Reproduction in dinoflagellates is most commonly asexual (vegetative). The nucleus divides mitotically. In unarmored forms, the rest of the cell may divide by simple fission. In thecate forms, cell division may be accomplished by the shedding of the theca, cell division, and the regeneration of thecae by the daughter cells, or it may occur by a diagonal division of the theca such that each daughter cell gets half of the original theca and regenerates the other half.

In photosynthetic dinoflagellates, vegetative reproduction is often related to the diurnal light cycle. Many dinoflagellates divide only at night, although the timing and frequency of division varies. In culture, vegetative reproduction may occur as often as every 8 hours.

Paleontologists are most likely to be interested in the sexual reproduction of dinoflagellates. After the gametes fuse, a cyst may be formed. Walls of the cyst may be made of cellulose (and therefore not readily preservable), sporopollenin, or calcium carbonate. One Paleocene form has a calcareous outer wall and an organic inner wall. Although no modern dinoflagellates are known to produce cysts made of silica, some contain silica. Siliceous fossil cysts may be primary composition or may result from secondary replacement.

With the exception of the genera *Polykrikos* (Fig. 7.12C) and *Noctiluca* (Fig. 7.12B), the normal vegetative, motile stage of dinoflagellates is haploid. The life-cycle of dinoflagellates has many variations, but a basic pattern is shown in Figure 7.15. Haploid vegetative cells become or produce haploid gametes. Gametes fuse and a zygote forms. The zygote often has two longitudinal flagella. The zygote enlarges, its cell wall may thicken, and it loses its motility. This stage is called the hypnozygote. The cytoplasm contracts, red bodies may develop, and a resistant-walled cyst may form inside the theca. As the cyst becomes part of the sediment, the thecal plates break apart or are destroyed. After a period of dormancy (from hours to years), during which the first and sometimes the second meiotic cell divisions occur, the cell (or cells) emerges through an opening determined by predeveloped structural weakness, the archeopyle. New cells rapidly form new thecae, and the cycle continues.

The specific details of the sexual cycle vary considerably within the members of the Dinophyceae. Gametes may be of the same size as the vegetative cells or may be considerably smaller. Gametes may be all of one size or of two greatly different sizes. Gametes from a single clone

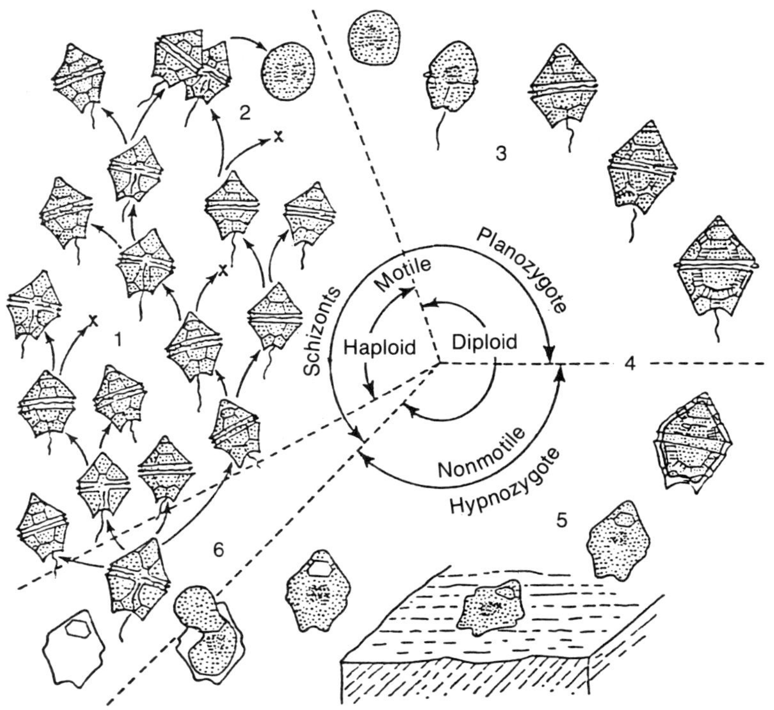

Figure 7.15 Schematic life-cycle of a cyst-producing dinoflagellate. 1, Growth and population expansion by vegetative division (x's represent unfigured vegetative cells). 2, Gametes fuse and form zygote. 3, New theca forms on motile zygote. 4, Cyst forms, flagella are lost, reddish bodies appear in protoplast. 5, Thecal plates lost or destroyed and the cyst behaves as a sedimentary particle. 6, Following dormancy and meiosis (timing varies), protoplasm excysts, cyst remains in sediment, new thecae form. (Reproduced from Evitt WR. *Sporopollenin Dinoflagellate Cysts. Their Morphology and Interpretation*. Austin: American Association of Stratigraphic Palynologists Foundation, 1985:11.)

may be able to fuse and reproduce, or two different mating types may be necessary. Zygotes may have a motile stage (the planozygote) before the nonmotile stage or there may be no motile zygote. The two meiotic divisions may occur during the motile stage or the encysted stage, or one or both meiotic divisions may even occur after germination.

In culture, sexual reproduction can be induced by nutrient depletion or by reducing the temperature or period of light. In nature, the triggers are likely to be far more complex. Sexual reproduction is known to occur in the late summer and fall and in the late stages of natural blooms.

Living dinoflagellates may form temporary cysts at times in their life-cycle other than during sexual reproduction. Most, if not all, fossilized cysts are believed to be hypnozygotes, rather than such temporary cysts.

Comparisons of modern, motile dinoflagellates with fossil dinoflagellate cysts are complicated. Only a few dozen sexual life histories have been documented. Not all modern dinoflagellates form any kind of cyst and only about 20% of modern taxa produce acid-resistant, geologically preservable cysts. Only about one-third of the known modern cysts have been correlated with their motile stages. Similar appearing cysts are known to germinate into very different motile stages, and seemingly identical motile stages have produced very dissimilar cysts.

Toxicity

About 20–30 extant species of dinoflagellates produce water-soluble or lipid-soluble toxins. These species are all photosynthetic, are all estuarine or neritic, and differ little from nontoxic forms. Both planktic and benthic species produce toxins. Toxic planktic dinoflagellates probably all produce benthic, sexual resting cysts, which are important in initiating toxic blooms. Toxic dinoflagellates form nearly monospecific blooms with very high concentrations (up to millions of cells per liter). Blooms develop in discrete water masses that differ in salinity and chemical composition from the surrounding water, such as zones of upwelling, tidal fronts, and estuaries affected by rainfall and runoff. Some areas experience blooms on an annual basis. Blooms can be transported by currents considerable distances from where they originate and can be detected on satellite images.

Blooms of toxic dinoflagellates can cause poisoning of marine animals, such as fish, shellfish, and filter-feeding mollusks. Humans who eat animals in which toxins have accumulated may suffer from a variety of reactions including diarrhea, facial numbness, general numbness, dizziness, muscular weakness, and respiratory failure that has caused over 200 deaths since the 1600s. Even sea-spray containing fragments of toxic dinoflagellates can cause coughing, sneezing, and other respiratory irritation — as space shuttle workers found out when they installed heat-shielding tiles on site at Cape Canaveral. Toxic dinoflagellates also cause massive fish kills, which can result in millions of dollars in clean-up costs and lost revenue in the seafood and tourist industries.

Bioluminescence

Dinoflagellates cause most of the bioluminescence in the surface waters of the oceans. Not all marine dinoflagellates are luminescent, and even within a species, some strains are luminescent, whereas others are not.

Bioluminescence in dinoflagellates may occur either as a flashing light or as a dim glow lasting up to several hours. The blue–green light emitted is the result of a chemical reaction of the enzyme luciferase oxidizing the substrate luciferin. In various tropical bays, the spectacular displays of bioluminescent dinoflagellates are popular tourist attractions.

Ecology

Dinoflagellates live in a variety of marine and nonmarine environments and are a major component of the food web. The majority of species are marine. Oceanic dinoflagellate species commonly have elongate horns or ornate wings (Fig. 7.1A, B), which increase surface area for flotation and nutrient uptake. Neritic dinoflagellates more typically have simple morphologies. Dinoflagellates live in tidepools and in the interstitial water in shore sands and muds. They are also common in freshwater lakes, ponds, rivers, and bogs, and blooms with concentrations high enough to color the water may occur. Red snow and ice in the Alps and Canada result from the accumulation of dinoflagellates or their cysts.

Most dinoflagellate species seem to be cosmopolitan or to be restricted to wide latitudinal bands. Although it is easier to think of species as preferring warm-water or cold-water environments, it is more accurate to think of species as being tolerant of a wide range of temperatures or of only a narrow range of temperatures. Within the genus *Ceratium* (Fig. 7.1C), temperature-related morphologic variation is common. Differences in the length of the horns and in the angle between horns reflect differences in temperature. Dinoflagellates tolerate a wide range in salinity. In culture, most marine species show optimal growth at salinities of 10–20 parts per thousand (‰), in contrast to the normal, open-ocean value of 35‰. One nonphotosynthetic species thrives in culture at salinities from 4 to 130‰.

The autotrophic dinoflagellates live within the photic zone. The factor limiting their productivity may be the availability of such nutrients as phosphorus and nitrogen. Thus, they reach high concentrations in the spring when nutrients increase, decline during the summer as nutrients are depleted, and may have a second time of increased productivity if there is a late summer increase in nutrients. They seem to prefer more stable and predictable environments. In contrast, heterotrophic dinoflagellates occupy a variety of other environments and may respond to quite different seasonal cues.

Biologists have mostly studied the motile stage of free-living planktic dinoflagellates from freshwater, brackish, and marine environments. Planktic dinoflagellates may or may not have a benthic (encysted) stage as an integral part of their life-cycle. With the exception of the actiniscians (see p. 128), paleontologists deal almost exclusively with the benthic stage in the dinoflagellate lifecycle. Unfortunately, few studies on dinoflagellate ecology have emphasized cysts.

Cyst-forming dinoflagellate species live in nearly all marine environments from the equator to the poles and from the intertidal zone to the deep oceans. They dominate in shallow coastal waters and their cysts may serve as seeds for oceanic populations. Assemblages of cysts in the sediments, however, do not reflect the overall dinoflagellate assemblages, because most living species do not produce resistant cysts. For example, more cyst-forming species live in shallow, temperate environments than in tropical oceans, although the total number of dinoflagellate species is much greater in the tropics.

Living cysts and the distribution of cysts in modern sediments show two general trends. The first is an inshore–offshore trend, in which specific assemblages are associated with estuarine, nearshore, neritic, or oceanic environments. The second trend is climatic, relating assemblages to present latitudinal bands. Ocean circulation patterns are also reflected in cyst-distribution patterns.

The morphology of dinocysts from recent sediments may be directly related to salinity. In the Black Sea and Baltic region, *Tectatodinium* and *Spiniferites* species are round in outline in normal salinities and are cruciform in low-salinity environments (Fig. 7.16). *Lingulodinium* and *Operculodinium* species may show greatly reduced processes in low-salinity environments.

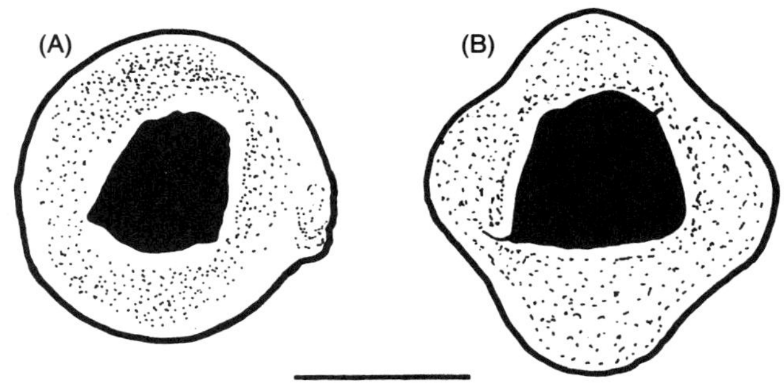

Figure 7.16 *Tectatodinium psilatum* Wall and Dale, from low-salinity sediments, Quaternary, Black Sea. A, Spherical variety. B, Cruciform variety. Scale bar, 20 μm. (Redrawn from photo in Wall D, Dale B, Harada K. *Micropaleontology* 1973; 19:18–31.)

Paleobiology

The life-cycle of the dinoflagellates presents problems to the paleontologist hoping to make paleobiologic inferences. The fossil record of the dinoflagellates represents one stage in the life-cycle, the cyst, and only a small minority of living species produce cysts that have geologically preservable walls. Among cyst-producing species, the number of cysts formed does not reflect the number of active motile cells. The number and variety of cysts in bottom sediments show no simple relationships to the number of cyst-producing species in modern plankton. Periods of high dinoflagellate productivity in present times and presumably in the geologic past are not necessarily reflected by abundant, preservable dinocysts in the sediments.

Many types of cyst that are abundant in the fossil record have no modern counterparts. Other types of cysts have apparent modern relatives that have habitats that are very different from what the habitats must have been in the past. The distribution of modern cysts in bottom sediments is also strongly influenced by ocean currents and may have little relationship to the living populations. Fossil cysts, likewise, may be far removed from their area and environment of formation.

Freshwater vs. marine environments

Few reliably identified freshwater deposits contain pre-Pleistocene dinoflagellate cysts. Preservable cysts, mostly of the *Peridinium* type, are common in many lake sediments today, so the presence of dinoflagellate cysts does not, as is often cited, indicate a marine influence. Generally, diverse dinocyst assemblages containing gonyaulacaceans as well as peridiniaceans probably indicate marine conditions. The paleontologist must determine that the marine dinocysts are of the same age as the rest of the sediments before inferring a marine environment. Reworked dinocysts can be extremely well preserved!

Inshore–offshore trends

Inshore–offshore trends can be recognized in the distributions of cysts in present-day bottom sediments. Variations in diversity, assemblage compositions, and dominances in fossil dinocyst distributions probably also indicate environmental trends, although in a less precise way. The most abundant fossil dinocyst assemblages are recovered from rocks from middle neritic to upper bathyal environments; relative abundance decreases in both inshore and offshore directions. Considerable study is needed, with corroborating sedimentologic and paleontologic evidence, before dinocyst-based paleoecologic inferences can approach the quality of those based on other, better documented fossil groups.

The distribution of the modern cysts allows inferences to be made concerning assemblages of fossil cysts containing many of the same species. In upper Miocene to Pleistocene deposits from the Gulf of Mexico, five neritic to oceanic paleoecologic categories can be recognized from their constituent dinocyst species. Co-occurrences of specimens of inshore and oceanic species can be an indication of transport by storms, upwelling, or other currents, most likely from shallow to deep water.

Distinctive assemblages of dinocysts recur in sediments older than Neogene, and sedimentologic and other evidence is used to relate these assemblages to paleoenvironments. Few or none of the cysts can be related to forms with known environmental tolerances, so only general paleoenvironmental interpretations result. The lower Tertiary units of southern England show four distinct assemblages of dinoflagellate cysts and acritarchs dominated by one or a few genera. The distribution of these associations corresponds to lithology, and perhaps the associations dominated by *Areoligera* (Fig. 7.9A) and *Spiniferites* (Fig. 7.9B) indicate an open-sea environment; the association dominated by *Micrhystridium* (an acritarch) is inner neritic, marking initial and closing stages of marine transgressions; and the association dominated by *Wetzeliella* (Fig. 7.18 27) and related genera is estuarine.

The ratio of the number of gonyaulacacean species to peridiniacean species may be an indicator of salinity in a Cretaceous deposit in Canada. High gonyaulacacean ratios indicate an open-marine environment; low ratios indicate a nearshore, reduced-salinity environment. The study from Canada backs up the paleoenvironmental interpretations of the associations of fossil cysts in England. *Areoligera* and *Spiniferites* are gonyaulacacean genera and *Wetzeliella* is a peridiniacean genus. Recent advances in the study of feeding strategies of living dinoflagellates may shed some light on these empirical observations. Most gonyaulacacean dinocysts may have been formed by autotrophic organisms whereas most peridiniacean dinocysts may represent heterotrophs.

Multivariate techniques applied to relative abundances of species can be used to define groups of cyst species (communities) which are then interpreted to reflect inshore and offshore paleoenvironmental trends. The percentage of reworked dinocysts can indicate proximity to ancient shorelines. Reworked dinocysts in Gulf of Mexico sediments, for example, vary directly with the clay : silt ratio and inversely with carbonate content and therefore are related to shoreline position.

Paleoclimate and provincialism

In Quaternary sediments, glacial and interglacial intervals are distinguished by assemblage composition and relative abundance of dinocysts. For example, in a North Sea core, cold, arctic-like deposits show low dinocyst diversity and recovery, much reworking, and dinocyst assemblages with a significant proportion of *Spiniferites elongatus* and round *Protoperidinium* species. Warm-temperate conditions are indicated by high dinocyst recovery, abundant *Operculodinium centrocarpum*, and a significant proportion of warm-water species such as *Spiniferites mirabilis* and pentagonal *Protoperidinium* species. Multivariate or other statistical techniques may allow paleoclimate studies using dinocysts to be extended to the Neogene.

In older sediments, dinocyst genera and species may show regionally restricted distributions. Boreal and Tethyan provinces can be recognized in Late Jurassic and Early Cretaceous dinocyst assemblages. The Boreal province shows considerably less taxonomic diversity than the Tethyan and also includes species not present in the Tethyan.

Detailed distribution patterns of Late Cretaceous peridiniaceans from North America, South America, Europe, and Africa show three diagnostic assemblages (Fig. 7.17). These assemblages seem to show regional differentiation rather than local paleoecologic control. Tropic and subtropic paleolatitudes are represented by the Malloy Suite, which is characterized by species of *Andalusiella*, *Cerodinium*, *Phelodinium*, and *Senegalinium*. The Williams Suite represents a warm-temperate assemblage and is dominated by species of *Chatangiella* (Fig. 7.18 no.18), *Isabelidinium*, *Manumiella* (Fig. 7.18

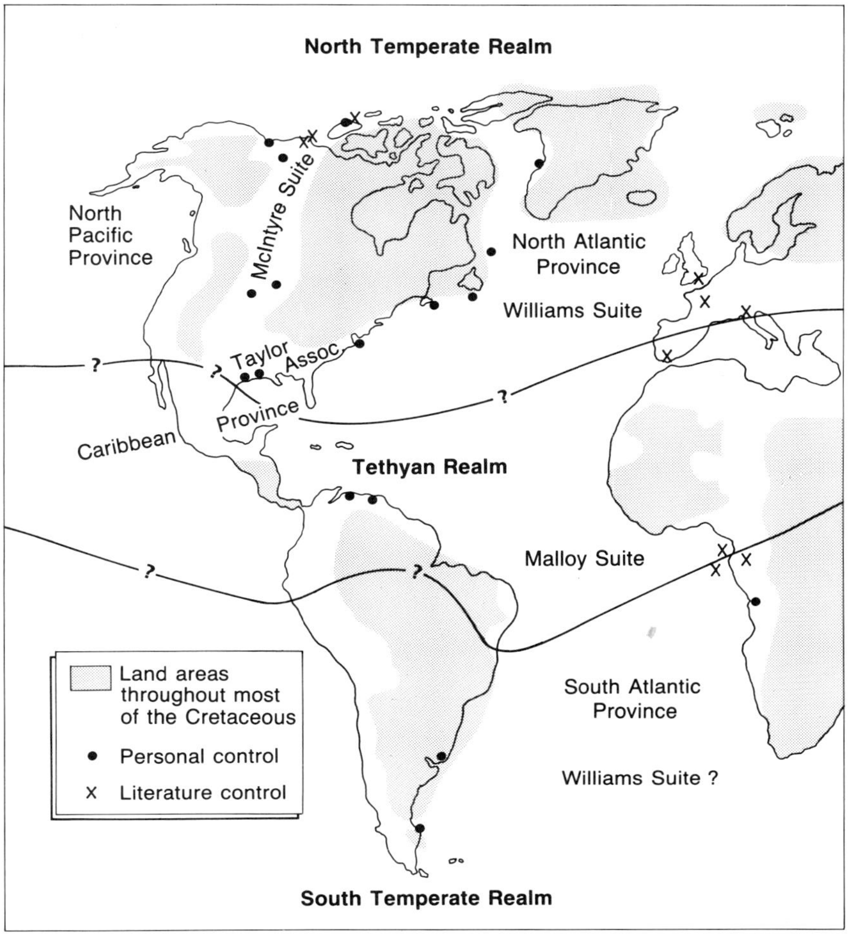

Figure 7.17 Late Cretaceous dinoflagellate cyst provincialism. (Reproduced from Lentin JK, Williams GL. *Amer Assoc Stratigraphic Palynologists Contrib Ser* 1980; 7:45.)

no.19), *Senegalinium*, *Spinidinium*, and *Trithyrodinium*. The McIntyre Suite is a boreal assemblage dominated by *Chatangiella* and *Laciniadinium* species.

Fossil red tides

Very abundant, monospecific dinocyst assemblages in the fossil record may have resulted from ancient dinoflagellate blooms. When associated with mass mortality of fossil invertebrates, the dinocyst assemblage probably represents a fossil toxic bloom. One probable fossil red tide occurs in the Lower Cretaceous of Denmark. Here the bivalve death assemblage is monospecific, the size–frequency and growth-ring distributions of the bivalves indicates a time delay between the growing season and the time of death, and the sediments and fossils indicate a quiet depositional environment and no transport after death. A single dinocyst species makes up more than 99% of the assemblage.

Biostratigraphy

Dinoflagellate biostratigraphy is based on the stratigraphic record of preserved cysts. Dinocysts allow detailed subdivision of many parts of the stratigraphic record and have been particularly useful for oil exploration in strata where calcareous microfossils are not abundant such as in Alaska and the North Sea. Although the lowest and highest occurrences of dinocyst taxa make excellent stratigraphic markers, these occurrences of cysts may not be evolutionary appearances or extinctions or even immigrations and emigrations of the organisms that produced them.

The oldest widely accepted fossil dinoflagellate cyst is *Arpylorus* in the Silurian. The single-known species shows what appears to be a precingular archeopyle, faint indications of paratabulation, and a general dinoflagellate-like appearance. No direct fossil record of the dinoflagellates exists between *Arpylorus* and the Triassic, although many of the Paleozoic acritarchs may actually be dinoflagellate cysts.

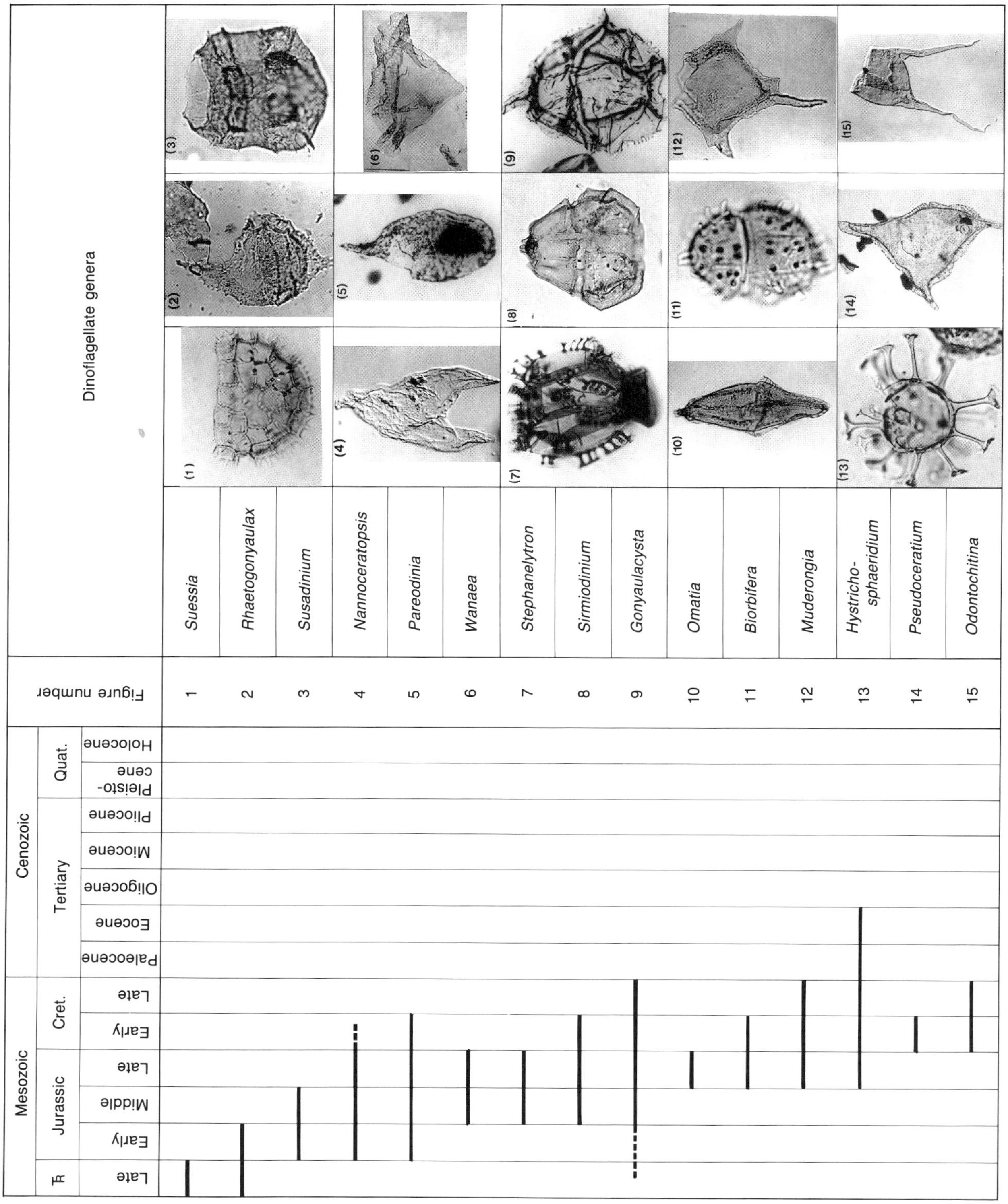
Dinoflagellate genera
Figure number
Suessia 1
Rhaetogonyaulax 2
Susadinium 3
Nannoceratopsis 4
Pareodinia 5
Wanaea 6
Stephanelytron 7
Sirmiodinium 8
Gonyaulacysta 9
Omatia 10
Biorbifera 11
Muderongia 12
Hystricho-sphaeridium 13
Pseudoceratium 14
Odontochitina 15
Cenozoic
Quat.
Holocene
Pleisto-cene
Tertiary
Pliocene
Miocene
Oligocene
Eocene
Paleocene
Mesozoic
Cret.
Late
Early
Jurassic
Late
Middle
Early
Ꞧ
Late

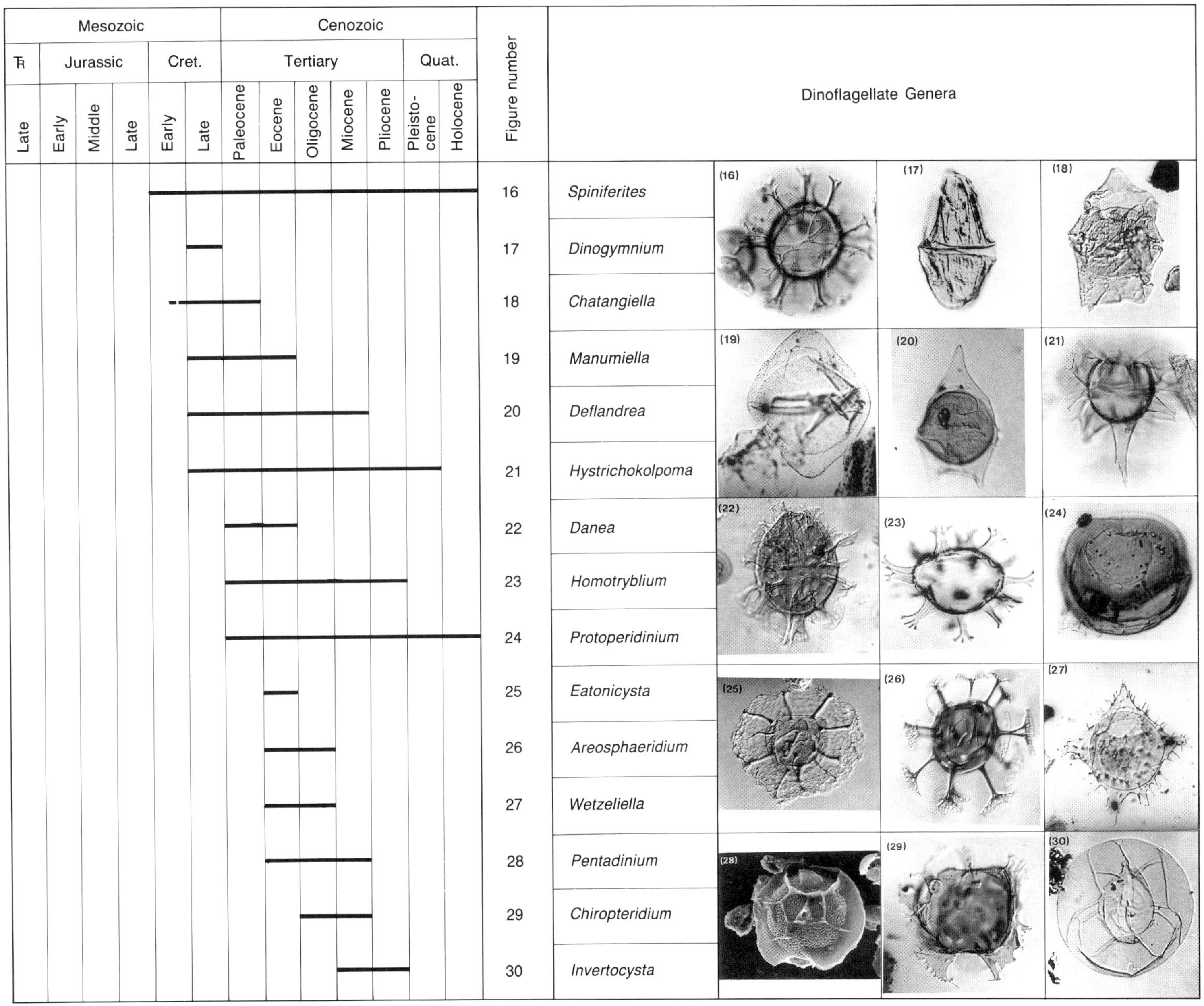

Figure 7.18 Some stratigraphically important dinocyst genera. (Photographs 1, 6, 7, 10, 12, 25, and 26 courtesy of Lewis Stover; 2, 3, 11, and 14 courtesy of Robin Helby; 8 courtesy of Nairn Albert.)

Late Triassic dinocyst assemblages have been reported from the Arctic, Australia, New Zealand, and Europe. They consist of forms with a relatively large number of paraplates, such as *Suessia* (Fig. 7.18 no.1) and *Rhaetogonyaulax* (Fig. 7.18 no.2).

The Early Jurassic was dominated by cysts of *Nannoceratopsis* (Fig. 7.18 no.4) and forms with disintegration archeopyles and tabulation patterns distinct from typical peridiniaceans and gonyaulacaceans. The genus *Gonyaulacysta* (Fig. 7.18 no.9) and its allies expanded in the Middle Jurassic. The ceratioid lineage (forms resembling modern *Ceratium*) appeared in the Late Jurassic but diversified during the Cretaceous. Chorate dinocysts first appeared in the Jurassic.

The Cretaceous was characterized by the radiation and the apparent end of the ceratioid fossils (such as *Muderongia*, Fig. 7.18 no.12, *Pseudoceratium*, Fig. 7.18 no.14, and *Odontochitina*, Fig. 7.18 no. 15). Large, cavate peridiniaceans such as *Chatangiella* (Fig. 7.18 no.18) were an important constituent of many Late Cretaceous assemblages. The distinctive fossil genus *Dinogymnium* (Fig. 7.18 no.17), which may be a preserved theca, was restricted to the Late Cretaceous.

In the Tertiary, a lineage including *Wetzeliella* (Fig. 7.18 no.27) and related genera first appeared in the Late Paleocene, dominated many Eocene assemblages, and then disappeared by the end of the Oligocene. Total dinocyst diversity began to decline in the Eocene and the decline continued to the present. Many modern peridiniacean species produce rather nondescript round cysts (Fig. 7.18 no.24).

No widely accepted dinoflagellate biozonation can be applied over large areas and long time-spans. For shorter time-spans, more than 50 published regional and local zonations allow very precise, relative-age determinations. For example, the Late Jurassic, a time interval of approximately 20 My, is divided into more than a dozen dinocyst zones and subzones in northwestern Europe. Many ongoing efforts in dinocyst biostratigraphy are focused on first and last appearance datums rather than on a worldwide zonation. Datums are then placed into the framework of other biozonations (such as those of ammonites, calcareous nannofossils, planktic foraminifera), sea-level curves and, indirectly, a numerical time-scale.

The dinoflagellate record across the Cretaceous–Tertiary boundary

The Cretaceous–Tertiary (K–T) boundary was a time of major extinction, especially among many marine plankton groups. Detailed examinations of the organic-walled dinocyst floras across the K–T boundary in Scandinavia, Antarctica, the southeastern USA, and elsewhere show successive appearances and disappearances but no major floral break.

For most calcareous and siliceous plankton, the K–T boundary marks a time of major faunal and floral change. The lowest Paleocene calcareous flora consists largely of *Thoracosphaera*, which are probable dinocysts with calcareous walls. Whatever major destruction of the oceanic flora and fauna occurred at the K–T boundary, the cyst-forming mode was an adaptation that aided survival.

Biostratigraphic record of zooxanthellae

Most modern reef-building corals contain symbiotic algae. Where original aragonite remains, geochemical studies of fossil corals can possibly detect the former presence of zooxanthellae. In older rocks, morphology, life habits, and geographic distribution of fossils provide indirect evidence for the presence of zooxanthellae. Silurian and Devonian tabulate corals, but not Devonian and Mississippian rugose corals, have growth forms and corallite size–frequency distributions that suggest the presence of zooxanthellae. The rapid diversification of the scleractinian corals at the end of the Triassic may be the result of the acquisition of algal symbionts.

In modern seas, the giant clam *Tridacna* has symbiotic zooxanthellae. Fossil rudistid bivalves and some Permian reef-dwelling brachiopods may have had them as well.

Evolutionary history

If the dinoflagellates are among the most primitive of the eukaryotes, why is their earliest record in the Silurian? Also, why is there a large stratigraphic gap—from Devonian to Jurassic—before abundant and diverse dinoflagellates are known from the fossil record?

At least two possible explanations exist for these phenomena and they are not mutually exclusive:

1 dinoflagellates did not produce cysts or the cysts they did produce did not withstand fossilization;

2 dinoflagellates produced fossilizable cysts but these cysts are not recognized as dinoflagellate fossils.

The modern genus *Ceratium* (Fig. 7.1C) forms rather fragile cellulosic cysts. Abundant and diverse *Ceratium*-like cysts are known from Late Jurassic and Cretaceous sediments but are unknown from the Tertiary. Apparently, members of this group stopped making fossilizable cysts about 70 Ma. If it was not for modern forms, the entire lineage would appear to have become extinct.

Some modern dinoflagellates produce cysts that do not show an obvious relationship to the dinoflagellates. For modern forms, the relationship is known from cultures. For fossil forms, only morphology can provide clues. Many of the fossils known as acritarchs were probably the cysts of dinoflagellates.

Thus, the features of the dinoflagellate life-cycle complicate the interpretation of their evolutionary history considerably. A first appearance in the fossil record may in fact be the earliest record of a lineage that has been living for millions of years without producing a fossilizable, recognizable cyst. Similarly, the highest occurrence of a fossil form may have nothing at all to do with its extinction.

Models for the evolution of the major orders of dinoflagellates must deal with armored and unarmored forms, armored forms with many plates and with few plates, and forms with anteriorly located flagella and with ventrally located flagella. Models for dinoflagellate evolution fall into three general categories:

1 a progressive increase in the number of thecal plates;
2 a progressive decrease in the number of thecal plates;
3 some combination of the above.

Evolution of the major lines of the dinoflagellates is a vigorously debated topic—there certainly is no consensus at the present time.

The plate increase model (Fig. 7.19, top) has the Prorocentrales (armored dinoflagellates, few plates, anteriorly located flagella) as the ancestral line. From this line arose the Dinophysiales (armored, relatively few thecal plates, ventrally located flagella). The remainder of the dinoflagellates arose from the ancestral line by the increase in the number of plates (other armored forms, such as the Peridiniales) and by the secondary loss of a rigid theca (unarmored forms, such as the Gymnodiniales).

The plate reduction model (Fig. 7.19, bottom, simplified) has an unarmored dinoflagellate, perhaps one that produced acritarch-like cysts, as the ancestral form. This ancestor gave rise to the Gymnodiniales. Following the development of a cellulosic wall, Late Triassic forms (such as Fig. 7.18 no.1 and no.2) with many polygonal plates on the cyst and presumably on the theca evolved later into forms with fewer plates: the Peridiniales, then the Dinophysiales. The number of plates could be reduced by the fusion of smaller plates or by the differential growth of some plates at the expense of others. A relatively recent change in swimming direction would convert ventrally located flagella to anteriorly located flagella giving rise to the Prorocentrales.

One combination model is shown in Fig. 7.20. Here, the Prorocentrales are considered to be primitive. A change in swimming direction and the differentiation of the theca into many plates is postulated to have occurred early in dinoflagellate history. The Gymnodiniales could have arisen from this ancestor by the loss of prominent thecal plates. As in the plate reduction model, the fossil record is used to infer that having many plates, as in such forms as *Suessia* and the Peridiniales, is a more primitive state than having few plates (the Nannoceratopsiales, the Dinophysiales).

Within the Peridiniales, a number of possible evolutionary lineages have been proposed based on morphology of fossil dinocysts. One important Paleogene lineage includes *Wetzeliella* and related genera (Fig. 7.21). At least three cyst groups diverged from *Wetzeliella* during the early part of the Eocene: in *Charlesdowniea*, the ends of the processes form platforms; in *Dracodinium*, the apical

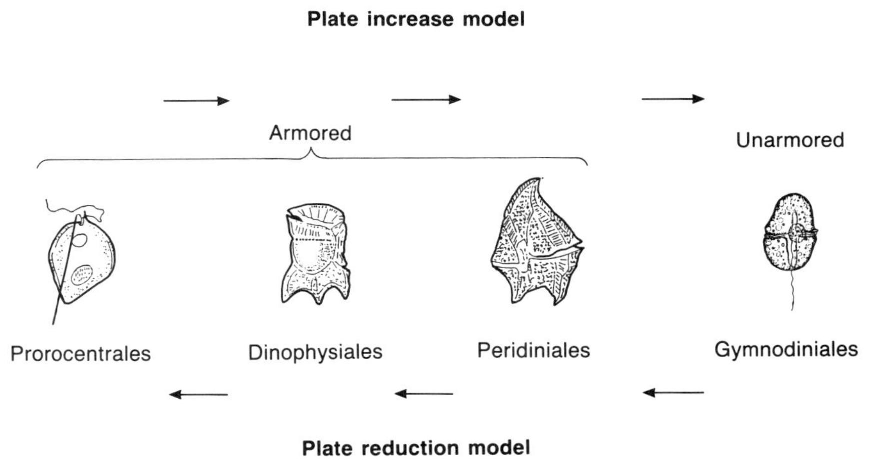

Figure 7.19 Highlights of two models of dinoflagellate evolution: the plate increase model (read from left to right) and the plate reduction model (read from right to left). (Models simplified from: Loeblich AR III. *J Protozool* 1976; 23:13–28; Dörhöfer G, Davies EH. *Evolution of Archeopyle and Tabulation in Rhaetogonyaulacinean Dinoflagellate Cysts*. Royal Ontario Museum, Life Sciences Miscellaneous Publications, 1980, Eaton G. *Paleontology* 1980; 23:667–88.)

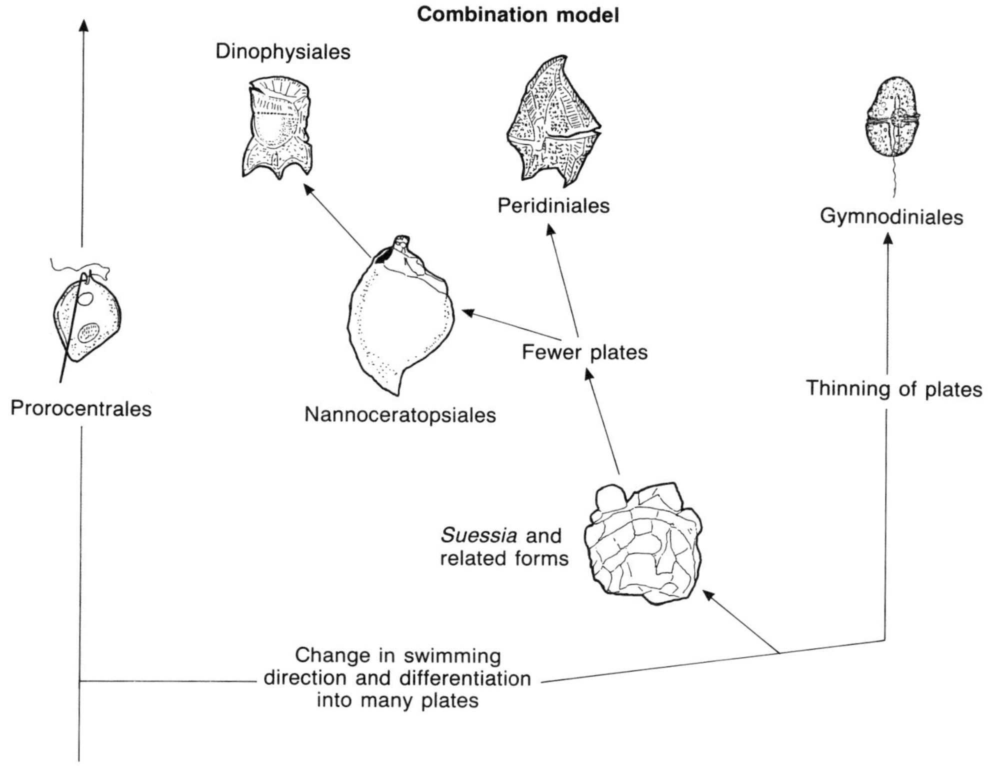

Figure 7.20 Combination model of dinoflagellate evolution. (As proposed by Bujak JP, Williams GL. *Can J Bot* 1981; 59:2077–87.)

horn nearly vanishes; and in *Rhombodinium* and *Gochtodinium,* the operculum is attached along the apical margin of the archeopyle. This and other proposed lineages will undoubtedly focus future studies to refine or refute the suggested phylogenetic relationships. On a smaller scale, possible evolutionary scenarios within one or a few cyst-based genera have been advanced. For example, species of *Areosphaeridium* (Fig. 7.18 no.26) probably evolved from *Eatonicysta* (Fig. 7.18 no.25) during Eocene time.

Acknowledgments

I thank Gregory Gaines for helpful comments, Nairn Albert, John Dodge, Robin Helby, Jane Lewis, and Lewis Stover for beautiful pictures, and the American Association of Stratigraphic Palynologists Foundation for permission to reproduce illustrations.

Supplementary reading

Dale B. Dinoflagellate resting cysts: "benthic plankton." In: Fryxell GA, ed. *Survival Strategies of the Algae*. Cambridge: Cambridge University Press, 1983: 69–136. This comprehensive introduction to the cysts of living dinoflagellates presents a benthic view of dinoflagellate ecology.

Dodge JD. *Atlas of Dinoflagellates*. London: Farrand Press, 1985. This slim volume contains exquisite scanning electron micrographs of modern dinoflagellates—mostly thecae, some cysts.

Evitt WR. *Sporopollenin Dinoflagellate Cysts. Their Morphology and Interpretation*. Austin: American Association of Stratigraphic Palynologists Foundation, 1985. Evitt presents the insights of two decades of short courses on fossil dinoflagellate cysts; his emphasis is clearly on morphology.

Harland R. Dinoflagellates, their cysts and Quaternary stratigraphy. *New Phytol* 1988; 108:111–20. Harland illustrates the use of dinocysts in detecting climate changes in the Quaternary marine record.

Noe-Nygaard N, Surlyk F, Piasecki S. Bivalve mass mortality caused by toxic dinoflagellate blooms in a Berriasian-Valanginian lagoon, Bornholm, Denmark. *Palaios* 1987; 2:263–73. The authors present well-documented studies of three massive bivalve kills in the fossil record, including one kill that is most probably due to toxic red tide.

Sarjeant WAS. *Fossil and Living Dinoflagellates*. London: Academic Press, 1974. Written by a paleontologist, this book was the first to bring together information on the living organism and on the fossils.

Spector DL, ed. *Dinoflagellates*. London: Academic Press, 1984. Sixteen authors describe various aspects of dinoflagellates—emphasis

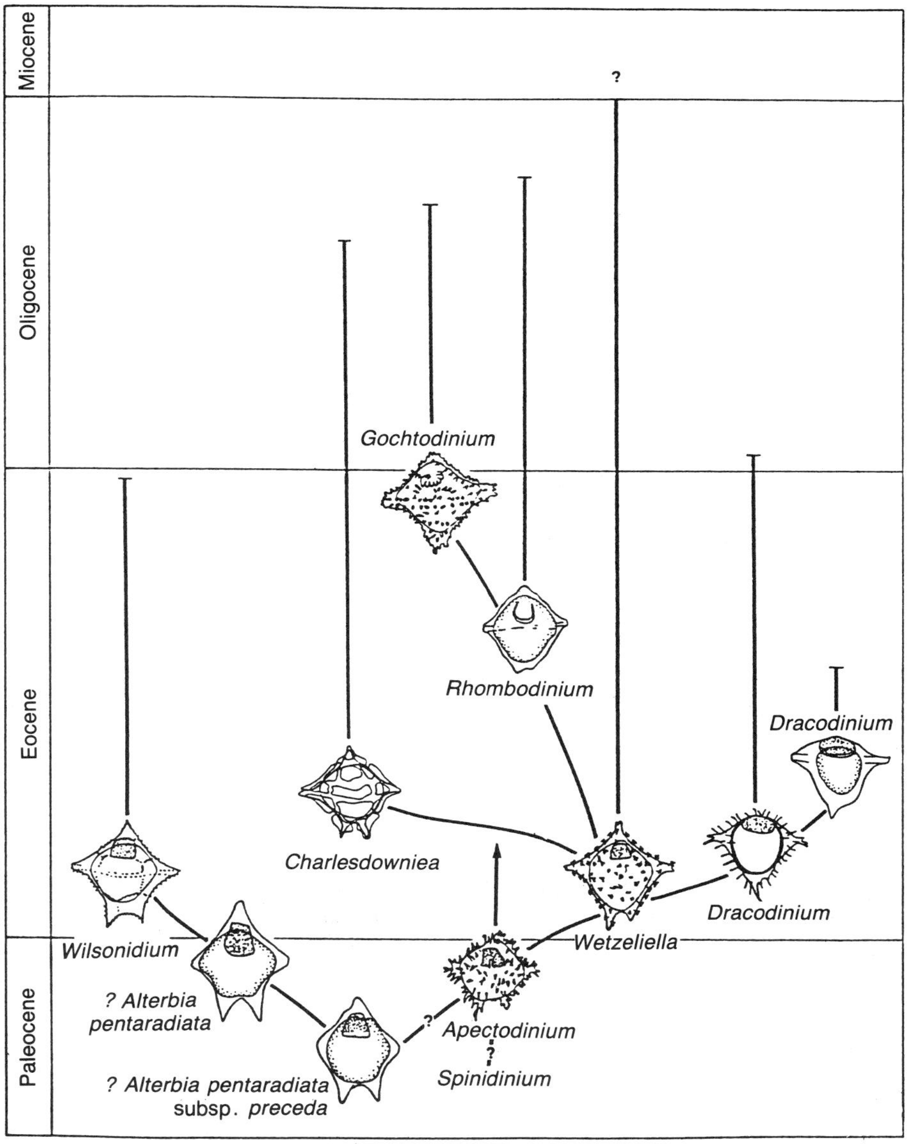

Figure 7.21 Possible phylogenetic relationships in the Wetzelielloideae. (Reproduced from Bujak JP, Davies EH. *Amer Assoc Stratigraphic Palynologists Contrib Ser* 1983; 13:73.)

is on the biology of living dinoflagellates, one chapter on dinocysts and one on evolution.

Stover LE, Evitt WR. Analyses of pre-Pleistocene organic-walled dinoflagellates. *Stanford Univ Public Geol Sci* 1978; 15:1–300.

Stover LE, Williams GL. Analyses of Mesozoic and Cenozoic organic-walled dinoflagellates 1977–1985. *Amer Assoc Stratigraphic Palynologists Contrib Ser* 1987; 18:1–243. These two works contain descriptions of all valid fossil organic-walled dinoflagellates genera published to date in a consistent format that makes comparisons easy.

Tappan H. *The Paleobiology of Plant Protists.* San Francisco: WH Freeman, 1980. Chapter 4 is a massive synthesis of knowledge of the biology and paleobiology of dinoflagellates.

Taylor FJR, ed. *The Biology of Dinoflagellates. Botanical Monographs, 29.* Oxford: Blackwell Scientific Publications, 1987. Eighteen authors detail the biology of dinoflagellates—a comprehensive chapter on fossil dinoflagellates by Dave Goodman.

Williams GL. Dinocysts. Their classification, biostratigraphy and palaeoecology. In: Ramsay ATS, ed. *Oceanic Micropalaeontology 2.* London: Academic Press, 1977: 1231–1325. This book chapter is a major synthesis from the paleontologic perspective and an attempt to propose a worldwide dinocyst zonation.

Williams GL, Bujak JP. Mesozoic and Cenozoic dinoflagellates. In: Bolli HM, Saunders JB, Perch-Nielson K, eds. *Plankton Stratigraphy.* Cambridge: Cambridge University Press, 1985: 847–964. This chapter summarizes dinoflagellate biostratigraphy. A range chart (accompanied by 23 plates of pictures) plots range against numerical age for nearly 400 species.

Williams GL, Sarjeant WAS, Kidson EJ. A glossary of the terminology applied to dinoflagellate amphiesmae and cysts and acritarchs: 1978 edition. *Amer Assoc Stratigraphic Palynologists Contrib Ser* 1978; 2A:1–121.

ENDOSKELETAL DINOFLAGELLATES

John J. Ernissee, Kevin McCartney

The endoskeletal dinoflagellates are an obscure and poorly understood group of dinoflagellates which produce a solid internal skeleton. Of the several groups known in the modern plankton, only one, containing *Actiniscus*, has a significant fossil record. They have received little attention among specialists in the fossil organic-walled plankton because the siliceous spicules dissolve in the hydrofluoric acid used in sample preparation. Instead, these fossils are commonly described and illustrated by those working on the silicoflagellates and ebridians, with which they are associated.

Actiniscus and similar genera are characterized by solid siliceous spicules consisting of bars or "rays" with triradial, cross, or T-shaped cross-section and numerous alveoli, reticulations, or spines upon them. They are, however, much simpler in overall morphology than are the ebridians, with which they have been classified. They also do not have windows or connecting elements. The most abundant species, *Actiniscus pentasterias*, possesses a strongly concave, pentaradial skeletal pattern (Figs 7.22, 7.23E), and varies considerably in the degree of silicification. Other species of *Actiniscus* have either radial or bilateral symmetry, while most of the other genera tend to be bilateral.

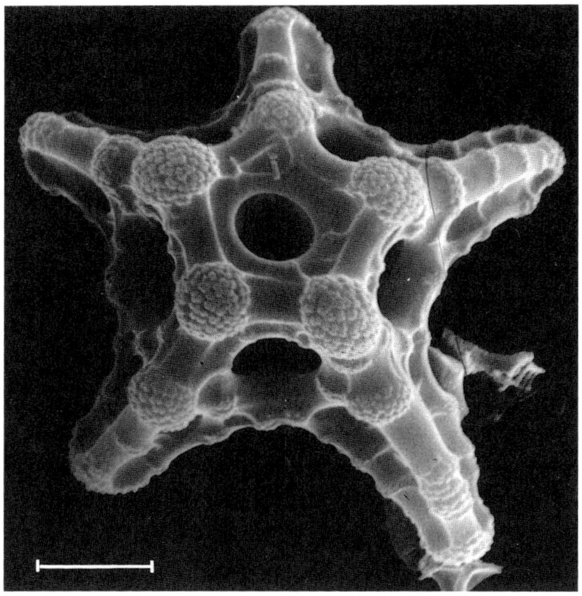

Figure 7.22 *Actiniscus pentasterias* from the subarctic Pacific Ocean. Note the enlarged, rough nodes surrounding the central knob which indicates hypersilicification. Compare with a more normal specimen in Figure 7.23E and with the same development on *Foliactiniscus folia* in Figure 7.23B. The causes of variations in degree of silicification have not been established. Scale bar, 10 μm. (Photograph courtesy of Dr Kozo Takahashi, Woods Hole Oceanographic Institution.)

Actiniscus was first described by Ehrenberg in 1840 and is the most widely distributed genus in both the modern plankton and in the fossil record. Living *Actiniscus* has been described from most of the world's oceans as well as several freshwater localities in Canada, where it may be a relict occurrence. Concentrations of *Actiniscus* range from less than 15 cells per liter upwards to more than 100, demonstrating its relative sparsity. The genus shows a distinct seasonality in the subarctic Pacific Ocean.

The second genus to be described was *Cardufolia* (Fig. 7.23C, D), which Hovasse described in 1932, based on material from fossil materials from Karand (Romania) and Oamaru (New Zealand). No additional actiniscian genera were recognized until 1973, when Dumitrica described two genera, *Foliactiniscus* (Fig. 7.23A, B) and *Cinctactiniscus*, in deep-sea sediments from the southwestern Pacific Ocean that he believed were clearly endoskeletal dinoflagellates, and another, *Calicipedinium* (Fig. 7.23G) from the Radiolarian Schist of Romania, whose affinities were uncertain.

Fossil *Actiniscus* are widespread in Miocene diatomites, and range from Oligocene through the present. With few exceptions, most *Actiniscus* species are relatively long ranging and thus are not biostratigraphically useful. However, since the group has received so little biostratigraphic or paleoecologic attention, their ultimate utility cannot be determined. Clearly, much remains to be learned about this enigmatic group.

Supplementary reading

Dumitrica P. Cenozoic endoskeletal dinoflagellates in southwestern Pacific sediments cored during Leg 21 of the DSDP. In: Burns RE, Andrews JE, van der Lingen G, *et al.* eds. *Initial Reports of the Deep Sea Drilling Project*. Vol. 21. Washington: US Government Printing Office, 1973: 819–35. The indispensible reference to endoskeletal dinoflagellates, with excellent drawings of the new genera and species Dumitrica describes.

Perch-Nielsen K. Late Cretaceous to Pleistocene archaeomonads, ebridians, endoskeletal dinoflagellates, and other siliceous microfossils from the subantarctic southwest Pacific, DSDP, Leg 29. In: Kennett JP, Houtz RE, Andrews P, *et al.* eds. *Initial Reports of the Deep Sea Drilling Project*. Vol. 29. Washington: US Government Printing Office, 1975: 873–907. A well-illustrated paper and an expansion of the results from Dumitrica's Leg 21.

Perch-Nielsen K. Eocene to Pliocene archaeomonads, ebridians, and endoskeletal dinoflagellates from the Norwegian Sea, DSDP Leg

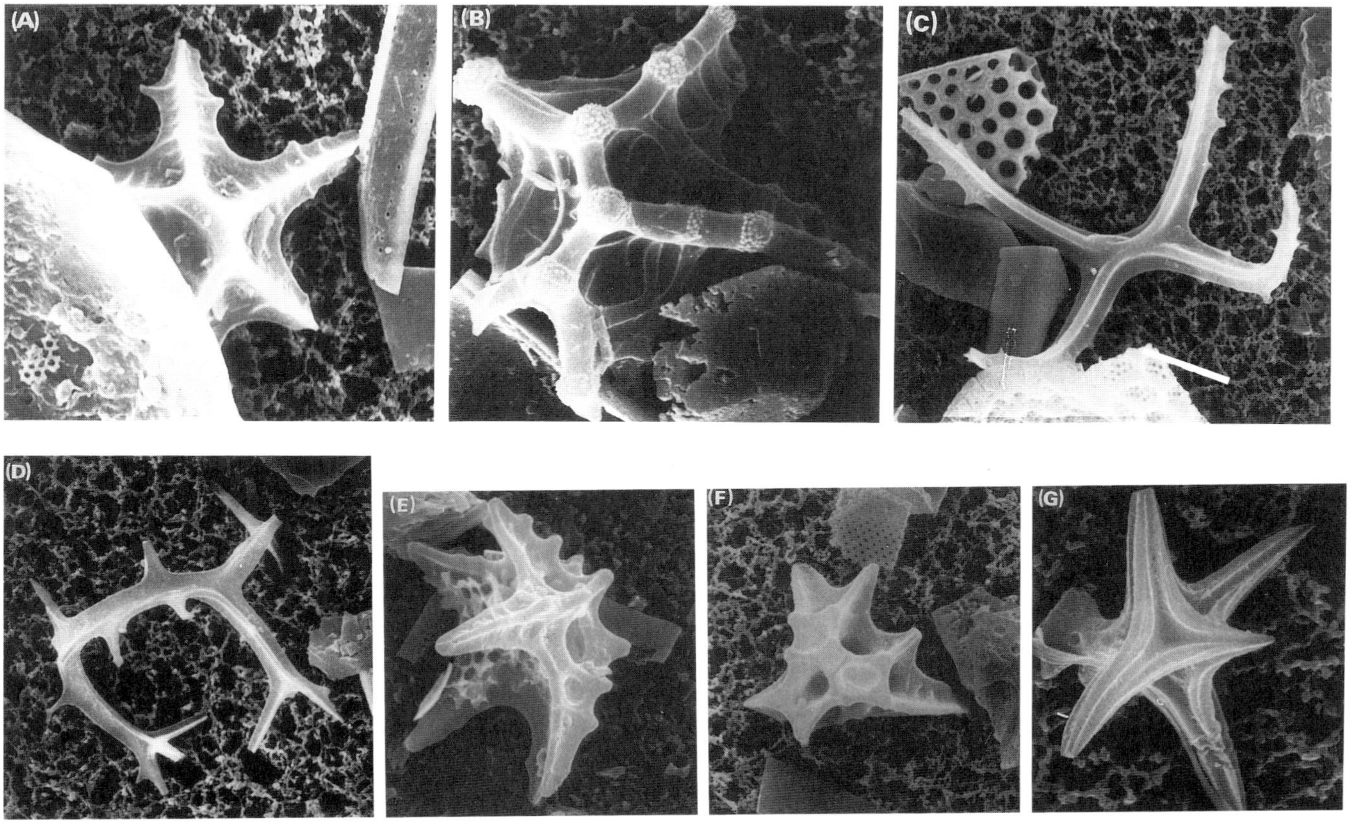

Figure 7.23 Illustrations of some common fossil endoskeletal dinoflagellates. A, *Foliactiniscus folia*. ×1900. B, *F. folia*. Note the callosities at the nodes indicating hypersilicification. ×3040. C, *Cardufolia gracilis*. ×1292. D, *C. onopordoides*. ×1650. E, *Actiniscus pentasterias*. ×1650. Compare with the hypersilicified form in Figure 7.22. F, *A. atlanticus*. ×1650. G, *Calicipedinium* cf. *C. hexastylus*. Miocene Pungo River Formation. ×3960.

38. In: Talwani M, Udintsev G, Bgorklund K, *et al.* eds. *Initial Reports of the Deep Sea Drilling Project*. Vol. 38. Washington: US Government Printing Office, 1976: 147–75. A valuable companion to Perch-Nielsen's Leg 29 paper and better illustrated.

Tappan H. *The Paleobiology of the Plant Protists*. San Francisco: WH Freeman, 1980. The biology of the endoskeletal dinoflagellates is discussed and many forms are illustrated on pp 250–6 of Chapter 4.

Ebridians

8

John J. Ernissee, Kevin McCartney

Introduction

The ebridians are a somewhat obscure group of small planktic marine organisms with two flagella and a siliceous skeleton made of solid elements. They occur in all oceans throughout most of the Tertiary, but have been in decline since the Miocene, with only three living species known.

History of study

The study of ebridians began in 1844 with Christian G. Ehrenberg's description of *Dicladia clathrata*. This and other ebridians were considered by Ehrenberg and his contemporaries as diatoms or radiolarians. However, after an initial flurry of activity by nineteenth-century micropaleontologists, little ebridian work was done until the 1930s. At that time, largely because of the work of Georges Deflandre and Raymond Hovasse, new species descriptions proliferated for about two decades. Renewed interest in ebridians came about as the result of deep-sea drilling by the Glomar Challenger. This work greatly improved our understanding of ebridian distribution and variability. Still, fundamental questions of taxonomy, basic biology, paleobiogeography, paleoecology, and skeletal variation remain, making the study of ebridians a rich area for future research.

The general lack of scientific interest in ebridians, either by biologists or paleontologists, is partly due to their general scarcity. They seldom constitute more than a few percent of the siliceous plankton and when they are abundant, they are not diverse. Moreover, they tend to be rather small, usually less than 25 μm, and can easily be overlooked during preliminary analyses. Although a thorough compilation of the descriptions of the genera and species was published by Loeblich and colleagues in 1968, few of the descriptions were in English, which has been an additional obstacle to their study. The early papers contained mostly line drawings lacking detail or failing adequately to show the three-dimensional character of the ebridian skeleton. Finally, the complicated skeleton of the ebridians has been burdened by equally complicated terminology (Fig. 8.1) — enough to dissuade scientists from routinely including them when working with other groups.

Skeletal morphology

The ebridian skeleton is similar in both size and general construction to those of silicoflagellates. The skeleton is composed of bars of opaline silica that are generally 3–4 μm in diameter and have a triradial, cross or T-shaped cross-section. Ebridians differ from the silicoflagellates in lacking the basal ring and hollow elements. They also do not have the hemispherical construction found in silicoflagellates and seldom have a radially symmetric series of spines. The ebridian skeleton usually has internal elements which pass through the interior of the skeleton, whereas silicoflagellates do not. The possession of internal skeletal members is similar to radiolarians, but ebridian skeletons are much smaller and generally much simpler.

The central or interior ebridian skeleton has a tetraxial (e.g., *Hermesinum*) or triradial (e.g., *Ammodochium*) symmetry with the skeletal elements intersecting at angles close to 120°. The skeletal elements form a simple framework with large external openings which, as with the silicoflagellates, are referred to as windows. With some ebridians, such as *Ebriopsis*, the windows are often of equal size, but in others, the window size may vary considerably. The exterior elements are generally bowed outwards and have a reticulate surface, while interior elements are usually straighter and have a smooth surface. As with the silicoflagellates, the general form and surface ornamentation show remarkable variability. Skeletal elements are either robust or fragile in appearance. Spines may be found on some ebridians.

The overall form of the skeleton is commonly circular, elliptical, or rectangular in outline (Figs 8.2, 8.3). Commonly, one end of the skeleton is somewhat enlarged,

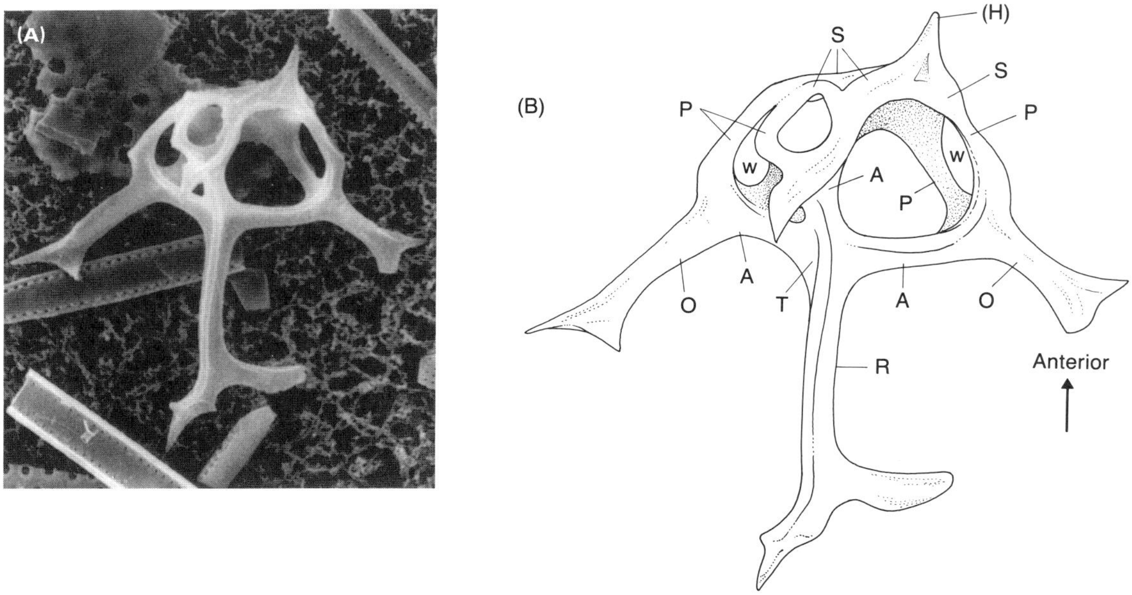

Figure 8.1 Skeletal morphology of a typical ebridian, *Hermesinum*. A, Scanning electron micrograph. ×1200. B, Labeled drawing. The initial spicule (T) may be tetraxial, as in this *Hermesinum* sp., or a flattened triradial one as in *Ammodochium*. Because some of the rear elements are obscured in this orientation, only the upper ones have been labeled. The anterior horn (H) is formed on the anterior ring produced by the synclades (S). R, rhabde; A, actine; O, opisthoclade; P, proclade; S, synclade; W, window. Miocene, Pungo River Formation, North Carolina.

and it presumably contains the cell nucleus. This enlarged end is considered anterior when orienting the skeleton. In many species (e.g., *Ammodochium*), this distinction cannot be made easily and the designation of anterior or posterior is arbitrary, making a consistent system of skeletal description difficult.

The skeletal terminology for the ebridians was one proposed by Deflandre in 1934, and was based on that of sponge spicules. Bars that radiate from the center are called **actines**, while the lower or axial one, if present, is called a *rhabde*. Elements at the exterior of the skeleton, which generally connect adjacent actines, are called **clades**, which are subdivided into *proclades*, *opisthoclades*, and *synclades*. Proclades are directed towards the anterior end, and commonly are connected with synclades to form a nuclear ring which sits atop an open enclosure which, presumably, held the nucleus. Opisthoclades are directed posteriorly from the ends of the actines and may join the rhabde. In general, the opisthoclades tend to be less completely developed and may be reduced to a short spine at the end of an actine.

Ebridians may also have a **lorica**, which is an inflated, chamber-like structure which generally develops at the anterior end of the skeleton atop the nuclear ring (Fig. 8.4). In the case of *Micromarsupium*, the lorica encases the entire skeleton. Initially, the lorica has both a basal and apical opening, but the basal opening tends to close as the lorica develops. The lorica is not considered to be a cyst, since it forms exterior to the main skeletal elements of the ebridian; the apical opening is relatively large and the lorica can often be perforate. The surface of the lorica is often reticulate.

The loricate stage is generally rare; however, it tends to be more widely developed in early Tertiary forms, particularly in *Ebriopsis, Hovassebria, Podamphora, Pseudammodochium*, and especially on the double and triple skeletons of *Ammodochium*. The lorica has not been observed on living ebridians. A similar feature called a cyst occurs in *Ammodochium* and *Parathranium*, where the globular structure forms within the framework of the skeleton and has only a tiny pore. As with the lorica, the function of the cyst is not understood, although it has been suggested that either may act as protective resting stages during unfavorable conditions.

Biology and systematics

Ebridians have two flagella (Fig. 8.5) which are dissimilar in size and in the manner in which they beat the water. The flagella are located near a large nucleus; at least one of the flagella is anterior. The organism does not appear to have pseudopodia. The group's name (from the Latin *ebrius*, "drunken") derives from the irregular helical pattern which the organism makes as it swims. The ebridians

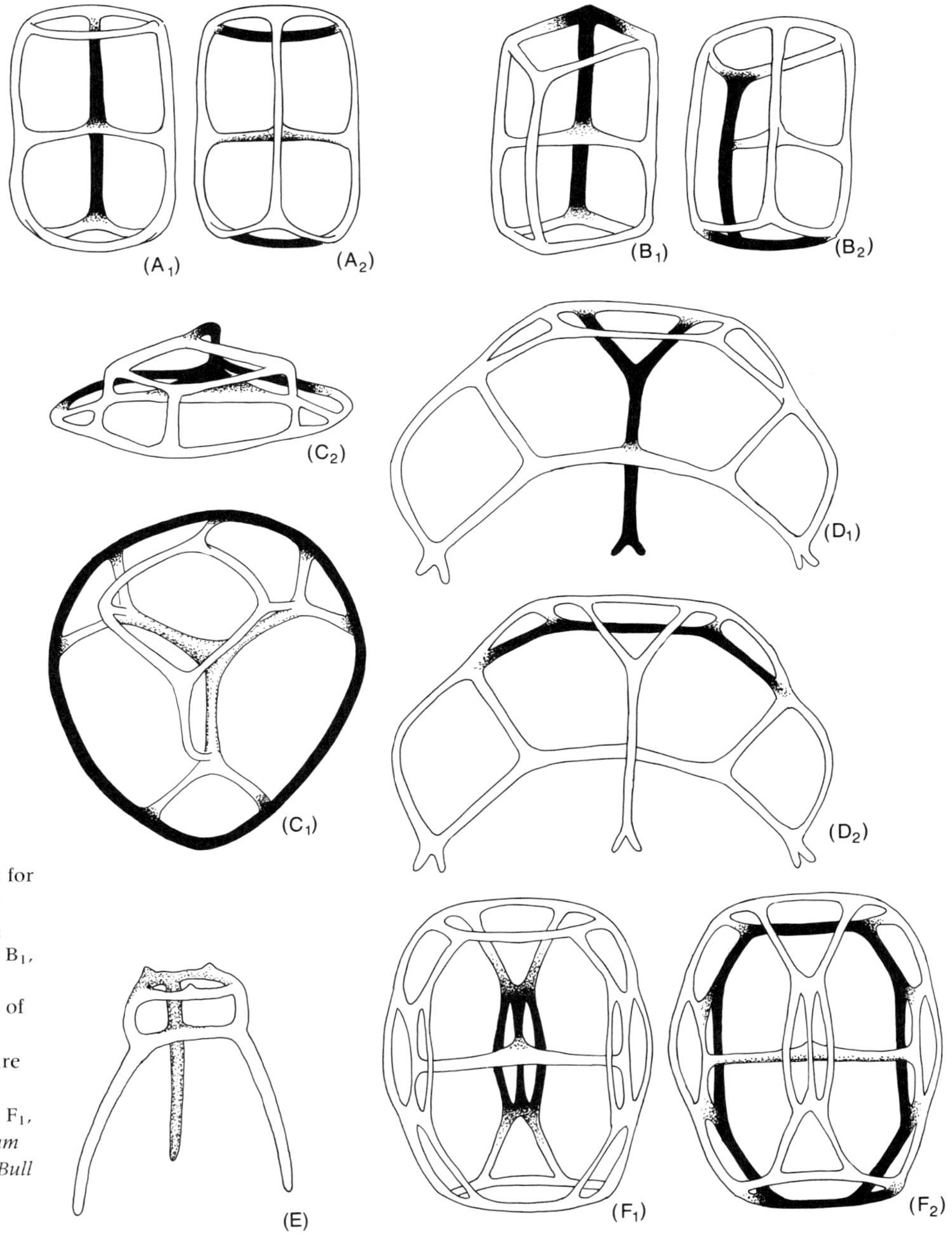

Figure 8.2 Schematic skeletal diagrams for some common ebridians which have an initial triradial spicule. A_1, Front and A_2 back views of *Ammodochium rectangulare*. B_1, Front and B_2, back views of *Ebrinula paradoxa*. C_1, Top and C_2, oblique views of *Ebria tripartita*. (Compare with the less complete skeleton of *E. tripartita* in Figure 8.7.) D_1, Front and D_2, back views of *Ditripodium fenestratum*. E, *Parathranium*. F_1, Front and F_2, back views of *Ammodochium speciosum*. (Modified from Deflandre G. *Bull Biol France Belgique* 1951; 85:1–84. E Courtesy of P. Dumitrica.)

are heterotrophic, with diatoms being their major source of food.

Little is known of the reproductive cycle of living ebridians. They are believed to undergo simple **fission**. With fossil forms, however, the skeletal morphology has suggested to some workers that sexual reproduction also occurs in some Tertiary species. During asexual reproduction, a daughter skeleton slowly develops within the parental cytoplasm. Nuclear division normally occurs when the skeleton is fully developed, followed by fission, with each daughter cell receiving half the cell contents and one of the skeletons. If the daughter and parent skeletons fuse during their development, the cell division process is apparently inhibited, resulting in double and even triple skeletons. This is relatively common in *Ammodochium* and *Parathranium* (Fig. 8.6), but is rare or unknown in other genera.

Ebridians are predominantly neritic, but can also be found in estuaries and embayments, near river mouths, and in inland saline seas, such as the Black Sea and the Salton Sea of California. Temperature tolerances are wide, although most prefer temperate to cool waters, as do

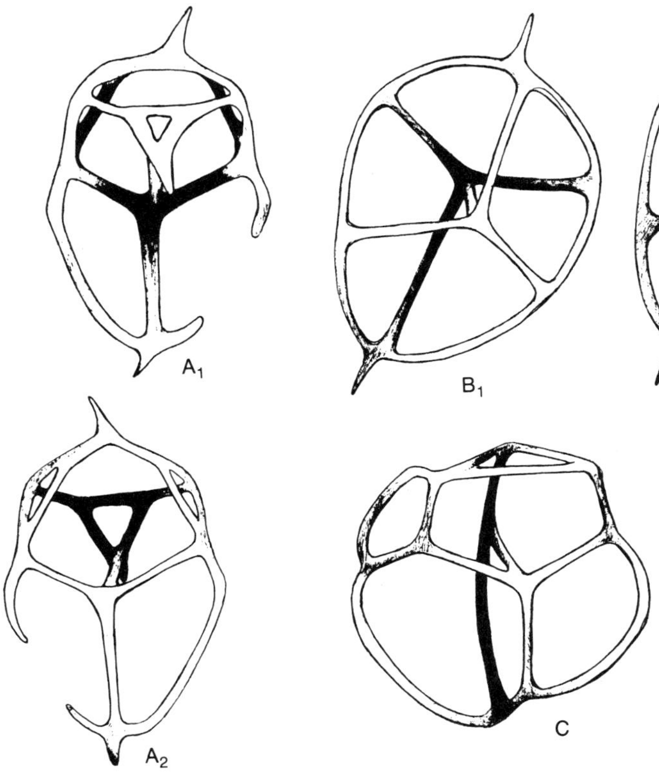

Figure 8.3 Schematic skeletal diagrams for some common ebridians which have an initial tetraxial spicule. A_1, Front and A_2, back views of *Hermesinum adriaticum*. B_1, Front and B_2, back views of *Ebriopsis antiqua*. C, *Podamphora tenuis*. Contrast the "idealized" form of *H. adriaticum* here with a typical fossil form in Figure 8.8E. (Modified from Deflandre G. *Bull Biol France Belgique* 1951; 85:1–84.)

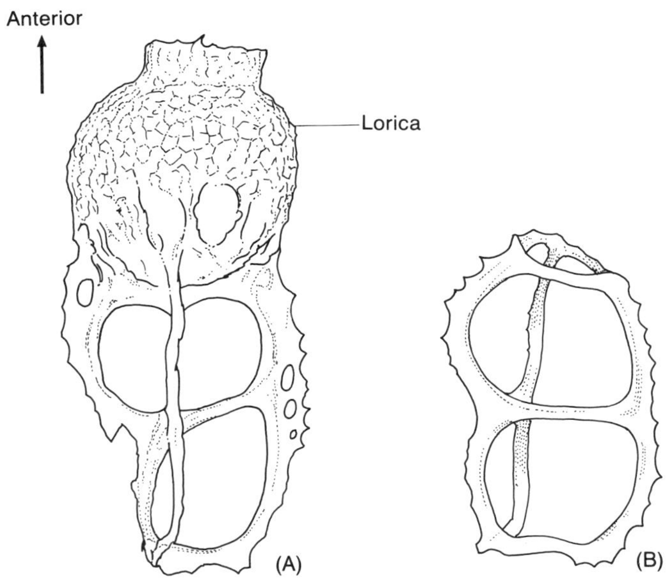

Figure 8.4 *Podamphora elgeri* with A, and without B, a lorica. The opening of the lorica is anterior. (Reproduced from Deflandre G. *Actualitiés Scient. et Ind.*, Paris: Hermann & Cie., 1936: no. 335: 1–95.)

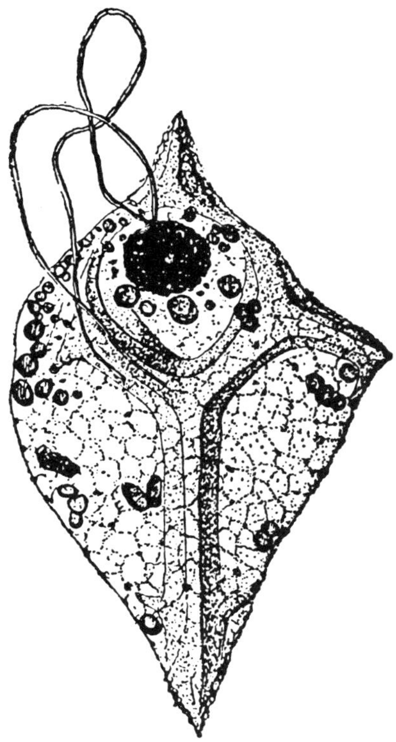

Figure 8.5 A sketch of a living *Hermesinum adriaticum* showing the two unequal flagella attached to the nucleus, which is, typically, located in the anterior portion of the skeleton which seems to surround and enclose it. (Reproduced from Hovasse R. *Acad Sci C R* 1934; 198:402–4.)

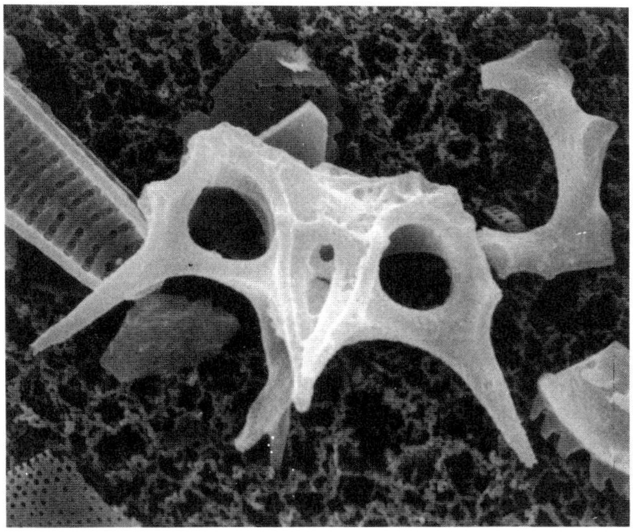

Figure 8.6 Scanning electron micrograph of a double skeleton of *Parathranium intermedium*. Miocene, Pungo River Formation, North Carolina. ×202.

siliceous plankton. *Ebria* has an optimum temperature of about 10°C, and is particularly widespread, while *Hermesinum* peaks at higher temperatures, roughly 20–25°C, which is reflected in its somewhat latitudinally restricted range.

While ebridians generally make up a very minor portion of the living plankton, blooms occur in both *Ebria* and *Hermesinum*. In summer, *E. tripartita* can reach more than 49 000 cells/liter in Long Island Sound and *H. adriaticum* constituted more than 43% of the total phytoplankton during one bloom at the mouth of the Nile River. Such occurrences, however, are sporadic. While concentrations of up to 235 000 cells/liter occurred off Rhode Island in August of 1970, none at all were found on the same date a year later. Living ebridians may, therefore, be considered opportunistic species which undergo population explosions when environmental conditions are appropriate. The nature of these ecologic controls is not at all understood and represents an area which needs further study.

The classification of the ebridians has been uncertain, since they have been considered both protozoans (subphylum Sarcomastigophora) and phytoplankton (phylum Chrysophyta or Pyrrophyta). A protozoan interpretation is understandable, since the organism is free-swimming and nonautotrophic, and the siliceous skeleton is similar to that in some chrysophytes (Fig. 8.7). However, the ebridians share several important traits with the dinoflagellates: two flagella of different lengths, a similar pattern of motion, a dinokaryon nucleus, and the colorless, yellow, or rose-colored cytoplasm typical of the heterotrophic dinoflagellates. The solid skeleton and mode of formation of the elements during mitosis is similar to that of *Actiniscus* (an endoskeletal dinoflagellate). The skeletons of the daughter cells begin as miniatures of the adult skeleton and reach full size and complete form upon maturity. These similarities led Hovasse to place the ebridians in the suborder Ebriida within the dinoflagellates. Deflandre believed that the dinoflagellates and ebridians belonged to separate classes, although he did not erect a class name for them. The most recent classification system and the one used here is from Tappan, reproduced in simplified and slightly modified form in Table 8.1.

Geologic occurrence and paleobiogeography

Many of the early papers described fossil species from poorly documented material, with the result that both the locations and ages of many of these species are in doubt. However, broad trends in generic diversity and

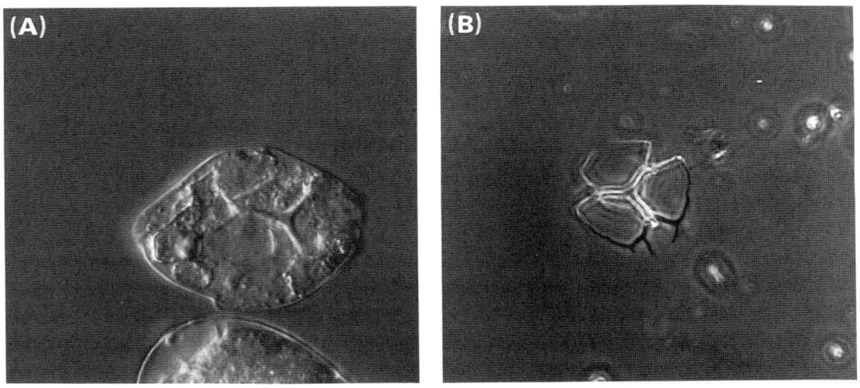

Figure 8.7 A living cell of *Ebria tripartita* (A), and a skeleton (B), both in phase contrast. Compare the frail-looking and incomplete skeleton with the "idealized" schematic in Figure 8.2C_1–C_2 (Photographs courtesy of P. Hargraves, Graduate School of Oceanography, University of Rhode Island.) ×459.2.

geologic ranges do show up in the record. Pre-Cenozoic ebridians are unknown, yet in the Paleocene, eight genera are recorded, representing three of the four families, hinting at extensive but as yet undiscovered Mesozoic development. From the Eocene through Miocene, they reach their peak diversity. At the end of the Miocene, they decline to only six genera, three of which survive to the present (Fig. 8.10).

The paleobiogeography of ebridians is poorly known, partly due, no doubt, to an inconsistent taxonomy. A mutual exclusivity in the paleogeographic distribution of *Parathranium tenuipes* and *P. intermedium* was suggested, but when their distributions are plotted (Fig. 8.11), including more recent information, no clear pattern emerges.

Biostratigraphy

The most useful fossils for biostratigraphic correlation have widespread occurrences over short time-intervals. Ebridians are not among them because:

1 many of the ebridian taxa are rather long ranging, which makes recognition of high-precision biostratigraphic zones very difficult;
2 their occurrence is both very sporadic and inconsistent;
3 their skeletal morphology is highly variable so that subdivision into species is difficult.

Another major problem plaguing ebridian study is that the early work emphasized naming and describing new species, with little effort to record the precise stratigraphic horizons from which the taxa were recovered. As a result, only the broadest time assignments are possible and some of these are geographically imprecise. The situation is improving as a result of the **Deep Sea Drilling Project** (DSDP) and the **Ocean Drilling Program** (ODP), but many more sections need to be examined before anything but the most general age assignments based on ebridians are possible. Moreover, reexamination of many of the type localities will be necessary to eliminate the uncertainties mentioned above. Clearly, even old fields may yield new harvests in ebridian studies.

Few attempts have been made to utilize ebridians in biostratigraphic zones. For example, on DSDP Leg 19, two zones for the northern Pacific were established: the *Ebriopsis antiqua* (spineless form) Zone, from the last appearance of *Cannopilus hemisphaericus* (a silicoflagellate) to the last appearance of *E. antiqua*, and the *Ammodochium rectangulare* Zone, from the last appearance of *E. antiqua* to the last appearance of *A. rectangulare*. These zones occur in the late Early Pliocene and the Late Pliocene, respectively. These zones were subsequently recognized

Table 8.1 Tappan's classification of ebridians. (From Tappan H. *The Paleobiology of the Plant Protists*. San Francisco: WH Freeman, 1980:480)

Class Ebriophyceae (Loeblich III 1970)

ORDER EBRIALES (Honigberg *et al.*, 1964)

Family Hermesinaceae (Fig. 8.8C, E–I)

Syn. Podamphoraceae, Hermesinidae, Craniopsidae, Ebriopsidae

Skeleton with triaene initial spicule, or atrophied rhabde resulting in triode. Rhabde axial to slightly excentric in position; proclades simple or bifurcating, connected by synclades to form nuclear ring; opisthoclades simple or lacking. Double and multiple skeletons and loricate stage known in fossils

Paleocene to Recent

Craniopsis, Ebriopsis, Haplohermesinum, Falsebria, Hermesinella, Hermesinum (syn. *Bosporella*), *Hovassebria, Micromarsupium, Parebriopsis, Podamphora* (syn. *Parammodochium, Parebria, Podium*), *Podamphoropsis, Spongebria.* (Note: *Pseudmicromarsupium* Busen and Wise 1976 may be a silicoflagellate; it is included here for completeness)

Family Ditripodiaceae (Fig. 8.8A, B, D)

Syn. subfamily Thraniinae, Tripodiidae (part), Ditripodiidae

Triode initial spicule; simple divergent proclades connected by synclades to form triangular or rounded nuclear ring; opisthoclades divergent, terminally free, pointed, or forked; mesoclades may occur. Double skeletons known; cyst known in *Parathranium*

Miocene to Pliocene

Ditripodium, Parathranium, Thranium, ?Dicladia (note: considerable controversy surrounds this taxon which some consider to be a synonym of *Parathranium*)

Family Ammodochiaceae (Fig. 8.9A, B)

Syn. subfamily Ammodochiinae, Tripodiidae (part), Ammodochiidae

Initial spicule triode or atrophied or absent; proclades and opisthoclades parallel to divergent, simple or bifurcate, connected by synclades to form nuclear and antapical rings, respectively; mesoclades may occur. Entire skeleton may be replaced by ovid perforated shell. Double skeletons, loricate stage, and cysts known

Paleocene to Pliocene

Ammodochium (syn. Tripodium), *Pseudammodochium*

Family Ebriaceae

Syn. Ebriidae, tribe Ebrioidae, Ebriacidae

Triode initial spicule; simple proclades perpendicular to triode, connected to rounded nuclear ring by anterior synclades; opisthoclades simple or bifurcating, connected by posterior synclades; opposite synclades may be connected by a rod

?Paleocene to Recent

Ebria (syn. *Ebriella, Proebria*), *Ebrinula*

Some genera that were considered to be ebridians have been placed elsewhere. *Cardufolia* is considered an actiniscid, while *Semantebria* and *Spyrebria* are probably radiolarians.

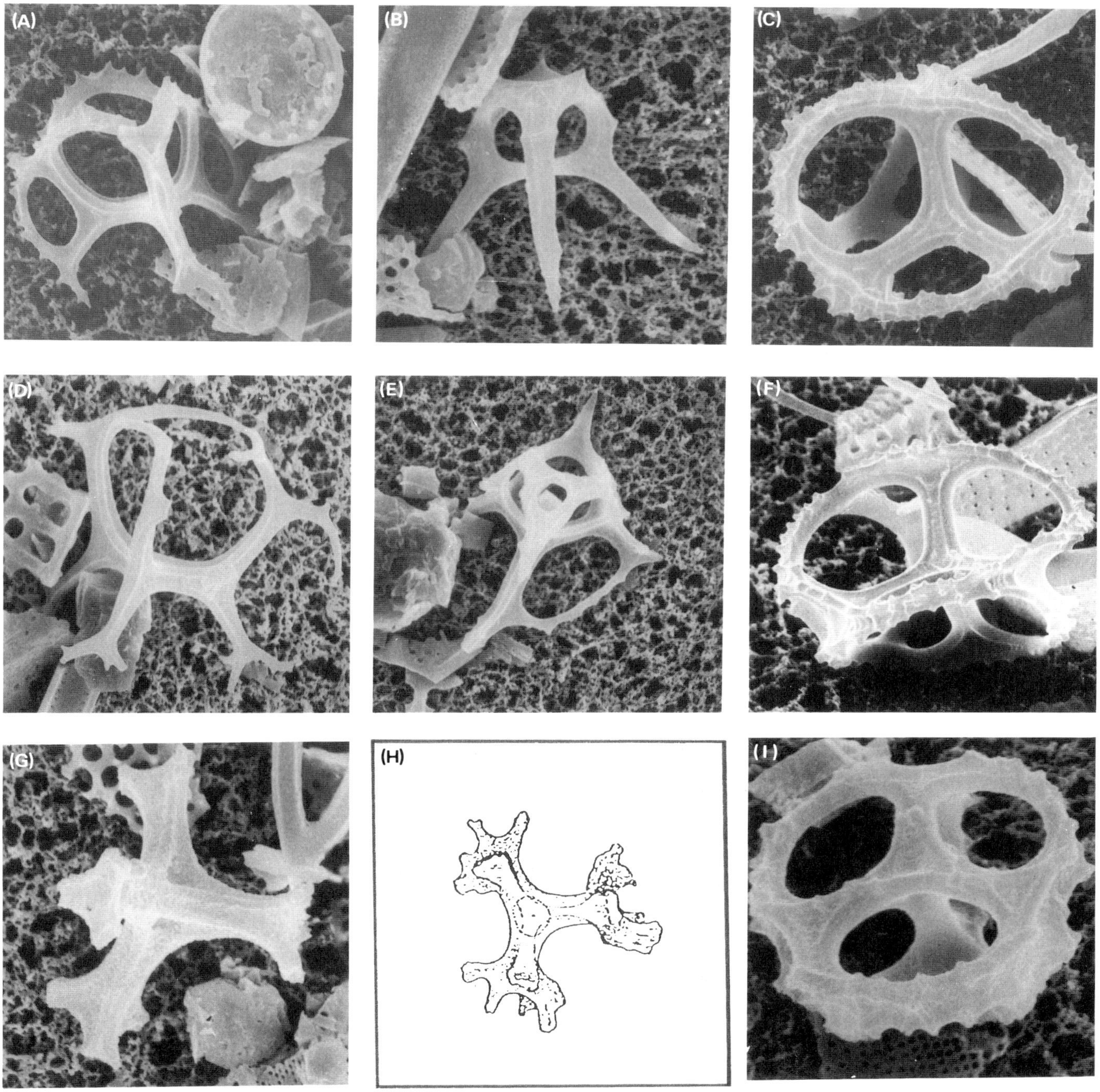

Figure 8.8 Illustrations of some common ebridians. A, *Ditripodium latum*. ×1800. B, *Parathranium intermedium*. ×1800. C, *Hermesinum schulzii*. ×2250. D, *Ditripodium latum* (note the outer element or mesoclade is frequently broken off, as in this specimen, giving the appearance of a *Thranium*). ×1800. E, *Hermesinum adriaticum*. ×1350. F, Another view of the specimen in C. The reticulated external surface of the skeleton is in strong contrast with the much smoother surfaces of the interior elements. G, *Falsebria* sp. This may be an incomplete or immature initial tetraxial spicule of another ebridian. However, the lack of intermediates between this form and any other more complete skeleton indicates that this is a legitimate species. ×2250. H, *Falsebria brachiolata*. I, *Haplohermesinum simplex*. ×2250. (All specimens from the Miocene Pungo River Formation, North Carolina. H, Reproduced from Deflandre G. *Bull Biol France Belgique* 1951; 85:1–85.)

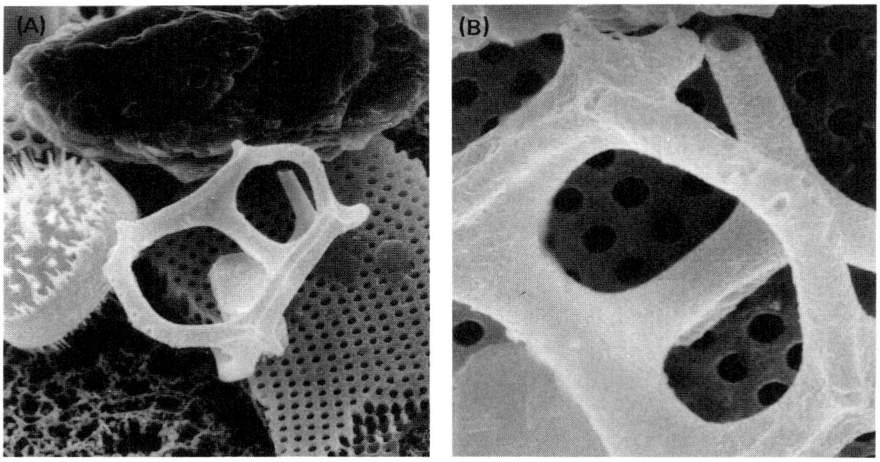

Figure 8.9 Two views of *Ammodochium pyramidale*. In B, the upper synclade is apparently hollow, indicating that ebridians do not universally have solid skeletal elements. A, ×1800; B, ×5400. Miocene Pungo River Formation, North Carolina.

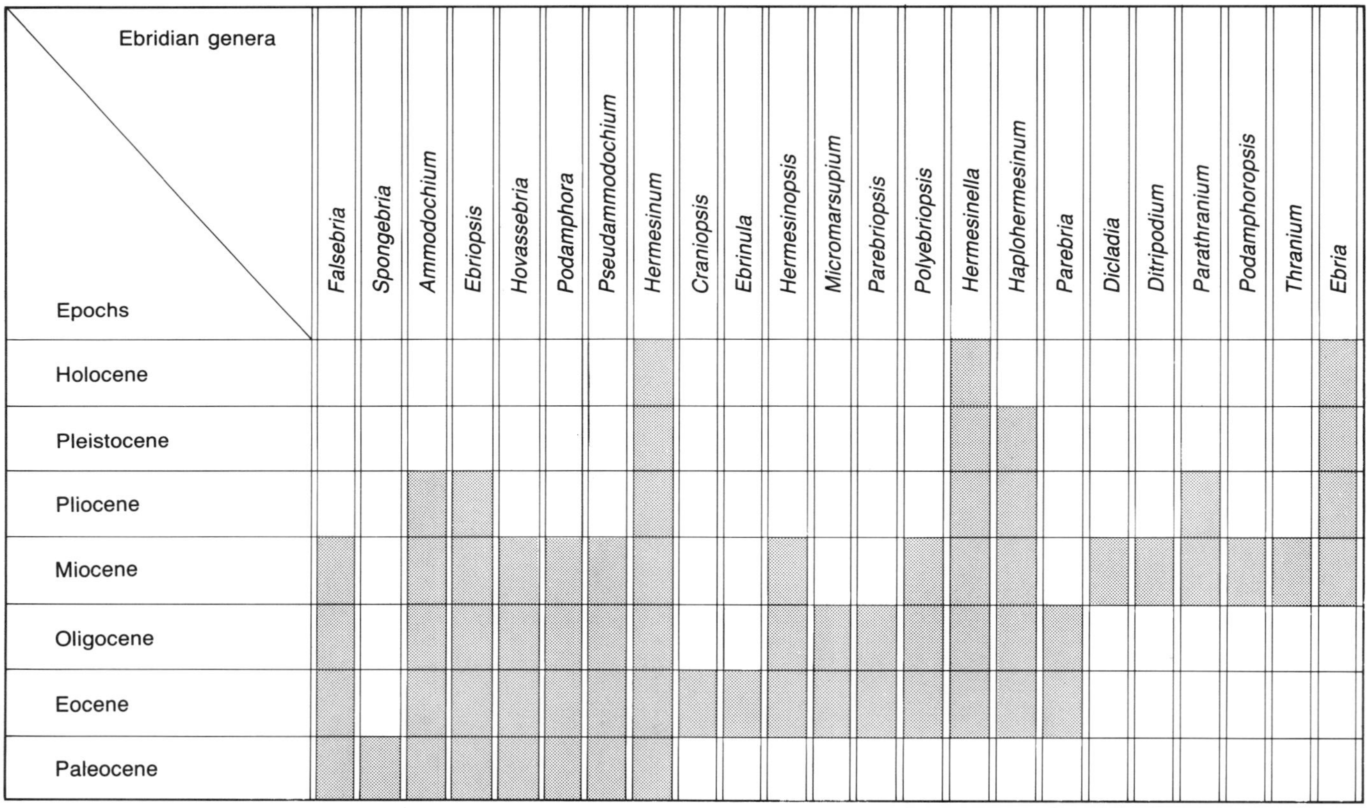

Figure 8.10 Geologic ranges of ebridian genera. (Modified from Loeblich AR III, Loeblich LA, Tappan H, Loeblich AR Jr. Memoir 106. Boulder: Geological Society of America, 1968:19.)

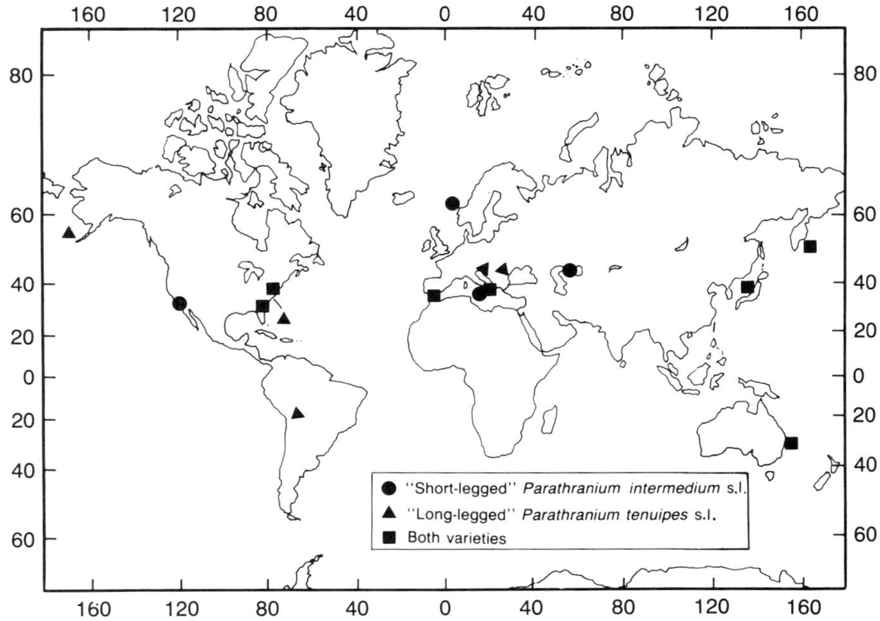

Figure 8.11 Geographic distribution of *Parathranium tenuipes* s.l. ("long-legged" species) and *P. intermedium* s.l. ("short-legged" species). No provinciality appears in the distribution of these species. The relatively small number of sites examined for ebridians can be gauged from this map.

in cores taken from the Sea of Japan on DSDP Leg 55 and on DSDP Leg 18, off Cape Mendocino, California. It would appear that these zones will be generally applicable to the northern Pacific, in a broad arch from northern California to the Sea of Japan.

A more ambitious scheme was proposed by Locker and Martini for the southwestern Pacific on DSDP Leg 90. It consists of five zones which use both actiniscids and ebridians (Fig. 8.12). Despite some taxonomic reassignments, the youngest two zones are essentially the same as those described from the North Pacific.

While these two zonal schemes may prove useful in the Pacific Ocean basin, they may not be useful in other regions. For example, *Thranium crassipes* has been reported from the Pungo River Formation of North Carolina, which has been assigned to the Middle Miocene on the basis of several other groups of microfossils. This is much earlier than the Upper Miocene assignment for the *T. crassipes* Zone in the Pacific. Obviously, caution is necessary in applying ebridian zones outside the ocean basin in which they were defined. Moreover, this illustrates the need for a concerted effort to establish ebridian zones in the Atlantic Ocean and elsewhere.

Epoch	Nannoplankton zones	Ebridian-actiniscidian zones
Upper Pliocene	NN16	?
Lower Pliocene	NN15	*Ebriopsis cornuta* zone
	NN14	
	NN12	*Hermesinum obliquum* zone
Upper Miocene	NN11B	
	NN11A	
		Thranium crassipes zone
	NN10	
Middle Miocene	NN9	*Hermesinella conata* zone
	NN8	
	NN7	
	NN6	
	NN5	
		Actiniscus elongatus zone
	NN4/5	?

Figure 8.12 The most extensive attempt thus far to utilize ebridians for biostratigraphic zonation. However, the *Thranium crassipes* Zone does not appear to be valid in the Miocene of North Carolina. (Modified from Locker S, Martini E. *Initial Reports of the Deep Sea Drilling Project, Leg 90*. Vol. 90, part 2. Washington: US Government Printing Office, 1986:939–51.)

Supplementary reading

Loeblich AR III, Loeblich LA, Tappan H, Loeblich AR Jr. *Annotated Index of Fossil and Recent Silicoflagellates and Ebridians, with Descriptions and Illustrations of Validly Proposed Taxa*. Boulder: Geological Society of America, 1968: 319. The essential ebridian reference for material published prior to 1967. All descriptions appear in their original languages.

Perch-Nielsen K. Late Cretaceous to Pleistocene archaeomonads, ebridians, endoskeletal dinoflagellates, and other siliceous microfossils from the subantarctic southwest Pacific, DSDP, Leg 29. In: Kennet JP, Houtz RE, *et al.*, eds. *Initial Reports of the Deep Sea Drilling Project*. Vol. 29. Washington: US Government Printing Office, 1975:873–907. A well-illustrated paper.

Perch-Nielsen K. Eocene to Pliocene archaeomonads, ebridians, and endoskeletal dinoflagellates from the Norwegian Sea, DSDP, Leg 38. In: Talwani M, Udintsev G, *et al.*, eds. *Initial Reports of the Deep Sea Drilling Project*. Vol. 38 (suppl.), Washington: US Government Printing Office, 1978:147–75. A valuable companion to Perch-Nielsen's Leg 29 paper and better illustrated.

Tappan H. *The Paleobiology of the Plant Protists*. San Francisco: WH Freeman, 1980:1028. Chapter 5 is devoted to the ebridians. This is the best general reference on the biology of the two groups and is well illustrated.

Chrysophytes

9

Jere H. Lipps, Kevin McCartney

OVERVIEW

Jere H. Lipps

Chrysophytes are algae grouped together chiefly because they possess anterior flagella at some stage in their life-cycle and silicified cysts, scales, or rods enclosing the cell. Only one group among the chrysophytes, the marine silicoflagellates, has a well-known fossil record. The remaining taxa in this group are reported infrequently as fossil cysts in marine rocks or in freshwater bog, pond, and lake deposits.

Living chrysophytes are mostly free-swimming single cells (Figs 9.1, 9.3) of quite small size (2.5–20 μm). A number of the species form colonies of cells, some of which are very complex. All species contain golden–yellow **plastids** and have a single nucleus. They possess a long and a short flagellum with mastigonemes at the anterior end of the cell. The cell wall is cellulosic and may form a **lorica** that holds the cell. The lorica may be mineralized with silica or calcium carbonate in some species. Commonly, the mineralized portion of the lorica is composed of tiny, complex scales. Silicoflagellates have a skeleton of hollow siliceous rods (Fig. 9.5), while many other species form silica cysts at some stage in their life-cycle. Silicoflagellate skeletons and cysts are commonly fossilized.

In the botanical classifications, the chrysophytes are placed in a single division, the Chrysophyta. Zoologic classifications separate many of the chrysophytes in different orders of phytoflagellates.

Another division, the Xanthophyta, contains fewer than 100 species; the group is poorly known at best. They resemble chrysophytes as some xanthophytes also produce siliceous scales but they have a very sparse fossil record in the Neogene.

CHRYSOPHYTE CYSTS

Jere H. Lipps

Although figured by Ehrenberg as early as 1854, fossil chrysophyte cysts, perhaps because of their small size, were first earnestly studied only in the 1920s and 1930s. G.D. Hanna in California and G. Deflandre in France described several genera from marine rocks, and J. Frenguelli described freshwater, as well as marine, taxa. In 1932, Deflandre proposed the enduring family Archaeomonadaceae for the marine forms. Later paleontologists, such as L. Rempi, E.J. Tynan, W.C. Cornell, M. Hajos, and K. Perch-Nielsen, described Cretaceous and Tertiary cysts chiefly from siliceous sedimentary rocks. D.P. Adam and A.D. Mahood pointed out the usefulness of these cysts as environmental indicators in fossil freshwater deposits.

Chrysophyte cysts are commonly abundant in marine and freshwater deposits, Antarctic sea ice, bogs, and damp places. In living forms, the cyst encloses the granular protoplast, the plastids, nucleus, and food reserves. Whether or not the encystment is an obligatory part of the life-cycle or a response to adverse environmental conditions is unknown. The marine cysts were originally assigned to the family Archaeomonadaceae but this is a false, although useful, division based solely on their marine occurrence. Fossil freshwater cysts are referred to the family Chrysostomataceae. Archaeomonads closely resemble cysts of freshwater chrysophytes but their relationships are uncertain, as the systematics of the chrysophytes is based on the living cell not the cyst.

Chrysophyte cysts are probably quite abundant but unobserved because of their small size (5–10, rarely 25, μm in diameter). They are most commonly hollow and spherical in shape although they vary from oblate to prolate or even may be lobed (Fig. 9.2). Each cyst has a single aperture that varies from a simple hole to rather

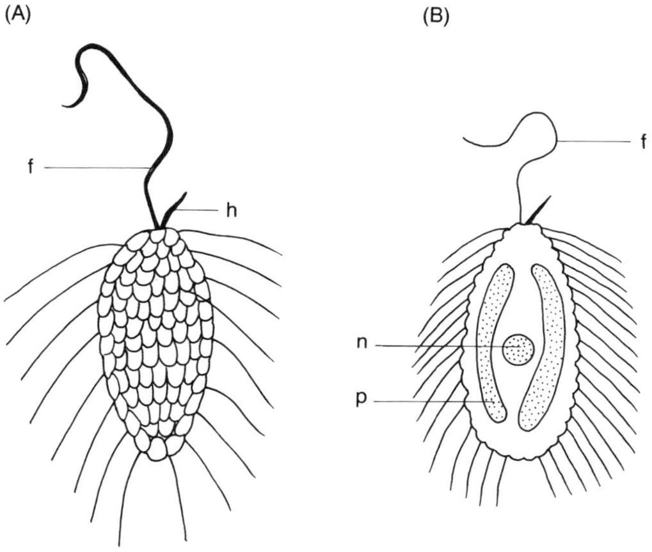

Figure 9.1 A living solitary chrysophyte, *Mallomonas* sp., showing the nucleus (n), plastids (p), haptonema (h), and flagellum (f). A, Cell showing external siliceous scales and spines. B, Cross-section of the cell. The cell is 50 μm in length.

complex types involving necks, expanded collars, raised areas, and flanges. The aperture is closed by an unsilicified organic plug. The cyst walls may be smooth, spinose, pitted, reticulate, ridged irregularly, or in parallel.

The fossil occurrences of chrysophyte cysts are revealing because they provide information on both marine and freshwater microphytoplankton. The fossil cysts occur in Cretaceous to Recent marine sedimentary rocks and in Quaternary freshwater deposits. In a few deposits, cysts are so abundant that the rocks have been called "archaeomonadites." In the siliceous Cretaceous Moreno Formation of California, archaeomonads number up to 11.8 million cysts/g rock.

Chrysophyte cysts appear to be more abundant in freshwater than in marine deposits. They occur in damp areas, including bogs, meadows, marshes, and swamps, where different cyst assemblages may occur over distances smaller than a meter. The cysts are a common component of lake sediments, particularly of high altitudes where freezing occurs. Similarly, modern marine cysts are abundant in fresh sea ice, although they form in the water column below the ice and are transported to and trapped in the ice. Freezing seems to be an important factor in cyst formation today, but in the Tertiary and Cretaceous, other factors must have induced cyst formation. Almost all fossil marine occurrences of cysts are in deposits rich in biogenic silica. The cysts may be associated with eutrophic conditions or with conditions favoring the preservation of silica on the sea floor during deposition.

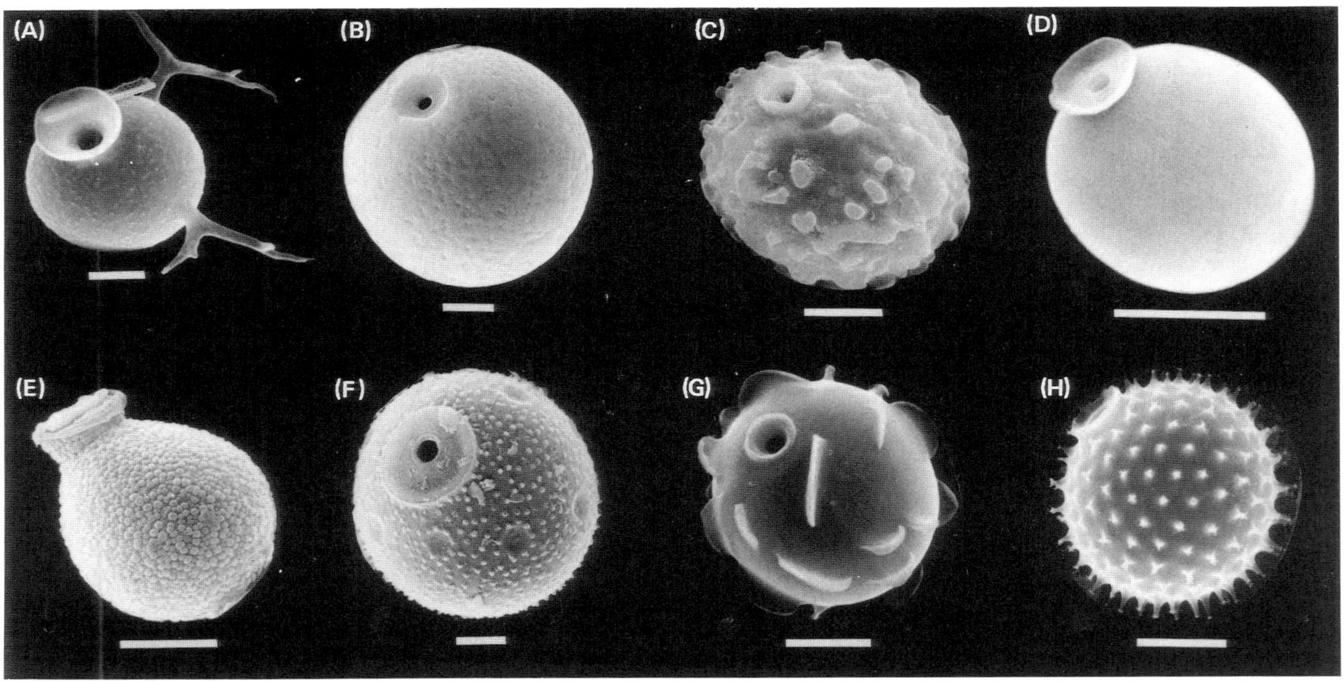

Figure 9.2 Scanning electron micrographs of chrysophyte cysts from freshwater environments in California and Washington State. A, G, H, from lake sediments; B, D, from a forested bog; C, from a wet meadow; E, from a spring; F, from a stagnant pond. Scale bars, 3 μm. (Reproduced from Adam DP, Mahood AD. *Geol Soc Am Bull* 1981; 92:839–44.)

Chrysophyte cysts are not very useful in biostratigraphic applications. The Late Cretaceous taxa differ from Tertiary ones, and the ranges of some species may be restricted to certain stratigraphic intervals. Eocene cysts are more diverse than those of the Oligocene, with diversity increasing in the Miocene but not to the Eocene levels. In addition to cysts, scales closely resembling those of modern chrysophytes are known from the latest Precambrian, a little more than 550 My in age. Possibly, chrysophytes or related algae were common but not preserved during the entire Phanerozoic.

Supplementary reading

Adam DP, Mahood AD. Chrysophyte cysts as potential environmental indicators. *Geol Soc Am Bull* 1981; 92:839–44. An excellent summary of the usefulness of terrestrial chrysophyte cysts in paleoenvironmental analyses.

Allison CW, Hilgert JW. Scale microfossils from the Early Cambrian of northwest Canada. *J Paleontol* 1986; 60:973–1015. Twenty-six unusual species of scale-bearing microfossils, some thought to be chrysophytes, are described and illustrated from rocks now known to be Late Proterozoic in age.

Cornell WC. Late Cretaceous chrysomonad cysts. *Palaeogeog Palaeoclimatol Palaeoecol* 1972; 12:33–47. Description of the occurrence of marine cysts.

Cronberg G. Chrysophycean cysts and scales in lake sediments. In: Kristiansen J, Andersen RA, eds. *Chrysophytes: Aspects and Problems*. Cambridge: Cambridge University Press, 1986:281–315. A complete review of chrysophytes, cysts, and scales found in fossil and modern deposits, especially in lakes.

Deflandre G. Archaeomonadaceae, une famille nouvelle de Protistes fossiles marins à loge siliceuse. *Comp Rend Acad Sci* 1932; 194: 1859–61. The original description of archaeomonads.

Mitchell JG, Silver MW. Archaeomonad (Chrysophyta) cysts: ecological and paleoecological significance. *BioSystems* 1986; 19:289–98. Description of archaeomonads in Antarctic sea ice and the means of their transportation to the sea floor. The authors propose that archaeomonad deposition has shifted from sediments underlying anoxic waters in the early Tertiary to those under sea ice in the late Cenozoic.

Tappan H. *The Paleobiology of Plant Protists*. San Francisco: WH Freeman, 1980:1028. A very complete and authoritative treatment of the biology and paleontology of xanthophytes and chrysophytes in Chapter 6.

Tynan EJ. Geologic occurrence of the archaeomonads. In: Farinacci A, ed. *Proceedings of the II Plankton Conference, Rome, 1970*. Vol. 2. Roma: Edizioni Tecnoscienza, 1971:1225–30. Summary of the stratigraphic occurrence of marine chrysophyte cysts.

SILICOFLAGELLATES

Kevin McCartney

Introduction

Silicoflagellates are planktic, microscopic algae with a single flagellum and a delicate, siliceous tubular skeleton. The flagellum is relatively long and anterior (Fig. 9.3). Silicoflagellates are both photosynthetic and heterotrophic. An additional nutrient source may be endosymbiotic blue–green algae, which have been seen with silicoflagellates in culture. They are generally considered among the Chrysophyceae or golden algae by botanists on the basis of pigmentation, mitochondrial and chloroplast structure, and siliceous skeleton. Protozoologists consider them to be an order in the Phytomastigophora.

Silicoflagellates first appear in the Early Cretaceous and are abundant in the Late Cretaceous. They have existed in all oceans since then, although with marked changes in their relative abundance and species diversity. They are exclusively marine and generally restricted to the euphotic zone. Although widely distributed, they occur in relatively low numbers, usually comprising only 1 or 2% of the siliceous component of marine sediments, and are consequently not as well studied as other microfossil groups such as the diatoms. Silicoflagellates are commonly found in areas of high diatom concentration, and fluctuations in abundances of diatoms are accompanied by similar changes in the abundance of silicoflagellates. Like the diatoms, silicoflagellates are most abundant in areas of upwelling and in equatorial waters but are also abundant in high latitudes.

Silicoflagellates are fairly useful in biostratigraphy, especially in higher latitudes and deeper waters where calcareous microfossils are less abundant or dissolve. Their biostratigraphic value, however, is somewhat limited because the silicoflagellate assemblages change considerably with latitude, and their morphologies appear to be influenced by environmental factors. These changes indicate that silicoflagellates may have considerable potential as environmental indicators.

History of study of silicoflagellates

Silicoflagellates were first described by the prolific C.G. Ehrenberg, who wrote some 20 papers on them among his work on many other microfossil groups. Ehrenberg first described silicoflagellates from Sicily and erected the

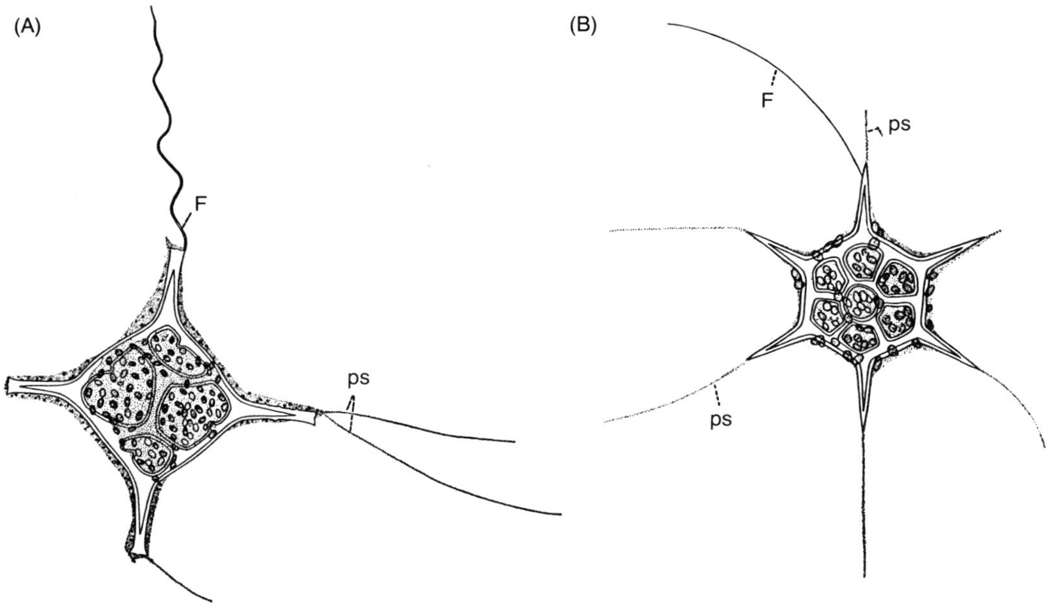

Figure 9.3 Living silicoflagellates showing the flagellum (F) and pseudopodia (ps). A, *Dictyocha fibula* with four basal sides. B, *Distephanus speculum* with six basal sides. The flagellum (F) is anterior (top) with respect to the direction of movement. (Reproduced from Marshall SM. *The Silicoflagellate and Tintinnoinea*. British Museum, Great Barrier Reef Expedition 1928–29, Science Report; modified by Loeblich AR III, Loeblich LA, Tappan H, Loeblich AR Jr. Memoir 106. Boulder: The Geological Society of America, 1969:4,6.)

genus *Dictyocha*. Most micropaleontologists of the nineteenth century, including Ehrenberg, did not recognize the silicoflagellates as a separate taxonomic group, generally considering them as diatoms or radiolarians. Not until 1890 did A.H.C. Borgert introduce the term *silicoflagellate* and propose the order Silicoflagellata for them. They were finally placed in the plant kingdom in 1894 by Haeckel because of their brownish-yellow chromatophores. Finally, in 1901, E. Lemmermann separated the silicoflagellates and ebridians, which had previously been considered together.

Knowledge of living microplankton expanded considerably with the great oceanographic exploration, beginning with the famous **HMS Challenger** expedition of 1872–76. Interest in fossil microplankton increased when in 1928 Schulz published a large compilation of silicoflagellate and ebridian species from many places around the world. Classification of the silicoflagellates was originally based on skeletal morphology, but during the 1940s and 1950s J. Frenguelli and G. Deflandre departed from these systematic principles by emphasizing the variability of the silicoflagellate skeleton. This variability was further illustrated by more recent studies of modern silicoflagellates, including those of Shirley Van Valkenburg, who raised silicoflagellates in laboratory cultures.

Our knowledge of fossil silicoflagellates increased dramatically because of the **Deep Sea Drilling Project** (DSDP, 1968–83) and **Ocean Drilling Program** (ODP, 1985 to present). These provided drilled core samples of deep-ocean sediment and sedimentary rock, allowing detailed paleontologic and paleogeographic interpretation. The development of biostatigraphic zonations based upon a variety of microfossil groups, including silicoflagellates, was an important part of this work. This research has greatly increased the number of known silicoflagellate taxa. How the various silicoflagellate morphologies are related taxonomically, however, remains unsettled. Some workers emphasize the skeletal variability of the organism while others continue the taxonomic description of various skeletal morphologies of silicoflagellates.

Skeletal morphology and systematics

Morphology

The silicoflagellate skeleton is a simple geometric framework of small size (20–60 μm, rarely up to 150 μm) that encloses most of the protoplasm. The general structure of the skeleton is a latticework of linear to somewhat curved elements (also called bars or rods) that form a simple dome; there are no internal skeletal members and the number of elements that make up the skeleton are relatively few (Fig. 9.4). Most commonly, the skeleton consists of a polygonal **basal ring** with three to eight or more sides and an **apical structure** that forms a dome above the basal ring. The apical structure generally consists of elements, called **struts**, that attach to the basal ring and whose apical ends can either meet at an apex or support an **apical bridge** (Fig. 9.4A, C) or ring (Fig. 9.4B). Spines generally occur at the corners of the basal ring, although some forms have rounded corners without spines or, rarely, enlarged knobs; spines also occur on the apical structure of some forms.

Despite their variability, the basic silicoflagellate skeletal morphotypes appear to have certain consistent characters. The number of struts, for example, is equal to the number of **basal sides**, and there is one strut per side. Similarly, the apical and basal rings have the same number of sides, and the entire skeleton generally has some form of rotational symmetry. Also, the skeletal elements are joined only in triple junctions, and the angle formed by the juncture of two elements is seldom acute and usually close to 120°. Triple junctions also occur where two elements meet a spine, as at the corners of the basal ring and the apex of *Dictyocha aculeata* (Fig. 9.4A, C). Exceptions to these general observations are unusual and commonly appear to be aberrant.

Systematics

Silicoflagellate skeletons exhibit remarkable variety that has spawned a taxonomy, particularly in paleontologic usage, that may not indicate actual biologic relationships. Skeletons collected in nature or grown in culture show extreme variability. While biologists have a broad silicoflagellate species concept, paleontologists usually emphasize the distinctness of the skeletal designs, and recognize many species.

The number of silicoflagellate genera recognized by

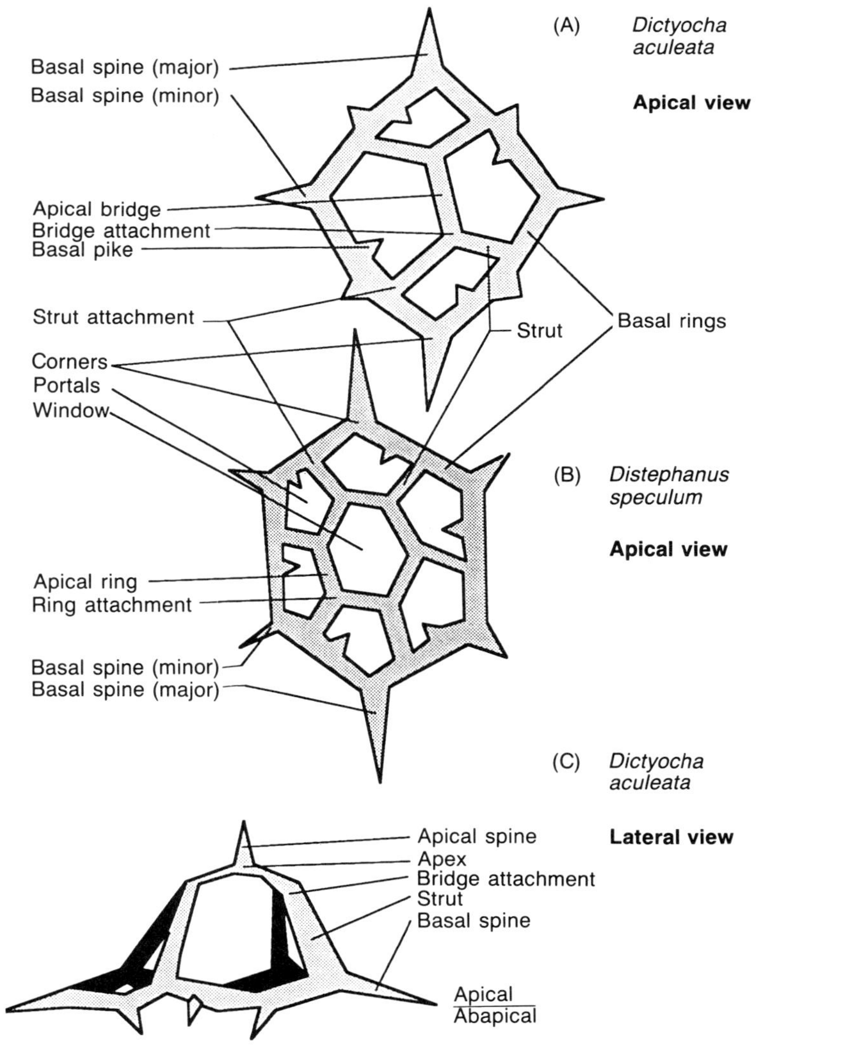

Figure 9.4 Silicoflagellate descriptive terminology. A, C, Apical and lateral view of *Dictyocha aculeata*. B, Apical view of *Distephanus speculum*. (Reproduced from McCartney K, Loper DE. *Paleobiology* 1989; 15:283–98.)

paleontologists varies, but six basic morphologic groups are in the Cenozoic and four problematic groups are from the Cretaceous. These groups are commonly known by their generic names, *Corbisema, Dictyocha, Distephanus, Cannopilus, Bachmannocena*, and *Naviculopsis* (Fig. 9.5), all of which belong to the family Dictyochaceae, and the problematic Cretaceous *Vallacerta, Lyramula, Cornua*, and *Variramus*.

Corbisema (three-sided forms) (Figs 9.5A, 9.6A–C)
This group has a basal ring with three sides and three struts. The struts attach to the basal ring at or very near the midpoint of each side and meet at their opposite ends at a single point or, rarely, form an apical ring or plate. The apex of the apical structure is positioned directly above the center of the basal ring. Spines may be at the corners of the basal ring. Most **corbisemids** have basal rings that are roughly equilateral, though *C. hastata* (Fig. 9.6C) has one side that is somewhat shorter than the other two, giving it a bilateral symmetry. Although the corbisemids have a basal ring with three corners by definition, two- and four-cornered variants are known. The corbisemids are generally dominant in the Cretaceous and much of the Paleogene. Their importance declines in the Oligocene and only the relatively small *C. triacantha* (Fig. 9.6A) is found in any abundance in the Miocene.

Dictyocha (four- and more-sided forms without apical rings) (Figs 9.5B, 9.6D–F)
Members of this group most commonly have a four-sided basal ring with a spine on each corner. Five- and six-sided forms are relatively uncommon or have limited geologic ranges. The struts meet to support an apical bridge. The bridge can be more or less parallel to the basal plane, or it can arch upwards with a spine at the apex (Fig. 9.4A, C). **Dictyochid** forms in which the four struts meet without a bridge (called medusid, after *D. medusa*) are rare in the Neogene.

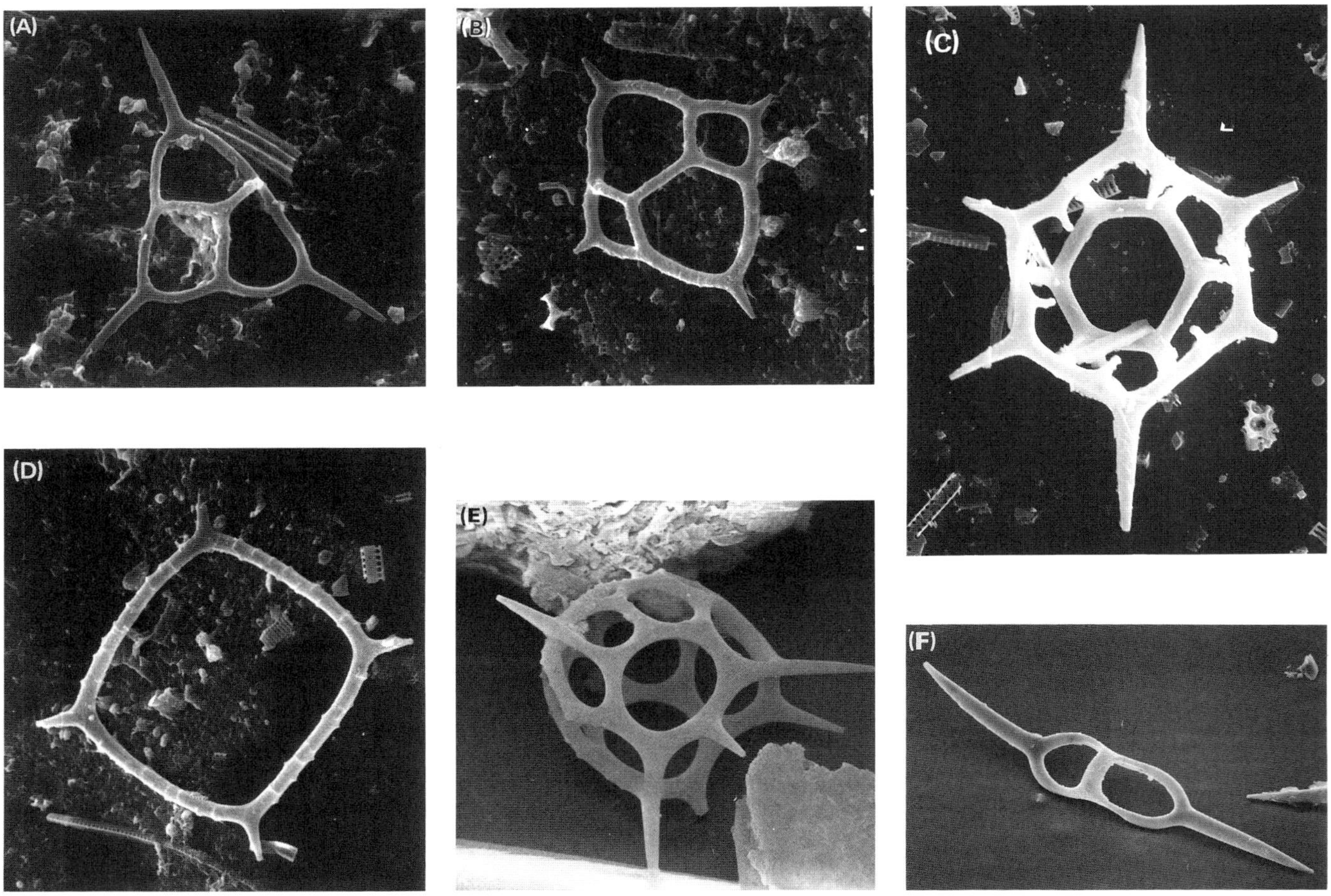

Figure 9.5 Silicoflagellate genera found in Cenozoic deposits. A, *Corbisema* (Middle Miocene, southwest Pacific). B, *Dictyocha* (Upper Miocene, southwest Pacific). C, *Distephanus* (Pliocene, Antarctic Ocean). D, *Cannopilus* (Miocene, Norwegian-Greenland Sea). E, *Bachmannocena* (Lower Pliocene, southwest Pacific). F, *Naviculopsis* (Middle Eocene, California).

The basic four-sided dictyochid design varies considerably. The bridge may be aligned with the major axis of the basal ring (fibulid, after *D. fibula*; Fig. 9.6C), with the minor axis (asperid, after *D. aspera*; Fig. 9.6D), or be diagonal with struts of unequal length (here called aculeatid, after *D. aculeata*; Fig. 9.6F). The asperids are most abundant through most of the Cenozoic but fibulids rapidly dominate by the upper Miocene (Fig. 9.7). The aculeatids dominate in the upper Pliocene and Pleistocene. In addition, some asperids have a bridge that extends across the minor axis without struts; this morphology occurs sporadically in the upper Miocene and lower Pliocene and is called *Neonaviculopsis* by some workers.

Distephanus (forms with apical rings)
(Figs 9.5C, 9.6G–H)
Silicoflagellates with apical rings may have from three to eight (or more) sides, with the four- (Fig. 9.6G) and six-sided (Fig. 9.6H) forms being most important. The apical ring typically has the same number of sides as the basal ring, and forms a polygonal opening called a window. The six-sided *D. speculum* is probably the best known of the ringed silicoflagellates, although the four-sided *D. crux* was abundant during the Oligocene to middle Miocene. Three-sided silicoflagellates with apical rings are commonly considered within *Corbisema*. Skeletons with divided apical rings forming multiple windows (Fig. 9.6J) technically belong to the genus *Cannopilus*, but some are taxonomically placed among the **distephanids** on the basis of associations with single-windowed forms.

Distephanids of any type may show considerable variability, especially in the general size of the basal ring and the relative size of the apical ring. When the basal ring is somewhat elongate, the spines that are associated with the major axis are longer than the other spines. The apical ring is commonly rotated with respect to the basal ring, especially where the basal ring is elongate. This

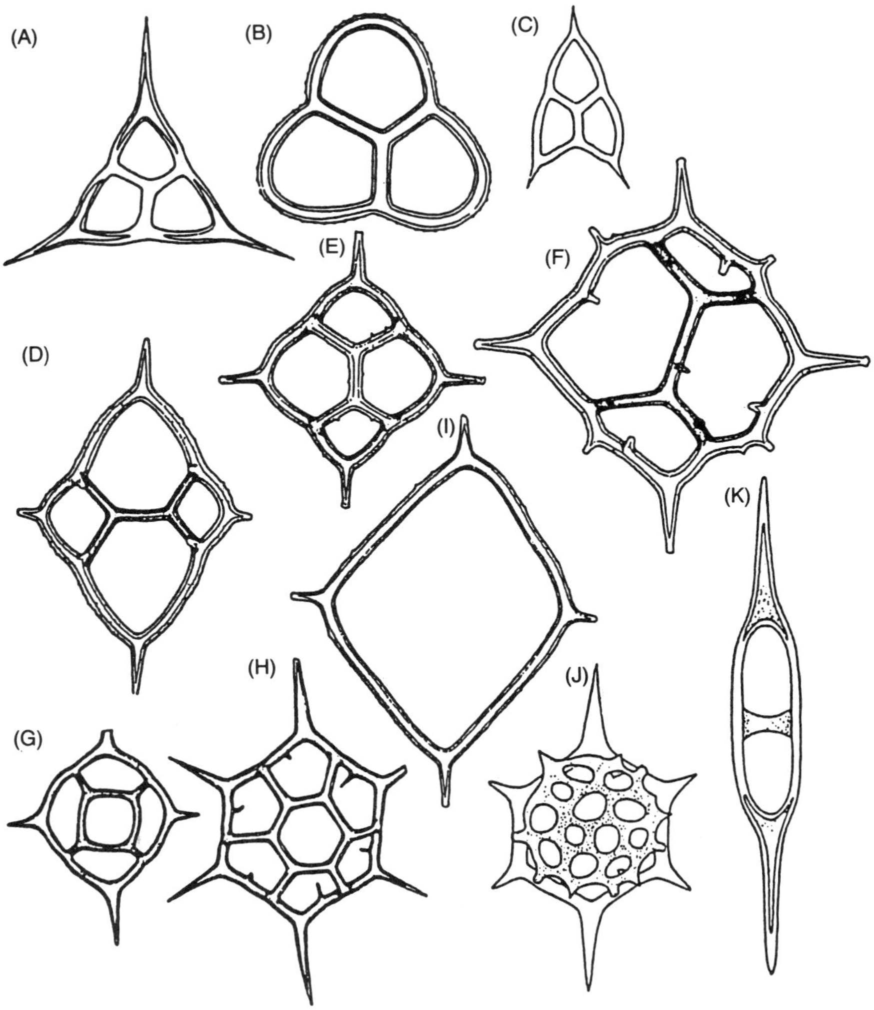

Figure 9.6 Representative silicoflagellate species from the Cenozoic. A, *Corbisema triacantha*; B, *C. inermis*; C, *C. hastata*; D, *Dictyocha aspera*; E, *D. fibula*; F, *D. aculeata*; G, *Distephanus crux*; H, *D. speculum*; I, *Bachmannocena elliptica*; J, *Cannopilus hemisphaericus*; K, *Naviculopsis biapiculata*. (A–C, J–K reproduced from Schulz P. *Botanischen Archiv* 1928; 21:225–92. D–I reproduced from Dumitrica P. *Initial Reports of the Deep Sea Drilling Project* 1973; 13:902–33.)

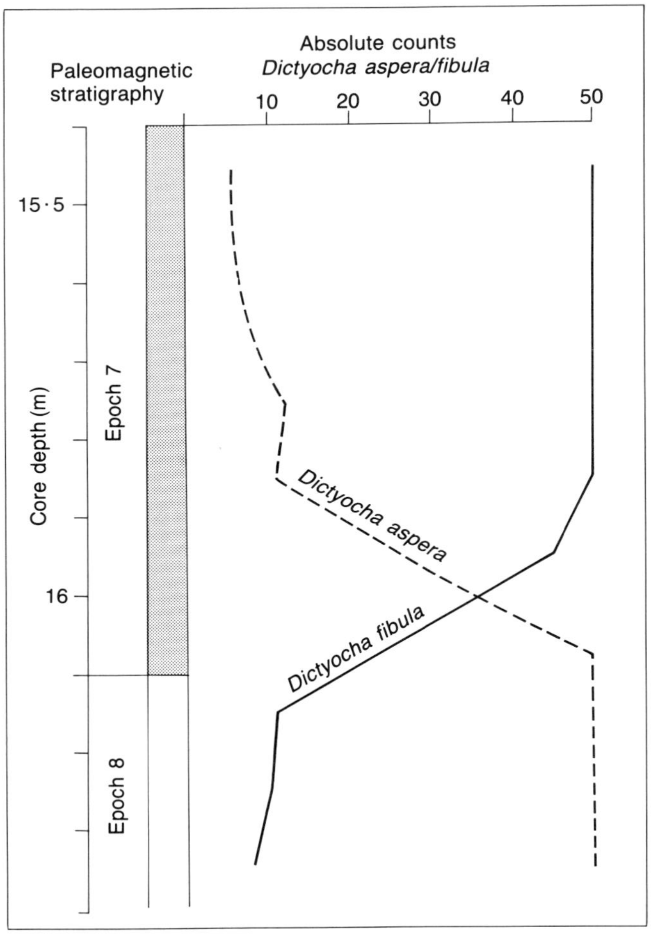

Figure 9.7 Change in dominance from *Dictyocha aspera* to *D. fibula* in the middle Late Miocene of an equatorial Pacific core. (Reproduced from Burkle LH. *Micropaleontology* 1981; 27:332–34.)

rotation is almost always sinistral (counter-clockwise) when seen from apical view. The six-sided distephanids are especially common in the late Cenozoic, but forms with eight or more sides also occur. These generally tend to have apical rings with diameters almost as large as the basal ring, while distephanids with six or fewer sides usually have relatively smaller apical rings. One group of eight-sided forms, until recently considered as the separate genus *Octactis*, has a relatively large apical ring with very thin apical elements.

Cannopilus (multiwindowed forms) (Figs 9.5D, 9.6J)
The **cannopilids** have an apical ring that is subdivided by additional elements to form multiple windows. The basal ring varies from a distephanid-like hexagonal ring and six basal spines to a nearly circular ring and many windows that can, in extreme cases, give it a radiolarian appearance. Skeletons with less than six basal sides are relatively rare. The six-sided and few-windowed forms are considered as members of *Distephanus* by some workers on the basis of associations with one-windowed types and from evidence supplied by double skeletons. This genus is called *Caryocha* by some workers.

Bachmannocena (forms without apical structures) (Figs 9.5E, 9.6I)
Three- to many-sided **bachmannocenids** consisting only of a basal ring without an apical structure, were, until recently, placed in the genus *Mesocena*. The three- to six-sided forms may be quite similar in size, spine length, and ornamentation to other silicoflagellates that are associated with them in the same rock. These may be ecophenotypic variants of *Corbisema* or *Dictyocha*. The many-sided forms, however, are distinct and are not similar to any other types; they are sometimes referred to in the separate genus *Paradictyocha*.

Naviculopsis (elongate forms without struts) (Fig. 9.6K)
Naviculopsids are very elongate with the minor axes connected by a bridge. Major-axis spines are usually present and relatively long, but minor-axis spines are absent. The apical structure consists of a strutless bridge or plate that arches across the width of the skeleton. The apical bridge varies in width and stratigraphic position and so is commonly used as a taxonomic character. The group is locally very abundant in Eocene rocks but declined through the Oligocene and became extinct in the early Miocene.

Most of the major groups discussed here have forms with a skeletal pattern of one group but which apparently are more closely related to another. For example, some silicoflagellates have the naviculopsid shape but are considered within the genus *Corbisema* and *Dictyocha*. *Dictyocha* sometimes occur without struts, and strutted *Naviculopsis* are also found. Some workers consider such unusual skeletal morphologies as separate genera, but these morphologies are generally similar to standard ones that co-occur in the same rocks. Such transitory configurations have caused taxonomic problems; they also provide important insight into the variability and evolution of the silicoflagellate skeletal design.

Vallacerta (basal rings that have apical domes without portals or windows) (Fig. 9.8A)
Vallacerta has a basal structure that is generally similar to some of the Tertiary silicoflagellate groups. The basal elements are tubular and basal spines occur at each corner. The apical structure, however, is different from that of any Tertiary silicoflagellate in having a solid domal

plate without portals or windows. Typical late Cretaceous morphologies have a basal ring with five sides, although variants with more or fewer sides also occur. Late Cretaceous *Vallacerta* also lack basal pikes. The group is known to occur in the Early Cretaceous, where seven-sided forms with basal pikes appear to be prevalent.

Lyramula (Y-shaped skeletons) (Fig. 9.8B)
Lyramula is commonly considered a silicoflagellate because of its tubular construction and its resemblance to an incomplete basal ring. The skeleton generally has two relatively long skeletal members, called limbs, and a shorter third member that resembles a basal spine. Specimens with more than two limbs occur uncommonly and thus may be aberrant. This group can be extremely common in siliceous sedimentary rocks of the later Cretaceous.

Cornua (triradiate skeletons without basal rings) (Fig. 9.8C)
This group has three radiating skeletal elements that bifurcate at each end. It is very poorly known and rarely found. Quite possibly these are aberrant corbisemids whose basal elements do not connect at the corners to form a ring, and their rare occurrence in coastal waters may support this conclusion. *Cornua* is almost unknown from deep-sea sedimentary rocks, which is also where aberrant silicoflagellates are less abundant. The recent discovery of *Variramus* in the Early Cretaceous, however, might indicate that *Cornua* is an intermediate between *Corbisema* and more primitive morphologies.

Variramus (branching skeletons without a particular symmetry or basal rings) (Fig. 9.8D)
Variramus consists of an apical structure that lacks a basal ring. The general morphology of the skeleton, which can include spines and **pikes**, is often similar to the apical structures of four- or more-sided *Dictyocha*, although more complex morphologies also occur and may be conspecific with the simpler skeletons. The genus is characterized by extreme variability, even by silico-flagellate standards; indeed, no two specimens of this group appear to be alike.

Biology

Silicoflagellates are unicellular, with clear protoplasm surrounding a relatively large and centrally located nucleus. The skeleton marks the approximate limits of the cell, but is not easily seen in living specimens. The flagellum is relatively long, anterior with respect to the direction of movement, and usually positioned at the base of a long spine. The flagellum is generally rigid with only the distal tip being motile. Thin pseudopodia may extend from the other corners. These pseudopodia become more prominent during skeletal secretion.

Only asexual reproduction by cell division in silicoflagellates is known. Paired or double skeletons, representing a stage in asexual division, connected at the abapical face occur seasonally and in fossil material. These skeletons are usually mirror images of each other, although variations occur, especially in the apical structure. Silicoflagellates are not known to reproduce sex-

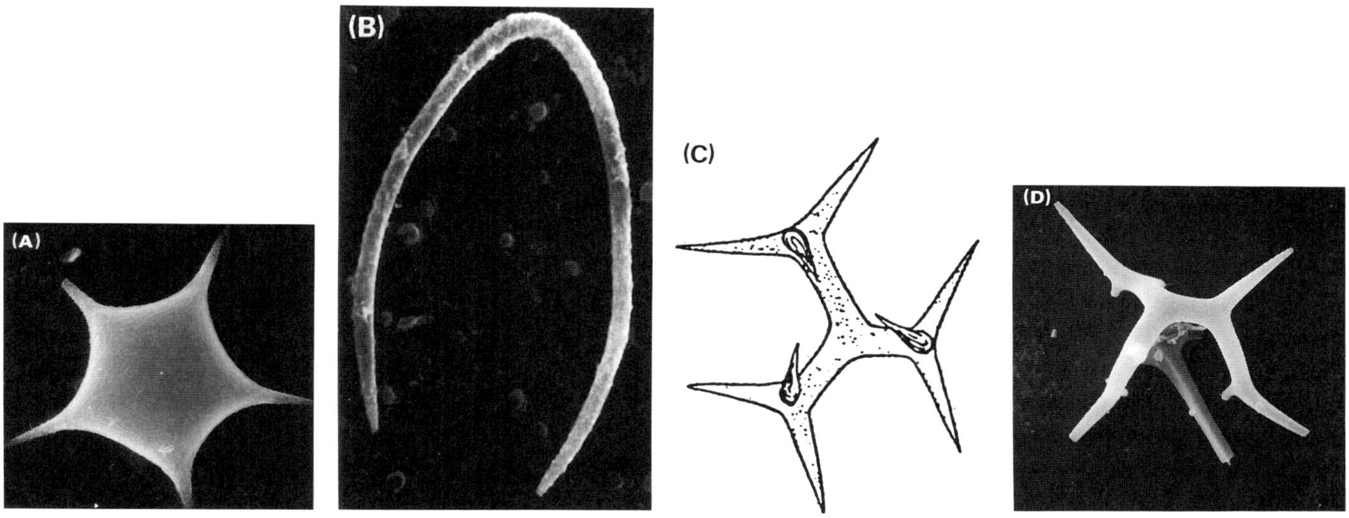

Figure 9.8 Cretaceous silicoflagellate genera. A, *Vallacerta* (Late Cretaceous, Arctic Ocean). B, *Lyramula* (Upper Cretaceous, California). C, *Cornua*. (Reproduced from Schulz P. *Botanischen Archiv* 1928; 21:225–92.) D, *Variramus* (Lower Cretaceous, Antarctic Ocean).

ually, but the great stability of the fossil and Recent forms, given their known variability, indicates that silicoflagellates possess temporary phases of sexual reproduction.

Silicoflagellates are exclusively marine, but live in the Black Sea where salinities are as low as 10‰. Both laboratory and field data suggest that the silicoflagellates do well in reduced light and prefer moderate to high nutrient levels. They may be more affected by temperature and water stability than by salinity or water density. The skeleton appears to be more variable in coastal environments.

Some silicoflagellates occur within definite temperature limits. *Distephanus speculum* generally occurs in water cooler than 15°C and is restricted to high latitudes. *Dictyocha*, on the other hand, is found in water warmer than about 10°C and thus predominates at lower and middle latitudes (Fig. 9.9). The transition in dominance from one morphology to another can occur quite rapidly, such as occurs in the Sea of Japan. The preference by *Distephanus* for waters colder than those dominated by *Dictyocha* occurs also in the geologic record, causing numerous workers to propose *Distephanus* : *Dictyocha* ratios as a useful indicator for past water temperature.

While more complex skeletal morphologies, such as *Distephanus*, are associated with colder water, the ratio of *Distephanus* : *Dictyocha* may be useful in local areas such as the Gulf of California, but not in other regions. Such ratios also have limited geologic utility because the distephanids have only been abundant in the later Cenozoic.

Unusual silicoflagellate morphologies have potential as paleoenvironmental indicators. For example, *Distephanus speculum* morphologies that lack an apical ring are found in high latitudes near where ice-rafted sediments cease to be deposited and they may result from rapid changes in temperature or nutrient supply. Variation in the size of silicoflagellates and the length of the basal spines may also be environmentally related. For example, the size of *Distephanus speculum* decreases in the South Atlantic between the tropical and polar regions. The length of the basal spines may decrease the rate of sinking, which can be important to an organism that lives in the photic zone.

The species problem

Exactly how many silicoflagellate species exist in the Recent or at times in the geologic past is a perplexing question. Most biologists believe that only two silicoflagellate species live today and that the four-sided dictyochids and six-sided distephanids are conspecific. Paleontologists, however, tend to subdivide the skeletal morphologies taxonomically. Both viewpoints are

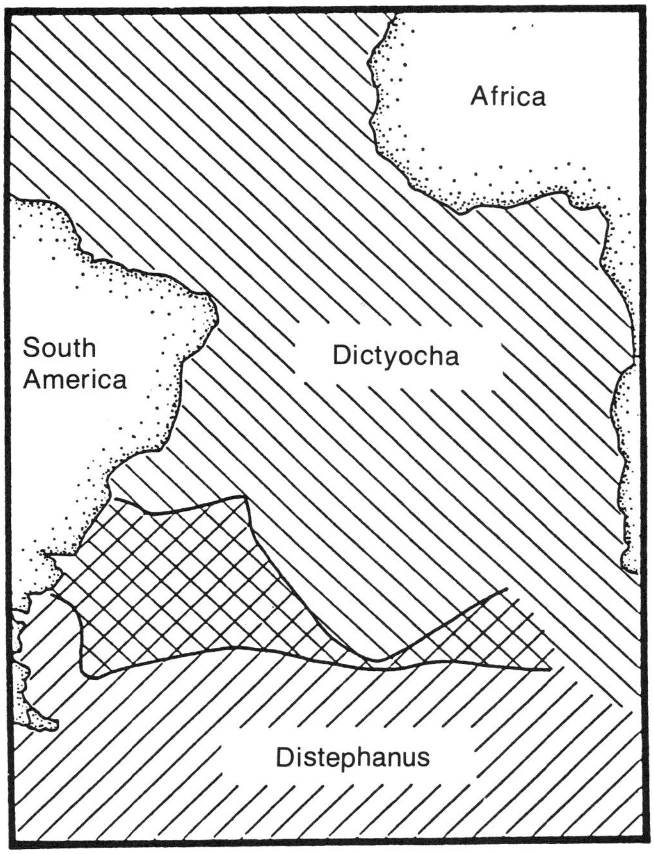

Figure 9.9 Distribution of the Recent silicoflagellates *Dictyocha* and *Distephanus* in the South Atlantic. (Reproduced from Lipps JH. *Proceedings of the North American Paleontological Convention, Chicago*, 1970; 2:965–93.)

supported because the skeletal morphology may be remarkably regular but at other times or places may be exceptionally variable. Some morphotypes occur frequently and have a uniform appearance over broad geographic areas or time intervals, but other morphologies are much less common, occur sporadically, and might very well be variants.

Exceptional variability of silicoflagellates occurs commonly in both modern and fossil assemblages. A single population can exhibit variation crossing species or generic boundaries as defined by the present taxonomy. Laboratory cultures of cloned *Dictyocha fibula* produced skeletons representing three genera and four species according to some taxonomic classifications.

Aberrant silicoflagellates are relatively common, although they usually make up less than 3% of a sample. They give insight into the variability of the silicoflagellate skeletal morphology and indicate that deviations from the regular morphology may not be lethal to the organism. Common abnormalities include disruption in the regular symmetry of the skeleton, failure to close the basal or apical ring, or struts that are forked or fused. Unusual numbers of exceptionally grotesque specimens are sometimes found, such as specimens collected off the northern California coast. Aberrant specimens are less common in the deep sea than in coastal areas. Explanations for unusual variation include high chlorinity, phytoplankton bloom conditions, genetic weaknesses, or other abnormal ecologic conditions.

Biostratigraphy

Silicoflagellate biostratigraphy is especially well studied in tropical and subtropical waters (Fig. 9.10) and has considerable potential in higher latitudes and deeper waters where calcareous plankton do not thrive or are not preserved. The simplicity of silicoflagellate skeletons make identification and age determination much easier than for the calcareous nannoplankton, planktic foraminifera, or radiolarians, although the silicoflagellate zonations are not as detailed. The geologic ranges and relative abundance of most silicoflagellate species, however, vary markedly with latitude and nutrient levels. Thus silicoflagellate zonations are useful only for local or regional correlations, especially in higher latitudes. Direct correlations with coccolith and foraminiferal zonations are not always possible.

Evolution

Interpretations of evolution of the silicoflagellates are complicated by their variability. The time when one morphology evolves from another is difficult to determine because a morphology may disappear and reappear several times in the geologic record. The genera *Dictyocha* and *Distephanus*, for example, appear to give rise to one another several times. However, skeletal complexity appears to increase throughout the Cenozoic. The number of basal sides has generally increased, and the apical structure has developed from the simple corbisemid design to an apical bridge, ring, and divided ring; skeletons with more complex designs generally appear first in high latitudes.

The ancestor of silicoflagellates is unknown. The presence of a flagellum suggests that the silicoflagellates are not closely related to the diatoms, while their cytology is distinct from the dinoflagellates. The earliest silicoflagellates are poorly understood and have morphologies that are generally quite different from the Cenozoic skeletal morphologies. They first appear as two distinct groups, *Variramus* and *Vallacerta*, in Lower Cretaceous

Silicoflagellates		
Age	Zone	Subzone
Quaternary	*Dictyocha aculeata aculeata*	
Quaternary	*Bachmannocena quadrangula*	
Quaternary–Pliocene	*Dictyocha stapedia stapedia*	*Dictyocha perlaevis delicata*
		Dictyocha perlaevis ornata
Pliocene–Miocene	*Dictyocha fibula*	
		Dictyocha neonautica
Miocene	*Dictyocha brevispina*	
Miocene	*Corbisema triacantha*	*Distephanus stauracanthus*
Miocene	*Naviculopsis ponticula*	
Miocene	*Naviculopsis quadrata*	
Miocene	*Naviculopsis lata*	
Oligocene	*Naviculopsis biapiculata*	*Distephanus speculum haliomma*
Oligocene–Eocene	*Corbisema apiculata*	
Eocene	*Dictyocha hexacantha*	
Eocene	*Naviculopsis foliacea*	*Dictyocha spinosa*
		Naviculopsis robusta
Eocene–Paleocene	*Naviculopsis constricta*	
Paleocene	*Corbisema hastata*	
Cretaceous	*Lyramula furcula*	

Figure 9.10 Silicoflagellate biostratigraphic zonation for tropical and subtropical regions. (Modified from Bukry D. *SEPM Special Publication* 1981; 32:433–44.)

rocks near Antarctica (Fig. 9.11). Similarities between *Variramus* and the apical structures of *Corbisema* and *Dictyocha* suggest an evolutionary relationship, possibly with *Cornua* as an intermediate. The *Corbisema* and *Dictyocha* lineages survived the Mesozoic and are the ancestors of the Cenozoic silicoflagellates. *Vallacerta* and *Lyramula* appear to be distinct, and their relationship to each other or the other morphologies is uncertain; both became extinct near the Cretaceous–Tertiary boundary.

Development through the Cenozoic

The Deep Sea Drilling Project provided a wealth of information on relative abundances of silicoflagellates, especially in the Cenozoic, and permits a synoptic view of the trends in silicoflagellate development during this time. The most important groups of strutted silicoflagellates contrast their geologic ranges and relative abundances through the Cenozoic (Fig. 9.12, cannopilids are not included).

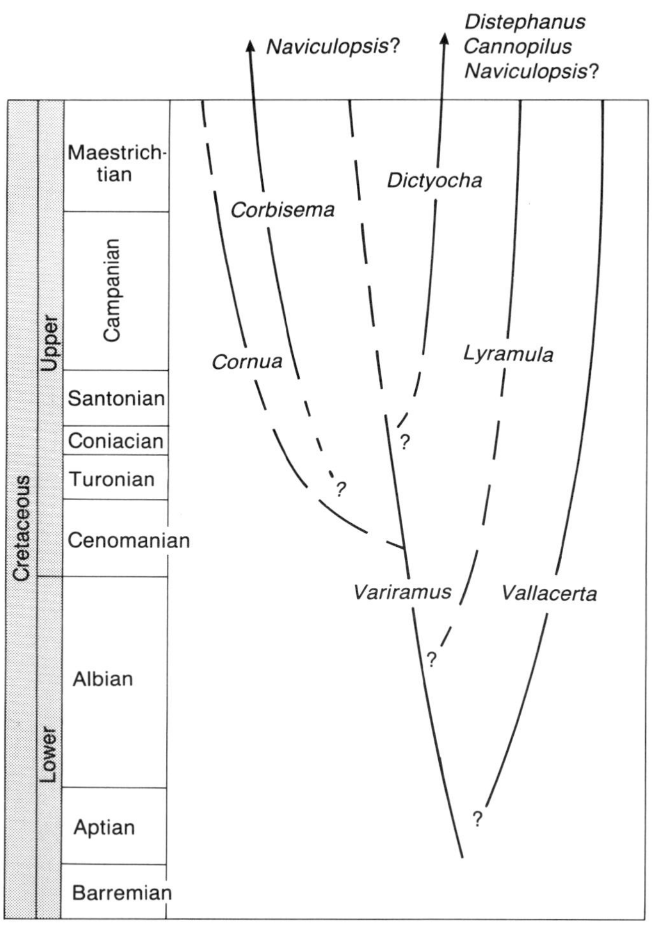

Figure 9.11 Silicoflagellate phylogeny during the Cretaceous. (Reproduced from McCartney K, Wise SW Jr, Harwood DM, Gersonde R. *Initial Reports of the Deep Sea Drilling Project*. Washington: US Government Printing Office, 1990; 113:427–42.)

Paleogene

The Paleocene and Early Eocene were characterized by predominate *Corbisema*, although some dictyochid morphologies, including unusual five- and six-sided forms, also occurred. Three-, four-, five-, and six-sided distephanid morphologies made sporadic appearances at high and middle latitudes during the Eocene, but did not occur at lower latitudes. By late Eocene and early Oligocene *Corbisema* still predominated, but *Dictyocha* (four-sided forms only) become more abundant. Many exceptionally robust *Dictyocha* existed during this time, some with very thick elements, broad plates, or unusual spines; such morphologies occurred rarely after the Early Oligocene. Through the rest of the Paleogene, *Dictyocha* and *Distephanus* became increasingly abundant and began to predominate, especially at high latitudes, while *Naviculopsis* and *Corbisema* declined. The dominant *Distephanus* were quadrate morphologies.

The general trend during the Paleogene was for changes of dominance from corbisemid and naviculopsid to dictyochid to distephanid forms and a change to more complex apical structures with changes occurring at higher before lower latitudes. By the end of the Paleogene the dictyochids and distephanids were both abundant at high latitudes, while corbisemids were still common and distephanids uncommon at lower latitudes. The only corbisemid that survived the Paleogene in any abundance was a small form with a relatively high apex (*Corbisema triacantha*).

Neogene

During the Miocene *Corbisema* and the four-sided *Distephanus* became very rare. The six-sided *Distephanus* became increasingly abundant, especially at high latitudes, where multiwindowed forms appeared. The cannopilids were most abundant and complex in Late Pliocene high latitudes and then became extinct. The distephanids became increasingly abundant throughout the rest of the Cenozoic and generally make up nearly 100% of the silicoflagellates in Recent high-latitude sediment.

Among the four-sided *Dictyocha*, both asperid and fibulid forms were common in the same rocks, although asperids predominated, prior to the late Miocene. The asperids, however, rapidly declined in the late Miocene

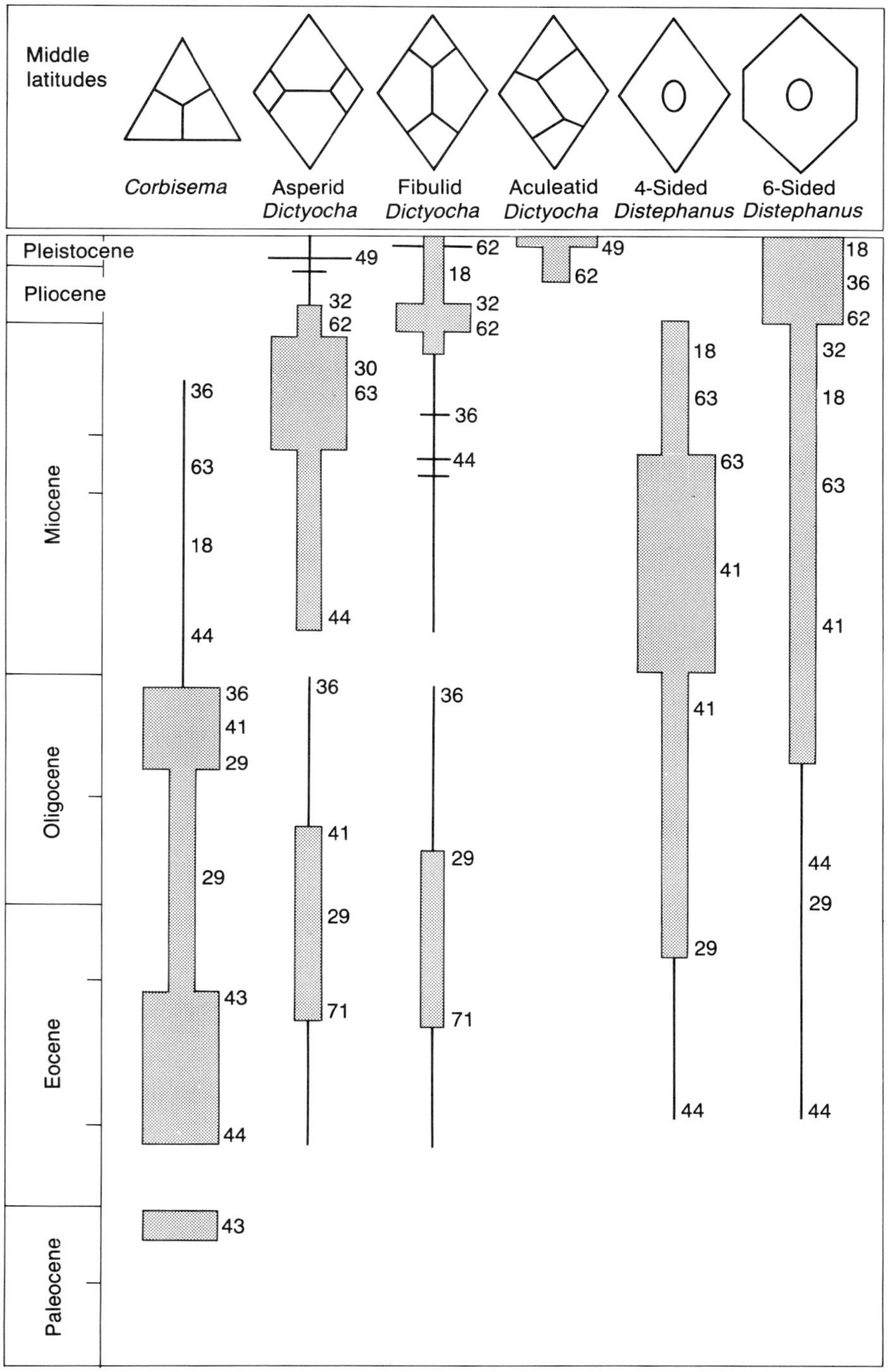

Figure 9.12 The geologic range and relative abundance in middle latitude (23.5–50°) of silicoflagellate morphotypes. Wide band indicates the relative abundance is commonly greater than 50% of strutted silicoflagellates. Narrow band denotes 10%; line indicates regular presence. The numbers refer to the Deep Sea Drilling Project leg from which this information was obtained.

and were seldom common in younger deposits. The change in dominance occurs very rapidly (Fig. 9.7) and has been used as a zonal boundary, although it is not synchronous over a broad area. The fibulids first became commonly abundant at high latitudes and they began to develop longer bridges (*D. longa*) and forms with a slightly diagonal bar (*D. perleavis, D. stapedia*). The bridge appears to become progressively more inclined (*D. subaculeata, D. aculeata*) in younger sediments.

In Recent waters six- and seven-sided distephanids predominate at high latitudes, while four-sided dictyochids with inclined bars predominate at lower latitudes. Middle latitudes usually contain a mixture of the two groups. The *D. speculum* populations move toward the equator in autumn. An unusual seven- to nine-sided form previously known as *Octactis*, but now considered synonymous with *D. speculum*, is especially abundant in recent nearshore environments.

Supplementary reading

Lipps JH. Evolution and ecology of silicoflagellates. In: Yochelson EL, ed. *Proceedings of the North American Paleontological Convention, Chicago*. 1970; 2:965–93. Review article with interesting speculation on ecology and functional morphology.

Locker S, Martini E. Silicoflagellates and some sponge spicules from the southwest Pacific, Deep Sea Drilling Project, Leg 90. In: Kennett JP, von der Borch C, *et al.*, eds. *Initial Reports of the Deep Sea Drilling Project*. Vol. 90. Washington: US Government Printing Office, 1986:887–924. Includes an excellent interpretation of the silicoflagellate species, as determined from fossil specimens; typical of the best of the deep sea silicoflagellate studies.

Loeblich AR III, Loeblich LA, Tappan H, Loeblich AR Jr. *Annotated Index of Fossil and Recent Silicoflagellates and Ebridians with Descriptions and Illustrations of Validly Proposed Taxa*. Boulder: The Geological Society of America, Memoir 106, 1969; 319. An essential guide to all silicoflagellate taxa described before 1968; includes all original descriptions of valid taxa (in original language).

McCartney K, Loper DE. Optimized skeletal morphologies of silicoflagellate genera *Dictyocha* and *Distephanus*. *Paleobiology* 1989; 15:283–98. Models mathematically some of the silicoflagellate skeletal morphologies.

McCartney K, Wise SW Jr, Harwood DM, Gersonde R. An enigmatic Lower Cretaceous (?) silicoflagellate from ODP Site 693: progenitor of the order Silicoflagellata? In: *Initial Reports of the Deep Sea Drilling Project*. Washington: US Government Printing Office, 1990; 113:427–42. Discusses the relationship of the Cretaceous morphologies, including the new *Variramus*.

Perch-Nielsen K. Silicoflagellates. In: Bolli HM, Saunders JB, Perch-Nielsen K, eds. *Plankton Stratigraphy*. Cambridge: Cambridge University Press, 1985:811–46. Summarizes the biostratigraphic utility of silicoflagellates and includes photographs of many species.

Tappan H. *The Paleobiology of Plant Protists*. San Francisco: WH Freeman, 1980:1028. Chapter 7 provides the best review in English of silicoflagellate biology.

Diatoms

10

John A. Barron

Introduction

Diatoms are unicellular, golden-brown algae that are characterized by an external two-part skeleton called a **frustule** of opaline silica. Frustules are commonly very intricate and varied, and the patterns and types of ornamentation of the frustule form the basis of nearly all diatom taxonomy. The size of diatom frustules ranges from less than 1 μm to more than 1000 μm, but most frustules range in size from 10 to 100 μm. Thus, diatoms are comparable in size to dinoflagellates but generally larger than calcareous nannofossils. Unlike calcareous nannoplankton and dinoflagellates, however, the vegetative cell of diatoms lacks flagella, although many diatoms produce flagellated gametes during their brief period of sexual reproduction. Consequently, planktic diatoms are dispersed passively by currents in surface layers of the ocean. Adaptations that promote flotation include frustule shapes and processes that increase the ratio of surface area to volume, formation of colonies, and storage of fats or oils in the cell, which reduce the specific gravity of the cell.

Diatoms are found practically everywhere that light and moisture occur, including in virtually all marine, brackish, and freshwater environments as well as in soils, ice, and attached to rocks and other substrates within spray or splash zones near water. Of the approximately 175 living and fossil diatom genera, 70% are exclusively marine, 17% are exclusively freshwater, and the remaining 13% are either predominantly marine or freshwater with only a few species crossing from one into the other.

In the modern oceans, diatoms are important **primary producers**, and a high abundance of diatoms typifies the most fertile areas of the oceans. Generally, high abundance of diatoms correlates strongly with waters of high nitrate and phosphorus concentrations, whereas diatoms can be quite rare in waters with low concentrations of these nutrients. Beneath the fertile, nutrient-rich waters sediments on the sea floor consist of abundant diatoms and are commonly referred to as diatom **oozes**.

With lithification, the purer diatom oozes become **diatomites**, which are mined commercially. Most of the commercial uses of diatomite take advantage of the low density and high porosity caused by the diatom frustules. The predominant use of diatomite in the USA is as a filtering agent (e.g., swimming pool filters and filtration of impurities by breweries, wineries, and domestic water works), but it is also used as a mineral filler (e.g., as a paint flattening agent), a fine abrasive, an insulating material, and as an absorbant.

Like calcareous nannoplankton, diatoms are widespread in the ocean's photic zone (water depths typically <100 m), but they also have various benthic lifestyles within the photic zone. Unlike calcareous microfossils, diatoms can be very abundant and diverse at higher latitudes. Diatoms either live singly or form colonies. Many diatoms secrete a mucilaginous material that covers the siliceous wall and serves as a means of attachment of cells in colonies. The mucilaginous material can also be used to attach benthic diatoms to a substrate and its extrusion is used as a means of propulsion along a substrate in some benthic diatoms.

History of diatom studies

Study of the varied and ornate diatom frustules has been popular with microscopists since the early nineteenth century. Diatom taxonomy was founded in the publication of a number of thick monographs containing numerous hand-drawn illustrations of diatoms, which were compiled by such European workers as P.T. Cleve, C.G. Ehrenberg, A. Grunow, A. Schmidt, and H. Van Heurck in the latter half of the nineteenth century. One such example is the so-called Schmidt atlas, published between 1874 and 1959, which contains 480 large, mostly hand-illustrated plates of both marine and nonmarine diatoms. Geologic studies utilizing diatoms began in 1919 with G.D. Hanna, and later K.E. Lohman described many Cretaceous to Pliocene marine diatom assemblages preserved in strata from North America. At the same time, F. Hustedt (1927–66) began a compilation of diatom taxonomy and ecology that remains today as a key reference. Detailed stratigraphic studies using diatoms were not initiated until the late 1950s, and only with the collection of deep-sea cores by oceanographic institutions and the Deep Sea Drilling Project (DSDP) in the 1960s and 1970s did marine diatoms begin to receive widespread application for biostratigraphy and paleoecology. Use of the scanning electron microscope (SEM) in recent years has provided considerable impetus to the study of diatom morphology and has resulted in numerous changes of diatom taxonomy at generic and higher levels.

Morphology and systematics

Diatoms are considered a separate algal class, the **Bacillariophyceae**, and are placed along with Chrysophyceae and Xanthophyceae in the division Chrysophyta. These algal groups all form similar endoplasmic cysts, secrete silica at some stage, store oils rather than starch, and possess a bipartite cell wall at some stage. Diatoms have also been elevated by some workers to a separate division, Bacillariophyta, based on differences, mainly in the structure of the diatom frustule, among these groups.

Diatoms are divided into two orders, the Centrales, which tend to have radial symmetry with the structural center formed by a point and the Pennales, which have basically bilateral symmetry with the structural center formed by a line (Fig. 10.1, Table 10.1). Almost all diatom taxonomy is based on the external opaline skeleton or frustule. Standardization of diatom terminology has been proposed by international committees through the International Society for Diatom Research, which meets every 2 years.

The silica elements of the diatom cell wall constitute the frustule, which includes two overlapping valves, the **epivalve** and the **hypovalve**, and the girdle, consisting of the epicingulum and the hypocingulum (Fig. 10.2). The epivalve and epicingulum are derived from the parent cell and overlap the hypovalve and hypocingulum, which are created internally during vegetative division. The siliceous valve wall may be single layered or double layered so that it forms a honeycomb-like cross-section with chambered pores termed areolae (Fig. 10.3).

Much of diatom classification is based on the types and pattern of ornamentation that appear on the valve's face. Pores (areolae), processes, marginal and submarginal spines, hyaline or clear areas, and other specialized structures are important. Descriptions of centric diatoms are commonly based on the presence, type, and number of any processes, spines, or specialized structures, the type of perforation or areolar pattern, the number of areolae per 10 μm and any variation in that number along the valve's diameter, and any interruptions of the areolar pattern by raised or lowered regions or by hyaline areas (Figs 10.4, 10.5).

In describing pennate diatoms, the valve shape, the presence of raphes, or longitudinal slits which penetrate the valve's surface, the areolar pattern and pattern of any hyaline fields, and the presence of such other specialized structures as spines or processes on the apices are important. Pennate diatoms may also possess transverse thickenings of the valve wall or costae (Fig. 10.4), which may be used for taxonomy. Those pennate diatoms that are capable of movement along a substrate are distinguished by the presence of a raphe, which is a longitudinal slit or pair of slits through the valve wall along the apical axis (Fig. 10.6). In most cases, the raphe is present on both valves of the pennate diatom, but in some genera (e.g., *Acnanthes, Cocconeis*) the raphe occurs on only one valve; the opposite valve has a hyaline axial area or pseudo-

Figure 10.1 The two basic groups of diatoms, centrics and pennates. In centric diatoms the structural center is formed by a point. In pennate diatoms the structural center is formed by a line. A, *Coscinodiscus*. B, *Triceratium*. C, *Cocconeis*. D, *Rhaphoneis*. E, *Denticulopsis*. Scale bars, 10 μm.

Table 10.1 Classification of diatoms. (After Simonsen R. *Bacillaria*. Vol. 2. Braunschweig: J. Cramer, 1979:9–71)

Order Centrales—structural center formed by a point, auxospore formation by oogamy

SUBORDER COSCINODISCINEAE—valves with a ring of marginal processes, symmetry primarily without development of polarities
- Family Thalssiosiraceae
- Family Melosiraceae
- Family Coscinodiscaceae
- Family Hemidiscaceae
- Family Asterolampraceae
- Family Heliopeltaceae

SUBORDER RHIZOSOLENIINEAE—valves primarily unipolar, strongly elongated in direction perpendicular to the plane at which the two valves are joined in the frustule
- Family Pyxillaceae
- Family Rhizosoleniaceae

SUBORDER BIDDULPHIINEAE—valves primarily bipolar, secondarily tri- to multipolar to circular
- Family Biddulphiaceae
- Family Chaetoceraceae
- Family Lithodemiaceae
- Family Eupodiscaceae

Order Pennales—structural center normally formed by a line, auxospore formation not by oogamy

SUBORDER ARAPHIDINEAE—valves without a raphe
- Family Diatomaceae
- Family Protoraphidaceae

SUBORDER RAPHIDINEAE—valves with a raphe
- Family Eunotiaceae
- Family Achnanthaceae
- Family Naviculaceae
- Family Auriculaceae
- Family Epithemiaceae
- Family Nitzschiaceae
- Family Surirellaceae

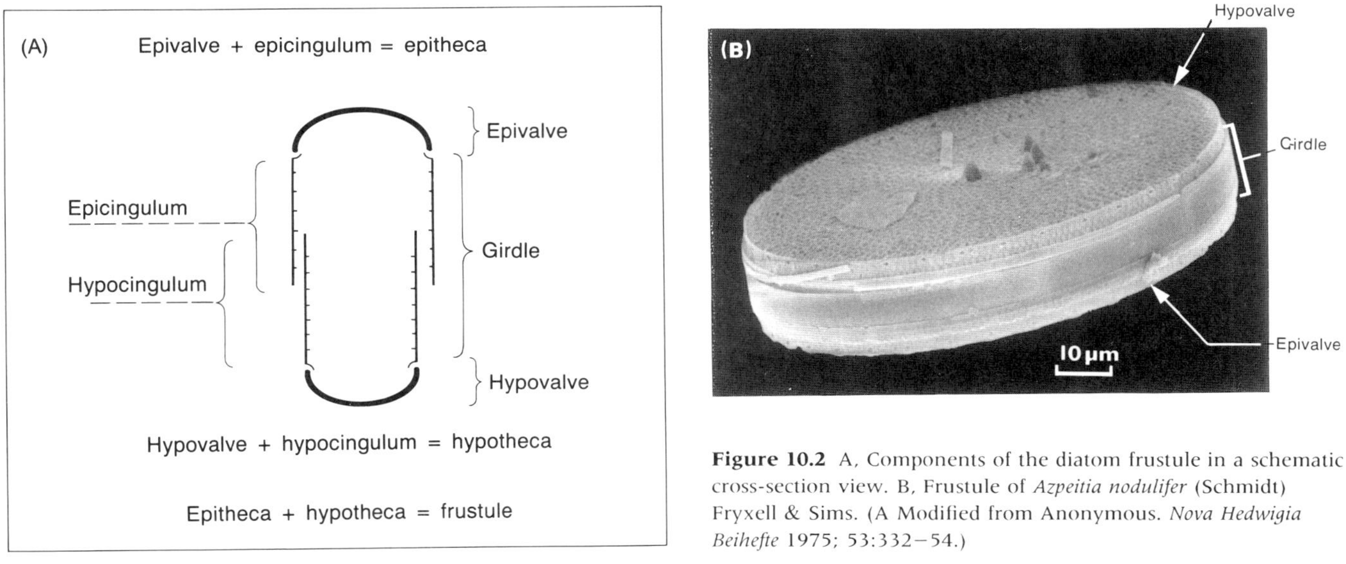

Figure 10.2 A, Components of the diatom frustule in a schematic cross-section view. B, Frustule of *Azpeitia nodulifer* (Schmidt) Fryxell & Sims. (A Modified from Anonymous. *Nova Hedwigia Beihefte* 1975; 53:332–54.)

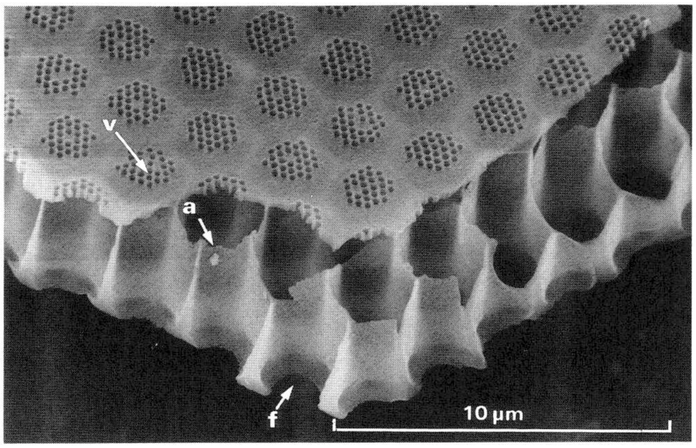

Figure 10.3 Cross-section of a two-layered wall of the diatom valve of *Coscinodiscus* showing the chamber-like areola (a), the sieve plate or vellum (v), and the unobstructed opening or foramen (f) opposite the vellum.

raphe. The raphe is a site for the secretion or channeling of a mucilaginous material, which adheres to the substrate and causes locomotion when it is actively extruded.

Many diatoms possess specialized structures, which are best studied with the SEM (Fig. 10.7). Labiate processes are tubes or openings through the valve wall with an internal flattened tube or longitudinal slit often surrounded by two lips. The number, distribution, and type of labiate processes are very useful in subdividing diatoms at the subordinal and familial levels.

Biology

In contrast to the other fossil protist groups, living diatoms are also actively studied by many biologists and botanists. Diatoms are easy to culture, so they are popular with cell physiologists, and their sensitivity to physical–chemical

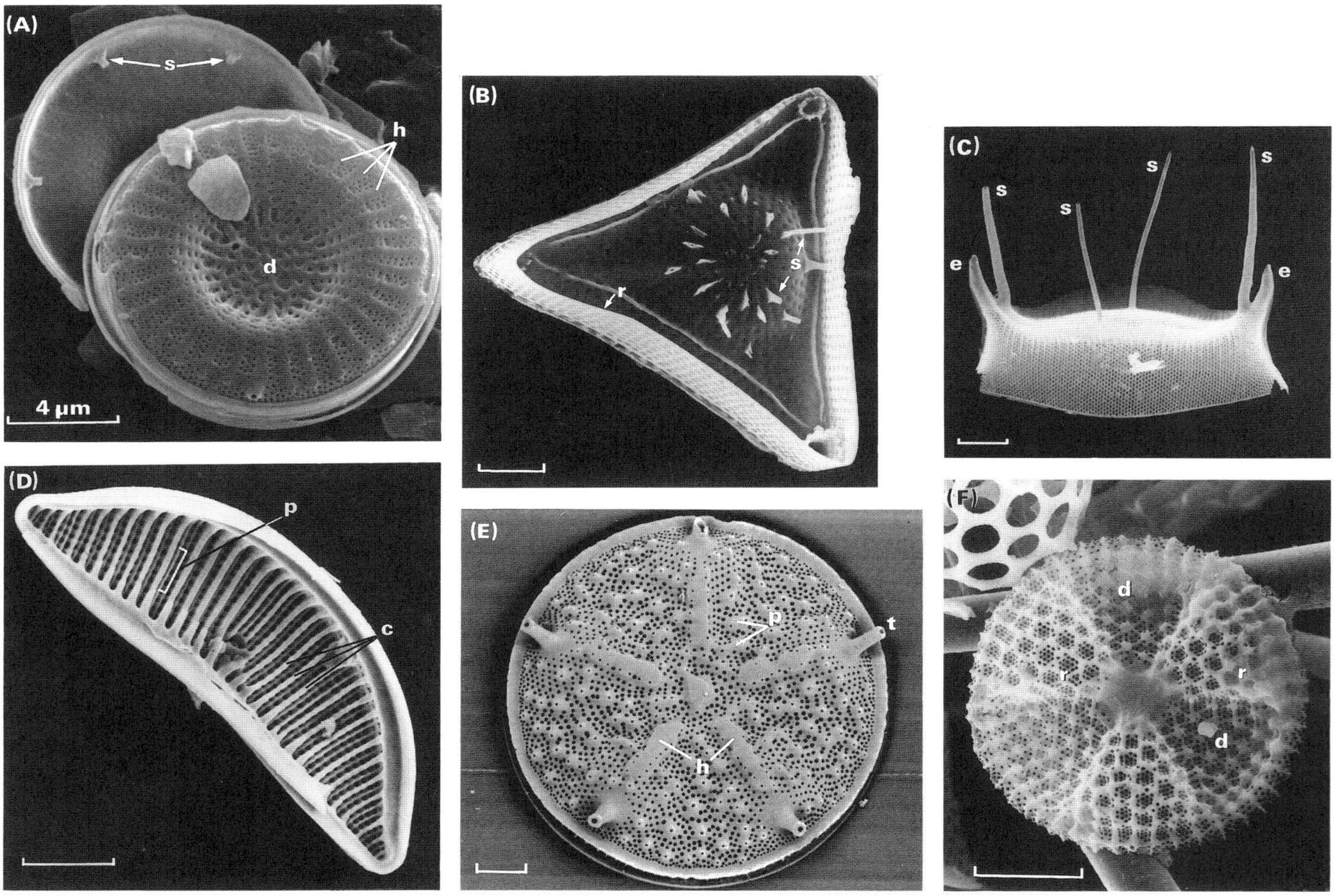

Figure 10.4 Specialized structures and variations on the valve faces of diatoms that are useful in description and classification. A, Two valves of *Stephanodiscus* showing external and interval views; (s), strutted process; (d), depressed center; (h), hyaline or clear areas. Scale bar, 4 µm. B, Valve of *Trinacria* showing a marginal ridge (r) and spines (s). If such spines are hollow tubes which are open to the interior of the valve, they are termed labiate processes. Scale bar, 10 µm. C, Marginal view of the valve of *Odontella* showing elevations (e) and spines (s). Scale bar, 10 µm. D, Valve of *Epithemia* showing costae or ribs (c) and punctae (p). Scale bar, 10 µm. E, Valve of *Aulacodiscus* showing tubes (t) or possibly labiate processes, punctae (p), and hyaline areas (h). Scale bar, 10 µm. F, Valve of *Actinoptychus* showing raised (r) and depressed (d) sectors. Scale bar, 10 µm.

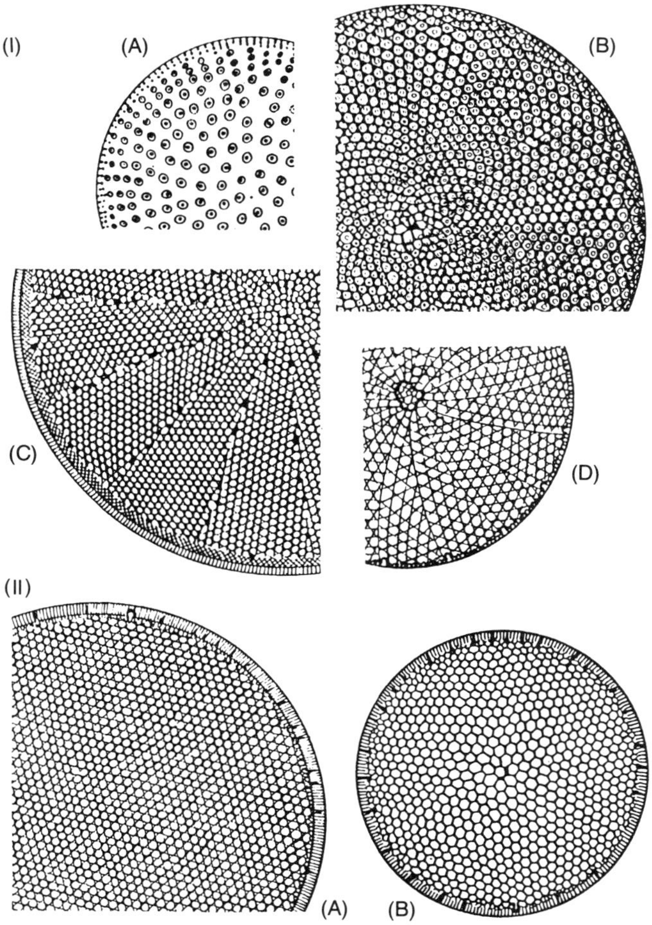

Figure 10.5 Areolation patterns in centric diatoms. I, Radial areolation: A, radial, single areolae; B, with secondary rows in spirals; C, D, radial fasciculate (pinwheel-like) areolation. II, Tangential areolation: A, straight rows (lineate type); B, rows concave towards the margin (eccentric type). (Reproduced from Anonymous. *Nova Hedwigia Beihefte* 1975; 53:332–54.)

conditions makes them popular for studies of water quality.

The protoplast of diatoms consists of a cytoplasmic layer that lines the interior of the frustule and surrounds a large central vacuole. A diploid nucleus, which is commonly spherical or oval, and two to several pigment-bearing plastids lie within the cytoplasmic layer. The plastids are the site of photosynthesis, and their pigments include chlorphyll *a*, c_1, and c_2 as well as accessory carotenoids and xanthophylls.

Reproduction

Diatoms have both sexual and vegetative modes of reproduction. Simple binary fission may occur from one to eight times a day and is regulated by intensity and duration of light and concentrations of nutrients. During cell division, the protoplast increases in size, the nucleus divides mitotically, and other organelles are divided or duplicated. The cytoplasm then divides in half parallel to the valve face, and two new silica valves form in the center of the old cell. Each parent valve becomes an overlapping **epivalve** for the daughter cells, and two newly formed daughter valves each become overlapped **hypovalves** for the daughter cells. Consequently, after cell division, the cell that retains the epivalve of the parent is the same size as the parent cell, while the one in which the hypovalve of the parent becomes the epivalve of the daughter is slightly smaller. Thus, with successive divisions, one of the daughter cells will always be smaller than the parent cell and the average size of successive generations will become progressively smaller (Fig. 10.8). Eventually, however, the size of the cells will become so small that they cannot live, and maximum cell size will be restored when gametes are produced and the diatom enters a phase of sexual reproduction. Various modes of gamete formation and various gamete morphologies occur in diatoms. In centric diatoms the female gamete or egg is nonmotile, and the male gametes are uniflagellate; in pennate diatoms the gametes are morphologically alike and nonflagellate. Whatever the gamete type, fusion of male and female gametes results in the formation of a specialized zygote cell termed an **auxospore**. A diatom

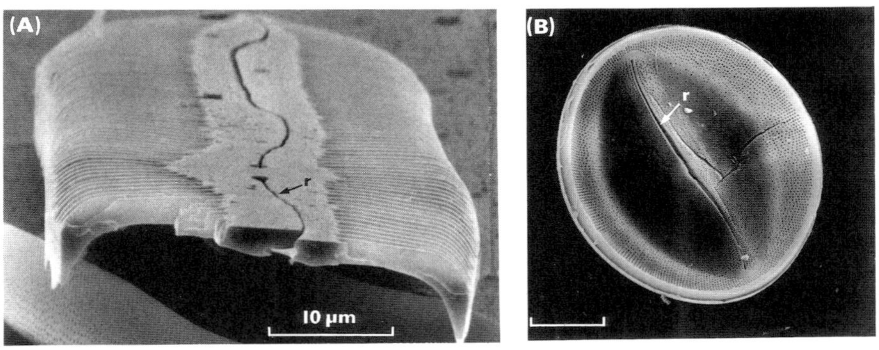

Figure 10.6 The raphe (r)—longitudinal slits through the valve wall—of benthic, pennate diatoms. A, *Pinnularia*; B, *Cocconeis*. Scale bars, 10 µm.

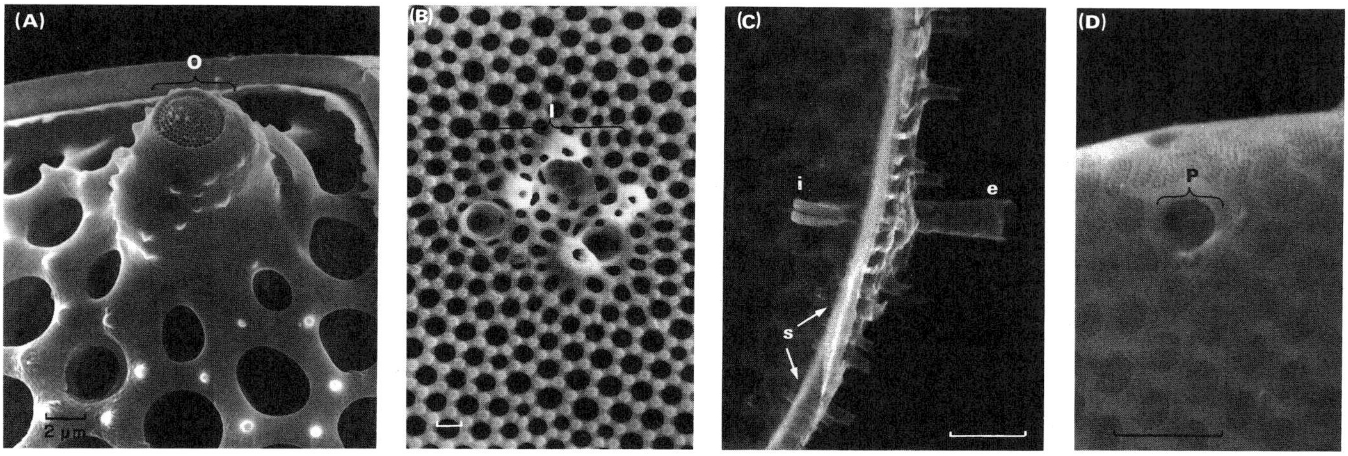

Figure 10.7 Close-up views of specialized structures of diatoms as seen in the scanning electron microscope. A, Ocellus (o) of *Odontella* showing fine pores. Mucilage, which is secreted through the fine pores, aids in attachment to substrate. B, Multiple central labiate processes (l) of *Azpeitia*. C, Marginal labiate process of *Thalassiosira* showing internal (i) and external (e) tubes. Numerous marginal strutted processes (s) are also visible. D, Pseudonodulus (p) of *Actinocyclus*.

cell with a maximum-sized frustule is then produced within the auxospore.

Resting spores

Many neritic planktic diatoms have a dimorphic life-cycle, with the normal planktic vegetative form alternating seasonally with a thicker-walled resting form or cyst that may sink to the bottom. Siliceous resting spores or **statospores** commonly form after a period of active vegetative reproduction when nutrient levels have been depleted or reduced. Statospores have two overlapping valves like the frustule, but they lack a girdle band, and the two valves are dissimilar; the epivalve is commonly spinose and the hypovalve smooth. Statospores may remain within the parent cell or may be isolated from it. Restoration of higher nutrient levels coupled with the proper length of day causes the statospore to germinate and the cell returns to its normal vegetative state. If the statospore has been retained within the parent frustule, the cytoplasm expands to conform to the size of that frustule. If the statospore has been isolated from the parent cell, the cytoplasm expands greatly, rounds off somewhat like an auxospore, and produces a new frustule with a diameter about twice that of the original parent. Seasonal upwelling may be an active mechanism for returning statospores or their vegetative products to levels higher in the water column where photosynthesis is more effective.

Ecology

Diatoms are photosynthetic and normally require light and CO_2 for growth and reproduction. A few diatoms, however, are facultative heterotrophs and are capable of using preformed organic compounds for growth in the dark or in light in the absence of CO_2. Concentrations of nitrogen, phosphorus, and silica are important limiting factors to diatom growth and reproduction, and diatom distribution and abundance in the modern world's oceans

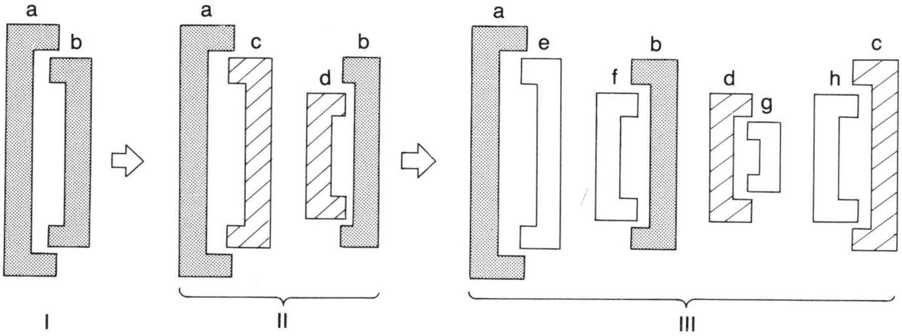

Figure 10.8 Cross-section of diatom frustule through two asexual reproductions showing that the large epivalve is inherited from the parent by the daughter and that the mean size of the population decreases with successive divisions. Roman numerals indicate generations. Arrows indicate asexual reproductions. Letters identify the valves through this process.

show a direct correspondence to areas of higher phosphate concentration in surface waters. Individual species typically also have specific salinity and temperature tolerances, so that diatom assemblages change as characteristics of water masses change. Additional elements such as sulfur, boron, barium, and manganese, as well as vitamin B_{12} have been shown in culture studies to be necessary for growth in various diatom species.

Seasonal changes in temperature, light, and nutrient concentrations commonly result in a seasonal succession of dominant planktic diatom species. Intense blooms of selected species may be caused by seasonal upwelling, which enriches surface waters within the photic zone with nutrients that are normally entrained in deeper waters. In coastal marine environments, such diatoms as *Chaetoceros* and *Skeletonema* grow rapidly (one to three cell divisions per day) in response to the onset of upwelling until they dominate the phytoplankton; during blooms, the abundance of diatoms in a given volume of sea water may increase 100-fold. Blooms last on the order of 2–3 weeks and terminate when nutrient supplies are depleted, causing a rapid decrease in the abundance of the blooming species. Immediately following the bloom, the blooming species commonly forms abundant resting spores, and different species more tolerant of lower nutrient concentrations increase in abundance. Further changes occur in succession in the local diatom assemblage, as seasonal changes in local oceanographic conditions occur; for example, an offshore oceanic assemblage may move into the coastal areas as upwelling and seasonal coastal currents diminish. Freshwater diatoms also experience seasonal blooms when seasonal winds or temperature changes result in mixing of stratified lakes bringing nutrients to the surface, where they can be used by planktic diatoms.

Biogeography

Areas of high diatom production in the modern oceans correspond closely with those areas that contain high surface water concentrations of nitrates and phosphates (Fig. 10.9). Nutrients are rapidly depleted from surface waters by phytoplankton, and a renewed supply of nutrients must be brought up from deeper waters by upwelling in order to sustain high levels of biologic productivity and abundant diatom growth. Coastal areas affected by eastern boundary currents along the coasts of California, Peru, and southwestern Africa are especially prone to upwelling. Upwelling may also be caused by diverging surface currents, such as in the eastern equatorial Pacific. In colder areas of the oceans, near Antarctica and in the North Pacific north of 40° N, surface-water temperatures do not differ significantly from deep-water temperatures, and nutrients can more easily rise to the surface from deeper waters without wind-enhanced upwelling and allow high diatom productivity.

Three major belts of diatomaceous sediments are found in the modern oceans:

1 a southern belt circling the globe between 45° and 65° S;

2 a northern belt within the Pacific including the Sea of Okhotsk, Sea of Japan, and Bering Sea, and in the Atlantic developed somewhat in the Norwegian Sea;

3 an equatorial belt which is well defined in the Pacific and Indian Oceans and less well defined in the Atlantic Ocean (Fig. 10.10).

These belts of sediment coincide with regions of abundant diatoms in the surface waters (Fig. 10.9), although dilution by carbonate sediments occurs in equatorial regions, and dilution by terrigenous debris occurs near the continents.

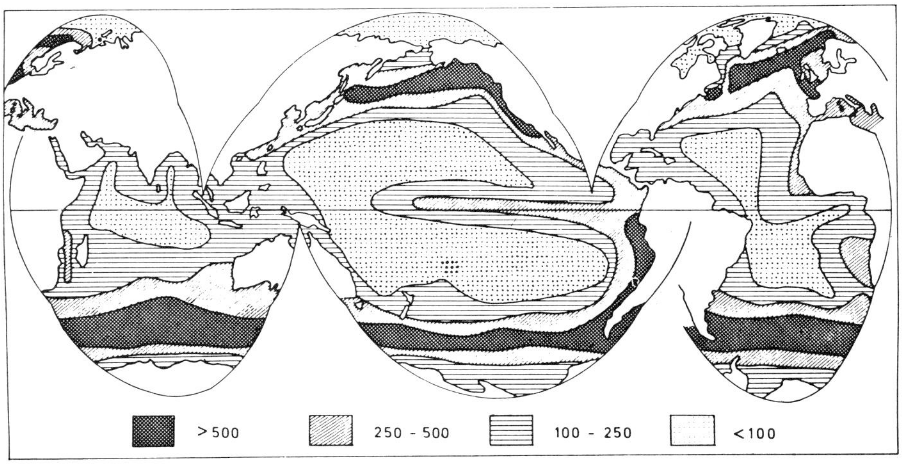

Figure 10.9 Global variation of the extraction of dissolved silica (g SiO_2/m^2 per year) by phytoplankton in near-surface ocean waters. This represents relative diatom production in near-surface waters, because diatoms are overwhelmingly the dominant siliceous phytoplankton group in the oceans. (Reproduced from Calvert SE. *Pelagic Sediments on Land and Under the Sea. Special Publication No. 1, International Association on Sedimentology.* Oxford: Blackwell Scientific Publications, 1974:279.)

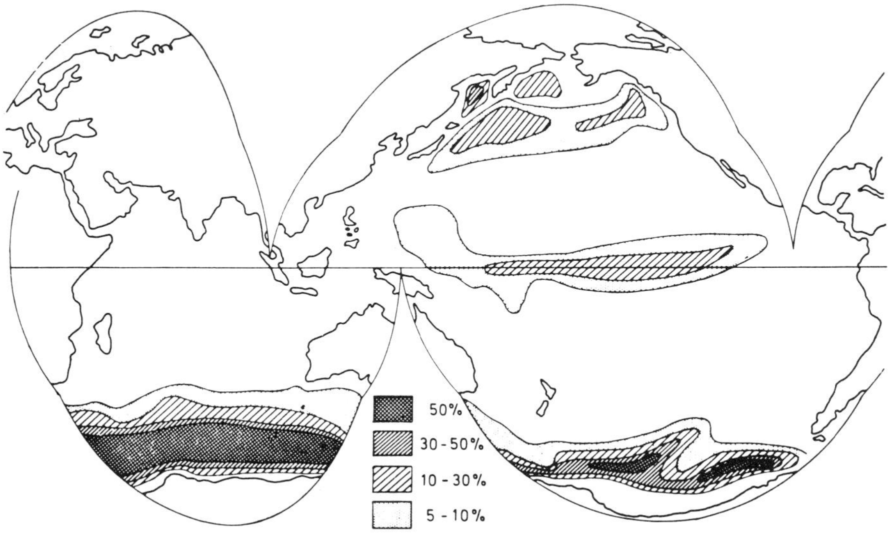

Figure 10.10 Distribution and concentration of bigenous opal in surface sediments of the Pacific and Indian Oceans on a $CaCO_3$-free basis. Diatoms dominate in the high-latitude areas, whereas radiolarians dominate by weight percent (but not necessarily by numbers) in the tropics. (Reproduced from Calvert SE. *Pelagic Sediments on Land and Under the Sea. Special Publication No. 1, International Association on Sedimentology.* Oxford: Blackwell Scientific Publications, 1974:282.)

Distinct planktic diatom assemblages are found in regions of differing surface water masses. Within the North Pacific, equatorial, subtropical, subarctic, northwestern marginal, and transitional diatom assemblages coincide with the various water masses and result from differences in salinity and water temperature. Many diatom species are endemic to a given region; for example, certain forms are not found outside of the Antarctic region. Other diatom species are more cosmopolitan in their distribution and can be found in varying numbers throughout the world's oceans.

Paleoecology

Diatoms can be excellent indicators of paleoenvironments and relative paleotemperatures. The distribution of living species is influenced by differences in temperature, nutrients, salinity, and other physical–chemical parameters. Good preservation of diatom frustules, however, requires both rapid burial and the lack of exposure to burial temperatures in excess of 50°C or to more alkaline conditions in sediment pore waters (pH >7). Sea water itself is undersaturated with respect to silica so that, after death, bacterial action causes decay of the organic coating of mucus surrounding the diatom frustule, and dissolution of the frustule begins. Thus, after death, the longer a marine diatom frustule is exposed to sea water, the more likely is it to be dissolved. Only 1–5% of the living assemblage of diatoms in surface plankton is estimated to compose the death assemblage found in sea-floor sediments. Those diatom valves that are preserved in the sediments are either robust—they commonly include the thickly silicified resting spores of neritic diatoms—or they have settled rapidly through the water column by settling in aggregates or by incorporation into the **fecal pellets** of zooplankton.

Ratios of warm- to cold-water diatoms have been used to estimate paleotemperatures. Tertiary paleoecologic studies typically involve quantitative analysis of diatom assemblages and require inference of the environmental tolerances of extinct species. Quaternary paleoclimatic studies combine factor analysis of modern assemblages with correlation of the factored assemblages to modern temperature and salinity conditions. Down-core quantitative studies are used to determine fluctuations in the assemblages through time, and a paleoceanographic model is developed that can explain changes in the distribution of water masses through time and may allow the estimation of paleotemperatures.

Fossil diatoms can also be used to estimate paleosalinity, recognize glacial or interglacial stages in the Quaternary, and trace Antarctic Bottom Water characterized by the presence of endemic antarctic diatoms in the south Atlantic and Indian Oceans. Nonmarine diatoms are blown thousands of miles into the equatorial north Atlantic Ocean from dry lake beds in the Saharan region of Africa, and their presence in marine cores records fluctuations in the degree of aridity in northwestern Africa and the strength and distribution of winds.

Diatoms are also used extensively in paleolimnology. They are sensitive to physical and chemical changes in lakes, and changes in lacustrine diatom assemblages accurately tract environmental changes in lakes. Varved lake sediments provide a year-by-year record, so that the effects of human settlement and industrialization on lake conditions can be monitored through studies of diatoms.

As diatoms are photosynthetic and live within the **photic** zone, they have limited application as paleodepth indicators. Nevertheless, a transect from the strand line to the open ocean passes from assemblages dominated by benthic diatoms through assemblages dominated by diatoms which spend part of their life-cycle in the plankton to neritic planktic assemblages, and finally, to **holoplanktic** open ocean assemblages. Consequently, the relative distance to the strand line can be estimated by analyzing fossil assemblages, allowing for variations in down-slope transport.

Biostratigraphy

Diatoms offer many advantages for biostratigraphy over other microfossils. They are abundant and diverse at high latitudes where calcareous microfossils generally are relatively sparse or low in diversity. Thus, diatoms are the primary means of correlating high-latitude deep-sea sediments. Unlike calcareous microfossils, siliceous diatom frustules may be well preserved in sediments deposited in 5000 m or greater water beneath areas of high productivity. As small microfossils (generally 10–100 μm), diatoms are relatively easy to process, and large numbers are generally present in small sample sizes of 1 cm^3 or less.

The relative instability of the opaline silica of the diatom frustule is the primary limitation of diatoms for biostratigraphic studies. Opaline silica (opal-A) begins to convert to cristobalite (opal-CT) as buried sediments are exposed to temperatures in excess of 50°C causing destruction of the diatom frustule. Consequently, diatom frustules are normally removed by dissolution from sediments that have been deeply buried. Diatoms may not be dissolved if they are incorporated into dolomitic concretions early in their diagenetic history.

Recent advances in marine diatom biostratigraphy have largely been due to the DSDP (now, the Ocean Drilling Program). Recovery of deep-sea cores of relatively terrigenous-free sediment that has been subjected to little diagenesis has allowed the construction of a biostratigraphy for most of the Tertiary. Naturally, Neogene and Quaternary diatom biostratigraphies are considerably more refined than Late Cretaceous and Paleogene biostratigraphies, because more material with a greater geographic distribution is available. In addition, many Neogene and Quaternary diatom biostratigraphic events (species appearances and extinctions) have been correlated directly to magnetic stratigraphy, so that correlation of diatom biochronologies across paleobiogeographic boundaries is easier. Magnetic stratigraphy also establishes which species appearances and disappearances are diachronous across latitude, so that migrational events can be distinguished from evolutionary events. Middle Eocene to Oligocene zonations for both low and high latitudes have recently been proposed, but reference sections are few, and correlation between different paleobiogeographic provinces is difficult. Cretaceous to Early Eocene diatom biostratigraphy is still preliminary, and the effects of paleoenvironment on species ranges are hard to assess because of the scarcity of long, continuous reference sections.

Marine diatom biostratigraphy typically employs holoplanktic diatoms, because they are more widespread than benthic or **neritic** planktic diatoms. In shallow-water sections, however, local zonations utilizing neritic planktic and benthic diatoms as marker taxa may be necessary because of the scarcity of holoplanktic diatoms; correlation of such shallow-water diatom zonations to open-ocean zonations is, therefore, often difficult.

Both high- and low-latitude zonations have been developed, as diatoms are diverse and abundant in both climatic regions, and the assemblages show a high degree of provincialism, especially in post-Eocene sedimentary rocks. In low-latitude regions abundant diatoms are restricted to more fertile regions of the oceans (e.g., the eastern equatorial Pacific), and diatoms are not used as extensively for biostratigraphy as calcareous microfossils. Nevertheless, Late Cenozoic low-latitude diatom biostratigraphy is as refined as calcareous microfossil biostratigraphy and is well correlated to magnetic stratigraphy. The stratigraphic ranges of important planktic diatom genera are illustrated in Figure 10.11.

Generally, separate zonations have been proposed for the southern ocean (Antarctic) and the middle-to-high-latitude regions of the North Pacific and North Atlantic. Diatoms are especially useful for correlations in the southern ocean and in the North Pacific north of 40° N, because of the scarcity or low diversity of calcareous microfossils and the lack of well-developed radiolarian biostratigraphies in those regions. Although many diatom species have bipolar distributions, few show synchronous appearances and extinctions between the northern and southern hemispheres, and a number of species are endemic to one region or another, especially since the latest Miocene (ca. 6 Ma).

Age resolution within the Late Cretaceous to Oligocene is typically of the order of 1–3 My. Within the Miocene to Quaternary where numerous stratigraphic sections of diatoms have been studied and where correlation with paleomagnetic stratigraphy is fairly good, age resolution can approach 100 000–300 000 years. Late Cretaceous to

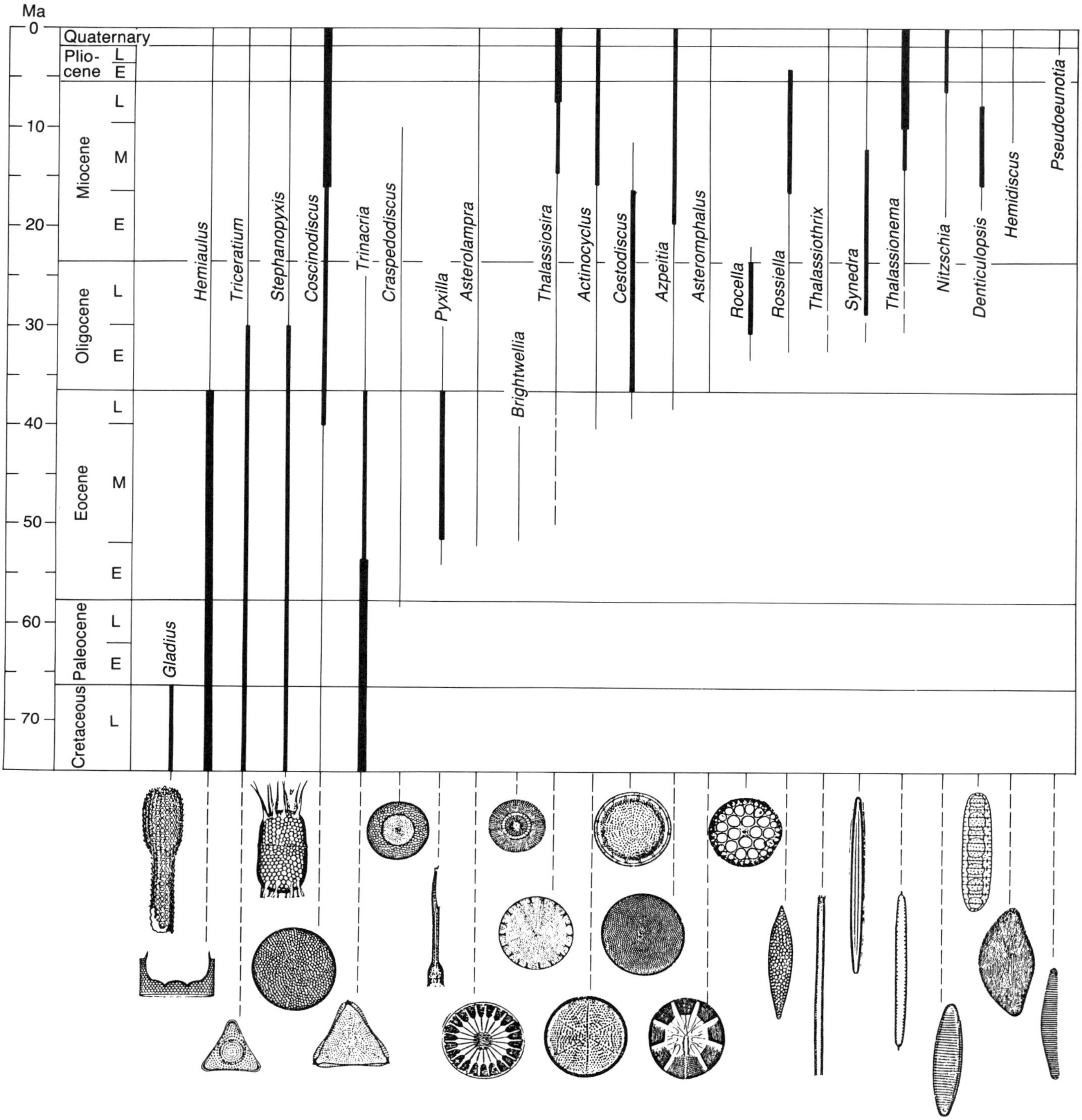

Figure 10.11 Stratigraphic ranges of important genera of planktic marine diatoms. Width of bar indicates the relative importance (abundance) of the genus during a given time interval. Recent subdivisions of genera based on SEM study suggest that *Triceratium* (as shown) includes the genus *Sheshukovia*, *Stephanopyxis* includes the genus *Pyxidicula*, and *Denticulopsis* includes the genera *Crucidenticula* and *Neodenticula*.

Eocene biostratigraphic zonations mainly use species events of *Hemiaulus*, *Triceratium*, *Pyxilla*, and *Trinacria*, whereas Oligocene to Quaternary zonations rely heavily on species events of *Rocella*, *Coscinodiscus*, *Actinocyclus*, *Nitzschia*, *Denticulopsis*, and *Thalassiosira* (Fig. 10.11).

Recent trends in marine diatom biostratigraphy include the use of quantitative data as a biostratigraphic tool. Quantitative diatom biostratigraphy is more commonly employed in the study of Quaternary sediments, where changes of abundance in detailed sampling intervals are referenced to oxygen isotope stratigraphy, paleomagnetic stratigraphy, or carbon-14 dating. Within a paleoceanographic province, abundance changes may be induced by abrupt and widespread environmental changes that may approach isochroneity. Size changes in the tropical Pacific diatom *Azpeitia nodulifer* can be used to recognize the last glacial maximum (oxygen isotope stage 2) and the preceding major interglacial event (oxygen isotope stage 5e). Small-sized individuals of *A. nodulifer* are relatively more abundant than larger-sized individuals during glacial intervals, whereas larger individuals are relatively more common during interglacial periods.

The development of nonmarine biostratigraphy has lagged considerably behind that of marine biostratigraphy because:

1 nonmarine deposits tend to be of limited areal extent so that patterns of succession are difficult to compare;
2 the deposits typically are poorly dated by means of other fossils;
3 individual deposits represent relatively brief periods of geologic time;
4 few specialists study fossil nonmarine diatoms.

Centric planktic diatoms are more reliable in nonmarine diatom biostratigraphy than pennate benthic taxa, and periods of diversification of centric nonmarine diatoms occurred during the early Middle Miocene, early Late Miocene, and Pliocene.

Evolutionary history

The oldest reliable record of diatoms is from the Early Jurassic (Toarcian stage), and reports of diatoms from older rocks have been shown to represent contamination. As stated above, the susceptibility of the opal-A of the diatom frustule to dissolution means that records of diatoms in rocks older than the Campanian (Late Cretaceous) represent very special conditions of preservation that include shallow burial, low-heat flow, and protection from dissolution processes in concretions early in the rock's diagenetic history.

The earliest record of abundant, well-preserved diatoms from the Aptian–Albian reveals that diatoms were already diverse with numerous genera with complex and differing morphologies. Diatoms must have had a much longer evolutionary history; the earliest diatoms may have been spherical and lacked a siliceous cell wall but possessed siliceous scales such as those found in related Chrysophyta. Such scales have been observed in the auxospores of several centric diatoms and their morphology is very much like that of the simplest valve structure of centric diatoms. Thus, the susceptibility of the opal-A of diatoms to dissolution and the possible lack of a well-developed siliceous frustule would explain why diatoms do not have an older fossil record.

Diatom generic extinction (23%) at the Cretaceous–Tertiary boundary is very low compared to generic extinction rates of coccolithophorids (73%), radiolarians (85%), and planktic foraminifera (92%) (Fig. 10.11). Many Late Cretaceous diatoms formed nonplanktic resting spore stages, and they might have survived short-lived catastropic events such as the global-darkening event which is postulated at the Cretaceous–Tertiary boundary.

No mass extinctions of marine diatoms have been observed during the Cenozoic, but times of relatively rapid turnover in marine diatom assemblages occurred near the Early Eocene–Middle Eocene boundary, in the middle part of the Oligocene, and in the middle part of the Middle Miocene. Each of these events represents the gradual extinction of a number of diatom species that had characterized assemblages for millions of years and their gradual replacement by newly evolved species. Other periods of change were at the end of the Middle Eocene and at the Oligocene–Eocene boundary, where the dominant species of low-latitude assemblages show abrupt abundance changes. In the Late Eocene to earliest Oligocene increased provincialism developed, so that high- and low-latitude diatom assemblages become difficult to correlate.

Another time of major change was the latest Miocene (6–5 Ma) during which provincialism increased in the North Pacific, and a number of species of *Thalassiosira* evolved that were to characterize the Pliocene. Interestingly, planktic nonmarine diatoms also showed major changes in the latest Miocene with the appearance of *Stephanodiscus* and *Cyclostephanos*, genera that radiated in the Pliocene and Pleistocene.

The onset of northern hemisphere glaciation at about 2.5 Ma had a major effect on global paleoceanography, and a number of extinctions of both high- and low-latitude diatoms are concentrated in the Late Pliocene between 2.5 and 2.0 Ma. The latest Pliocene and Pleistocene are characterized by diatom assemblages that differ very little from modern assemblages, although in some areas of the ocean individual species may show large fluctuations of abundance that correspond to changes between glacial and interglacial conditions. Each of these times of high turnover in marine diatom species coincides with periods of major rapid cooling at high latitudes. These successive cooling periods were part of an overall Cenozoic cooling trend, which was interrupted by more stable periods that contained some relatively brief intervals of warming.

Presumably, relatively stable paleoceanographic conditions were disrupted by these cooling steps, resulting in changes in the diatom assemblages. At various times, notably in the late Eocene and latest Miocene, a steepening of latitudinal thermal gradients caused an increase in provincialism. High-latitude cooling may also have resulted in an increased surface-water circulation and increased equatorial upwelling, which, in turn, resulted in changes of diatom assemblages in the low-latitude oceans; for example, during the middle Miocene. As diatoms are accurate indicators of surface-water conditions, major changes in Cenozoic marine-diatom assemblages typically correspond with the major climatic cooling steps of the Tertiary.

Late Cretaceous to Eocene diatom assemblages are dominated by relatively robust genera such as *Gladius, Hemiaulus, Triceratium, Trinacria, Stephanopyxis,* and *Pyxilla* (Fig. 10.11). During the early Oligocene these robust forms begin to decline and are gradually replaced by more finely silicified genera such as *Coscinodiscus, Cestodiscus, Thalassiosira, Synedra,* and *Thalassionema* (Fig. 10.11). This trend toward more finely silicified forms is intensified in the middle and late Miocene as the genera *Thalassiosira, Thalassionema, Nitzschia,* and *Denticulopsis* increase in diversity and numbers in both the high- and the low-latitude oceans. Many of the more finely silicified forms that expand during the late Cenozoic are more elongated in shape compared to the more rounded outline of early Cenozoic forms (Fig. 10.11). Even more finely silicified and elongated diatoms such as *Chaetoceros* and *Skeletonema* dominate modern coastal upwelling assemblages in temperate regions of the world, but the frustules of these forms dissolve rapidly upon death and they are poorly represented in the sediment record.

This Oligocene to Quaternary trend toward more finely silicified forms coincides with an increasingly more vigorous circulation of the ocean's surface and deep waters. Increased mixing of the oceans renews silica and other nutrients to the surface waters of the oceans especially in regions of upwelling. Seasonal upwelling along with the proper duration and intensity of light promotes rapid diatom growth. In order to reproduce rapidly and exploit the high levels of nutrients initially available during seasonal upwelling, it would be an advantage for a diatom species to use relatively small amounts of silica in the construction of its frustules. Upon death of individual diatoms, finely silicified frustules would also dissolve more readily than more heavily silicified frustules and reintroduce silica to the environment where it can be used by growing diatoms.

The earliest freshwater diatoms are pennate forms known from the Paleocene of Russia, but no nonmarine diatoms are known prior to the Late Eocene in North America. Freshwater diatoms may have had a separate evolutionary history from marine diatoms since at least the Cretaceous, but poor taxonomic studies and the lack of preserved diatoms in marginal environments hinder studies on the origin of freshwater diatoms. The invasion of freshwater habitats by diatoms may have occurred more than once, as different freshwater genera show affinities to a number of marine genera.

In any case, freshwater diatoms first diversified during the Oligocene, but planktic freshwater diatoms exhibited other periods of expansion during the middle Miocene and the Pliocene and Pleistocene. Radiation during the middle Miocene might have been due to increased volcanism that supplied silica for diatom growth; and it may have been aided by the relatively high global sea level of the middle Miocene, which might have eased the invasion of nonmarine environments through the flooding of the continents. Increased middle Miocene volcanism also created conditions that favored the preservation of nonmarine diatomites in isolated basins with little terrigenous input and layering of soft diatomaceous sediments between relatively resistant volcanic rocks. Diversification of planktic freshwater diatoms in the Pliocene and Pleistocene was largely confined to the family Thalassiosiraceae, and it may reflect increased seasonality and the ability of this group of diatoms to form resting spores to survive changing physical—chemical conditions in lakes.

Supplementary reading

Barron JA. Miocene to Holocene planktic diatoms. In: Bolli HM, Saunders JB, Perch-Nielsen K, eds. *Plankton Stratigraphy.* Cambridge: Cambridge University Press, 1985:763—809. Detailed biostratigraphy using diatoms for the marine Neogene.

Burckle LH. Marine diatoms. In: Haq BU, Boersma A, eds. *Introduction to Marine Micropaleontology.* New York: Elsevier/North Holland, 1978:245—66. A good overview of diatom paleontology.

Calvert SE. Deposition and diagenesis of silica in marine sediments. In: Hsu KJ, Jenkyns H, eds. *Pelagic Sediments on Land and Under the Sea. Special Publication of the International Association on Sedimentology No. 1.* Oxford: Blackwell Scientific Publications, 1974:273—99. Review of siliceous sedimentation and diagenesis focussing particularly on diatoms.

Fenner J. Late Cretaceous to Oligocene planktic diatoms. In: Bolli HM, Saunders JB, Perch-Nielsen K, eds. *Plankton Stratigraphy.* Cambridge: Cambridge University Press, 1985:713—62. Detailed diatom biostratigraphy for the marine Late Cretaceous and Paleogene.

Hustedt F. Die Kieselalgen Deutschlands, Osterreichs und der Schweiz. In: Rabenhorst L, ed. *Krypogamen-flora von Deutschland, Osterreich und der Schweiz*. Vol. 7. Leipzig: Akademie Verlasges, 1927–66: part 1, pp. 1–920, figs 1–542, 1927–1930; part 2, pp. 1–845, figs 543–1179, 1931–1959; part 3, pp. 1–816, figs 1180–1788, 1962–1966. A key reference on diatom taxonomy and ecology.

Ricard M, ed. *Proceedings of the Eighth International Diatom Symposium, Paris, August 27–September 1, 1984*. Koenigstein: Koeltz Scientific Books, 1984:781. Many recent studies of modern and fossil diatoms.

Schmidt A, *et al. Atlas der Diatomaceenkunde*. Leipzig: OR Reisland, 1874–1959:480 pp. A catalog with mostly hand-drawn illustrations of marine and nonmarine diatoms.

Simonsen R. The diatom system: ideas on phylogeny, *Bacillaria*. Vol. 2. Braunschweig: J. Cramer, 1979:9–71. A classification and suggested phylogeny for diatoms.

Tappan H. *The Paleobiology of Plant Protists*. San Francisco: WH Freeman, 1980:1028. Excellent detailed review of diatom biology and paleontology in Chapter 8.

Vanlandingham SL. *Catalogue of the Fossil and Recent Genera and Species of Diatoms and their Synonyms*. 8 volumes. Lehre and Vaduz: J. Cramer, 1967–79. A useful compilation of diatom species.

Werner D, ed. *The Biology of Diatoms*. Botanical Monographs, Vol. 13. Berkeley and Los Angeles: University of California Press, 1977:498. A complete discussion of the biology of diatoms by a variety of specialists.

Calcareous nannoplankton

11

William G. Siesser

Introduction

Calcareous nannoplankton are minute, algal protists that first appeared in the Late Triassic, and whose descendants live in great abundance in the surface waters of today's oceans. Living nannoplankton cells are oval to elongate and range in size from about 15 to 100 μm. During one stage of their life-cycle, calcareous nannoplankton secrete tiny (average length about 8 μm) platelets of calcite called **coccoliths**. The coccoliths interlock to form a spherical external skeleton, the **coccosphere** (Fig. 11.1), which surrounds the living cell.

Studies of nannoplankton have increased exponentially in the last 25 years, and fossils of these protists are now among the most important microfossils used in Mesozoic and Cenozoic biostratigraphic investigations. They are useful in biostratigraphy because:

1 they are extremely abundant in marine sedimentary rocks;

2 as planktic organisms they have a wide geographic distribution;

3 their rapid evolutionary changes allow recognition of narrow subdivisions of geologic time.

Living calcareous nannoplankton are assigned to the class Coccolithophyceae, and are therefore idiomatically referred to as **coccolithophores**. Coccolithophores are covered by circular or elliptical coccoliths (Figs 11.1, 11.2A). Many older rocks contain similar-sized calcareous objects of diverse shapes (e.g., shaped like stars, horseshoes, nails, baskets, etc., Fig. 11.2) which are found together with the circular or elliptical coccoliths. These noncircular, nonelliptical platelets are often called **nannoliths** to distinguish them from coccoliths. Some workers simply use the term coccolith in a broad sense, however, referring to all the calcareous fossils (i.e., the coccoliths sensu stricto and the nannoliths) that occur in the 2–25 μm size range, regardless of their shape. Coccolith will be used in the broad sense in this chapter.

The biologic characteristics of the extinct organisms that secreted nannoliths are unknown. Many nannoliths resemble coccoliths somewhat, and in all likelihood are related to coccolithophores. However, great differences in the shapes of some of the nannoliths suggest that they could belong to an entirely different class or even division. The term **calcareous nannofossils** has therefore gained wide acceptance when referring specifically to extinct forms, which may or may not actually be related to coccolithophores, but whose fossils occur with, and traditionally have been studied with, the coccoliths of the coccolithophores. Calcareous nannoplankton is best used as a collective term for the group when generally referring to both living and fossil forms.

History of study

C.G. Ehrenberg first recorded "coccoliths" in 1836 while examining chalk from the island of Rugen in the Baltic Sea. He considered the elliptical, flattened disks he saw to be of inorganic origin, believing them to have formed as spherulitic concretions.

T.H. Huxley recognized the same disk-shaped fossils in deep-sea sediments collected by HMS *Cyclops* in 1857. He named them "coccoliths" since they looked "...somewhat like single cells of the plant *Protococcus*...." Huxley nevertheless attributed an inorganic origin to the coccoliths, as had Ehrenberg. G.C. Wallich also found coccoliths in samples from the north Atlantic sea floor. In addition to isolated coccoliths, Wallich found coccoliths joined together to create minute spheres, which he named "coccospheres" (Figs 11.1, 11.2A). Unlike Ehrenberg and Huxley, Wallich believed that coccospheres were organic in origin, and speculated that they were the larvae of planktic foraminifers.

In 1861, H.C. Sorby found coccospheres in the English chalk. Sorby also believed coccospheres were the remains of some organism, but doubted that they belonged to a foraminifer. He based this conclusion on the differences in optical properties of the calcite composing coccosphere and foraminifer skeletons. After publication of Sorby's paper, most natural scientists concluded that coccoliths

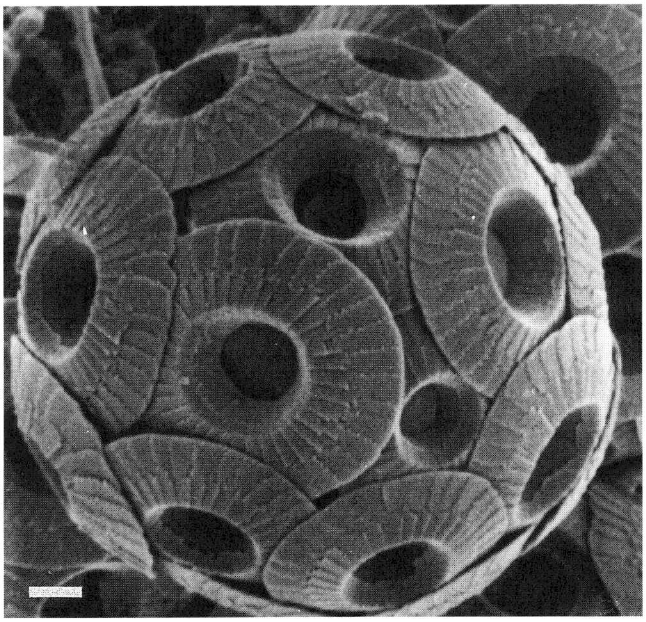

Figure 11.1 Scanning electron micrograph of a coccosphere of *Umbilicosphaera sibogae*. Interlocking circular coccoliths form a hard external skeleton around the soft cellular material of the protist. Scale bar, 1 μm. (Reproduced from Nishida S. *Atlas of Pacific Nannoplankton*. Micropaleontology Society of Osaka, 1979:5.)

and coccospheres were organic in origin. However, Ehrenberg, the venerable microscopist and discoverer of coccoliths, never accepted this opinion. As late as 1873, in one of his last papers, he still referred to them as inorganic structures.

In 1872, the British ship **HMS Challenger** sailed on the first major oceanographic expedition, a 4-year, globe-encircling cruise. John Murray was the biologist on board, and he and his coworkers found coccospheres floating in the ocean water. They eventually concluded that coccospheres were the skeletal parts of calcareous algae and noted that they occurred in surface waters ranging from the tropics to higher latitudes with temperatures as low as 7°C.

Intensive systematic work on calcareous nannoplankton began in the first decades of the twentieth century. This detailed work was initially done by only a few workers: H. Lohmann, C.H. Ostenfield, A.D. Arkhangelsky, and J. Schiller, who discovered, described, and named many living, and a few fossil, calcareous nannoplankton. Additional work on the biology and physiology of the group resulted in the discovery by Ostenfield of the nucleus of the cell in 1900; of the presence of two golden-brown chloroplasts on either side of the nucleus by A. Weber van Bosse in 1901; and of flagella on the organism by Lohmann in 1902. Lohmann coined the word nannoplankton for very small plankton which could pass through the normal mesh of a phytoplankton net (60 μm or less in size).

In the late 1920s and 1930s E. Kamptner in Austria, G. Deflandre in France, and T. Braarud and his colleagues in Norway described many living and fossil species. During the 1950s and 1960s, the Norwegian group investigated the cytology, physiology, and ecology of living nannoplankton and pioneered attempts at culturing them in the laboratory. M. Parke and her colleagues in Plymouth, England also studied the growth, calcification, and lifecycle of living nannoplankton during this period.

Improved microscopy techniques were developed during the early 1950s which markedly aided the study of nannoplankton. Kamptner applied cross-polarized light to nannoplankton study, and T. Braarud and E. Nordii first used phase-contrast illumination and transmission electron microscopy. The transmission electron microscope (TEM) was routinely used from the mid 1950s onward, and large numbers of TEM photographs of nannoplankton were published to supplement photographs made using the transmitted-light microscope.

In 1954 M.N. Bramlette and W.R. Riedel first demon-

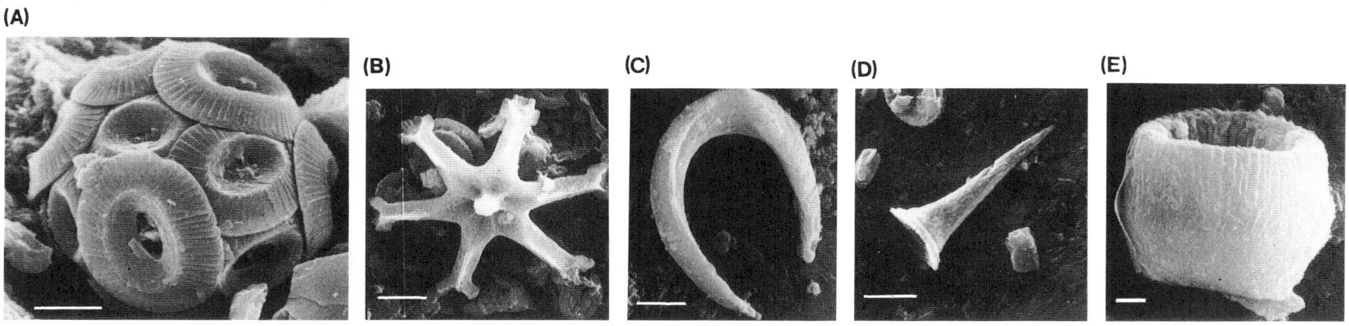

Figure 11.2 Scanning electron micrographs showing diverse coccolith and nannolith shapes. A, Coccosphere of *Coccolithus pelagicus*, built of elliptical coccoliths. Note the overriding and imbrication of adjacent crystal elements making up each coccolith. B–E, Nannoliths. B, *Discoaster surculus*. C, *Amaurolithus delicatus*. D, *Blackites spinosus*. E, *Scyphosphaera apsteinii*. All scale bars, 4 μm.

strated the usefulness of calcareous nannofossils in biostratigraphy. The initiation of the **Deep Sea Drilling Project** (DSDP) in 1968 made the stratigraphic value of calcareous nannofossils apparent to the larger scientific community. Stratigraphic ages could be determined accurately and quickly. Abundant, well-preserved nannofossil assemblages contained in the relatively continuous DSDP cores soon led to the development of refined nannofossil zonations. The scanning electron microscope (SEM) replaced the TEM as an investigative tool in the late 1960s and early 1970s, and soon became widely used for nannoplankton illustration. The transmitted-light microscope equipped with cross-polarizers and phase-contrast illumination is, however, still the preferred tool for routine biostratigraphic work.

Nannoplankton research reached a mature stage in the 1970s and 1980s. Highly refined biostratigraphic zonations are now available for the Jurassic, Cretaceous, and Cenozoic. Nannofossils have been used in detailed taxonomic studies, paleobiogeographic mapping, paleoclimatic and paleoceanographic reconstructions, evolutionary studies, biochronology, diagenesis, and stable-isotope investigations during this period.

Figure 11.4 Coccosphere of *Gephyrocapsa oceanica*, a heterococcolith. Crystal elements making up the shield, the bridge, and the grid across the central-area opening of each coccolith have different sizes and shapes. Scale bar, 1 μm. (Reproduced from Nishida S. *Atlas of Pacific Nannoplankton*. Micropaleontology Society of Osaka, 1979:12.)

Morphology and systematics

Morphology

Living coccolithophores are covered by a layer of organic scales on which rest small calcite platelets: the coccoliths. Coccoliths range in size from 1.3 to 50 μm on living coccolithophores, but most coccoliths found in the fossil record range from 2 to 25 μm; their average size is about 7 or 8 μm.

Coccoliths are divided into two general groups: **heterococcoliths** and **holococcoliths**. Heterococcoliths are constructed of crystals that differ in size and shape (Figs 11.3, 11.4), whereas the crystals forming holococcoliths are essentially identical in size and shape (Fig. 11.5) and are considerably smaller than the crystals forming heterococcoliths.

Most coccolithophores produce only heterococcoliths, and then only during the passive floating phase of their life-cycle. Living coccolithophores that do produce holococcoliths (about 30 species), do so only during their swimming phase. One species, *Coccolithus pelagicus* (Fig. 11.2A), is known to produce holococcoliths during its swimming phase and heterococcoliths during its floating phase.

Most heterococcoliths are formed of rhombohedral crystals; a few species are built of hexagonal prisms. The simplest arrangement of rhombohedral crystals on a heterococcolith is a single disk (Fig. 11.12A), with a raised wall of rhombohedral crystals at the circumference.

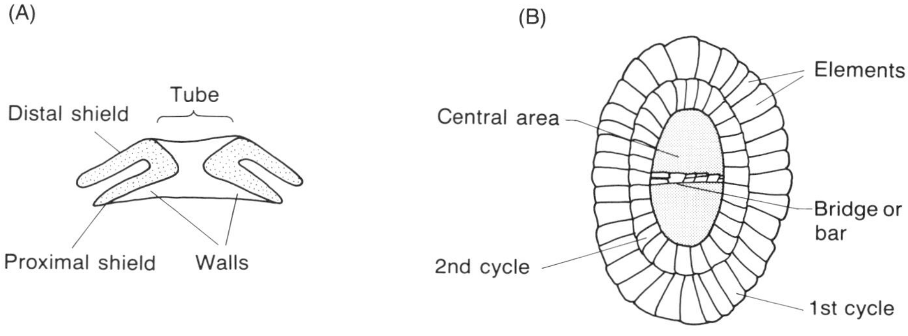

Figure 11.3 Diagrammatic morphology of a heterococcolith composed of two shields. A, Side view showing distal and proximal shields. B, Top view of distal shield.

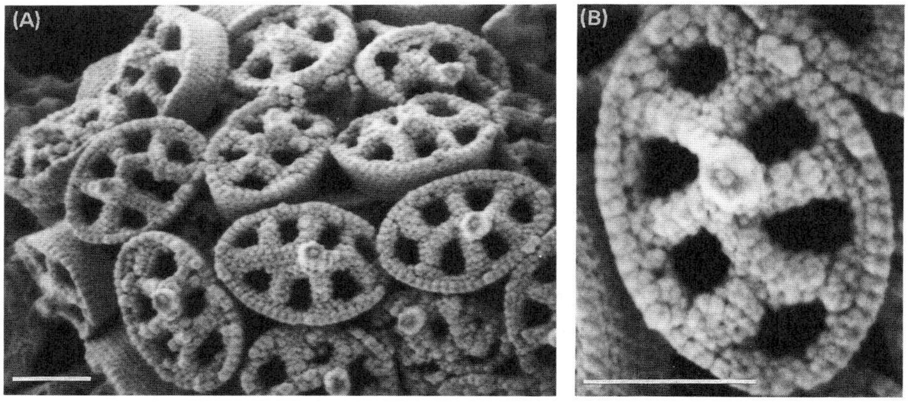

Figure 11.5 A, Coccosphere of *Helladosphaera* sp., a holococcolith. B, A single coccolith from *Helladosphaera*, made up of tiny crystal elements similar in size and shape. Both scale bars, 1 μm. (Reproduced from Nishida S. *Atlas of Pacific Nannoplankton*. Micropaleontology Society of Osaka, 1979:18.)

Other rhombohedral crystals grow inward from the base of the raised wall towards the center of the disk. Another, more complex arrangement of heterococcoliths consists of two circular or elliptical shields (Figs 11.3, 11.4, 11.6) joined by a central cylindrical column, but without a raised wall. The central column is either a solid pillar or a hollow tube with its wall composed of rhombohedral crystals, either joined like staves in a barrel, or spiraling obliquely down the wall of the tube (Fig. 11.3). The joined shields are denoted as proximal and distal. The proximal shield is on the inner (concave) side of the coccolith, and the distal shield is on the outer (convex) side of the coccolith (Fig. 11.6). Most heterococcoliths have, in addition, a distinctive closed or open central area (Figs 11.4, 11.6) from which crystal elements of the shield radiate outward. Shield elements taper near the central area in order to fit better around a circular center, whereas they expand towards the circumference, giving them a petal-like shape. The crowding together of elements often causes overriding and imbrication of adjacent elements (Figs 11.2A, 11.8J). A central process (a spine or knob) may project outward from the central area (Fig. 11.10C).

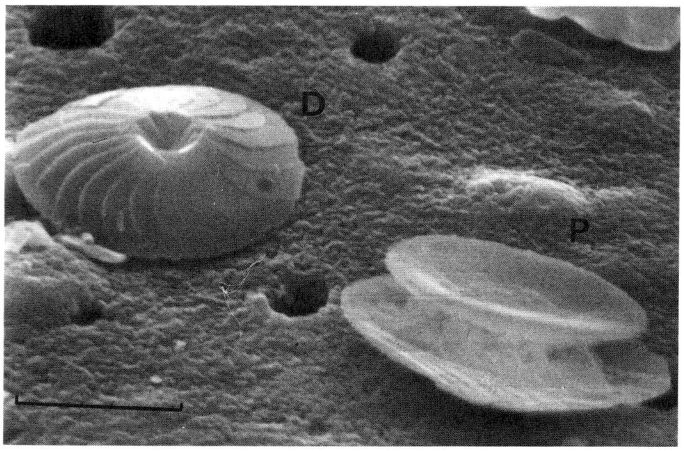

Figure 11.6 Detached coccoliths of *Calcidiscus leptoporus*, showing distal (D) and proximal (P) shields. On a coccosphere, the outer edge of a coccolith's distal shield wedges into the gap between the distal and proximal shields of adjacent coccoliths. Scale bar, 4 μm.

Holococcoliths are much less common than heterococcoliths among living coccolithophores; they are even less common in the fossil record. Their construction probably permits them to disintegrate more rapidly than the more robust heterococcoliths. Holococcoliths may be constructed of hexagonal prisms, rhombohedral crystals, or both. The crystal elements making up holococcoliths are commonly less than 0.1 μm in diameter. Their characteristic similarity in size and shape is shown in Figure 11.5.

The calcite crystals forming hetero- and holococcoliths often have differently arranged **optic axes**, resulting in a variety of **optic extinction patterns** seen between the crossed nicols of a polarizing microscope. Moreover, the distal shield of a heterococcolith may have an arrangement different from that of the proximal shield; or several cycles of crystals may be present on the same shield with axes arranged differently among the different cycles (Fig. 11.7). Collars, bridges, crossbars, spines, and other struc-

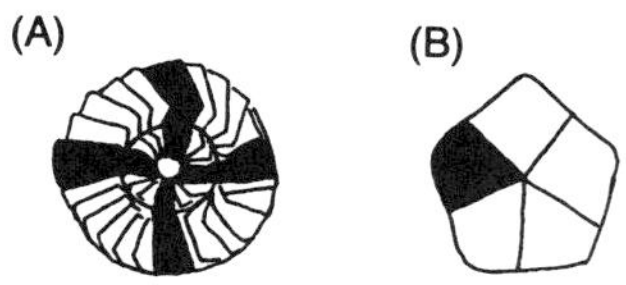

Figure 11.7 Diagrammatic illustration of coccolith optic extinction patterns between crossed nicols of a polarizing microscope. A, Dark lines swing across the distal shield of a placolith as the microscope stage is rotated. B, Successive segments of a pentalith darken as the stage is rotated.

tures of the central area may also have varying arrangements of axes. These varying structural arrangements produce optic extinction patterns which differ among species. Thus species are routinely identified on the basis of a particular pattern (Figs 11.28, 11.29). A few taxa are faint or entirely dark between crossed nicols. For example, the star-shaped genus *Discoaster* (Fig. 11.2B) normally lies flat on a microscope slide, with its optic axis (**c-axis**) perpendicular to the surface of the slide, causing it to appear dark between crossed nicols.

A number of terms have been introduced to describe the overall shapes which nannoplankton may take. These terms can replace more lengthy descriptive phrases. Table 11.1 is a list of the more common shape terms.

Classification

A characteristic of coccolithophores is that, like plants, they are autotrophic: they possess chloroplasts which are used to produce food photosynthetically. Coccolithophores swim, however, and a few are known to ingest bacteria and small algae. Despite these animal-like characteristics, most workers consider nannoplankton to be plant-like protists, based primarily on their ability to photosynthesize.

Coccolithophores possess two flagella and a third appendage, the **haptonema**, and thus are included in the division Haptophyta. Coccolithophores were originally considered to belong to the class Haptophyceae within the division Chrysophyta. The Haptophyta have recently been upgraded to division rank. Many living haptophytes develop only a scale cover and lack coccoliths. The class Coccolithophyceae was therefore created to include only those haptophytes that secrete coccoliths.

A number of interim and partial classification schemes for calcareous nannoplankton were published throughout the 1960s and 1970s. During the last decade, several comprehensive classifications have been presented, although there is still no universally applied classification, especially for the orders and families. Details of the composite classification used here (Table 11.2) may be found in publications listed at the end of this chapter. Type species or typical forms are illustrated for each family (Figs 11.8–11.11).

Classification of species

Calcareous nannoplankton are best classified under the **International Code of Botanical Nomenclature** rather than under the **International Code of Zoological Nomenclature**, because the botanical code provides

Table 11.1 Shape terminology of calcareous nannoplankton

Asterolith	star or rosette shaped; projecting arms are called "rays" (Fig. 11.9B)
Calyptrolith	holococcolith; basket shaped, open proximally (Fig. 11.8G)
Caneolith	disk shaped; lath-filled central area (Fig. 11.10L)
Ceratolith	horseshoe or wishbone shaped (Fig. 11.8H)
Cribrilith	disk shaped; numerous central-area perforations (Fig. 11.10A)
Cricolith	elliptical ring shape (Fig. 11.8K)
Cyclolith	circular ring shape (Fig. 11.8L)
Cyrtolith	basket shaped, commonly with a projecting central process (Fig. 11.10F)
Discolith	disk shaped; thickened, raised rim; may or may not have central-area perforations; definitions for this term may vary; synonymous in part with cribrilith
Helicolith	spiraling, overlapping marginal flange (Fig. 11.9F)
Lopadolith	basket, cup, or vase shaped, with high rim, opening distally (Fig. 11.10B)
Nannoconid	cone or cylinder shaped, with an axial canal (Fig. 11.9J)
Pentalith	five four-sided crystals joined to form a pentagon (Fig. 11.8D)
Placolith	two shields joined by a central tube (Fig. 11.8J)
Prismatolith	polygonal prisms, solid or perforate (Fig. 11.11A)
Rhabdolith	single shield, surmounted by a stem (Fig. 11.10E)
Scapholith	rhombohedral shape, with parallel laths in the central area (Fig. 11.8E)
Sphenolith	prismatic base formed of radial crystals, surmounted by a cone (Fig. 11.10I)
Stelolith	cylinder or column shaped (Fig. 11.9E)
Stephanolith	polygonal or cylindrical shape; hollow with radiating spokes (Fig. 11.10K)
Zygolith	elliptical ring with a crossbar supporting a central process (Fig. 11.11C)

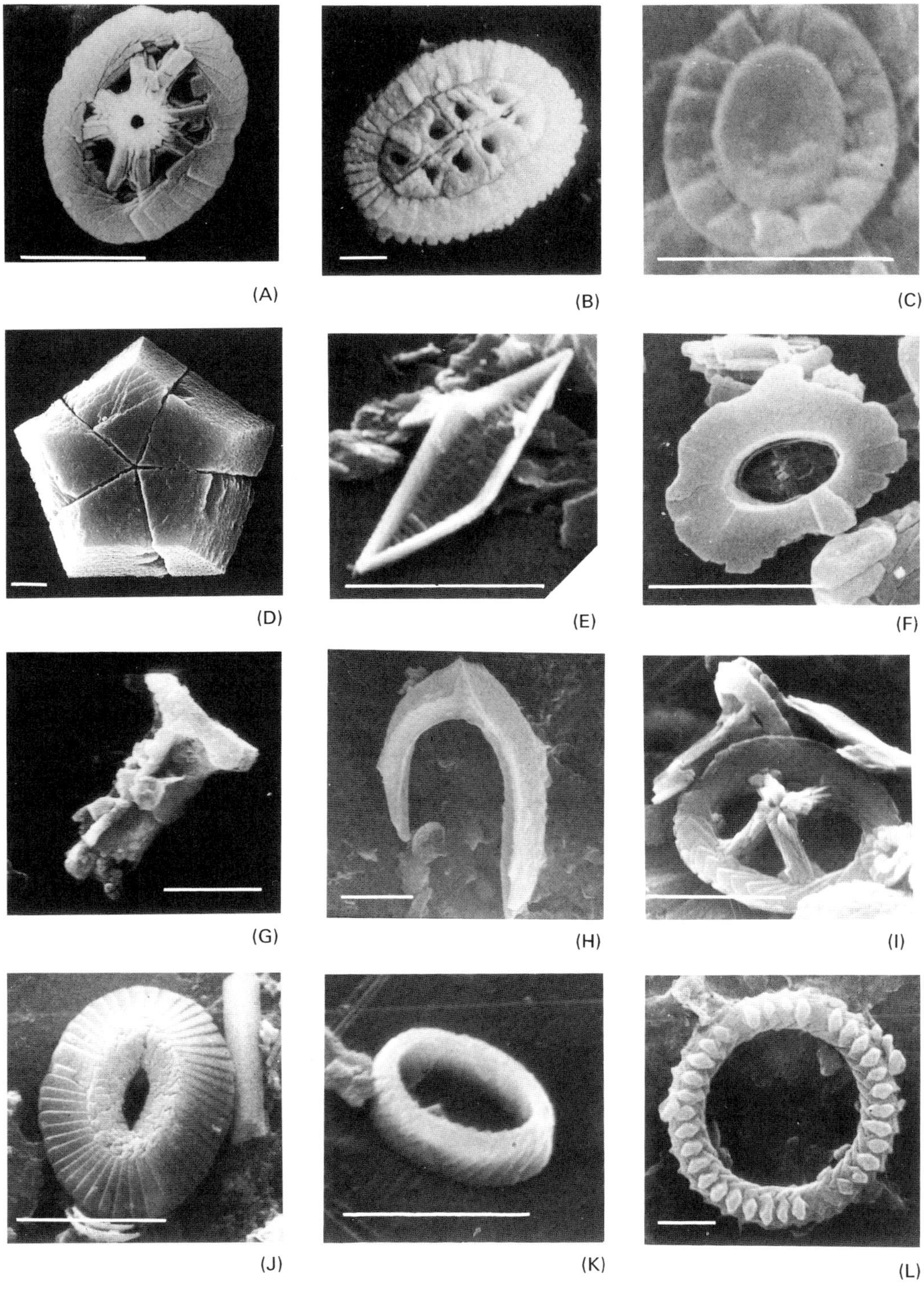

Figure 11.8 Scanning electron micrographs of the type species or a representative species of Coccolithophyceae families. A, Family Ahmuellerellaceae: *Ahmuellerella octoradiata*. B, Family Arkhangelskiellaceae: *Arkhangelskiella cymbiformis*. C, Family Biscutaceae: *Biscutum constans*. D, Family Braarudosphaeraceae: *Braarudosphaera bigelowii*. E, Family Calciosoleniaceae: *Anoplosolenia brasiliensis*. F, Family Calyculaceae: *Calyculus cribrum*. G, Family Calyptrosphaeraceae: *Zygrhablithus bijugatus*. H, Family Ceratolithaceae: *Ceratolithus cristatus*. I, Family Chiastozygaceae: *Chiastozygus platyrhethus*. J, Family Coccolithaceae: *Coccolithus pelagicus*. K, Family Coccolithaceae: *Pyrocyclus hermosus*. L, Family Coccolithaceae: *Coronocyclus nitescens*. Scale bars, 4 μm. (Photographs A, B, and F courtesy of A.R. Lord, Postgraduate Unit of Micropalaeontology, University College, London.)

Figure 11.9 Scanning electron micrographs of the type species or a representative species of Coccolithophyceae families: A, Family Crepidolithaceae: *Conusphaera mexicana*. B, Family Discoasteraceae: *Discoaster quinqueramus*. C, Family Eiffellithaceae: *Eiffellithus turriseiffelii*. D, Family Ellipsagelosphaeraceae: *Ellipsagelosphaera communis*. E, Family Fasciculithaceae: *Fasciculithus involutus*. F, Family Helicosphaeraceae: *Helicosphaera carteri*. G, Family Heliolithaceae: *Heliolithus riedelii*. H, Family Lithostromationaceae: *Lithostromation simplex*. I, Family Microrhabdulaceae: *Lithraphidites carniolensis*. J, Family Nannoconaceae: *Nannoconus steinmannii*. K, Family Podorhabdaceae: *Cretarhabdus conicus*. L, Family Polycyclolithaceae: *Eprolithus floralis*. Scale bars, 4 μm. (Photographs A, C, J, and K courtesy of A.R. Lord, Postgraduate Unit of Micropalaeontology, University College, London.)

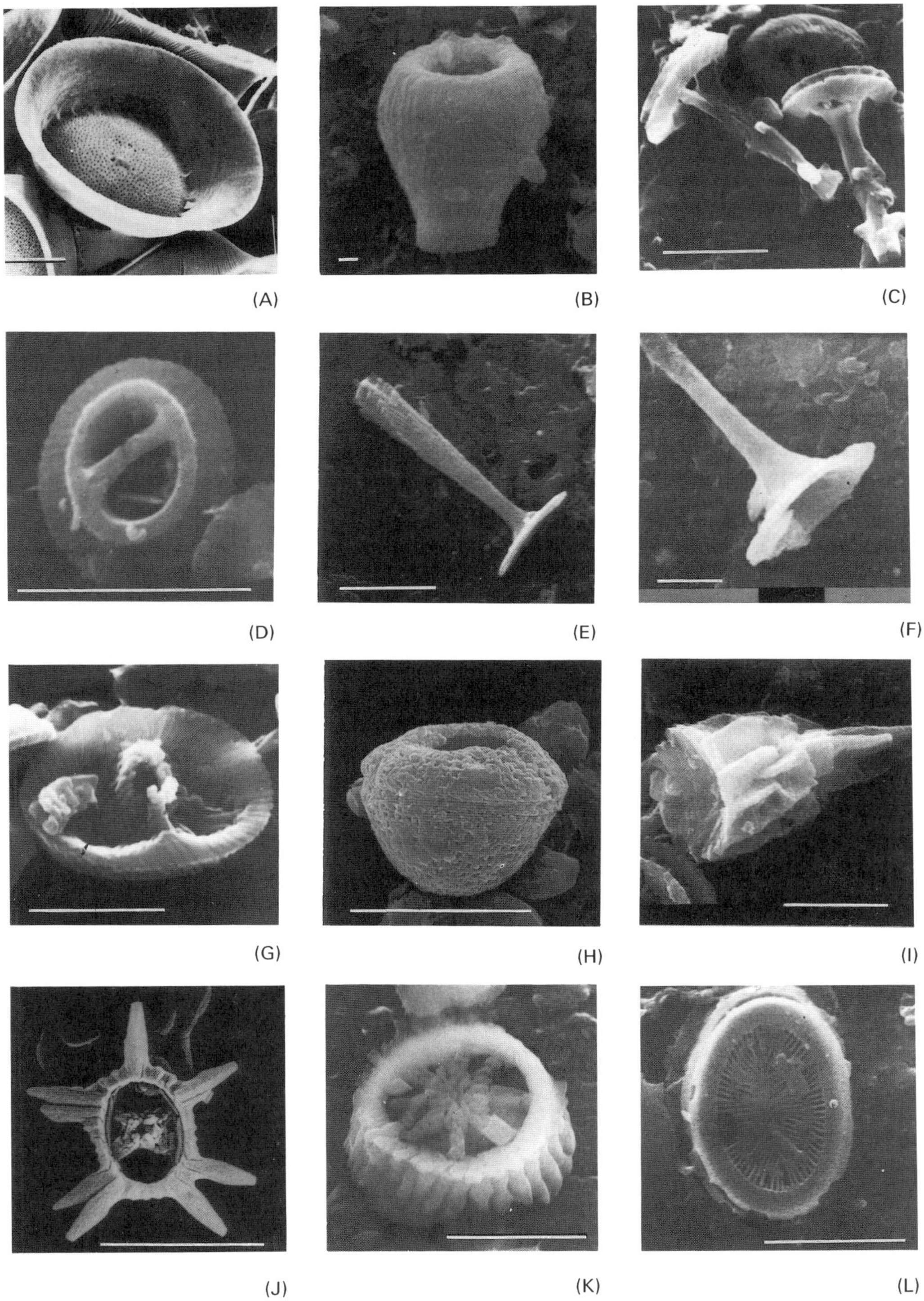

Figure 11.10 Scanning electron micrographs of the type species or a representative species of Coccolithophyceae families. A, Family Pontosphaeraceae: *Pontosphaera syracusana*. B, Family Pontosphaeraceae: *Scyphosphaera recurvata*. C, Family Prediscosphaeraceae: *Prediscosphaera cretacea*. D, Family Prinsiaceae: *Gephyrocapsa oceanica*. E, Family Rhabdosphaeraceae: *Rhabdosphaera claviger*. F, Family Rhabdosphaeraceae: *Discosphaera tubifer*. G, Family Rhagodiscaceae: *Rhagodiscus splendens*. H, Family Schizosphaerellaceae: *Schizosphaerella punctulata*. I, Family Sphenolithaceae: *Sphenolithus belemnos*. J, Family Stephanolithiaceae: *Stephanolithion bigotii*. K, Family Stephanolithiaceae: *Rotelapillus laffittei*. L, Family Syracosphaeraceae: *Syracosphaera pulchra*. All scale bars, 4 μm. (A Reproduced from Nishida S. *Atlas of Pacific Nannoplankton*. Micropaleontology Society of Osaka, 1979:11. Photographs H and J courtesy of A.R. Lord, Postgraduate Unit of Micropalaeontology, University College, London.)

Table 11.2 Classification of calcareous nannoplankton

Division Haptophyta (Hibberd 1972)

Syn. Prymnesiophyta (Hibberd 1976)

CLASS COCCOLITHOPHYCEAE (Rothmaler 1951)

Syn. Coccolithophorales (Lemmermann 1903), Coccolithophoridaceae (Fritsch 1935), Coccolithophorides (Stradner 1958), Coccolithophoriinae (Vekshina 1959), Haptophyceae (Christensen 1962), Coccolithophorea (Pitelka 1963), Coccolithophorineae (Wood 1965), Prymnesiophyceae (Hibberd 1976)

Family Ahmuellerellaceae (Reinhardt 1965)

Elliptical coccoliths with a wall of inclined crystal elements and a central area spanned by a cross aligned with the axis of the ellipse. Triassic/Early Jurassic–Late Cretaceous/Paleogene. Figure 11.8A

Family Arkhangelskiellaceae (Bukry 1969)

Elliptical coccoliths with a complex rim consisting of three to five tiers of elements. Two to four tiers are visible on the proximal side and two tiers on the distal side. Late Jurassic–Late Cretaceous. Figure 11.8B

Family Biscutaceae (Black 1971)

Circular to elliptical coccoliths consisting of two closely appressed shields. Each shield is constructed of radial, nonimbricate, petal-shaped elements. Early Jurassic–Paleogene. Figure 11.8C

Family Braarudosphaeraceae (Deflandre 1947)

Pentagon-shaped coccoliths. Early Cretaceous–Holocene. Figure 11.8D

Family Calciosoleniaceae (Kamptner 1927)

Rhomboidal coccoliths with calcite laths extending inward from the walls. Laths may meet along a central line. Early Cretaceous–Holocene. Figure 11.8E

Family Calyculaceae (Noël 1973)

Elliptical to subcircular coccoliths, each appearing like an elongate cup in side view. Central area is covered with a grid. Early–Late Jurassic. Figure 11.8F

Family Calyptrosphaeraceae (Boudreaux & Hay 1969)

Holococcoliths of widely varying design. Late Jurassic–Holocene. Figure 11.8G

Family Ceratolithaceae (Norris 1965)

Horseshoe-shaped coccoliths. Neogene–Holocene. Figure 11.8H

Family Chiastozygaceae (Rood, Hay & Barnard 1973)

Elliptical coccoliths with an X- or H-shaped central structure. Triassic/Early Jurassic–Paleogene. Figure 11.8I

Family Coccolithaceae (Poche 1913)

Circular or elliptical coccoliths with a distal shield of radiating, petal-shaped elements. The proximal shield is usually birefringent between crossed nicols, and may consist of one or two cycles of elements. The distal shield is larger and is not birefringent. The central area may be closed or open and possess or lack a central structure. Late Cretaceous–Holocene. Figures 11.8J–L

Family Crepidolithaceae (Black 1971)

Elliptical coccoliths consisting of a ring of elements without appreciable imbrication. A proximal floor and massive distal process may be present. Early Jurassic–Paleogene. Figure 11.9A

Family Discoasteraceae (Tan 1927)

Star- or rosette-shaped nannofossils. In plan view, they are not birefringent between crossed nicols. Paleogene–Neogene. Figure 11.9B

Family Eiffellithaceae (Reinhardt 1965)

Elliptical coccoliths with a narrow rim of inclined elements. An inner circle of plates covers most of the central area. A central structure sometimes supports a central process. Early–Late Cretaceous. Figure 11.9C

Family Ellipsagelosphaeraceae (Noël 1965)

Circular and elliptical coccoliths. Distal shield has slightly overlapping elements; proximal shield has radially arranged elements. Early Jurassic–Paleogene. Figure 11.9D

Family Fasciculithaceae (Hay & Mohler 1967)

Cylindrical nannofossils with a proximal column and a distal disk and cone. Paleogene. Figure 11.9E

Family Goniolithaceae (Deflandre 1957)
Pentagon-shaped coccoliths which have a wall composed of vertical elements enclosing a granular center. Late Cretaceous–Paleogene

Family Helicosphaeraceae (Black 1971)

Spiral-walled coccoliths, usually with a flange. Central area may be open, spanned by a bridge, or closed. Paleogene–Holocene. Figure 11.9F

Family Heliolithaceae (Hay & Mohler 1967)

Cylindrical nannofossils with a shortened proximal column and one or two distal cycles of elements. Paleogene. Figure 11.9G

Family Lithostromationaceae (Deflandre 1959)

Triangular, hexagonal or nearly circular nannofossils covered with symmetrically arranged depressions. Paleogene–Neogene. Figure 11.9H

(Continued)

Table 11.2 *Continued*

Family Microrhabdulaceae (Deflandre 1963)

Cylindrical, rod-, or spindle-shaped nannofossils. Late Jurassic–Late Cretaceous. Figure 11.9I

Family Nannoconaceae (Deflandre 1959)

Conical nannofossils composed of a thick wall of wedge-shaped elements perpendicular to, and spirally surrounding, an axial canal. Late Jurassic–Late Cretaceous. Figure 11.9J

Family Podorhabdaceae (Noël 1965)

Elliptical coccoliths with a narrow rim consisting of two to three cycles of elements. The wide central area is spanned by a variety of structures, commonly, but not always, supporting a hollow or solid central process. Early Jurassic–Late Cretaceous. Figure 11.9K

Family Polycyclolithaceae (Forchheimer 1972)

Cylinder-, block-, star-, or rosette-shaped nannofossils. Early–Late Cretaceous/Paleogene. Figure 11.9L

Family Pontosphaeraceae (Lemmermann 1908)

Coccoliths with a raised wall of varying height consisting of two cycles of elements and a large central area. Elements are generally arranged radially on the proximal side and concentrically on the distal side. The central area is usually perforated or contains depressions. Paleogene–Holocene. Figure 11.10A, B

Family Prediscosphaeraceae (Rood, Hay & Barnard 1971)

Circular or elliptical coccoliths, almost always with 16 elements in each of two shields. Specimens sometimes have a central process. Early–Late Cretaceous. Figure 11.10C

Family Prinsiaceae (Hay & Mohler 1967)

Circular to elliptical coccoliths. The distal shield is birefringent between crossed nicols, with a right-curving interference figure. Late Cretaceous–Holocene. Figure 11.10D

Family Rhabdosphaeraceae (Lemmermann 1908)

Nannofossils with a base consisting of a varying number of cycles of elements. A central process rises from the base. Paleogene–Holocene. Figure 11.10E, F

Family Rhagodiscaceae (Hay 1977)

Elliptical coccoliths with a wall composed of inclined elements and a granular central area. Late Jurassic–Late Cretaceous. Figure 11.10G

Family Schizosphaerellaceae (Deflandre 1959)

Nannofossils consisting of two overlapping hemispheres. Triassic–Late Jurassic. Figure 11.10H

Family Sollasitaceae (Black 1971)

Elliptical coccoliths with two shields and a large central opening occupied by a grid or bars, but without a central process. Early Jurassic–Paleogene

Family Sphenolithaceae (Deflandre 1952)

Nannofossils with a proximal shield or column, above which are tiers of radiating lateral elements, then a final elongate, sometimes bifurcating, distal structure. Paleogene–Neogene. Figure 11.10I

Family Stephanolithiaceae (Black 1968)

Circular, elliptical, or polygonal coccoliths. The outer wall has vertically arranged elements and the central area is spanned by variously arranged elements which may support a central process. Lateral spines may project from the outer wall. Early Jurassic–Late Cretaceous. Figure 11.10J, K

Family Syracosphaeraceae (Lemmermann 1908)

Coccoliths with a complex wall, and a central area partially closed by calcite laths. Neogene–Holocene. Figure 11.10L

Family Thoracosphaeraceae (Schiller 1930)

Spherical or ovoid nannofossils composed of interlocking polygonal elements. Late Jurassic–Holocene. Figure 11.11A

Family Triquetrorhabdulaceae (Lipps 1969)

Spindle-shaped rods constructed of three blades. Triradial cross-section. Paleogene–Neogene. Figure 11.11B

Family Zygodiscaceae (Hay & Mohler 1967)

Coccoliths with one or two cycles of inclined elements in the wall and a bridge aligned with the short axis of the ellipse. Early Jurassic–Paleogene. Figure 11.11C

for two taxonomic concepts—the **organ genus** and the **form genus**. These two concepts are recognized because plants are commonly fragmentary and the various parts cannot always be proved to belong together. The organ genus recognizes taxa that can be assigned to a family, whereas a form genus cannot be assigned to a family because its biologic relationships are unknown. The form-genus concept is useful in classifying nannofossils, as the original coccospheres usually disintegrate after death or reproduction, leaving a fossil record of only scattered, individual coccoliths.

Other problems also exist. Some species are **dimorphic**, possessing more than one kind of coccolith on a single coccosphere. *Syracosphaera pulchra* (Fig. 11.12A), for example, has spine-bearing coccoliths around the flagellar end of the coccosphere, but spineless coccoliths

Figure 11.11 Scanning electron micrographs of the type species or a representative species of Coccolithophyceae families. A, Family Thoracosphaeraceae: *Thoracosphaera saxea*. B, Family Triquetrorhabdulaceae: *Triquetrorhabdulus carinatus*. C, Family Zygodiscaceae: *Glaucolithus diplogrammus*. All scale bars, 4 μm.

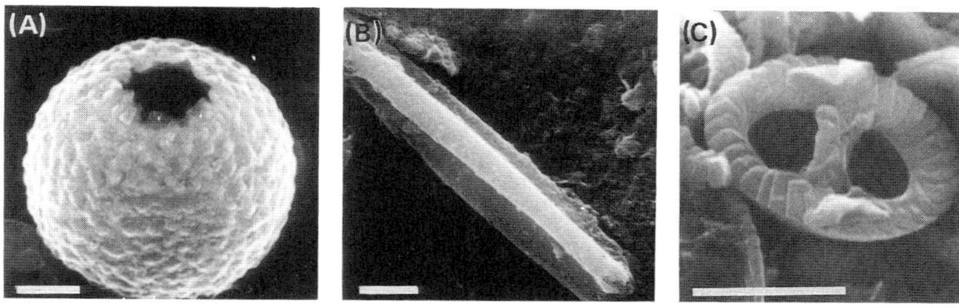

around the remainder of the coccosphere. Furthermore, the two species *Scyphosphaera apsteinii* and *Pontosphaera japonica* are now known to occur on the same coccosphere (Fig. 11.12B), as do *Helicosphaera carteri* and *H. wallichi* (Fig. 11.12C). In addition, some living coccolithophores exhibit dithecatism; i.e., they are constructed of a double layer of coccoliths, each layer of which may contain coccoliths of a markedly different shape. Figure 11.13 shows the **dithecate** genus *Syracosphaera*, which has an endothecal (internal) layer of caneoliths and an exothecal (external) layer of cyrtoliths.

When a coccosphere disintegrates, the coccoliths become separated. Because the paleontologist deals mostly with individual coccoliths, each dimorphic or dithecate variety may be erroneously assigned to a separate species, when in fact they actually belong to the same species. For stratigraphic applications this difficulty is not as serious as it may seem. The worst scenario is that two species are thought to exist, rather than only one; but both should have closely similar stratigraphic ranges.

Pleomorphism is yet another problem. For example, a heterococcolith-bearing phase alternates with a holococcolith-bearing phase in the life-cycle of *Coccolithus pelagicus*. Different coccoliths formed in different phases can result in the coccoliths being placed in two different genera or species, when the different coccoliths are simply variations in the life history of the same species. Less serious to classification are the subtle changes in appearance of some coccoliths in response to temperature or other ecologic variables. This has not hampered classification, and in fact has enhanced the usefulness of these nannofossils as paleoenvironmental indicators.

Biology

Coccolithophores have two distinct phases in their life-cycles: a motile, free swimming phase, and a nonmotile, floating phase. During the motile phase a cell has two smooth, whip-like **flagella** rising from one end of an oval to elongate body (Fig. 11.14). A third appendage, the

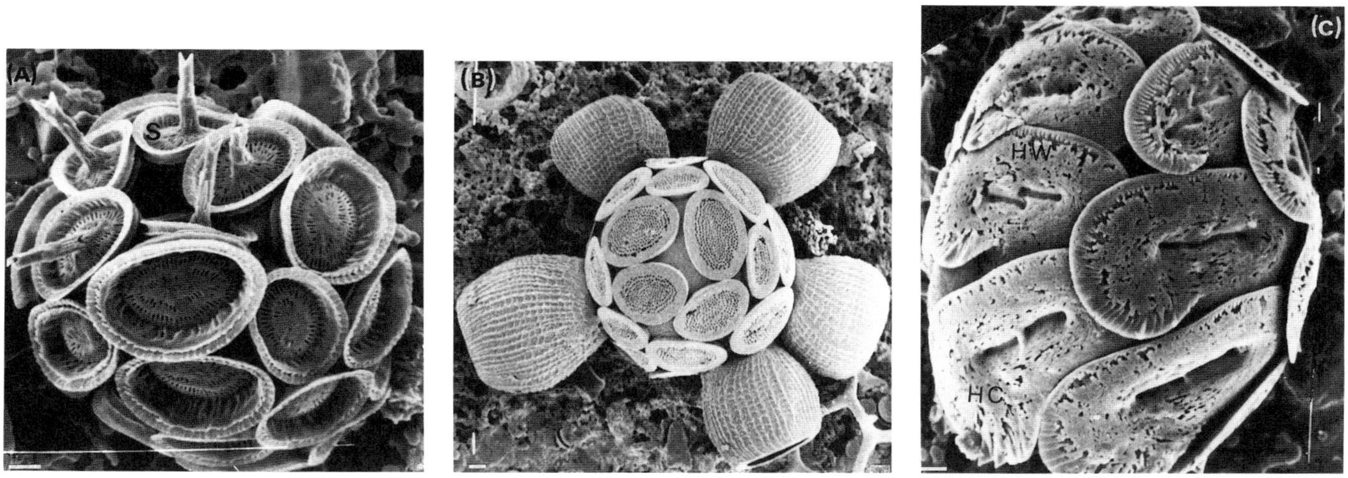

Figure 11.12 Coccospheres illustrating dimorphism. A, *Syracosphaera pulchra*, carrying both spine-bearing coccoliths (S) and spineless coccoliths. B, A single coccosphere bearing both the tall, vase-shaped *Scyphosphaera apsteinii* and the flattened, perforate "species" *Pontosphaera japonica*. C, A coccosphere bearing both *Helicosphaera carteri* (distinguished by two centrally aligned, slit-like pores (HC)) and *H. wallichi* (two offset slit-like pores (HW)). All scale bars, 1 μm. (Reproduced from Nishida S. *Atlas of Pacific Nannoplankton*. Micropaleontology Society of Osaka, 1979:6, 9, 11.)

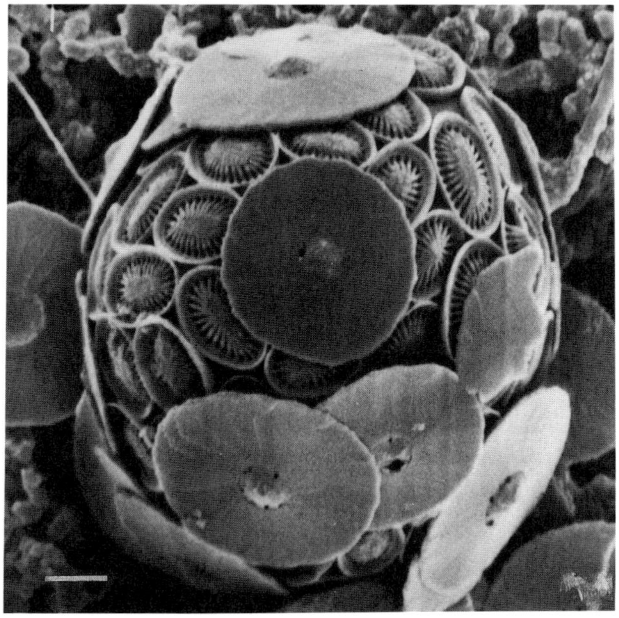

Figure 11.13 Coccosphere of *Syracosphaera variabilis* showing two layers of coccoliths (dithecatism). An outer layer of cyrtoliths (partially removed) encloses an inner layer of caneoliths. Scale bar, 1 μm. (Reproduced from Nishida S. *Atlas of Pacific Nannoplankton.* Micropaleontology Society of Osaka, 1979:8.)

haptonema, is attached between the two flagella (Fig. 11.14). The haptonema is generally coiled, although it can also be straight. It differs from the flagella in its internal structure and in the structure of its basal attachment. During the nonmotile phase the flagella disappear, although the haptonema may remain.

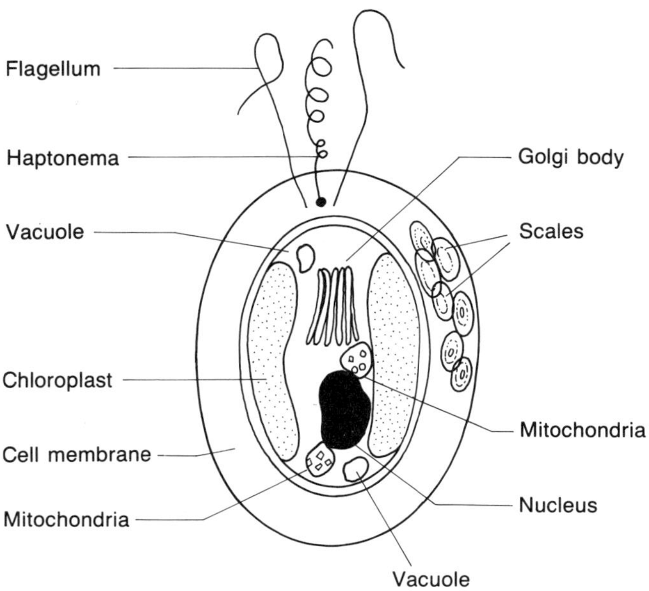

Figure 11.14 Diagrammatic sketch of a living coccolithophore cell. During the nonmotile stage, the flagella are absent and the cell is covered with calcite coccoliths.

Internally, the cell has a nucleus lying between two golden-brown chloroplasts (Fig. 11.14). The chloroplasts contain the chlorophyll used in photosynthesis and can change their shape and position within the cell in response to varying light intensity. Also within the cell are **mitochondria**, **vacuoles**, and the **Golgi body** (Fig. 11.14). The mitochondrial bodies are enclosed in membranes and have tubular extensions into the cell interior. Mitochondria contain enzyme systems that produce the energy needed for the cellular functions. Vacuoles are structures where waste products are stored prior to expulsion. The Golgi body (also known as the "dictyosome") consists of closely packed, swollen, sac-like structures called "cisternae." The Golgi body has a secretory function, first secreting organic scales that cover the cell, and later the coccoliths.

The protoplasm of the cell is enclosed within a double membrane that in turn is covered by an organic skin, the **pellicle**. Organic scales coat the outer cell membrane. These scales are oval, overlap, and have concentric ridges on their distal surfaces and radiating ridges on their proximal surfaces (Fig. 11.14). Scales act as organic bases for the calcite precipitation that forms the coccoliths.

Coccolith formation

The physiologic process of coccolith formation is complex and is not the same for all species. The Golgi body seems to be the site of coccolith secretion for most species, although a little-understood structure called a reticular body is instrumental in coccolith formation in some species. After scales and coccoliths are formed within the Golgi body, they are extruded to the cell surface where the coccoliths lock together to form the coccosphere.

In another method of coccolith formation, exemplified by *Emiliania huxleyi* (Fig. 11.15), the cell does not produce scales beneath the coccoliths. Calcite precipitation in living *E. huxleyi* occurs in an organic matrix adjacent to the reticular body. The coccolith begins to take its general shape owing to calcification beginning in the organic matrix. The base of the coccolith appears first, with mineralization of the shields advancing upwards and laterally. After calcification is essentially complete, coccoliths move to the surface of the cell where they form the external skeleton (Fig. 11.15).

A third type of coccolith formation occurs in *Coccolithus pelagicus* (Fig. 11.2A). The process begins with scales being produced within the Golgi body and then extruded to the cell's exterior. Unlike the previous two cases,

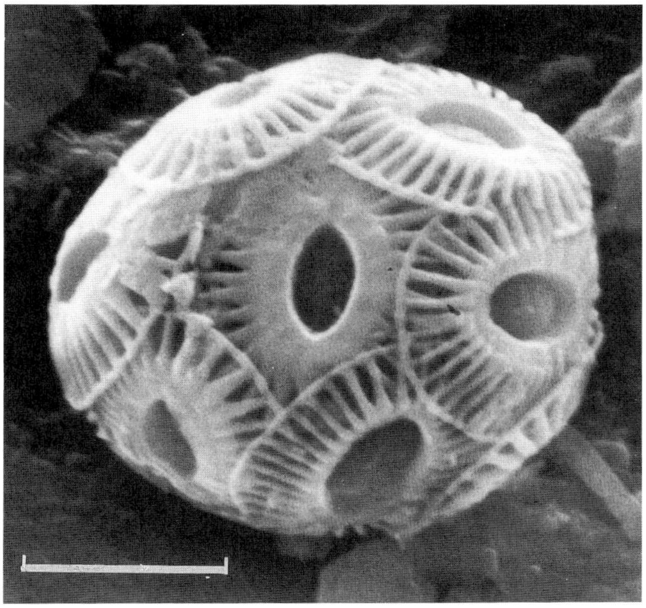

Figure 11.15 Coccosphere of *Emiliania huxleyi*, the most abundant coccolithophore in today's oceans. Scale bar, 2 μm.

however, *C. pelagicus* does not produce coccoliths within the cell interior. Coccolith precipitation takes place on the scales later, outside the cell membrane, but beneath the organic pellicle surrounding the cell.

The cell membrane of many living species holds the coccoliths loosely arranged around the cell surface. Coccoliths may hook together at their rims, then detach if disturbed, and hook together again later in a new position. This imparts a flexibility to the cells even when they are bearing coccoliths. In some species, however, calcite is deposited within the cell membrane, rigidly cementing the coccoliths in place. Cemented coccospheres are sometimes found in Holocene sediments as well as in older rocks.

Coccolith formation is apparently controlled by light. In darkened cultures, *E. huxleyi* stopped producing coccoliths. Coccolith production was renewed within half an hour after light was reintroduced. *E. huxleyi* then produced cells rapidly under the lighted conditions; individual coccoliths formed in about 1 hour, and a complete coccosphere formed in about 30 hours. The maximum rate of coccolith production continued for about 1 hour after the culture was placed in the dark again.

Coccolith composition and function

The small size of coccoliths has hampered detailed mineralogic and chemical analyses of their composition. Coccolithophores grown in laboratory cultures produced coccoliths composed predominantly of calcite, with small amounts of **aragonite** and **vaterite**. Analysis of fossil coccoliths, however, has shown them to be composed of low-Mg calcite (less than 4% $MgCO_3$), and not of aragonite or other minerals.

The purpose of the coccolith covering on coccolithophores is unknown. Some of the ideas proposed for coccolith function are:

1 as a shield to protect the cell against intense sunlight;
2 as the result of carbonate detoxification by calcium fixation;
3 as a metabolic byproduct;
4 as supporting and/or protecting armor;
5 as a light-gathering and concentrating organ;
6 as a ballasting or stabilizing mechanism.

Minor morphologic characteristics of coccoliths may also serve specific functions. For example, almost all coccoliths are perforate. The perforate nature of coccoliths probably has a physiologic function, perhaps to allow interaction between the cell and the surrounding water. Among coccolithophores, only *Braarudosphaera bigelowii* (Figs 11.8D, 11.16) is imperforate. However, the construction and ecology of *B. bigelowii* are also considerably different from other coccolithophores, and this species may not be a coccolithophore at all.

Many nannoplankton species have spines that probably have some functional purpose. Spines may act to control flotation of the organism, or have a defensive purpose: either a simple porcupine-like defense, or as a method to ensure rejection by specific filter-feeders.

Figure 11.16 Coccosphere of *Braarudosphaera bigelowii*, a species which prefers low-salinity, nearshore environments. Scale bar, 1 cm. (Reproduced from Nishida, S. *Atlas of Pacific Nannoplankton*. Micropaleontology Society of Osaka, 1979:16.)

Nutrition and locomotion

Coccolithophores produce carbohydrates photosynthetically from water and carbon dioxide. Several species are known to ingest and assimilate bacteria and algae, but this method of nutrition is an exception, and is always subordinate to photosynthesis. Solid food particles, when ingested, are taken in at the nonflagellar pole. Wastes are gathered into a vacuole and periodically expelled.

During the nonmotile stage of their life-cycle, coccolithophores are simple floating spheres. During the motile stage, they swim freely, using the pair of flagella for propulsion. Coccolithophores can swim at an average speed of 5–8 mm/minute (about 10–20 m/day). Since the rate of oceanic-current movement is generally faster than this, coccolithophores are considered to be essentially planktic, regardless of their swimming ability.

The haptonema may be several times the length of the flagella when it is fully extended. It has an adhesive tip, and it has been suggested that the haptonema may be used to anchor the cell to the substrate during a benthic phase in the life-cycle. Although possible for some near-shore species, substrate anchoring is clearly not likely for open-ocean species. Another possibility is that the open-ocean species may use the haptonema to attach themselves to other floating organic or inorganic particles.

Reproduction

Reproduction among coccolithophores is by single or double fission, sometimes accompanied by a swarm-spore stage. Two species have been studied in some detail. In cultures the holococcolith-covered form representing the motile stage in the life-cycle of *Coccolithus pelagicus* reproduced by simple **binary fission** first, forming two daughter cells. The daughter cells remained within the coccosphere until one of the coccoliths detached, providing an exit. The naked, nonmotile daughter cells escaped through the opening, and the empty coccosphere slowly sank to the bottom. The daughter cells then underwent double fission, resulting in four nonmotile cells. These cells grew in size and developed a heterococcolith covering in about 2 weeks. Motile, flagella-, and holococcolith-bearing cells were then freed from *C. pelagicus* and the cycle began again. This holococcolith-bearing, motile phase of *C. pelagicus* was once thought to be a separate species, *Crystallolithus hyalinus*.

Emiliania huxleyi is the other living species whose life-cycle is well known. Unlike *C. pelagicus*, *E. huxleyi* lacks organic scales and carries coccoliths only during the nonmotile stage. Reproduction takes two different paths. The nonmotile cell may divide by fission, with daughter cells sharing the original coccoliths. Alternatively, a naked swarm spore may escape from the nonmotile coccosphere and divide to produce other naked cells, which then produce coccoliths.

The speed of cell division probably varies among species; it can certainly vary within one species. *Emiliania huxleyi* achieves up to 4.8 cell divisions/day in the Black Sea, whereas the same species has a normal maximum of 1.2 divisions/day in the Atlantic. In cultures, *E. huxleyi* developed its highest reproduction rate between 18 and 24°C; in that temperature range, it reproduced almost four times as fast as it did at 7°C.

Very high rates of reproduction cause exceedingly rich local blooms of coccolithophore cells. For example, water samples collected north of Scotland contained 50 000–100 000 cells/liter, not an unusual number; local concentrations, however, exceeded 1 million cells/liter. Rich concentrations recorded off Oslo, Norway ranged from 14 million to 35 million cells/liter, and 30 million cells/liter of a single species, *Calcidiscus fragilis*, were found off Senegal after a flood.

Ecology

As with all organisms, coccolithophores can survive only within certain environmental tolerances. Light, salinity, and temperature are particularly important ecologic influences.

Coccolithophores depend on photosynthesis and are therefore restricted to oceanic surface waters within the reach of sunlight. The thickness of the photic zone varies from region to region, but is generally considered to be from the surface down to about 150 or 200 m. The maximum concentration of nannoplankton in the tropics is at a depth of about 50 m, and in more temperate regions the maximum concentration is between 10 and 20 m. This may be because many species cannot tolerate the intense sunlight of shallow depths, and prefer to live at depths where only the optimum amount of sunlight reaches them.

Most coccolithophore species prefer normal oceanic salinity, with the number of species becoming severely reduced at salinities higher than 38‰ or lower than 25‰. Several living species, however, are able to tolerate natural salinities considerably higher or lower than normal. *Emiliania huxleyi* has been found in canals with salinities of 15‰, in the Black Sea and Sea of Azov with salinities down to only 11‰, and in the Red Sea with a salinity of 41‰. Moreover, *E. huxleyi* has been grown in laboratory cultures at salinities between 16‰ and 45‰.

Braarudosphaera bigelowii also thrives in the low-salinity waters of the Black Sea, but is not found in the **hypersaline** waters of the Red Sea.

Other laboratory experiments suggest that salinity tolerances may be different for the motile and nonmotile stages of a coccolithophore. The nonmotile phase of *Cricosphaera* can tolerate salinities up to an astonishing 236‰ in laboratory cultures, whereas the motile phase can withstand a maximum salinity of 90‰. Such tolerances have not yet been documented in natural environments.

Many species of coccolithophores are indigenous to waters of a definite temperature range; a few have very narrow temperature tolerances. The observed temperature ranges of species have allowed delineation of biogeographic nannofloral zones (see "Biogeography" below).

Biogeography

Coccolithophores rival diatoms as the most abundant phytoplankton group in the oceans. Coccolithophores are most abundant in low latitudes and middle latitudes; they are not found in polar water masses. Greatest species diversity occurs in the equatorial latitudes, although the comparatively fewer species in higher latitudes are often more abundant in total number of cells/liter than at lower latitudes. *Emiliania huxleyi* has the widest biogeographic distribution (Figs 11.17, 11.18), and is probably also the most abundant living species. It makes up more than 50% of the total **nannoflora** in temperate waters, is less abundant in tropical waters, but makes up almost 100% in subpolar waters.

In the Atlantic, temperature-based nannofloral biogeographic zones have been established as follows: subarctic–subantarctic (2–6°C), transitional (6–12°C), subtropical (14–28°C), and tropical (>28°C). Figure 11.17 shows the location of the zones in the Atlantic, and lists the species characteristic of each zone.

Detailed studies of the **standing crop** of coccolithophores in the north and central Pacific show the variations of cells/liter with latitude and depth. In Pacific surface waters, coccolithophores were found to be most numerous in more northern latitudes. At 50° N, the most densely populated layer was at a depth of 5 m (170 000 cells/liter). At 45° N, it was at 50 m (65 000 cells/liter). Between 40 and 30° N, the greatest numbers were between 50 and 100 m, and at 17° 30′ N, the distribution was about the same throughout the upper 200 m water column (average about 3600 cells/liter).

In the entire upper 200 m, a maximum of 60 million

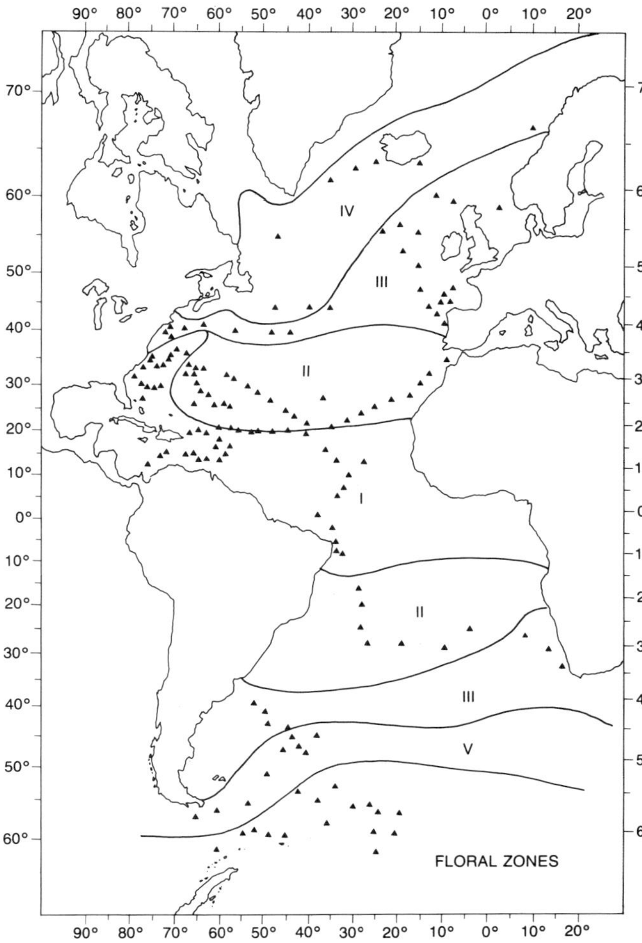

Figure 11.17 Coccolithophore nannofloral zones in the Atlantic Ocean. (Reproduced from McIntyre A, Be AWH. *Deep Sea Research* 1967; 14:561–97. Taxonomy in the accompanying species list has been updated by the writer.) Species of the Atlantic coccolithophorid floral assemblages arranged in descending order of importance within each group.

I Tropical: *Umbellosphaera irregularis, Calcidiscus annulus, Oolithotus fragilis, Umbellosphaera tenuis, Discosphaera tubifer, Rhabdosphaera stylifer, Helicosphaera carteri, Gephyrocapsa oceanica, Emiliania huxleyi, Calcidiscus leptoporus.*

II Subtropical: *Umbellosphaera tenuis, Rhabdosphaera stylifer, Discosphaera tubifer, Calcidiscus annulus, Gephyrocapsa oceanica, Umbilicosphaera sibogae, Helicosphaera carteri, Calcidiscus leptoporus, Oolithotus fragilis, Emiliania huxleyi.*

III Transitional: *Emiliania huxleyi, Calcidiscus leptoporus, Gephyrocapsa ericsonii, Rhabdosphaera stylifer, Gephyrocapsa oceanica, Umbellosphaera tenuis, Coccolithus pelagicus.*

IV Subarctic: *Coccolithus pelagicus, Emiliania huxleyi, Calcidiscus leptoporus.*

V Subantarctic: *Emiliania huxleyi, Calcidiscus leptoporus.*

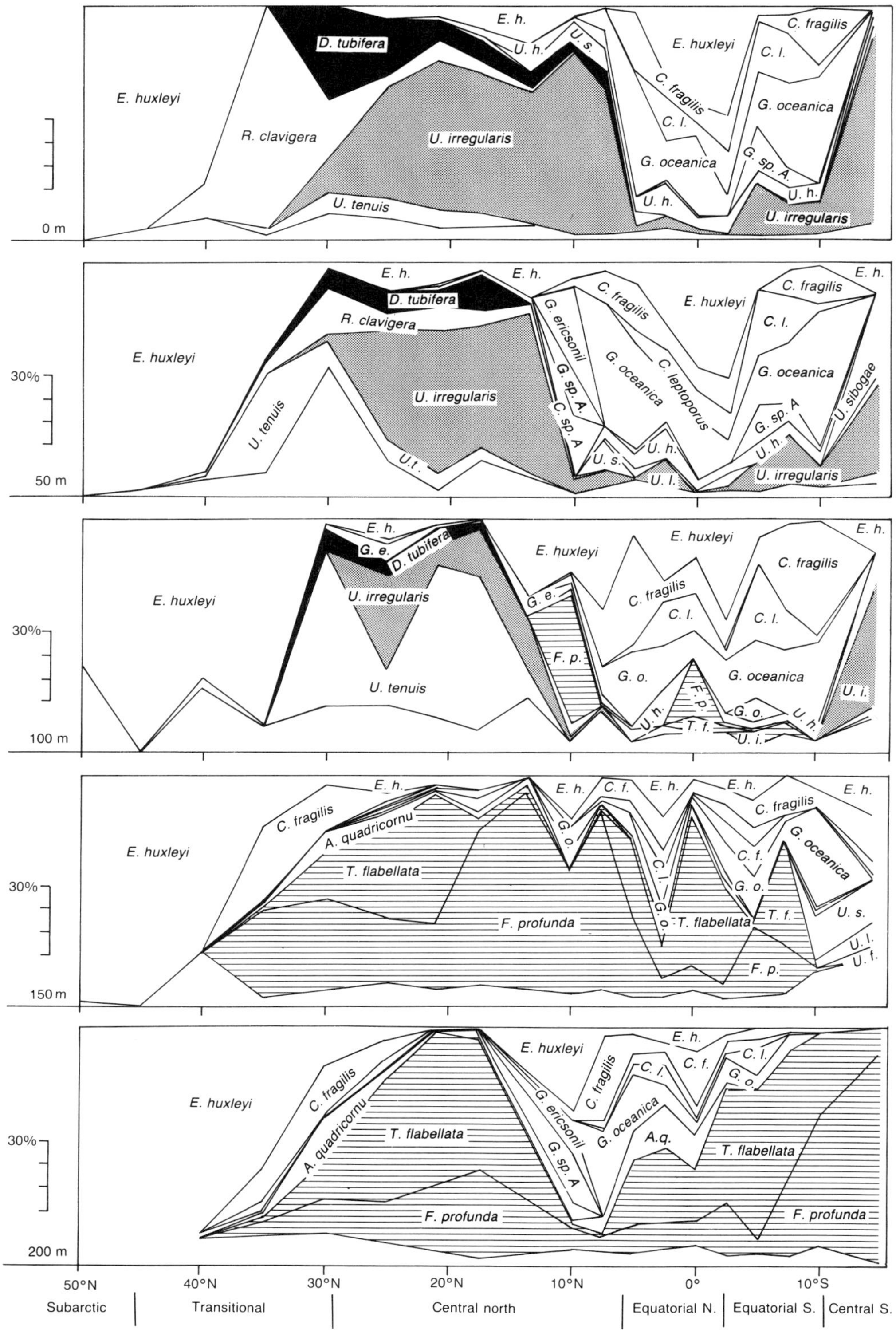

Figure 11.18 Relative abundance of different coccolithophore species at the surface and at depths of 50, 100, 150, and 200 m in the Pacific Ocean between 5° S latitude and 50° N latitude. (Reproduced from Okada H, Honjo S. *Deep Sea Research* 1973; 20:355–74.)

cells/m^3 occurred at 50° N, decreasing to 4 million at 25° N. The total number of cells increased again somewhat nearer the tropics in the Pacific. In the subphotic zone below 200 m coccolithophores were still found in suspension. More than 100 cells/liter occurred down to 4000 m in the equatorial area, but these were dead cells that had sunk from the surface.

The Pacific Ocean study also showed that some degree of species stratification occurs in the photic zone (Fig. 11.18). *Umbellosphaera irregularis* and *Rhabdosphaera claviger* were most abundant in the upper photic zone (0–100 m), whereas *U. tenuis* and *Oolithotus fragilis* were abundant in the middle photic zone (50–125 m), and *Florisphaera profunda* and *Thorosphaera flabellata* occurred only in the lower photic zone (100–200 m). *Emiliania huxleyi, Gephyrocapsa oceanica,* and *Calcidiscus leptoporus* were most numerous in the upper–middle photic zone (0–150 m), but ranged through the entire photic water column (Fig. 11.18).

A few taxa are more abundant in nearshore environments than in the open ocean. Various explanations have been proposed to explain this distribution, including:

1 nearshore turbidity and/or low salinity;

2 a restricting benthic stage in the nannoplankton life-cycle;

3 selective preservation in nearshore environments owing to greater amounts of dissolution at deeper depths.

The genera *Lanternithus, Pontosphaera, Scyphosphaera, Transversopontis, Zygrhablithus* and some *Helicosphaera* all seem to prefer nearshore conditions. Members of the family Braarudosphaeraceae, in particular, are conspicuously absent from open-ocean deposits, and are strong indicators of nearshore conditions.

Distribution in sea-floor sediments

Coccolith abundance decreases rapidly in a shoreward direction in Holocene sea-floor sediments. For example, study of the fine fraction of Holocene sediments off Australia showed that sediments underlying waters deeper than 100 m contained greater than 10% coccoliths. Between 100 and 40 m, sediments contained 2–10% coccoliths, and sediments shallower than 40 m had less than 2%. A similar pattern was reported from continental-shelf sediments off China. Coccoliths were rare shallower than 50 m, but both abundance and diversity increased progressively in a seaward direction across the shelf.

The concentration and diversity of species in sediments might be assumed to differ from their concentration and diversity in the overlying waters, since several conditions should act to obscure the surface water–sediment correlation. The first condition is seasonal fluctuation of coccolithophores in surface waters. Abundance and diversity of coccolithophores measured at any station in the ocean give values representative for only that instant in time. Marked seasonal variations occur in some species, notably *Emiliania huxleyi, Umbellosphaera irregularis,* and *U. tenuis*. In the North Atlantic, studies have shown that *U. irregularis* increased markedly in abundance in mid-summer, corresponding to a decrease in *E. huxleyi* at about the same time. *U. tenuis* bloomed in the spring. Coccolithophores in the western North Atlantic between 35 and 56° N were less abundant in the north during winter and in the south during summer. Individual species exhibited seasonal fluctuations in abundance as well. Seasonal fluctuation is not, however, a particularly important control on abundance and diversity when dealing with sediments. A thin lamination of sea-floor sediment records the surface-water assemblage over a long period of time, and seasonal and even longer cyclic variations tend to become unrecognizable.

Selective dissolution, which concentrates more-resistant species at the expense of less-resistant ones, poses a more serious difficulty. Numerous studies of differential solution have shown the relative resistance to solution among the more common coccolithophores is as follows (most resistant taxon listed first):

Coccolithus pelagicus
Gephyrocapsa sp.
Calcidiscus leptoporus
Helicosphaera carteri
Rhabdosphaera claviger
Umbilicosphaera sibogae
Umbellosphaera irregularis
Emiliania huxleyi
Discosphaera tubifer
Umbilicosphaera tenuis
Oolithotus fragilis.

Solution is apparently not important in sediments at depths shallower than about 3500 m, and some robust species are still present below 5000 m.

Long-distance transportation of discarded nannoplankton skeletons by oceanic currents is potentially the most serious problem. Individual coccoliths of *Emiliania huxleyi* settle through the water column at rates ranging from less than 1 μm/second to about 1.6 μm/second. At these rates, an individual coccolith would take 50–150 years to reach a 5000-m-deep sea floor, even disregarding current action. Oceanic currents would thus have ample time to transport coccoliths far from the water mass in which they lived. The transported coccoliths could eventually come to rest on sea floor underlying surface waters containing a much different nannoplankton assemblage.

Coccoliths do not sink as individual particles, however. Instead, **zooplankton** feed on coccolithophores and pack the coccoliths in their **fecal pellets**. A single fecal pellet may contain as many as 100 000 coccoliths, and as it is considerably larger than an individual coccolith, sinks much more rapidly and probably reaches the 5000-m level in 22–100 days. Enclosure in fecal pellets would, moreover, protect delicate coccoliths that might otherwise dissolve while descending to the sea floor. Once on the sea floor, the organic material in the fecal pellets decomposes, liberating the coccoliths.

The overall result is that the distribution of nannoplankton species in sea-floor sediments corresponds surprisingly closely to the distribution of the same species in the overlying surface waters (within about 200 km). This correlation between the assemblage in the surface waters and in the underlying sea-floor sediments thus provides a reliable basis for paleotemperature interpretations of oceanic waters based on cores taken from the sea floor (see "Paleoclimatology" below).

Paleobiology

Paleobiogeography

The paleobiogeography of calcareous nannofossils has been determined for many time intervals in the geologic past (e.g., Fig. 11.19). The paleobiogeography of Cenozoic taxa is better known than for Mesozoic taxa, owing to the more complete coverage by deep-sea cores for the Cenozoic. As an example, latitude-controlled assemblages for the Paleogene of the Atlantic Ocean have been recognized. Paleobiogeographic patterns of nannofossils of 12 Paleogene time intervals were mapped, allowing the recognition of paleoclimatic indicators among the species and the establishment of the relative latitude distributions of the major assemblages. The assemblages group into low-, middle-, and high-latitude floras. Five distinctive latitudinal shifts occur in the Paleogene distribution patterns. These have been interpreted as responses to major climatic fluctuations.

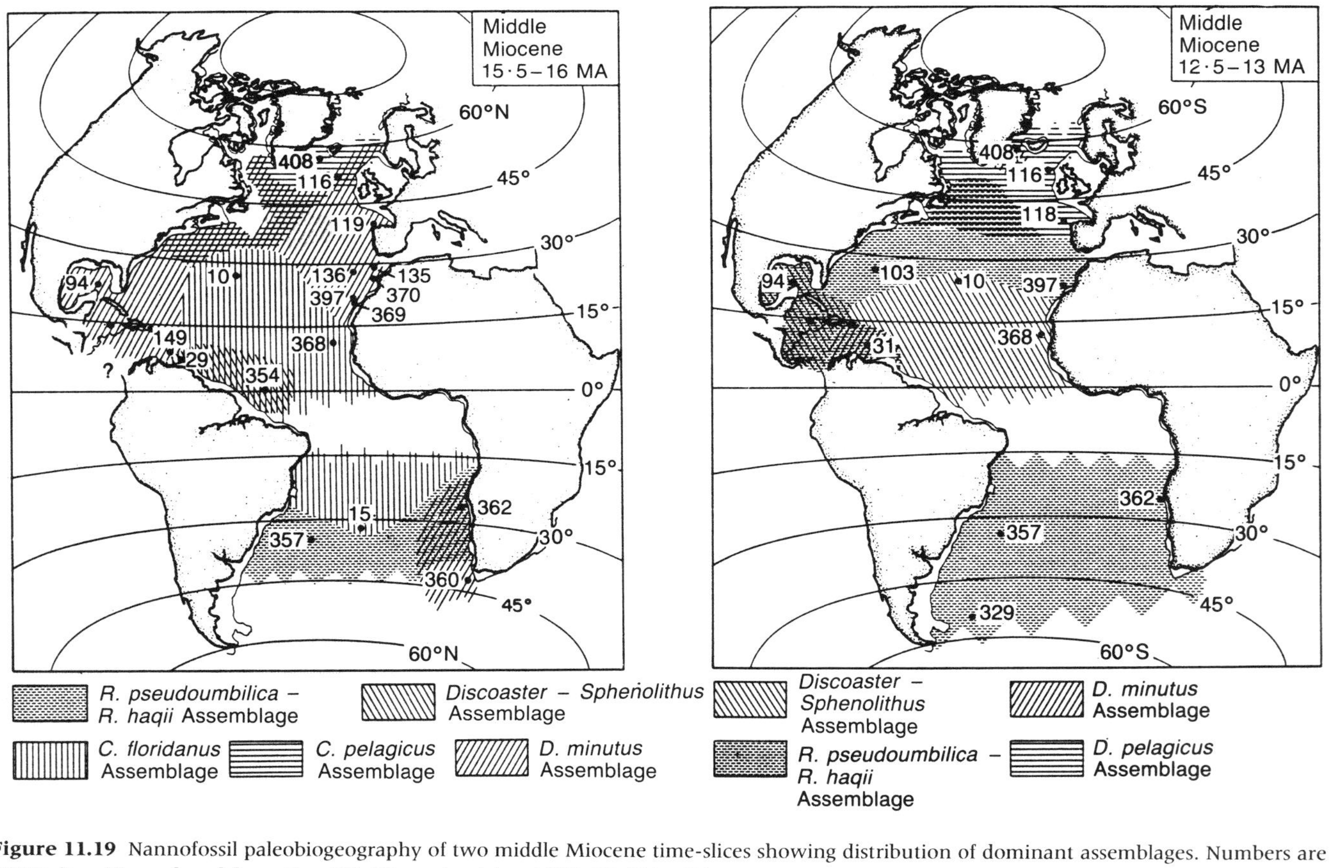

Figure 11.19 Nannofossil paleobiogeography of two middle Miocene time-slices showing distribution of dominant assemblages. Numbers are DSDP sites. (Reproduced from Haq BU. *Micropaleontology* 1980; 26:414–43.)

Paleoclimatology

Coccolithophores live in the photic zone of oceanic waters and are therefore under direct climatic control, a necessary requirement for paleoclimatic studies. Moreover, reliable paleoclimatic interpretations must rest on a firm foundation of paleoecologic and paleobiogeographic data. A considerable body of data is available on the relationships of modern coccolithophores to present-day water masses and latitudinal climatic gradients (Fig. 11.17). This has facilitated Pleistocene paleoclimatic studies, as essentially the same nannoplankton assemblage living today was living during the Pleistocene.

Assuming that the temperature requirements of coccolithophores have not changed since the Pleistocene, ancient water masses of a certain temperature range should have contained essentially the same species as similar water masses today. As Pleistocene glaciers waxed and waned, polar water masses moved correspondingly closer to, or farther from, the equator. The maximum cooling during the last glaciation some 18 000 years ago shifted the planktic populations and their host water masses an average of 15° of latitude southward in the North Atlantic. Latitudinally shifting water masses, each containing a slightly different species mixture, would cause a different mixture of planktic organisms to fall as sediment to the sea floor. A sediment core records these climatic fluctuations by the vertical changes from cool- to warm-water species.

Four methods have most frequently been used to determine relative paleotemperatures. The first is simply to investigate the relative abundances of certain species that are useful by themselves as paleoclimatic indicators. *Emiliania huxleyi* is one of these species, as it possesses recognizably different climatic morphotypes. The cold-water variety of *E. huxleyi* has a solid proximal shield, closed central area, and between 23 and 33 elements in the shield. The warm-water variety (Fig. 11.15) has an open proximal shield, a grill-like plate over the central area, and 30 to 40 elements. The changing ratio of cold-water/warm-water *E. huxleyi* morphotypes indicates changing climatic conditions.

The second method is to examine the entire nannofossil assemblage, rather than only a single species. Each species in the assemblage is designated as a cool- or warm-water species, and the general temperature of the ancient water mass is inferred on the basis of fluctuations in the ratio of cool-water/warm-water forms.

The third method is useful in pre-Pleistocene sediments. The genus *Discoaster* is more abundant than the genus *Chiasmolithus* (Fig. 11.29L) in warm waters and vice versa. The changing *Discoaster* : *Chiasmolithus* ratio can therefore be used as a guide to the changing paleotemperature of surface waters.

The fourth method is based on stable isotopes (see Chapter 4). Oxygen-isotope analysis of foraminifera has been used for some time as a means of inferring paleotemperatures of the sea water in which the foraminifera lived. Applying the $^{18}O:^{16}O$ ratio to calcareous nannofossils has only recently been adopted as a paleoclimatic tool. A major drawback to isotope analysis of nannofossils is the extremely small size of specimens, which precludes single-species analysis. Individual species of foraminifera can be picked from a sample and analyzed, whereas coccoliths can only be concentrated as a certain grain-size fraction, and that size fraction analyzed. Although the size fraction used may seem to consist of almost pure coccolith ooze, a small amount of carbonate material from other sources will nevertheless remain and influence the analyses. Moreover, nannofossil analyses will be of a polyspecific assemblage, rather than monospecific as in foraminiferal studies. This produces another source of potential error, because different species may fractionate oxygen isotopes by different amounts, which will influence the measured $^{18}O:^{16}O$ ratio.

Biostratigraphy

The greatest importance of calcareous nannofossils to geology lies in their application to high-resolution biostratigraphy. Nannofossils are very useful in stratigraphic correlation because of their abundance and diversity in marine sediments, their wide geographic distribution, and their rapid evolution, which allows a narrow zonal subdivision of the geologic time column.

The number of nannofossil specimens in even the smallest of samples may be enormous; oceanic oozes may contain 1 trillion specimens/cm^3 (Fig. 11.20), and shallower marine sediments as many as 1 billion specimens/cm^3. Even fine, sandy, inner-shelf sediments commonly contain thousands or tens of thousands of specimens/cm^3. A light-microscope smear slide with evenly spaced nannofossils usually contains 100 000 to 1 million specimens.

As floating marine organisms, nannofossils have had a very wide distribution in marine sediments from the Jurassic to the present. For example, an upper Eocene assemblage from North America will be remarkably similar to latitudinally equivalent upper Eocene assemblages from any other part of the world. Biozones can therefore be established and correlated that are essentially global in extent.

Figure 11.20 Even extremely small samples may contain enormous numbers of nannofossil specimens, as shown on this fracture surface of a middle Miocene chalk from DSDP Site 360, south Atlantic Ocean. Most of the specimens shown are the species *Cyclicargolithus floridanus*. Scale bar, 4 μm.

The evolutionary record of calcareous nannofossils shows generally rapid speciation. Although evolutionary changes appear to be gradual in some nannofossil taxa, changes are quite sudden in others. Reproduction by asexual cell division may explain why some new species appear and establish themselves suddenly, without clearly recognizable ancestors. Any successful mutation would be able to reproduce itself unaltered and in large numbers in a relatively short time. Evolutionary first occurrences and extinctions of species are unique events and are recognizable datum points in the geologic time continuum.

Nannofossils clearly have marked advantages for use in biostratigraphy, but there are also potential problems, the most important of which are ecologic control, diagenetic alteration, and reworking.

Ecologic control

Nannoplankton species may be controlled by environmental factors, especially temperature, as described in previous sections of this chapter. The appearance or disappearance of a predominantly low-latitude species in a high-latitude region could result from:

1 a real appearance or extinction;
2 a climatic change affecting water temperature;
3 changing current patterns.

The extinction of discoasters, for example, occurred in the mid-Pliocene in high latitudes, in contrast to their extinction at approximately the Pliocene–Pleistocene boundary in lower latitudes. The reason is that the warm-water discoasters were affected earlier in higher latitudes by the onset of glaciation.

Diagenetic alteration

Dissolution and secondary **overgrowths** are diagenetic effects that must be recognized for correct interpretation of a nannofossil assemblage. Many areas of the ocean floor are so deep (>4000 m) that dissolution has selectively removed some species, leaving behind only the more solution-resistant forms. Some age-diagnostic species may thus be preserved in shallow water, but diagenetically removed in deep water.

The effects of solution on some common nannofossils are as follows: rim margins of placoliths become raggedly incised; central-area structures are removed; central-area openings are irregularly enlarged; and distal and proximal shields are disconnected. Distal shields usually survive dissolution longer than the smaller proximal shields. Among the sphenoliths (Fig. 11.10I), the basal structures are commonly detached by solution. In discoasters, the central knob may be removed, and the small lateral spines on a discoaster's rays may also be destroyed. In general, however, discoasters are the most resistant nannofossils to solution, owing to their large size and crystal structure. Discoasters are also the nannofossils most susceptible to secondary overgrowth by calcite. Most of the calcite freed by dissolution within the sediment mass does not escape, but instead is reprecipitated on larger, more stable particles, such as discoasters (Fig. 11.21). The quantity of calcite accreted may double or triple the mass of a specimen. Overgrowth formation begins in and around the central area of a discoaster, then advances outward along the rays. Overgrowths can cause difficulty in identification of discoasters, especially when species-distinctive ray tips are obscured by overgrowths. Overgrowths on discoasters have resulted in the erroneous erection of some species names.

Notwithstanding the discussion above, nannofossils are still generally more resistant to solution than planktic foraminifera, and are therefore more useful for stratigraphic studies of deep-water sediments. Cellulose-like polysaccharide material is incorporated within the calcite of coccoliths, and this may be the reason for their greater resistance to dissolution. Organic coatings on calcite particles are known to inhibit precipitation of secondary calcite on the particles, and the organic pellicle around a

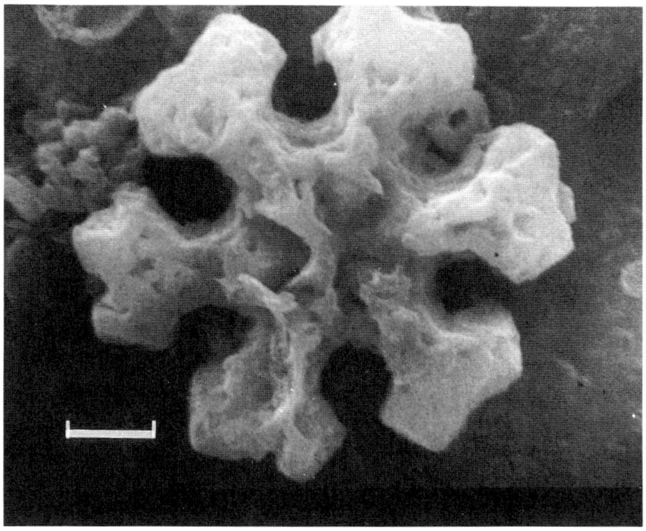

Figure 11.21 *Discoaster deflandrei*, heavily overgrown with secondary calcite. Scale bar, 4 μm.

coccolithophore's cell is likely to inhibit dissolution as well.

Reworking

The minute size of nannofossils makes them almost indestructible by mechanical abrasion. They are therefore able to be reworked, time and again, into progressively younger sediments. Almost any moderately rich sample will contain at least some reworked nannofossils, but fortunately the production of indigenous nannofossils is so rapid that reworked specimens will make up only a very small percentage of a normal assemblage. Exceptional cases are known, however. For example, the Bridlington Crag in England, whose Pleistocene age is well established on the basis of molluscs, contains an abundant, well-preserved, and fairly diverse assemblage of Jurassic and Cretaceous nannofossils, but no Cenozoic nannofossils. The nannofossil assemblage has been entirely reworked from Mesozoic rocks that crop out close to the depositional site of the Bridlington Crag. As another example, Cretaceous coccoliths have been extensively reworked from the Mississippi River drainage basin, ultimately mixing with Quaternary coccoliths in Gulf of Mexico sediments. Samples are easy to recognize as Quaternary, however, because the Cretaceous specimens are so clearly out of place.

Contiguous biozones differentiated by relatively few taxa are more difficult to recognize if reworking has occurred. The *Discoaster brouweri* (NN 18) Zone (Fig. 11.27) is one example. The top of this zone is defined by the last occurrence of *D. brouweri* (Fig. 11.29g). The base of the overlying *Pseudoemiliania lacunosa* (NN 19) Zone is also defined by the last occurrence of *D. brouweri*. A few reworked specimens of *D. brouweri* would extend the duration of the *D. brouweri* Zone and reduce the duration of the overlying zone. If *D. brouweri* were not present—owing to diagenetic or environmental causes—this zone would be indistinguishable from adjacent zones. Any of these situations could lead to errors in correlation.

Reworked specimens in a nannofossil assemblage can normally be recognized by comparison with the dominant species present. If rare or sparse species are obviously out of place, stratigraphically, with the rest of the assemblage, in all likelihood they have been reworked. Extensive overgrowths on some specimens in an assemblage of otherwise clean specimens also suggests reworking.

Calcareous nannofossil zonations

In 1954, Bramlette and Riedel first suggested that nannofossils could be used for biostratigraphic zonation, and many zonations for the Jurassic, Cretaceous, and Cenozoic were subsequently erected.

More zonations have been proposed for the Cretaceous than for the Jurassic. This is largely due to globally widespread Cretaceous marine deposits, both on the continents and in the oceans, and in part to the greater abundance and diversity of nannofossils during the Cretaceous and therefore greater potential for refined subdivision of Cretaceous rocks. Cenozoic zonations are even more refined than those of the Cretaceous for the same reasons (compare Figs 11.22–11.27).

All current Mesozoic and Cenozoic zonations use **interval zones** (biozones) whose boundaries are geologically synchronous **biohorizons** represented by unique biostratigraphic events: an evolutionary first occurrence or an extinction of a species. In theory, first occurrences should be more precise than last occurrences for defining zonal boundaries, since true last occurrences can be distorted by reworking. However, petroleum-company biostratigraphers working with borehole cuttings use last occurrences ("tops"), because of the problem of downhole contamination caused by sediment slumping from higher in the hole.

Reasonably abundant, cosmopolitan, solution-resistant species are obviously the best choices to define zonal boundaries. Currently used zonations use such species wherever possible, although certain zones are difficult to recognize because of the rarity of the zonal marker in some geographic areas or environments. Widely used nannofossil zonations for the Mesozoic and Cenozoic are

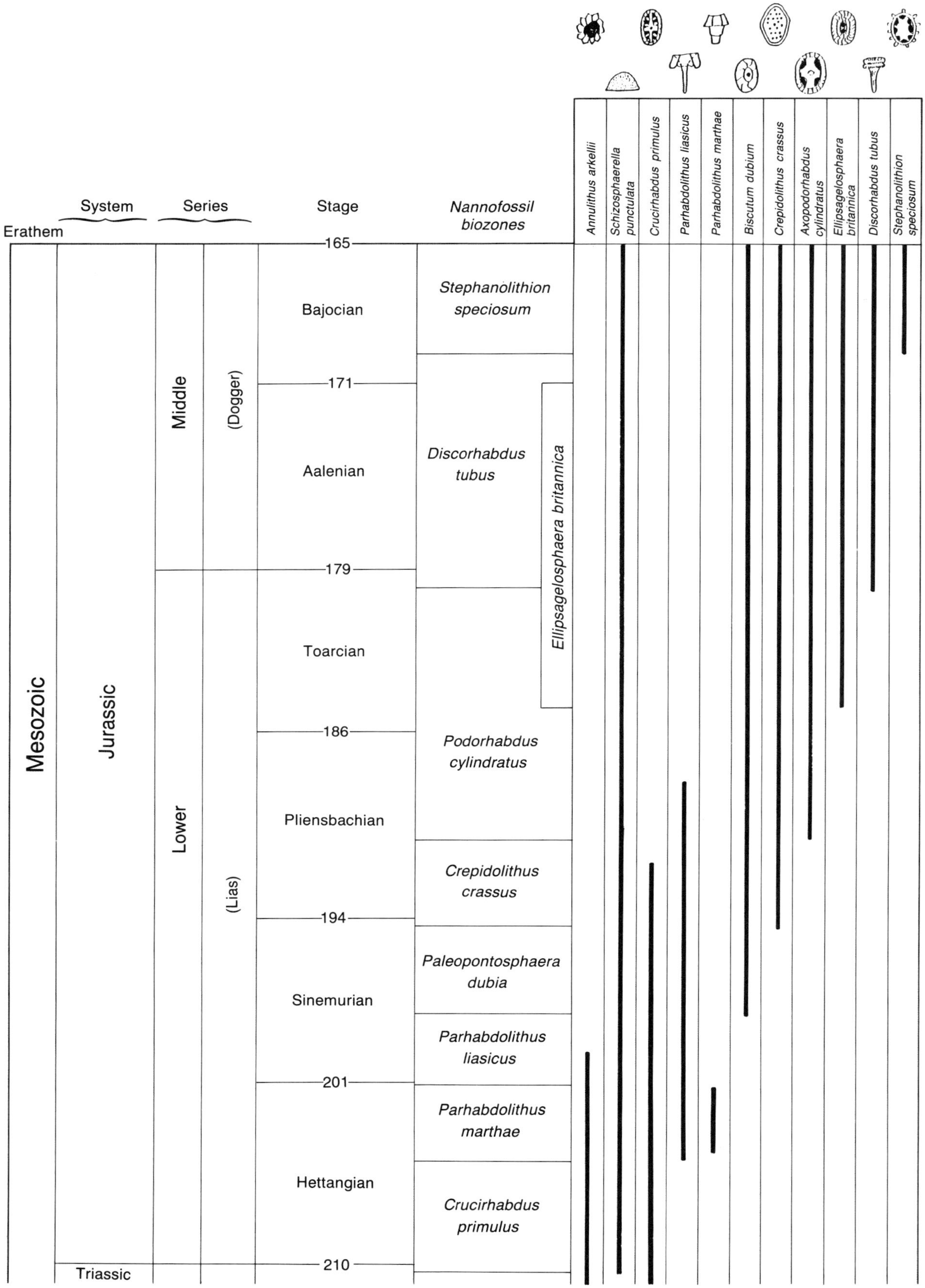

Figure 11.22 Lower and Middle Jurassic nannofossil biozones and stratigraphic ranges of species that define the zonal boundaries.

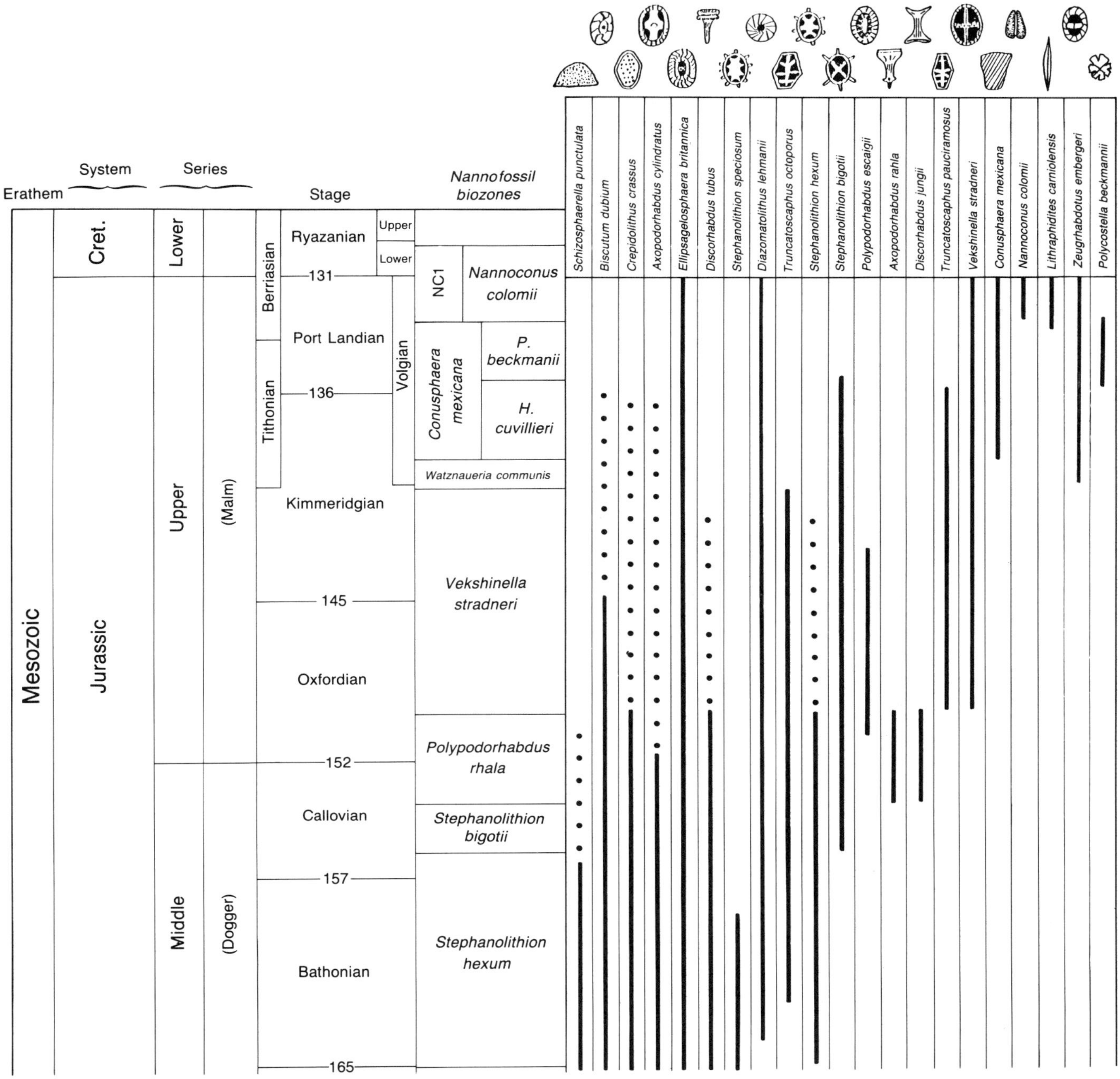

Figure 11.23 Middle and Upper Jurassic nannofossil biozones and stratigraphic ranges of species that define the zonal boundaries. Dots represent sporadic or questionable occurrences.

shown in Figures 11.22–11.27, plotted against chronostratigraphic units and against the stratigraphic ranges of the marker species whose first or last occurrences mark the boundaries of each biozone. Figures 11.28 and 11.29 show light photomicrographs of many of the marker species used in these zonations.

Evolutionary history

Paleozoic nannofossils have been reported, but these reports have not been confirmed. True calcareous nannofossils first appeared in the Late Triassic. Major extinction of nannofossils at the Triassic–Jurassic boundary does

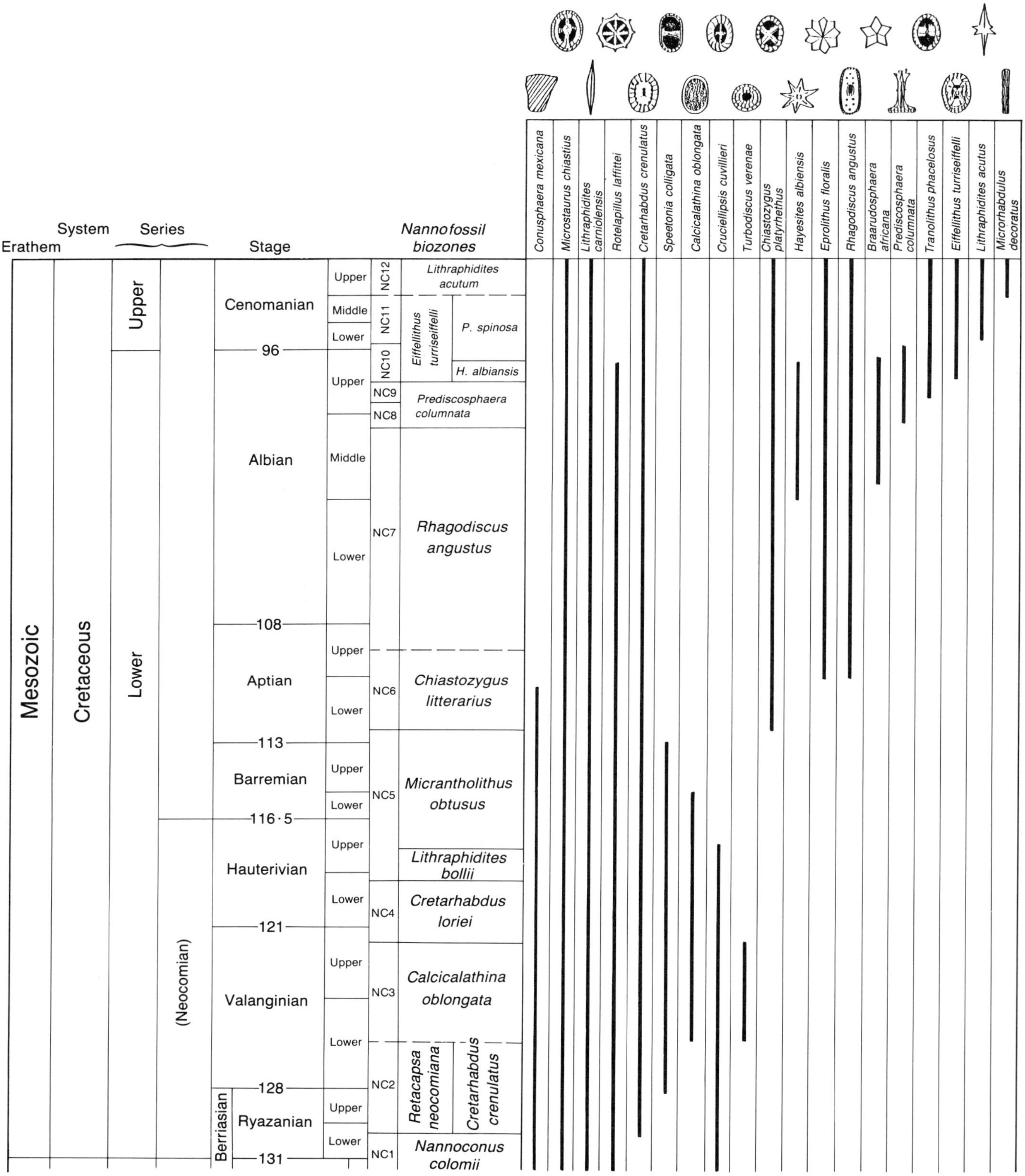

Figure 11.24 Lower Cretaceous nannofossil biozones and stratigraphic ranges of species that define the zonal boundaries.

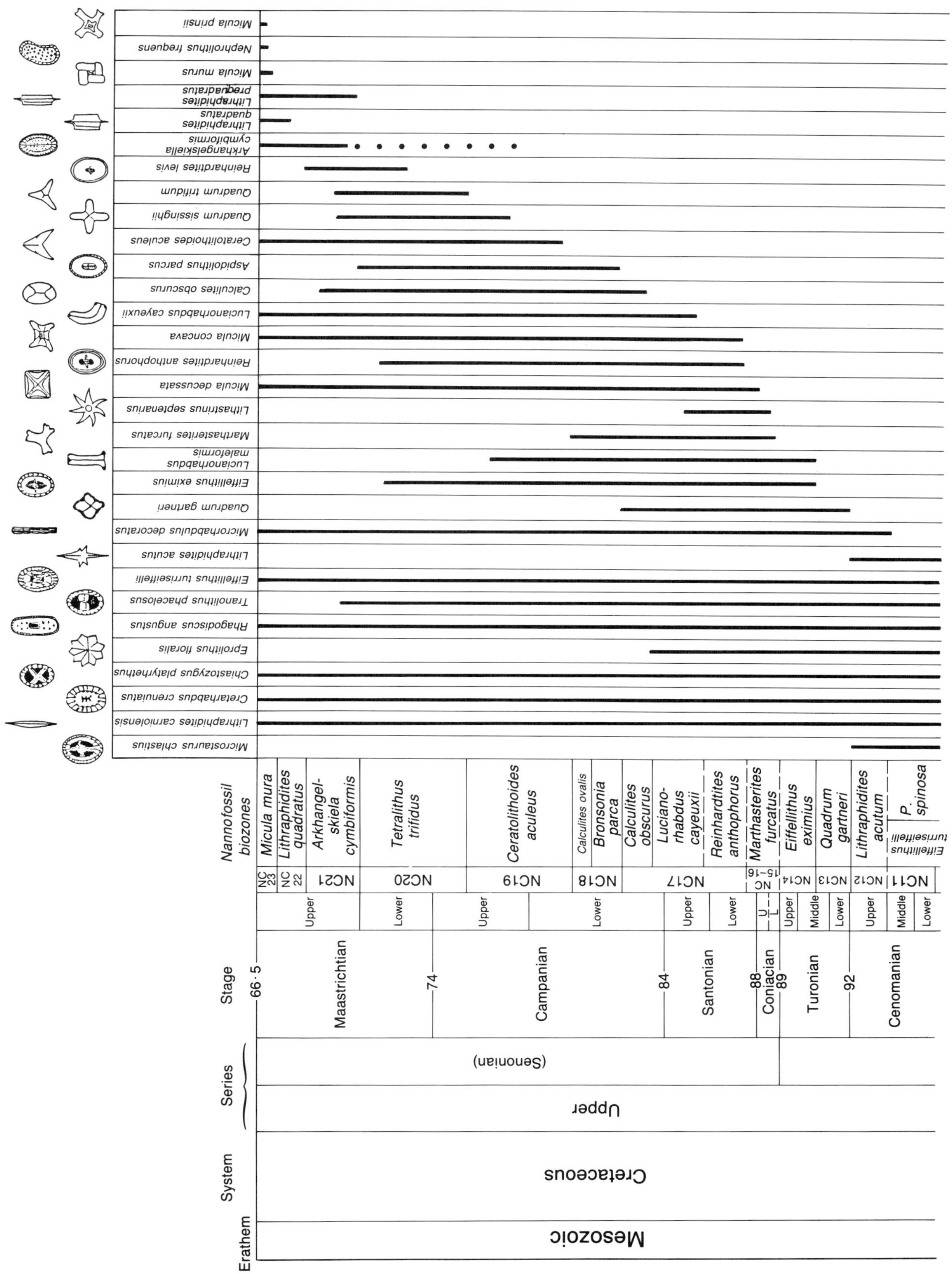

Figure 11.25 Upper Cretaceous nannofossil biozones and stratigraphic ranges of species that define the zonal boundaries. Dots represent sporadic or questionable occurrences.

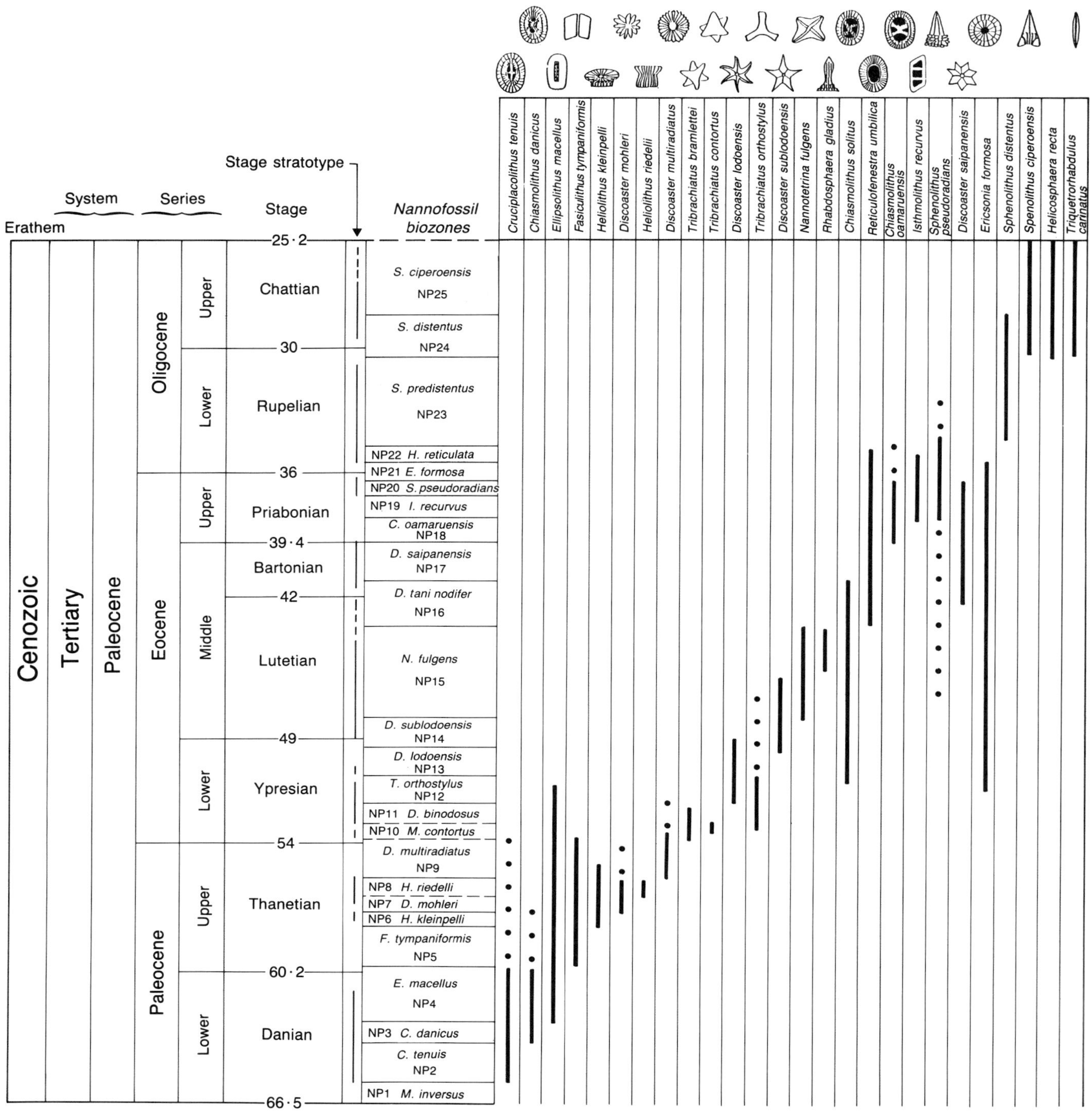

Figure 11.26 Paleogene nannofossil biozones and stratigraphic ranges of species that define the zonal boundaries. Dots represent sporadic or questionable occurrences.

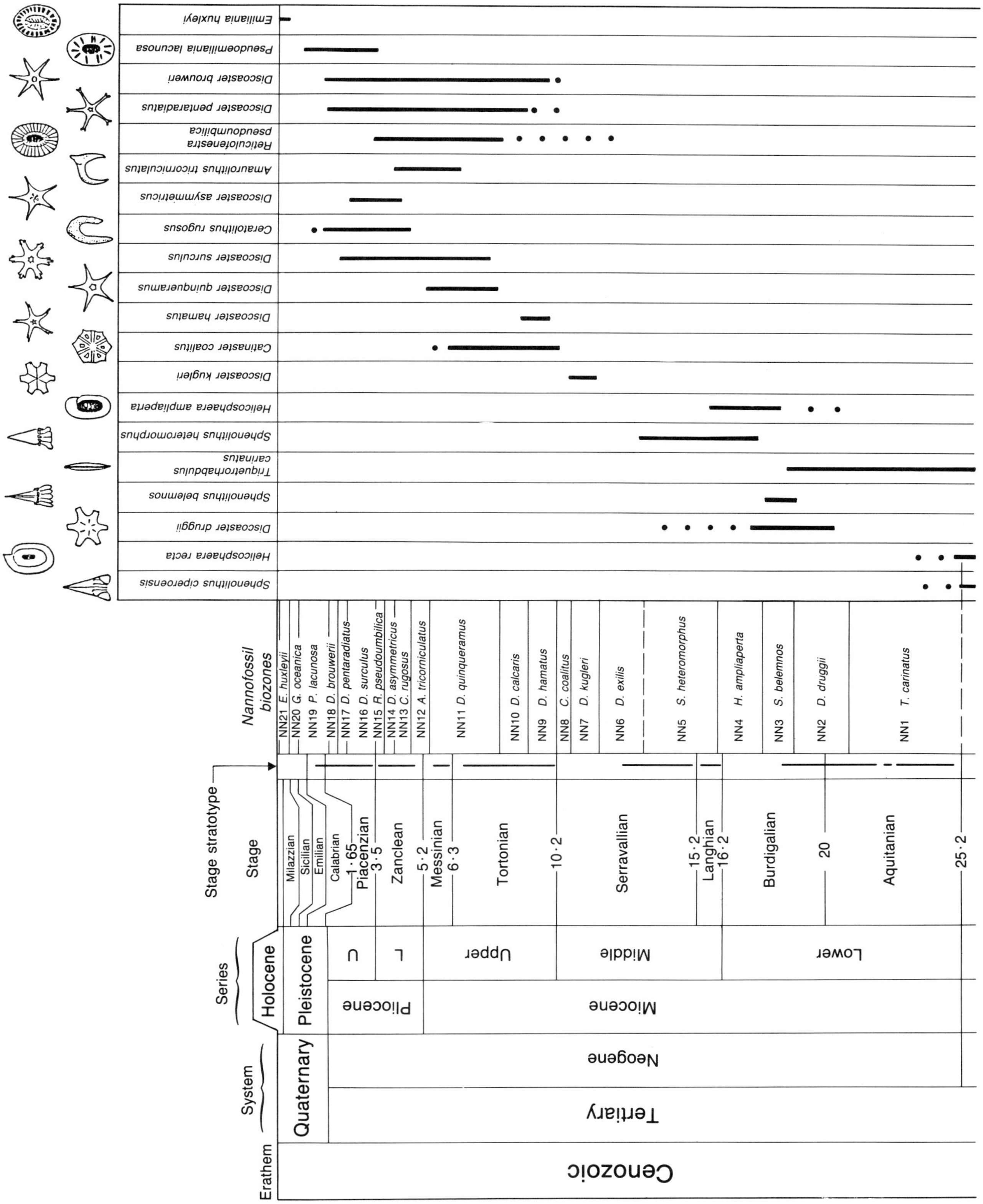

Figure 11.27 Neogene and Quaternary nannofossil biozones and stratigraphic ranges of species that define the zonal boundaries. Dots represent sporadic or questionable occurrences.

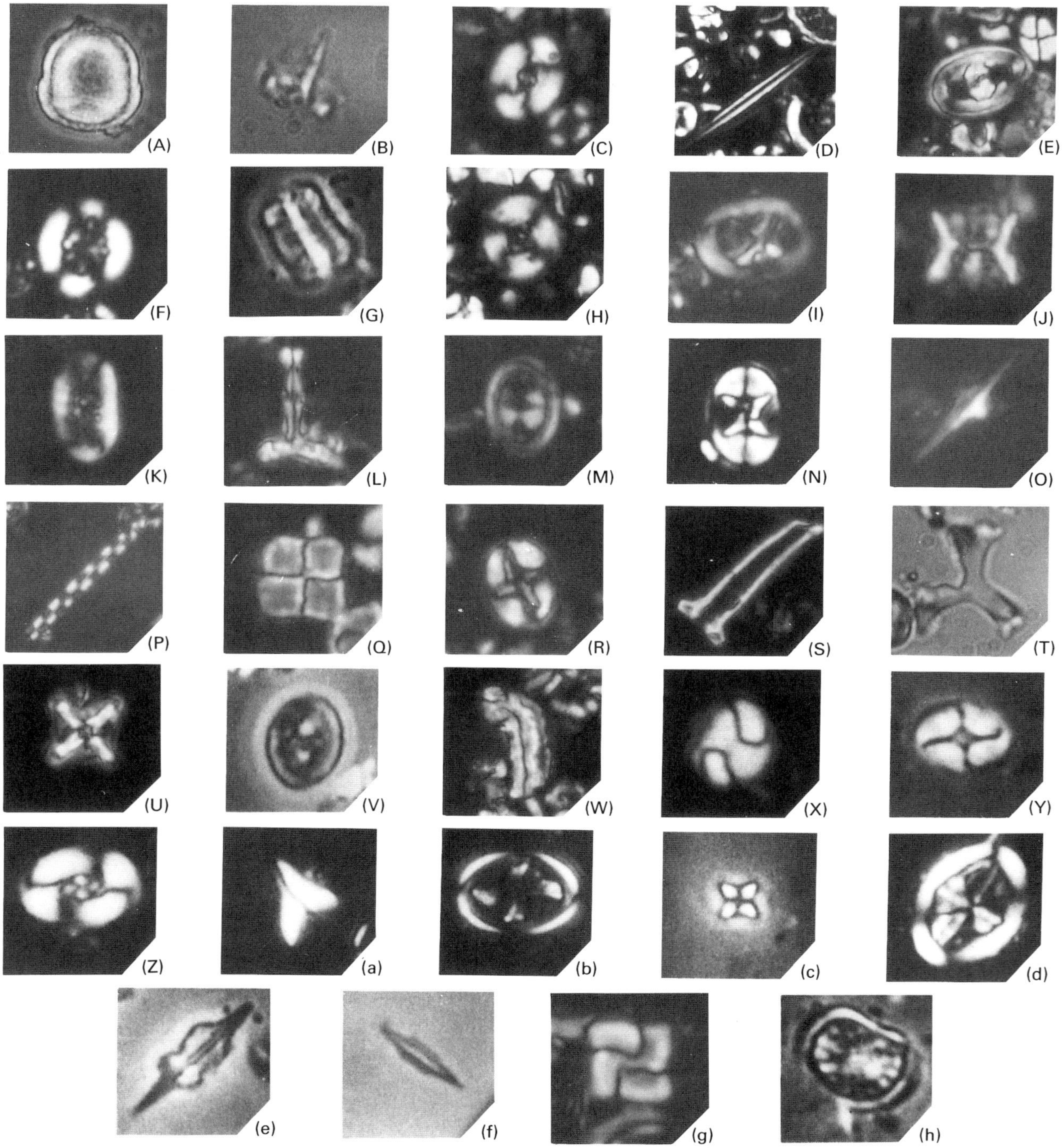

Figure 11.28 Light photomicrographs of some of the species that define the Mesozoic biozonal boundaries shown in Figures 11.22–11.25, and some other useful Mesozoic marker species. All figures were photographed between crossed nicols except A, B, G, T, V, c, e, f, and h, which were photographed using phase-contrast illumination. A, *Schizosphaerella punctulata*. B, *Parhabdolithus liasicus*. C, *Ellipsagelosphaera britannica*. D, *Lithraphidites carniolensis*. E, *Zeugrhabdotus embergeri*. F, *Cretarhabdus crenulatus*. G, *Nannoconus elongatus*. H, *Cyclagelosphaera deflandrei*. I, *Chiastozygus platyrhethus*. J, *Eprolithus floralis*. K, *Rhagodiscus angustus*. L, *Prediscosphaera cretacea*. M, *Tranolithus phacelosus*. N, *Eiffellithus turriseiffelii*. O, *Lithraphidites acutus*. P, *Microrhabdulus decoratus*. Q, *Quadrum gartneri*. R, *Eiffellithus eximius*. S, *Lucianorhabdus maleformis*. T, *Marthasterites furcatus*. U, *Micula decussata*. V, *Reinhardtites anthophorus*. W, *Lucianorhabdus cayeuxii*. X, *Calculites obscurus*. Y, *Calculites ovalis*. Z, *Aspidolithus parcus*. a, *Ceratolithoides aculeus*. b, *Gartnerago obliquum*. c, *Quadrum gothicum*. d, *Arkhangelskiella cymbiformis*. e, *Lithraphidites quadratus*. f, *Lithraphidites praequadratus*. g, *Micula murus*. h, *Nephrolithus frequens*. Approximate magnifications ×1000, except D, E, G, H, and W (×700), and B, K, O, a, e, and f (×1200).

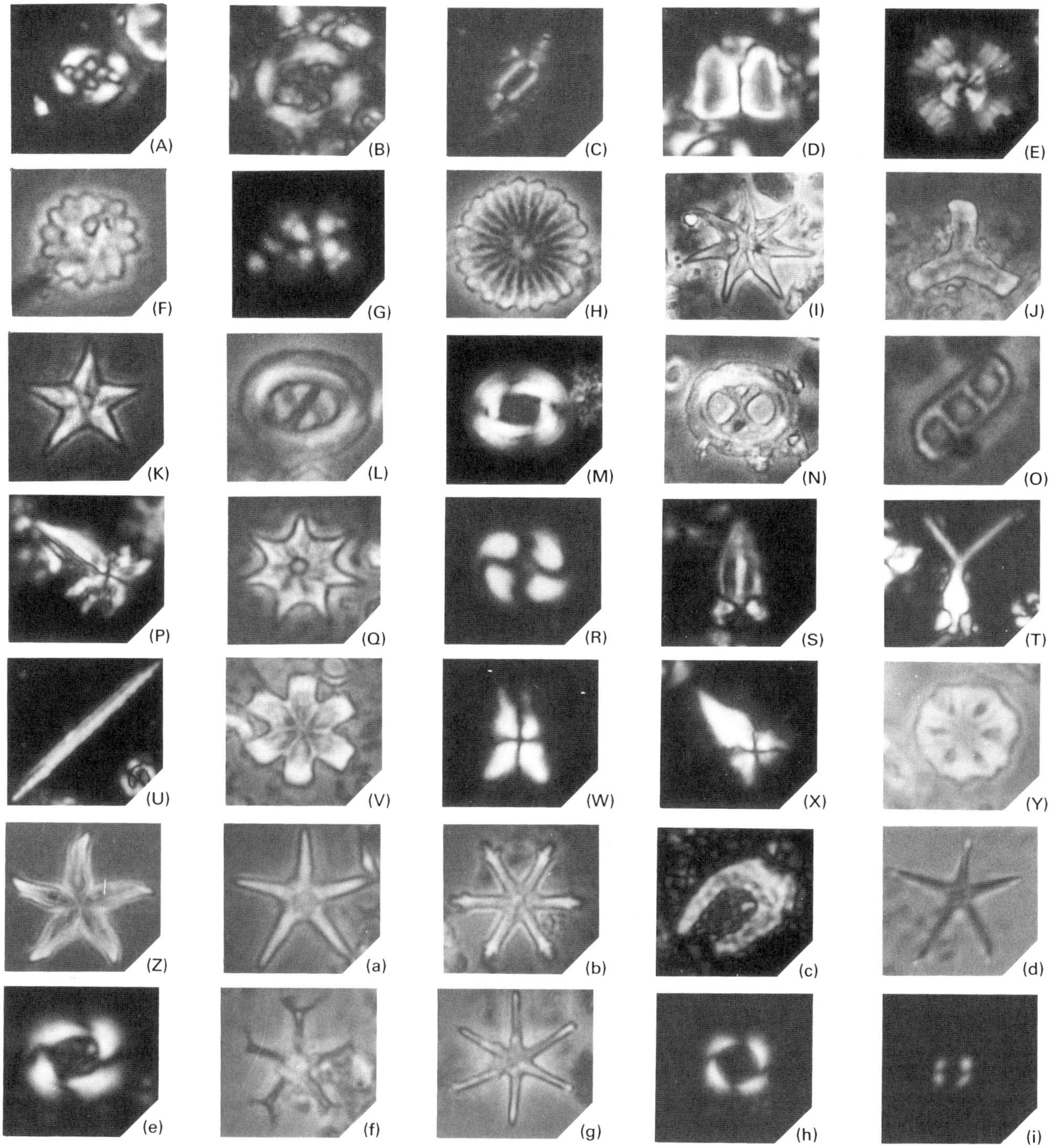

Figure 11.29 Light photomicrographs of some common species defining the Cenozoic biozonal boundaries shown in Figures 11.26 and 11.27. All species were photographed between crossed nicols except F, H, I, J, K, L, N, O, Q, V, Y, Z, a, b, d, f, and g, which were photographed using phase-contrast illumination. A, *Cruciplacolithus tenuis*. B, *Chiasmolithus danicus*. C, *Ellipsolithus macellus*. D, *Fasciculithus tympaniformis*. E, *Heliolithus kleinpellii*. F, *Discoaster mohleri*. G, *Heliolithus riedelii*. H, *Discoaster multiradiatus*. I, *Discoaster lodoensis*. J, *Tribrachiatus orthostylus*. K, *Discoaster sublodoensis*. L, *Chiasmolithus solitus*. M, *Reticulofenestra umbilica*. N, *Chiasmolithus oamaruensis*. O, *Isthmolithus recurvus*. P, *Sphenolithus pseudoradians*. Q, *Discoaster saipanensis*. R, *Ericsonia formosa*. S, *Sphenolithus distentus*. T, *Sphenolithus ciperoensis*. U, *Triquetrorhabdulus carinatus*. V, *Discoaster druggii*. W, *Sphenolithus belemnos*. X, *Sphenolithus heteromorphus*. Y, *Catinaster coalitus*. Z, *Discoaster hamatus*. a, *Discoaster quinqueramus*. b, *Discoaster surculus*. c, *Ceratolithus rugosus*. d, *Discoaster asymmetricus*. e, *Reticulofenestra pseudoumbilica*. f, *Discoaster pentaradiatus*. g, *Discoaster brouweri*. h, *Pseudoemiliania lacunosa*. i, *Emiliania huxleyi*. Approximate magnifications ×1000, except M, N, and Z (×700), and O, S, T, W, h, and i (×1200).

not occur. Calcareous nannofossils increased gradually in abundance and diversity throughout the Jurassic and during the Cretaceous until the Santonian. A rapid increase in diversity occurred during and after the Santonian, culminating in a Mesozoic diversity high (estimates of up to 240 different species) in the Maastrichtian (Fig. 11.30). Then came the well-documented massive extinctions at the termination of the Maastrichtian. Only an estimated 15–18 species survived into the early Danian. These few species gave rise to all the later Tertiary forms.

Diversity increased gradually again throughout the Paleocene and Eocene, culminating in an estimated Cenozoic high of up to 300 species in the early Eocene. Diversity then declined steadily until the close of the early Miocene, increased slightly in the middle Miocene, then declined again, resulting in less than 50 species during the Pleistocene (Fig. 11.30).

Estimating total species diversity in a stratigraphic interval is difficult, in large part because of differing taxonomic concepts among various nannoplankton specialists. Nevertheless, an estimated 500 to 550 valid nannofossil species are known from the Mesozoic fossil record, and an estimated 600 to 650 from the Cenozoic, which includes about 150 living species.

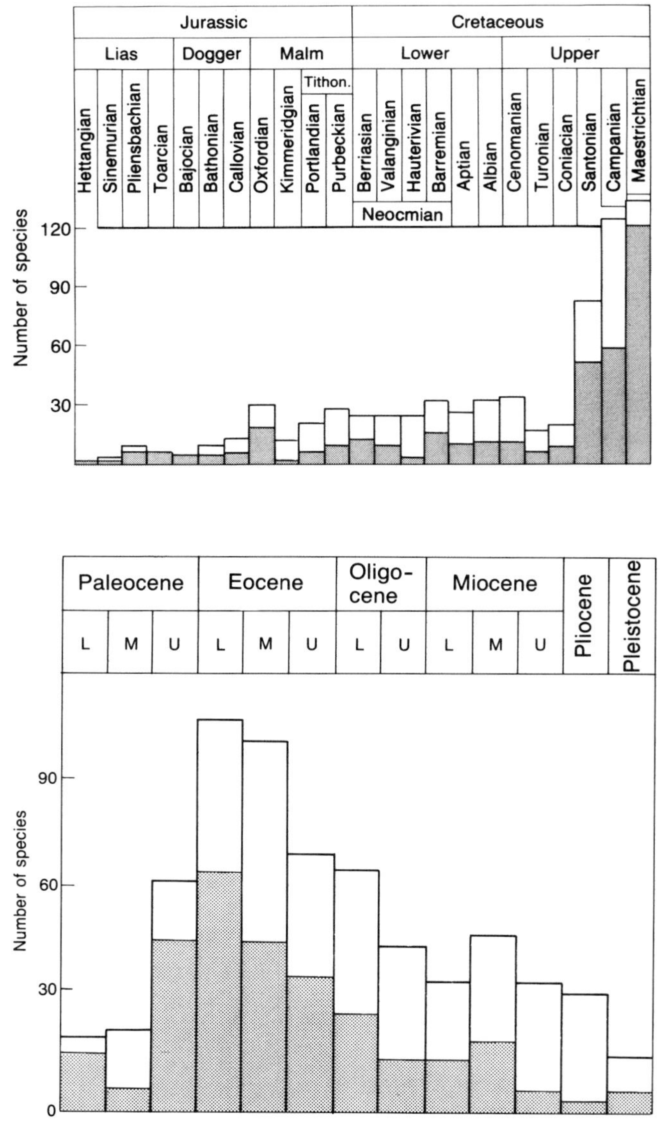

Figure 11.30 Nannofossil diversity during the Jurassic, Cretaceous, and Cenozoic. Shaded areas of bars represent new species. (Modified from Haq BU. *Marine Geology*. 1973; 15:M25–M30.)

Phylogenetic lineages

Modification of structural features among an evolving group of closely related taxa provides clues to ancestor–descendant relationships. Examples of several phylogenetic studies are described briefly in the following paragraphs.

In the Paleogene, two distinct lineages have been recognized within the genus *Chiasmolithus*, based in part on the construction and architecture of the crossbar (Fig. 11.31). In the lineage originating with *C. danicus* (lower half of Fig. 11.31), species have crossbars with axial sutures ("split bars"); in the other lineage (upper half of Fig. 11.31, originating with *Cruciplacolithus*), species' crossbars lack axial sutures ("nonsplit bars"). Further subdivision of *Chiasmolithus* is largely based on the shape of individual crossbars.

Two lines of evolution occur among the discoasters, one line arising from the Paleocene genus *Heliolithus* (Fig. 11.9G) and the other from *Fasciculithus* (Fig. 11.9E). *Discoaster* in general shows a trend towards less robust forms possessing fewer rays from the Paleocene until its extinction at the end of the Pliocene (Fig. 11.32).

Three evolutionary trends occur within the genus *Helicosphaera* during the Tertiary:

1 a change from ovoid, to rhomboid or broad-elliptical, to roughly rectangular, to long-elliptical;

2 a change in the shape of the terminal flange from compressed, to sharply pointed, to large-extended, to small-rounded flanges;

3 a simplification of the central-area construction with time (Fig. 11.33).

Study of lineages in the family Ceratolithaceae (Fig. 11.34) resulted in the recognition of a new genus, *Amaurolithus*, which shows faint or no birefringence in cross-polarized light. The other genus in the family,

Ceratolithus, is strongly birefringent. *Amaurolithus* first appeared in the late Miocene and evolved into *Ceratolithus* during the early Pliocene and is represented among the living species by *C. cristatus* (Fig. 11.8H) and *C. telesmus*.

Supplementary reading

Aubry MP. *Handbook of Cenozoic Calcareous Nannoplankton. Books 1–4*. New York: Micropaleontology Press, 1984:266, 1988:279, 1989:279, 1990:381. The first four of nine volumes planned as a comprehensive synthesis of all Cenozoic nannofossil species.

Bukry D. Biostratigraphy of Cenozoic marine sediment by calcareous nannofossils. *Micropaleontology* 1978; 24:44–60. Use of calcareous nannofossils in biostratigraphy, with emphasis on DSDP studies.

Gartner S. Calcareous nannofossils in marine sediments. In: Emiliani C, ed. *The Sea: The Oceanic Lithosphere*. Vol. 7. New York: John Wiley, 1981:1145–77. A summary treatment of nannofossils, including sections on morphology and biology, sedimentation and diagenesis, and evolution and biostratigraphy.

Haq BU. Calcareous nannoplankton. In: Haq BU, Boersma A, eds. *Introduction to Marine Micropaleontology*. Amsterdam: Elsevier, 1978: 79–107. An introductory-level chapter on the biology, ecology, classification, and biostratigraphy of calcareous nannoplankton.

Haq BU, ed. *Calcareous Nannoplankton, Benchmark Papers in Geology, 78*. Stroudsburg: Hutchison Ross, 1983:338. A collection of many of the classical papers on structural, biologic, biogeographical, evolutionary, and other aspects of calcareous nannoplankton.

Haq BU, ed. *Nannofossil Biostratigraphy, Benchmark Papers in Geology, 79*. Stroudsburg: Hutchison Ross, 1984:386. A collection of many of the classical papers on nannofossil biostratigraphy.

Hay WW. Calcareous nannofossils. In: Ramsay ATS, ed. *Oceanic Micropaleontology*. London: Academic Press, 1977:1055–1200. An overview of calcareous nannofossils, with detailed treatments of preparation and examination techniques and taxonomy.

Lord AR, ed. *A Stratigraphical Index of Calcareous Nannofossils*. Chichester: Ellis Horwood, 1982:192. A comprehensive stratigraphical treatment of calcareous nannofossils in the Mesozoic and Cenozoic of Britain, with emphasis on the Mesozoic.

Martini E. Standard Tertiary and Quaternary calcareous nannoplankton zonation. In: Farinacci A, ed. *Proceedings of the 2nd*

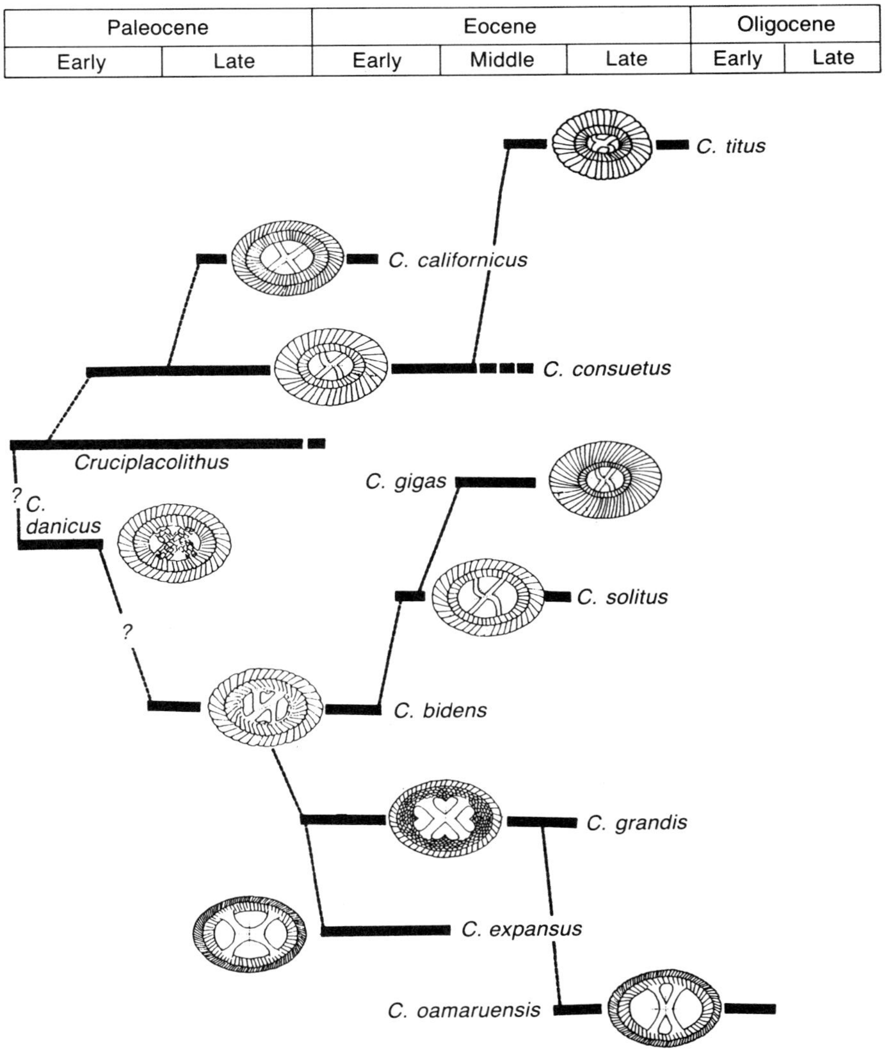

Figure 11.31 Phylogenetic lineage of the genus *Chiasmolithus*. (Reproduced from Haq BU. *Introduction to Marine Micropaleontology*. Amsterdam: Elsevier Science, 1978:98.)

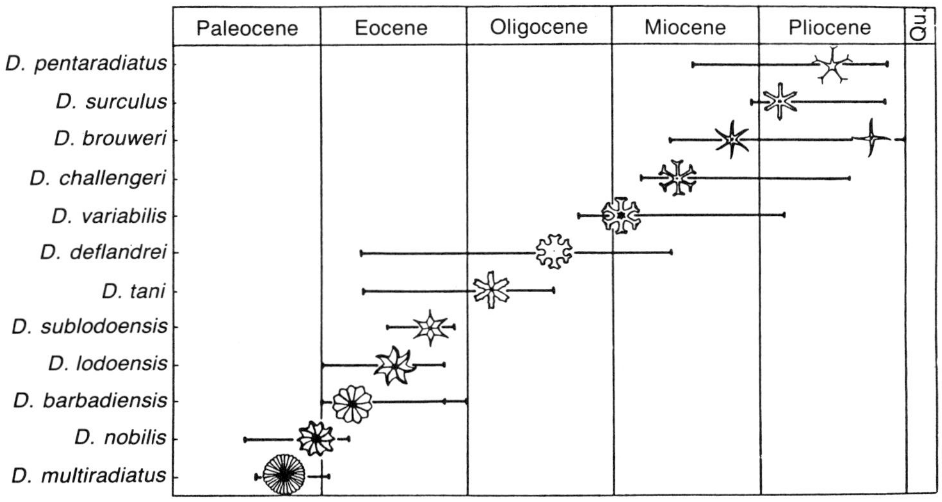

Figure 11.32 Evolutionary changes in the genus *Discoaster*, showing a Paleocene–Pliocene trend towards less robust bodies and fewer rays. (Reproduced from Martini E. *The Micropaleontology of Oceans*. Cambridge: Cambridge University Press, 1971:537.)

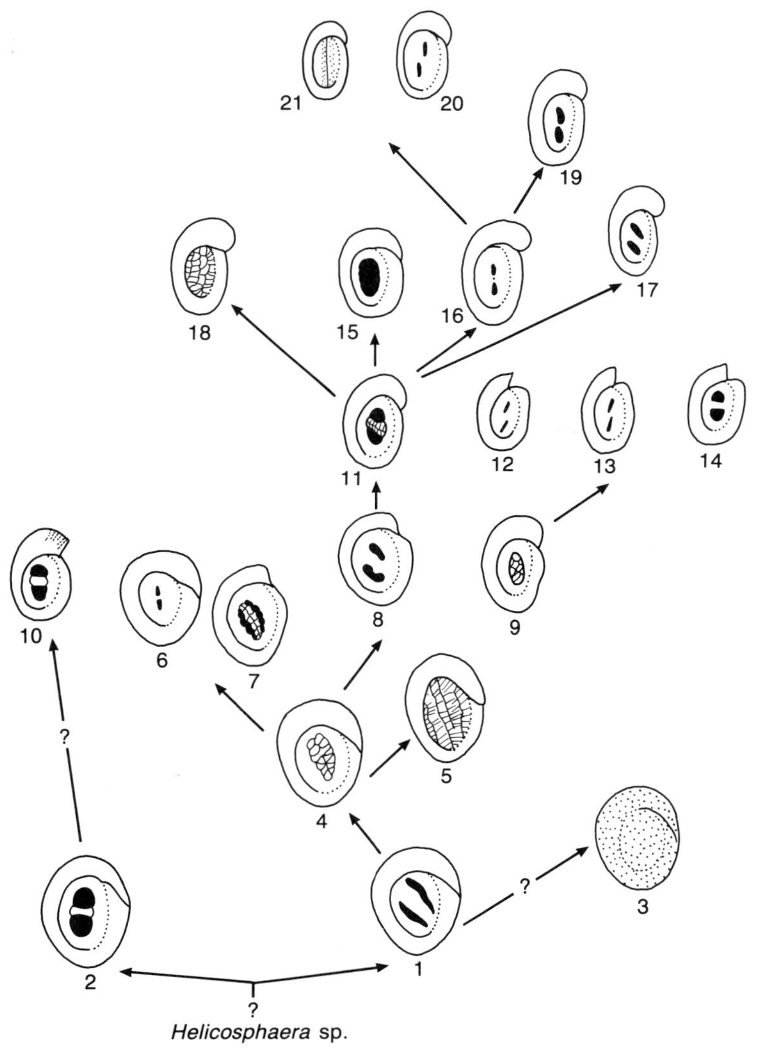

Figure 11.33 Phylogenetic lineage of the genus *Helicosphaera*. 1, *H. lophota*; 2, *H. seminulum*; 3, *H. papillata*; 4, *H. dinesenii*; 5, *H. heezenii*; 6, *H. compacta*; 7, *H. reticulata*; 8, *H. bramlettei*; 9, *H. euphratis*; 10, *H. wilcoxonii*; 11, *H. intermedia*; 12, *H. perch-nielseniae*; 13, *H. obliqua*; 14, *H. recta*; 15, *H. ampliaperta*; 16, *H. carteri*; 17, *H. rhomba*; 18, *H. granulata*; 19, *H. sellii*; 20, *H. wallichi*; 21, *H. hyalina*. (Reproduced from Haq BU. *Micropaleontology*. 1973; 19:32–52.)

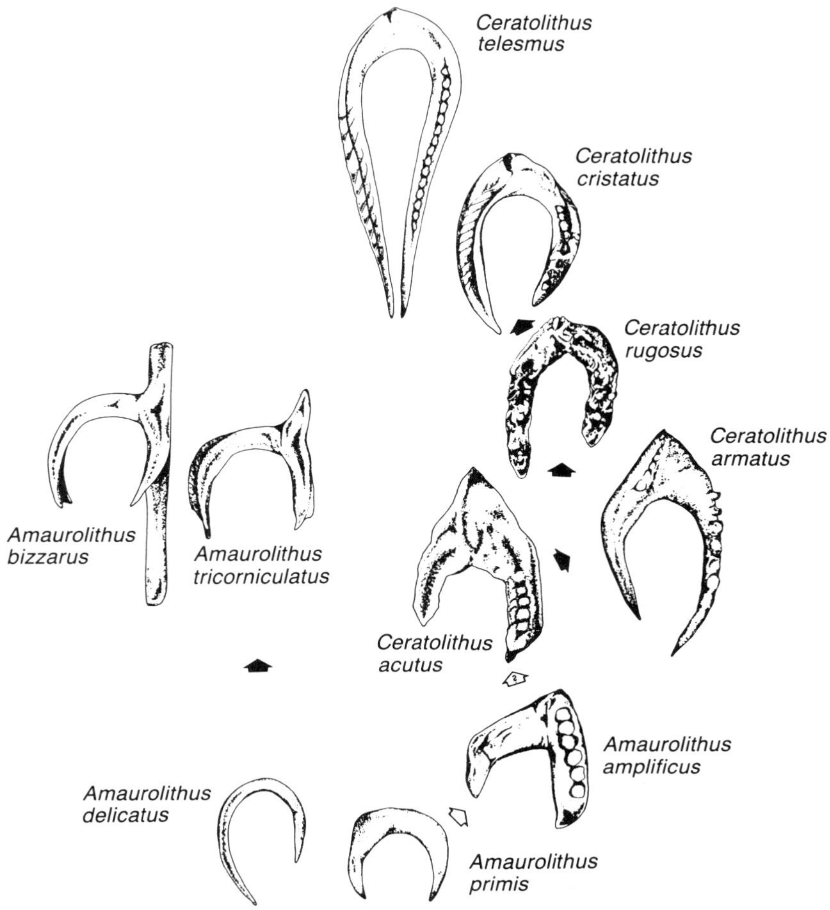

Figure 11.34 Phylogenetic lineage of the family Ceratolithaceae. (Reproduced from Gartner S, Bukry D. *J Res US Geol Survey* 1975; 3:451–65.)

Planktonic Conference, Roma, 1970. Rome: Edizioni Technoscienza, 1971:739–85. Definitions and zonal boundary indicators for one of the widely used Cenozoic biostratigraphical schemes.

Perch-Nielsen K. Mesozoic calcareous nannofossils. In: Bolli HM, Saunders JB, Perch-Nielsen K, eds. *Plankton Stratigraphy.* Cambridge: Cambridge University Press, 1985:329–426. A state-of-the-art taxonomical and biostratigraphical synthesis of Mesozoic nannofossils; all important species are illustrated by both LM and SEM photographs.

Perch-Nielsen K. Cenozoic calcareous nannofossils. In: Bolli HM, Saunders JB, Perch-Nielsen K, eds. *Plankton Stratigraphy.* Cambridge: Cambridge University Press, 1985:427–554. A similar taxonomical and biostratigraphical synthesis of Cenozoic nannofossils; also profusely illustrated.

Tappan H. *The Paleobiology of Plant Protists.* San Francisco: WH Freeman, 1980:1028. Chapter 9 is a summary of the biology, ecology, classification, and biostratigraphy of calcareous nannoplankton, containing more biologic and ecologic information than most other summaries.

Foraminifera

12

Stephen J. Culver

Introduction

Foraminifera are **testate** protozoans that have a wide variety of test composition and morphology (Fig. 12.1). They inhabit all marine environments from the intertidal zone to the deep ocean floor, they range from the poles to the tropics, and they can be planktic or benthic. Their abundance in sediment samples may reach tens of thousands per square meter, and their diversity in tropical environments may exceed 60 or 70 species in a sample of 300 individuals. They have a geologic range starting in the very earliest Cambrian through to the present day and showing varying patterns of evolution and extinction that are of considerable biologic and geologic significance.

For all of these reasons, foraminifera, studied by hundreds of industrial and academic paleontologists worldwide, continue to be the most important fossil group in both biostratigraphy and paleoenvironmental reconstructions. A voluminous literature concerns foraminifera, and it continues to increase. The *Treatise on Invertebrate Paleontology* on the foraminifera in 1964 recognized 2459 validly described genera. In less than 20 years this number had increased to 3558 and more than 60 000 species had been described. By 1988, 3620 described genera were recognized.

Most papers on living or fossil foraminifera are written by paleontologists. Nevertheless, they have immense biologic significance. The density of living benthic foraminifera can exceed $10^6/m^2$, with wet-weight biomass estimates ranging from 0.02 to more than $10\,g/m^2$. Hence, foraminifera constitute a major portion of the benthic biomass in shallow to deep marine ecosystems, and their many trophic interactions indicate their importance in benthic food webs. Planktic foraminifera, similarly, occupy an important place in ecosystems in the water column. Many species are opportunistic feeders, consuming whatever food particles they encounter, whether plant, animal, or organic detritus. They are themselves the selective or accidental prey of fish, various invertebrates, and perhaps other protozoans.

The major impetus for foraminiferal research, however, remains their unsurpassed value for solving geologic problems. Paleotemperatures can be estimated through isotope analysis of foraminiferal tests; **paleodepths** can be inferred from the composition of foraminiferal assemblages; the movement of water masses can be reconstructed through distributional studies; and movement of tectonic plates can be recognized using foraminiferal data. All of these patterns, recognizable in the stratigraphic record through studies of foraminifera, can be put into a biostratigraphic framework, again through the use of foraminifera. Twenty-eight biostratigraphic zones based on planktic foraminifera can be recognized in the Cretaceous, 22 in the Paleogene, and 20 in the Neogene. Similar detailed zonations are possible from the Late Paleozoic to the Recent using benthic foraminifera.

The scientific study of foraminifera has a long history stretching back to the sixteenth and seventeenth centuries when smaller foraminifera were illustrated, described, and variously classified as worms, corals, or molluscs. Foraminifera were first mentioned, however, in classical times. Herodotus (fifth century BC) noted the large benthic foraminifer *Nummulites* in the limestone of the Egyptian pyramids. Their true protozoan nature was first determined by F. Dujardin in 1835, 9 years after A. d'Orbigny (who considered foraminifera to be tiny cephalopods) produced his famous first classification of the foraminifera. Foraminifera continued to be described and illustrated throughout the nineteenth century. The classic 1884 monograph by H.B. Brady on the foraminifera dredged from the world's oceans by **HMS *Challenger*** remains a standard reference for foraminiferologists in addition to being an exquisitely illustrated work. Brady also wrote one of the first (1870) ecologically oriented studies of foraminifera when he noted low species diversity in brackish water environments.

Foraminiferal studies with an ecologic or paleoecologic emphasis began in earnest in the 1930s with the works of R. Norton and M.L. Natland in North America and W.A. Macfadyen in Europe. Such studies gained impetus and

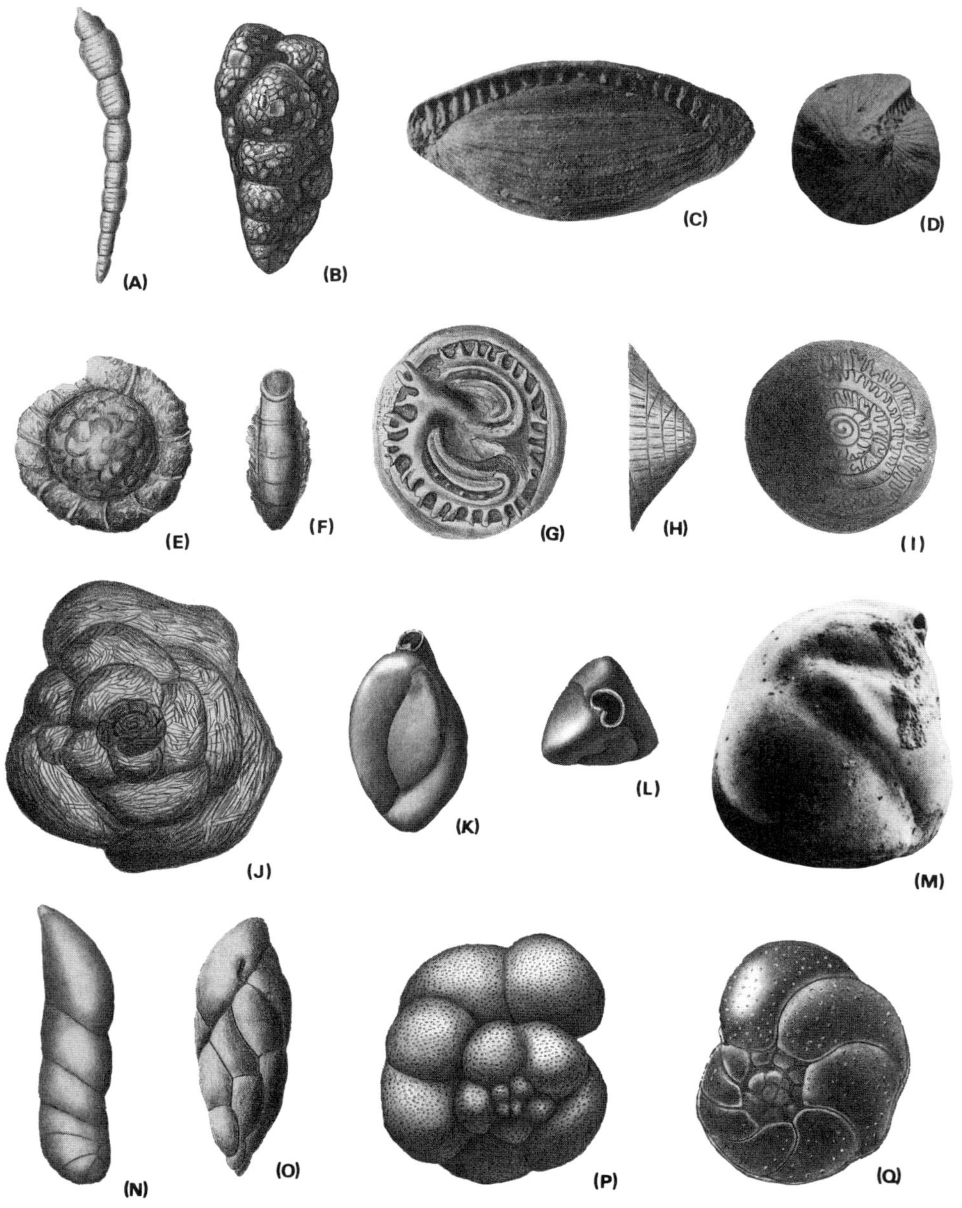

Figure 12.1 Representative genera of 12 foraminiferal suborders. A, Allogromiina, *Nodellum membranaceum*. Recent. Side view. ×36.5. B, Textulariina, *Eggerella advena*. Pleistocene. Side view. ×108. C, D; Fusulinina, *Schwagerina diversiformis*. Pennsylvanian. C, Front view. ×5.8. D, End view. ×5.8. E, F; Involutinina, *Involutina deslongchampsi*. Jurassic. E, Side view. ×54.8. F, Apertural view. ×54.8. G, H, I; Spirillinina, *Patellina corrugata*. Recent. G, Ventral view. ×110. H, Side view. ×110. I, Dorsal view. ×110. J, Carterinina, *Carterina spiculotesta*. Recent. Dorsal view. ×37. K, L; Miliolina, *Triloculina trihedra*. Pleistocene. K, Side view. ×67.9. L, Apertural view. ×67.9. M, Silicoloculinina, *Miliammellus legis*. Side view. Recent. ×131. N, Lagenina, *Marginulina dubia*. Miocene. Side view. ×62.1. O, Robertinina, *Robertina arctica*. Recent. Side view. About ×58.4. P, Globigerinina, *Neogloboquadrina eggeri*. Recent. Dorsal view. ×45.3. Q, Rotaliina, *Cibicides floridanus*. Recent. Ventral view. ×66. (A, E, F, G, H, I, J, N, O Reproduced from Cushman JA. *Foraminifera, Their Classification and Economic Use*. Cambridge: Harvard University Press, 1948. B, K, L Reproduced from Buzas MA. *Smithsonian Misc Collections* 1965; 145:1–30. C, D Reproduced from Dunbar CO, Henbest LG. *Illinois State Geol Surv* 1942; 67:1–218. M Reproduced from Resig JM, Lowenstam HA, Echols RJ, Weiner S. *Cushman Found Foraminiferal Res Spec Publ* 1980; 19:205–214. P, Q Reproduced from Todd R. *US Nat Mus Bull* 1965; 161:1–139.)

expanded in scope in the 1950s and 1960s with the extensive works, chiefly by F.L. Parker, F.B. Phleger, and O.L. Bandy, that documented the modern-day distribution of benthic foraminifera and the abiotic controls of those distributional patterns. More recently, the distributional patterns of foraminifera have been explained by both biotic and abiotic factors.

Biologic studies of foraminifera commenced around the turn of the twentieth century with work by J.J. Lister, F. Schaudinn, and L. Rhumbler, and have continued as a relatively minor part of foraminiferal research. Although the biology of foraminifera is still poorly understood, immense strides have been made recently in the study of both planktic and benthic foraminifera.

The importance of foraminifera to biostratigraphic studies was realized by the "father of micropaleontology," A. d'Orbigny, who used foraminifera in biostratigraphic studies of the Paratethyan Tertiary. However, the influence of members of the late nineteenth century English school of micropaleontology, who considered foraminiferal species to be geologically long ranging and quite variable in form, led to purely descriptive activities for more than half a century.

The need for more petroleum during World War I led to an increased utilization of foraminifera in biostratigraphy. Perhaps the most influential worker at this time was J.A. Cushman who, together with his many students, built upon the earlier classical European work and published 556 papers, mainly on the taxonomy and stratigraphic occurrence of foraminifera. Similar pioneering work on the Miocene of the American west coast was produced by R.M. Kleinpell. The search for petroleum spurred on research on foraminiferal biostratigraphy in the first half of this century. This phase of foraminiferal research dealt almost entirely with benthic foraminifera. Planktic foraminifera were disregarded for biostratigraphic studies as some species were considered to be extremely long-ranged, perhaps from Jurassic to Recent. In the decade prior to World War II detailed taxonomic and stratigraphic work by Subbotina in the Soviet Union showed the biostratigraphic potential of plantic foraminifera. The postwar search for petroleum in Trinidad, where faunas consisted almost entirely of planktic foraminifera, led to a revolution in biostratigraphy, commencing with the works of H.M. Bolli, P. Bronnimann, J.P. Beckmann, W.H. Blow, A.R. Loeblich, J.B. Saunders, and H. Tappan. Bolli's papers on Cretaceous and Tertiary planktic foraminifera of Trinidad published in 1957 set the standard for later biostratigraphic research.

Morphology and systematics

Systematic classifications are not static but change with increased knowledge of the organisms. Some 26 classifications of foraminifera have been published during the past 150 years. The evolution of foraminiferal classification parallels an explosion of information about cytoplasm and, in particular, test morphology. Although a classification of foraminifera should be based on data from many sources including morphology, reproductive biology, life habits, and geologic occurrences, the hard and soft part morphology still remain the most important bases for classification.

Foraminiferal skeletons or tests are built of chambers, which are cavities containing the cytoplasm with a surrounding firm wall (Fig. 12.2). Adjacent chambers are separated by **septa** but a connection is maintained by a hole or **foramen** for which the order Foraminiferida is named. The opening on the last chamber, through which pseudopodia extrude, is termed the **aperture**.

Tests may be single or multichambered. Multichambered forms exhibit **sutures**, the external line of junction between adjacent chambers that may be arranged in a variety of ways. When later chambers envelop earlier ones the test is called **involute**. When earlier chambers are visible the test is termed **evolute**. The test may coil in a single plane (**planispiral**) or in a spire (**trochospiral**). Each coil of the test is called a **whorl**. A basic terminology is used in describing the foraminiferal test (Fig. 12.2).

Cytology

Foraminifera are characterized by their finely granular pseudopodia (**rhizopodia**) and by the nuclear dimorphism occurring in the life-cycles of some species. Rhizopodia (Fig. 12.3) are cytoplasmic extensions beyond the test that are used in locomotion and feeding. They exhibit a variety of forms that can be related, in a general fashion, to mode of life. Benthic foraminifera, in general, produce fine strand-like projections that often anastomose into a net-like mass of granuloreticulopodia. Planktic foraminifera likewise produce numerous thin rhizopodia that repeatedly branch into a three-dimensional shroud of reticulopodia surrounding the test. The rhizopodia can exude directly from the test aperture or they can project outwards from an extramural sheath of cytoplasm that may surround the test.

Perhaps the most remarkable feature of the granular rhizopodia of foraminifera is the typical bidirectional cytoplasmic flow, first observed over a century ago by

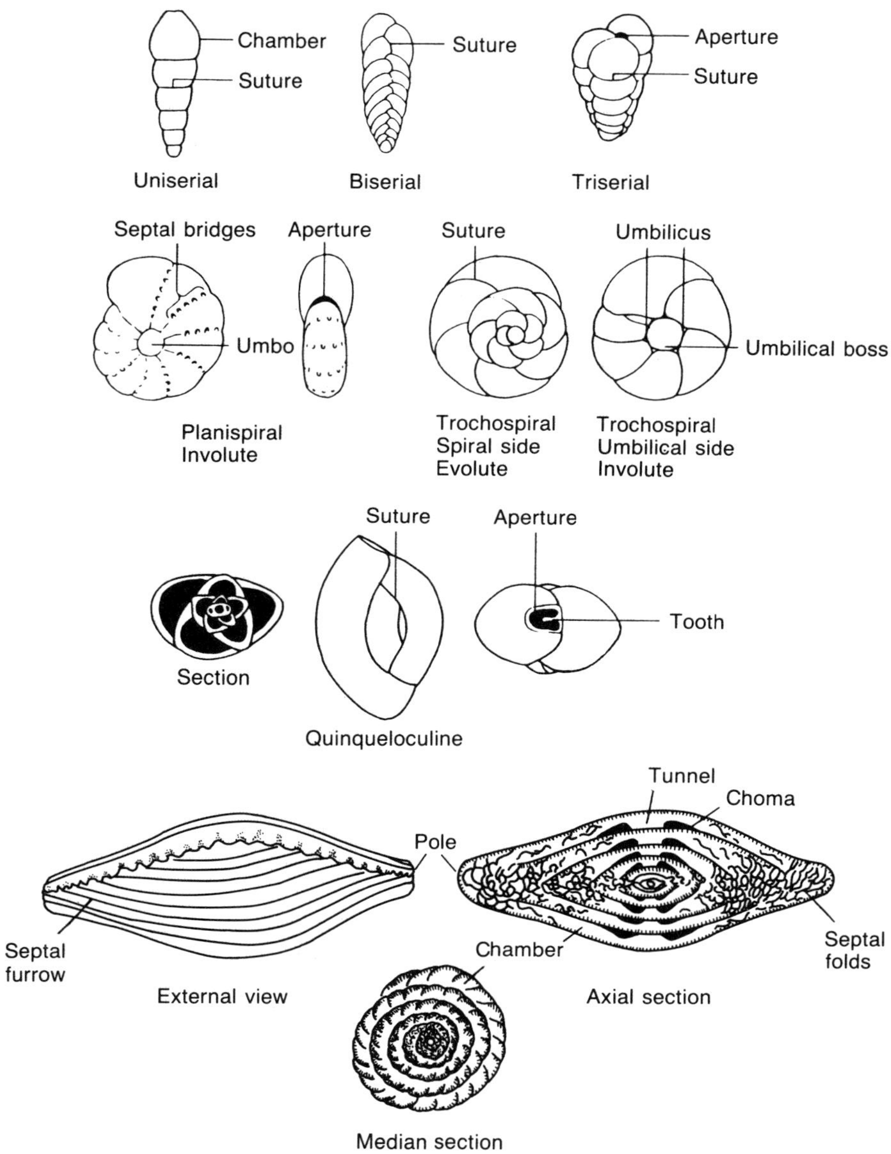

Figure 12.2 Terminology used in describing the foraminiferal test.

J. Leidy. Granules are transported in both directions simultaneously even in extremely fine rhizopodial strands. Although several models have been proposed to explain the bidirectional cytoplasmic flow, a convincing explanation for these streaming processes has yet to be documented.

The granules within the cytoplasm are of several kinds. They may be digestive vacuoles or vacuoles containing waste substances that are transported to the tips of rhizopodia, presumably for excretion. In the planktic foraminifera *Globigerinoides sacculifer*, the granular appearance is caused by dinoflagellate symbionts that are transported out to the distal parts of rhizopodia in the morning and are returned back into the test at night. Much of the granular material in rhizopodia is probably mitochondria.

Cytoplasm within the test is distinct from rhizopodial cytoplasm due to the higher density of ribosomes and organelles. The transition between the two can be sharp or gradual depending on the physiologic state of the individual. Among the organelles and inclusions noted in the cytoplasm of foraminifera are **Golgi bodies**, **mitochondria**, digestive vacuoles, microbodies, lipid droplets, **microtubules**, **xanthosomes**, and **endoplasmic reticulum**.

Composition and morphology of tests

The composition and morphology of tests form the major bases for foraminiferal classification at the subordinal and lower taxonomic levels. Three basic wall compositions

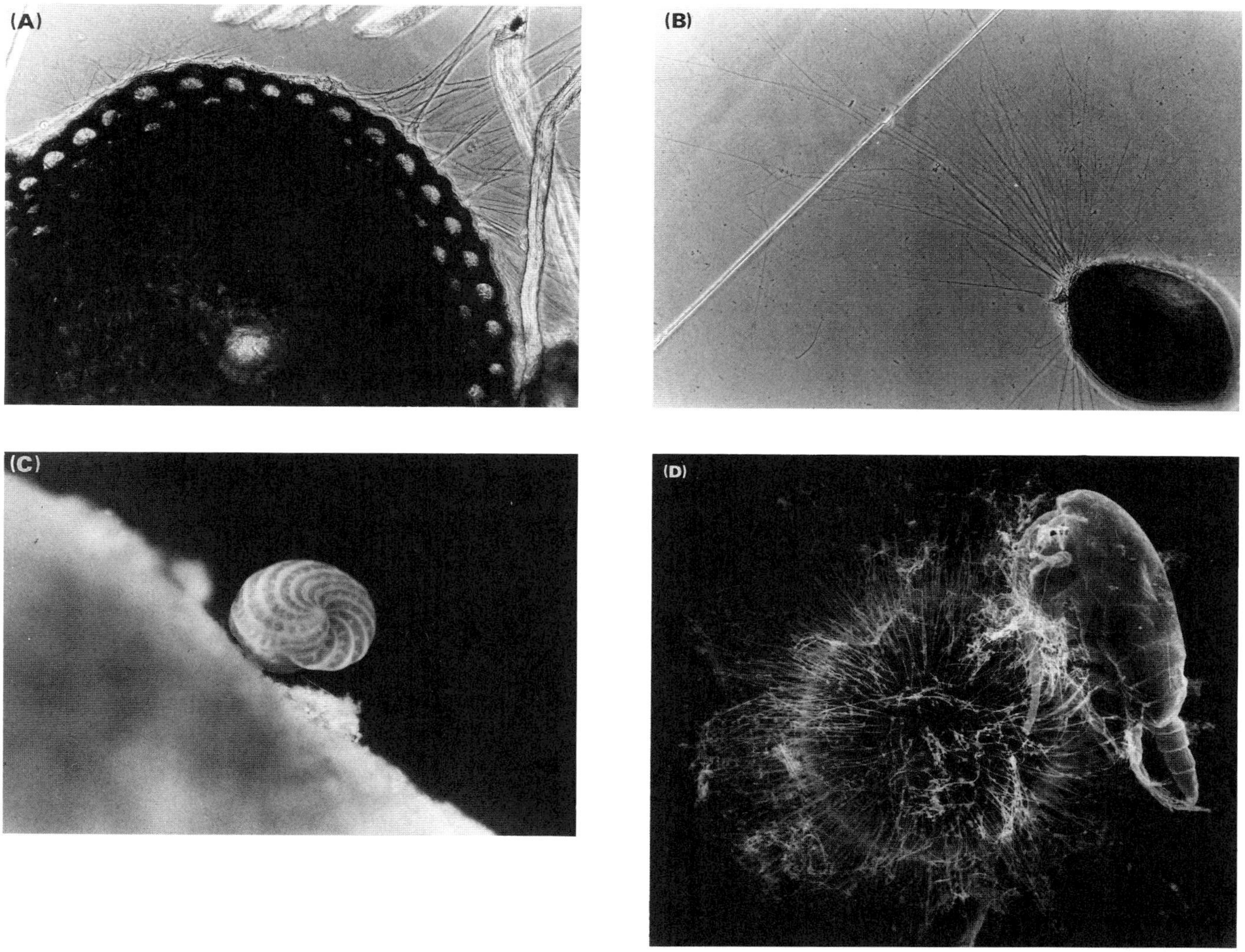

Figure 12.3 A, Pseudopodia extending from peripheral apertures of the large discoidal benthic foraminifera *Sorites marginalis*. Recent. About ×750. B, Pseudopodia radiating from the single aperture of the small benthic foraminifera *Quinqueloculina impressa*. Recent. About ×750. C, Benthic foraminifera *Archaias angulatus* moving across a stem of the marine alga *Penicillus*; pseudopodia extend across substrate. Recent. ×183. D, Spinose planktic foraminifera *Orbulina universa* feeding on calanoid copepod. The central calcareous test is surrounded by a network of flexible spines that are covered with sticky rhizopodial strands. Zooplankters such as copepods that inadvertently swim into the spine matrix are ensnared in the rhizopodial web and become stuck to the spines. The prey is slowly transported down the spines to the surface of the shell via streaming cytoplasm. Enzymes associated with the rhizopodial web gradually break down the prey tissue which is then transported through the apertures in the test to the internal body of cytoplasm. The food is then phagocytosed into vacuoles and subsequently digested. Recent. ×53. (A, B, C Courtesy of K. Wetmore, University of California at Berkeley. D Courtesy of H.J. Spero, University of California, Davis.)

have been recognized in the foraminifera: organic (membranous or tectinous), **agglutinated** (foreign particles held together by various cements), and secreted **calcium carbonate** or silica. In agglutinated and secreted tests an organic membrane forms the foundation for the overlying skeletal material.

The organic-walled test of foraminifera has been described as chitin or pseudochitin but is actually a proteinaceous mucopolysaccharide. Organic walls are thin and flexible and form **unilocular** tests that can change shape (Fig. 12.1A). Some species attach agglutinated particles on the organic wall foundation. These particles may cover the entire test or be restricted to just a portion of the test. Hence, a gradation exists from tests with purely organic walls to organic walls with some agglutinated particles to organic walls with loosely attached or firmly cemented agglutinated particles (Fig. 12.1B). Agglutinated walls can reflect the composition of the local bottom sediment or they can be composed of grains selected on the basis of composition, specific gravity, or size. Not only can particular grains be chosen for the test; the grains can be arranged in a manner that may be constant for all individuals of a particular species (e.g., coarse particles arranged around the aperture).

Chambers of most agglutinated genera have a smooth internal surface. Some genera, however, have a labyrinthic structure where portions of the inner wall project into the chamber resulting in a thickened test and a reduced volume within the chambers (Fig. 12.4). The microstructure of agglutinated walls can also vary. Pore-like structures have been reported in some genera, but other genera have solid walls. These variations in wall structure have, of course, been used in classifications.

Most calcareous foraminiferal walls are secreted, although some agglutinated walls are made of calcareous particles, especially in carbonate-rich tropical environments. Three major types of secreted calcareous walls are recognized: **microgranular**, **porcelaneous**, and **hyaline**. They form an important basis of foraminiferal classification. All are calcitic with the exception of a small group of foraminifera with aragonitic hyaline walls.

Microgranular walls are composed of equidimensional subspherical grains of crystalline calcite (Fig. 12.5). The grains are tightly packed without any evident cement. Most foraminifera with this wall type are from the Late Paleozoic (Fig. 12.1C, D); hence recrystallization is a problem, but the individual calcite grains were probably secreted by the organism.

Porcelaneous-walled foraminifera have variably translucent to opaque high magnesium calcite tests that are imperforate with the exception of the finely perforate proloculus. The wall is composed of thin inner and outer veneers enclosing a thick median layer of crystal laths (Fig. 12.5). The outer veneer can be composed of laths parallel to the surface, producing a smooth shiny appearance (Fig. 12.1K, L), or they can be arranged perpendicular to the surface to give a rough wall surface. Some porcelaneous taxa have a superficial coating of agglutinated particles that resembles the appearance of true agglutinated foraminifera. At least one species of deep-sea benthic foraminifera has test ultrastructure and gross morphology closely resembling calcareous porcelaneous foraminifera but is made of a secreted opaline silica (Fig. 12.1M).

Hyaline calcareous walls are of two main types; radial and granular. Under the **crossed nicols** of a polarizing microscope, optically radial forms exhibit a black cross with concentric rings of color, caused by crystals of calcite arranged with their **c-axes** normal to the test surface (Fig. 12.5). Optically granular forms, when similarly observed, show many small flecks of color but no extinction

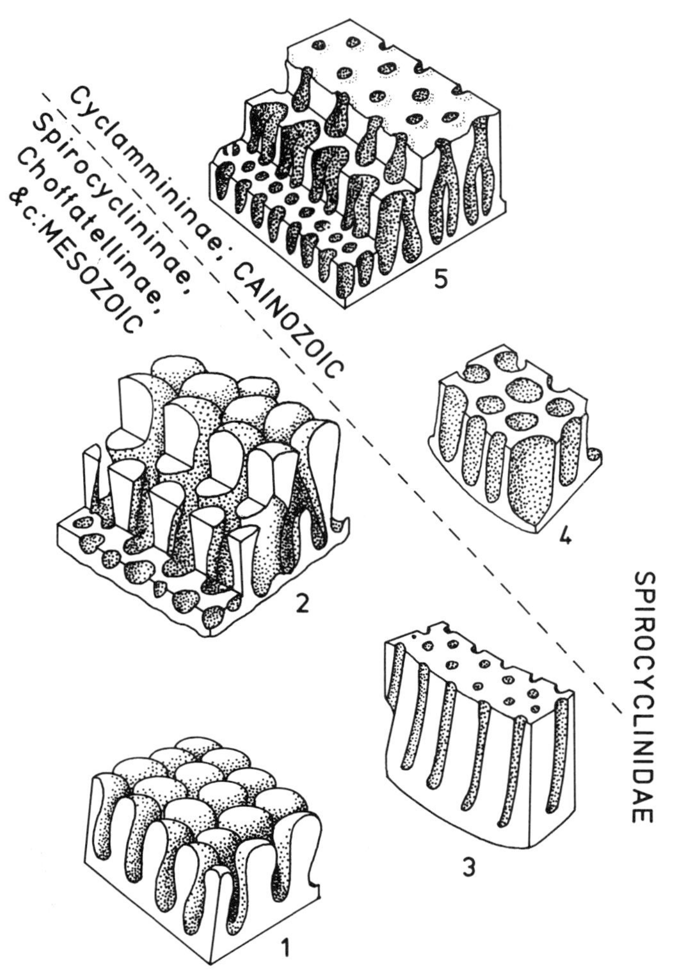

Figure 12.4 The complex wall structure of the Spirocyclinidae. 1, Labyrinthic; 2, labyrinthic and subepidermally alveolar; 3, 4, 5, alveolar. (Modified from Banner FT. *Rev Española Micropaleontol* 1970; 2:243–90.)

pattern. Transmission electron microscopy has shown that the granular appearance is due to microcrystals (ca. 1 μm diameter) oriented with their c-axes at varying angles to the test surface (Fig. 12.5). Combinations of the radial and oblique structure (Fig. 12.5) occur because crystallographically radial and oblique microstructures are very similar.

Hyaline foraminifera, in general, have a wall structure of several lamellae. A new lamella is added to the entire test when a new chamber is secreted, thus producing a test of many lamellae. Various types of lamellar wall structure have been recognized (Fig. 12.6).

The test wall of hyaline calcareous foraminifera is penetrated by fine pores some 1–15 μm in diameter (Fig. 12.1Q). The pores vary in size and distribution over the test of an individual; some trochospiral forms, for

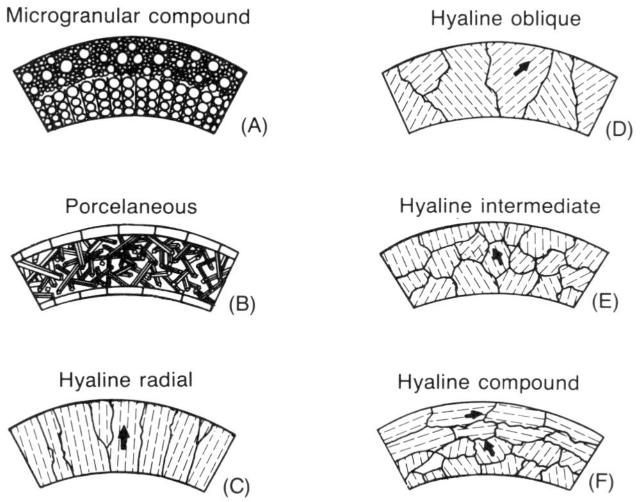

Figure 12.5 The main types of secreted calcareous walls. Arrows indicate orientations of c-axes of calcite microcrystals. A, Microgranular compound. Outer layer of irregularly packed calcite granules; inner layer of rows of calcite granules perpendicular to test surface. B, Porcelaneous. Smooth inner and outer layers of calcite laths arranged parallel to inner and outer wall surface; thick median layer of randomly oriented calcite laths. C, Hyaline radial (optically radial). Calcite microcrystals extend across wall; the optical c-axes of the crystals are perpendicular to test surface. D, Hyaline oblique (optically granular). Microcrystals and their c-axes extend obliquely across the wall. E, Hyaline intermediate. Small crystal units do not extend across wall; c-axes are almost perpendicular to wall surface in inner and outer zones; c-axes in a median zone are more oblique. F, Hyaline compound. Inner layer is of intermediate structure; outer layer has c-axes parallel to test surface. (Redrawn from Haynes JR. *Foraminifera*. New York: John Wiley, 1981:47.)

example, are perforate on the spiral side and imperforate on the umbilical side. Pores usually pass through the entire calcareous wall and end at an imperforate organic membrane that has micropores immediately beneath the pore. The function of pores is unclear but they probably do not function as pathways for rhizopodia. They are involved in formation of chambers and may perform an excretory or osmoregulatory function before being sealed off by the organic lining. The large pores in the planktic foraminifera *Orbulina* (Fig. 12.3D) may be exit sites for gametes. Pores may also help to increase buoyancy in planktic species or oxygen uptake in benthic species.

Two small groups can be distinguished from the other calcareous foraminifera by unique wall structure and composition. One group has a test constructed of an optically single crystal of calcite (Fig. 12.1G, H, I). These foraminifera are the only organisms known to have the entire test or skeleton composed of a single crystal. Another small group of hyaline foraminifera have aragonitic tests (Fig. 12.1O). Under the crossed nicols of a polarizing microscope these tests resemble radial, calcitic forms.

The genus *Carterina* fits into none of the test types discussed above. The test in *Carterina* is composed of spicules of calcite that might be secreted by the foraminifer. The spicules could also be the product of another organism that were later used by the foraminifer so that the test is actually agglutinated. Each fusiform spicule is a single crystal of calcite, embedded in a poorly cemented organic matrix and groundmass of small spicules (Fig. 12.1J).

The shape of a foraminiferal test depends on chamber shape and chamber arrangement. A large number of variations exist (Figs 12.7–12.12). Even unilocular forms have variable test shapes (Figs 12.7A, B, 12.9A, B, 12.11A, B), and **multilocular** species may exhibit several types of chamber shape and arrangement within a single test (Fig. 12.11E, J, M, N).

Foraminiferal tests may have **primary** and **secondary** (**supplementary** and **accessory**) **apertures** (Figs 12.13–12.15). The primary aperture is the main opening in the test and may be single or multiple and of various shapes (Figs 12.13, 12.14). Supplementary apertures, like the primary one, open directly into the chamber cavity (Fig. 12.15), whereas the accessory apertures open into cavities formed by various modifications of the aperture. Cyto-

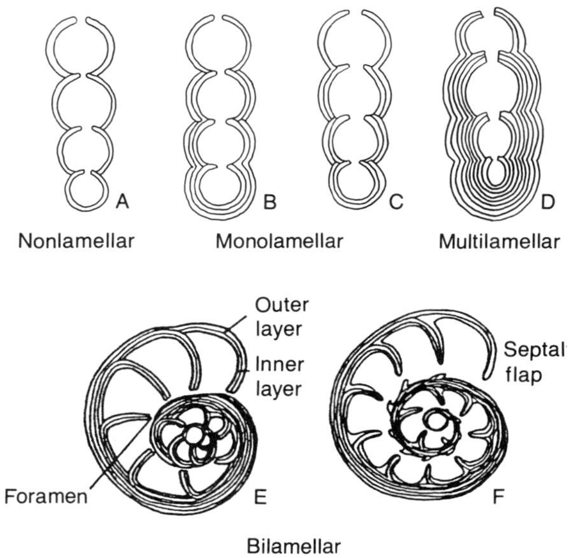

Figure 12.6 Diagrammatic sections of calcareous hyaline foraminifera showing variation in lamellar wall structure. A–D; Uniserial members of the Lagenina. A, Nonlamellar. B, Lamellar and fully investing. C, Partially lamellar. D, Multilamellar. E, F; Equatorial sections of low trochospiral or planispiral members of the Rotaliina. E, Bilamellar. F, Bilamellar with septal flap. (Redrawn from Haynes JR. *Foraminifera*. New York: John Wiley, 1981:49.)

plasm is extruded from the test through the apertures (Fig. 12.3A, B).

External modifications to the surface of the test are often referred to as ornamentation although many of these surficial features are functional. The functions probably include anchoring or stabilizing the test in sediment, deploying pseudopodia, mechanical breakdown of food particles, and prevention of predation. The test surface (Fig. 12.16) may vary from completely smooth to ornate sculpture on a single test. Agglutinated foraminifera show little surface sculpture, porcelaneous forms commonly show striae or costae while hyaline foraminifera show the greatest variety of ornamental modifications.

Systematics

Foraminifera are classified on the basis of test characteristics. Wall composition and microstructure are of primary importance in classification and are useful for the definition of suborders and superfamilies. Mode of chamber and septal addition (lamellar nature of hyaline test: see Fig. 12.6) are of secondary importance, while third in importance are apertural characters and their modifications. Finally, chamber form and arrangement and the free-living or attached nature of the animal are least important but still useful in classification.

Six hundred and eighty-four categories above the level of the genus have been recognized in the order Foraminiferida. These include 12 suborders, 74 superfamilies, 296 families, 302 subfamilies containing a total of 3620 genera.

Kingdom Protista
Subkingdom Protozoa

Phylum Sarcomastigophora Honigberg and Balamuth, 1963
Subphylum Sarcodina Schmarda, 1871

Superclass Rhizopoda von Siebold, 1845
Class Granuloreticulosea De Saedeleer, 1934

The most comprehensive modern classification was proposed in 1988 by A.R. Loeblich Jr and H. Tappan. They recognized 12 suborders in the Foraminiferida: their diagnoses are given in Table 12.1. The geologic range of each suborder is shown in Figure 12.17.

Biology and ecology

Reproduction

The tests of many species of foraminifera are **dimorphic**. This dimorphism may involve differences in size or in more obvious morphologic differentiation, and it is the result of **alternation of sexual and asexual generations**. This reproductive mode differentiates the foraminifera from other members of the Granuloreticulosea. Indeed, foraminifera are different from most animals in having an asexually produced uninucleate **haploid** generation (**gamont**) alternating with a sexually produced multinucleate **diploid** generation (**agamont** or **schizont**). Reproductive variations are possible; one generation may be secondarily repressed, asexual or sexual generations may be repeated, and in some taxa simple **fission** or **budding** may occur. Asexually formed specimens generally have more protoplasm initially and form a larger **proloculus**; hence, they are termed megalospheric. Specimens developing from gametes have less initial protoplasm and form a smaller proloculus; hence they are termed **microspheric** (Fig. 12.18).

The life-cycles of only 20 or so of the 4000 living species of foraminifera are known. These cycles exhibit considerable differences. For example, *Rotaliella heterocaryotica* has size differences but no morphologic differences between the two generations. The adult gamont is simply smaller than the adult agamont. In contrast, *Glabratella sulcata* exhibits both size and morphologic differences (Fig. 12.19). Two gamonts come together (Fig. 12.19) with an organic membrane joining the two tests; the appressed apertural faces are then dissolved and fertilization takes place inside the paired tests. **Zygotes** grow to about three chambers and are liberated following degradation of the organic membrane.

Smaller benthic foraminiferal tests may show the effects of reproduction (e.g., dissolution of the apertural face in *Glabratella*) and larger benthic foraminiferal tests also may undergo partial destruction during release of asexually produced **embryos** (Fig. 12.20). In spinose planktic foraminifera (e.g., *Hastigerina pelagica*), where reproduction takes place by **gametogenesis** deep in the water, the spines, septa, and apertural region of the test are progressively resorbed. In *Globigerinoides sacculifer*, a sac-like chamber is formed before dissolution of spines occurs and may serve to protect the expanding **multinucleate** cytoplasm during development of gametes. Finally, perhaps 300 000 flagellate gametes are released into the water through the aperture leaving the empty test to sink through the water column. Sinking is enhanced by ad-

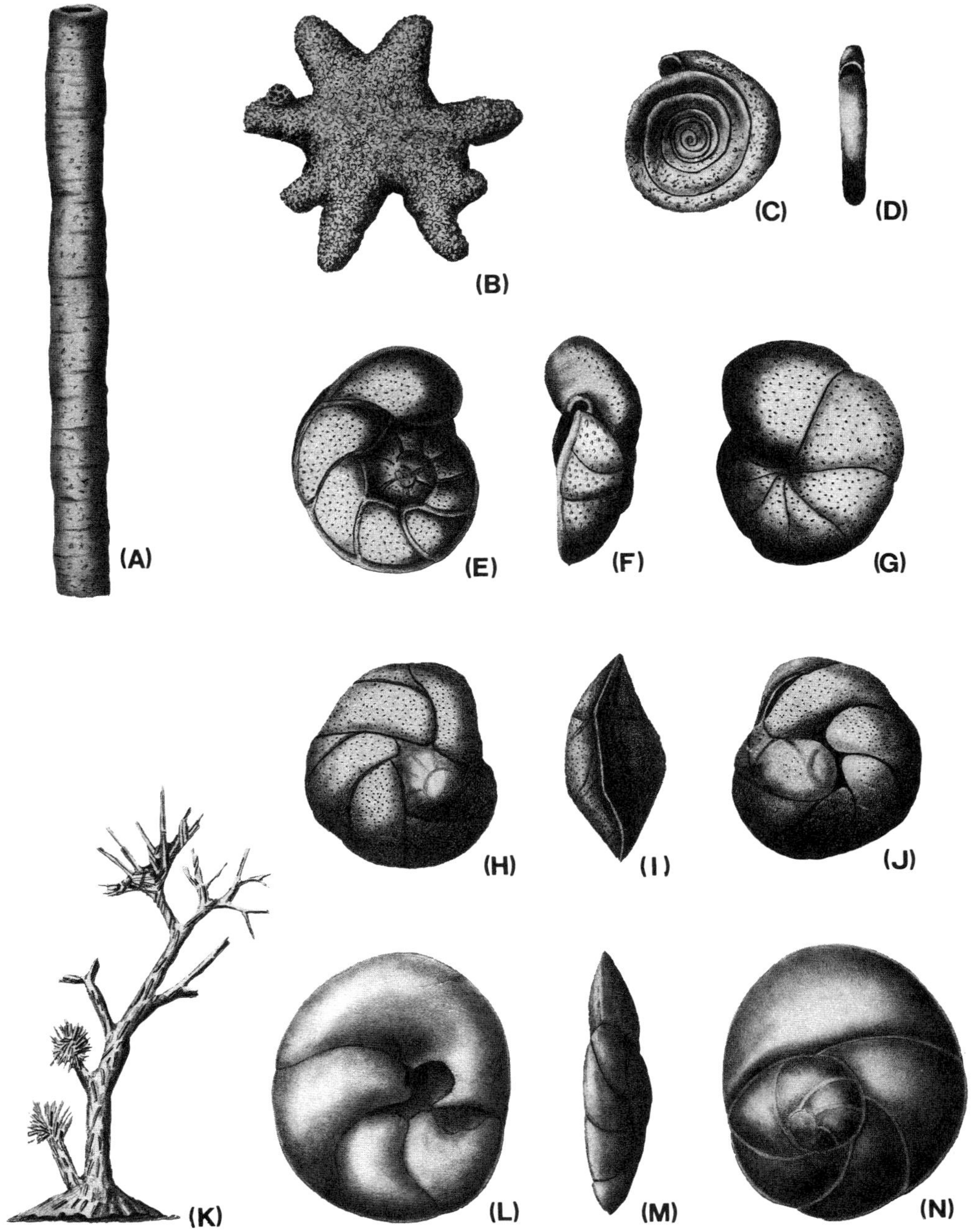

Figure 12.7 Examples of test shape. A, Tubular, *Bathysiphon rufescens*. Recent. Side view. ×12. B, Radiate, *Astrorhiza arenaria*. Recent. Side view. ×4. C, D; Discoidal, *Spirillina vivipara*. Recent. C, Side view. ×80. D, Apertural view. ×80. E, F, G; Umbilicoconvex trochospiral, *Cibicides lobatulus*. Recent. E, Dorsal view. ×48. F, Peripheral view. ×48. G, Ventral view. ×48. H, I, J; Lenticular, *Cassidulina teretis*. Recent. H, J, Opposite sides. ×60. I, Peripheral view. ×60. K, Arborescent, *Dendronina aborescens*. Recent. Side view. ×7.2. L, M, N; Spiroconvex trochospiral, *Rosalina williamsoni*. Recent. L, Ventral view. ×132. M, Peripheral view. ×165. N, Dorsal view. ×132. (A−N Reproduced from Cushman JA. *Foraminifera, Their Classification and Economic Use*. Cambridge: Harvard University Press, 1948.)

ditional calcification of the last-formed chambers. This may explain why thick-shelled forms are rarely collected in surface waters, whereas they are abundant on the sea bed.

Shell growth

The process of chamber formation has been described for several species of benthic and planktic foraminifera. In the calcareous hyaline benthic foraminifera *Discorbinella*

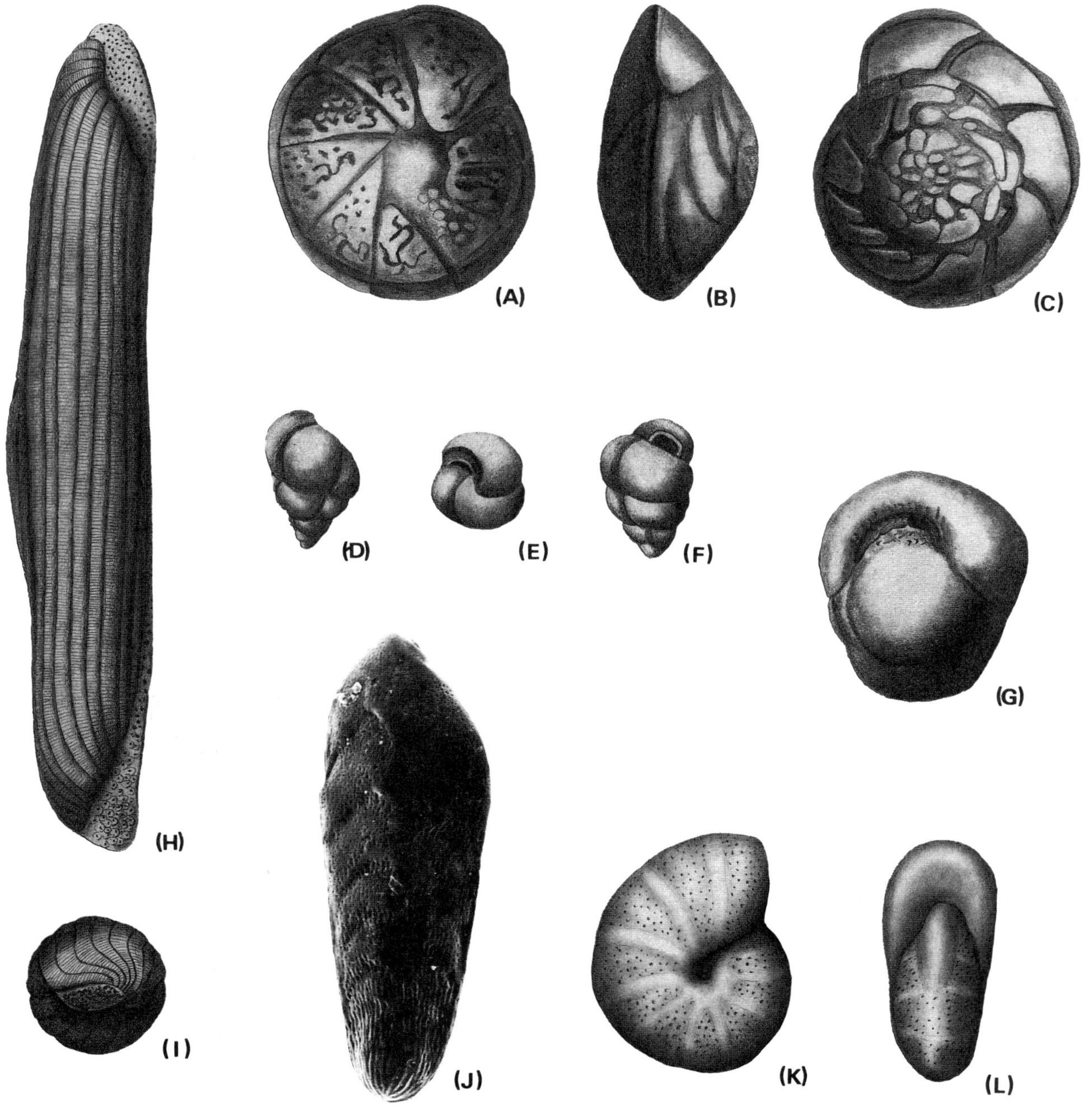

Figure 12.8 Examples of test shape. A, B, C; Biconvex, *Hoeglundina elegans*. Recent. A, Ventral view. ×35. B, Peripheral view. ×35. C, Dorsal view. ×35. D, E, F; Flaring, *Turrilina alsatica*. Oligocene. D, F, Opposite sides. ×90. E, Apertural view. ×90. G, Spherical/subspherical, *Pulleniatina obliquiloculata*. Recent. Apertural view. ×50. H, I; Fusiform, *Alveolinella quoyi*. Recent. H, Side view. ×15. I, End view. ×15. J, Lanceolate, *Bolivina punctatostriata*. Miocene. Side view. ×120. K, L; Biumbilicate, *Melonis barleeanus*. Recent. K, Side view. ×110. L, Apertural view. ×110. (A–I, K, L Reproduced from Cushman JA. *Foraminifera, Their Classification and Economic Use*. Cambridge: Harvard University Press, 1948. J Reproduced from Hayward BW, Buzas MA. *Smithsonian Contrib Paleobiol* 1979; 36:1–54.)

bertheloti, the pseudopodia extend from the aperture in a fan shape and form a protective **cyst** of organic matter. The pseudopodia retract to the location of a new chamber wall where an organic film is constructed followed by a covering of calcareous material. *Rosalina floridana*, also calcareous hyaline, forms a protective enclosure of algal fragments. Vesicular cytoplasm is extruded into its interior and a cytoplasmic template for the wall is formed. Membranes that will eventually be calcified are secreted on the template by pseudopodia that penetrate it. Crystallites

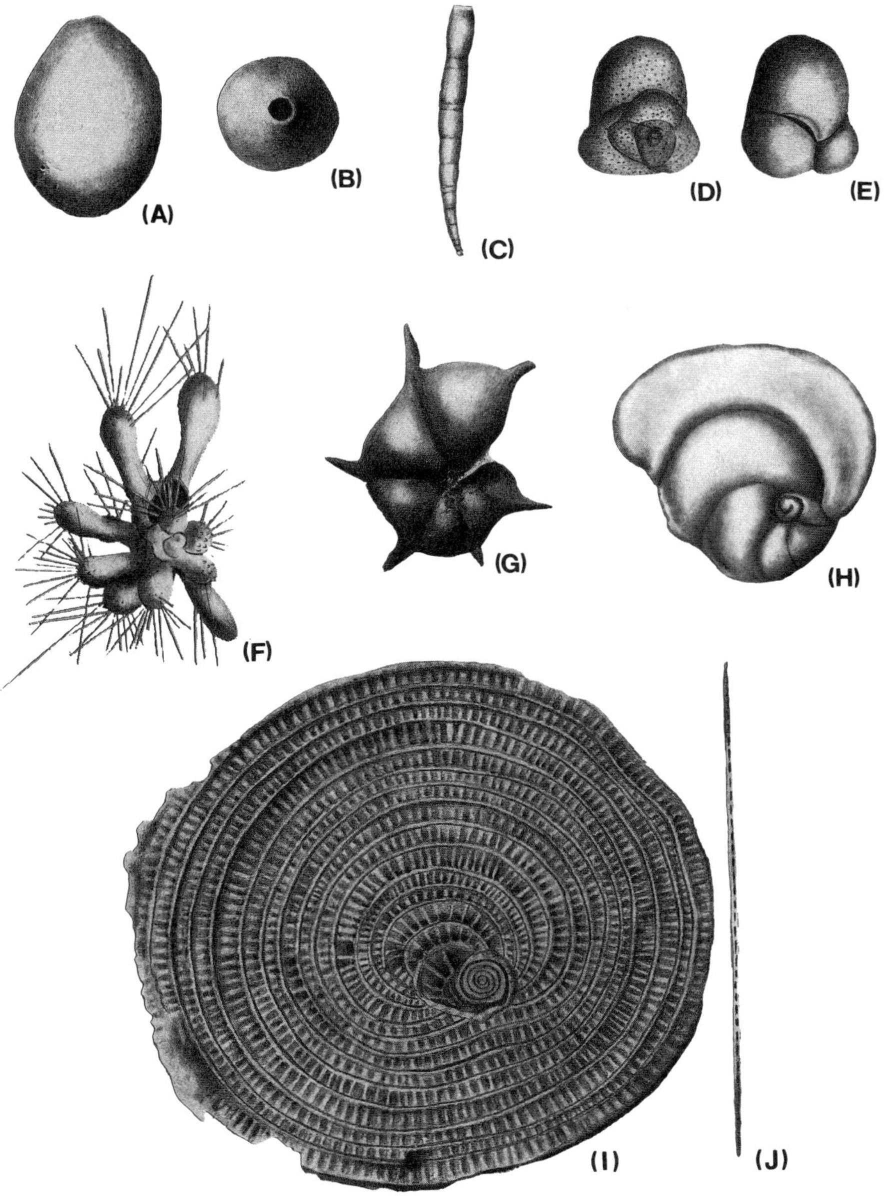

Figure 12.9 Examples of chamber shape. A, B; Globular, *Oolina globosa*. Recent. A, Side view. ×112.5. B, Apertural view. Recent. ×112.5. C, Tubular, *Earlandia perparva*. Pennsylvanian. Side view. ×22.5. D, E; Hemispherical, *Allomorphina pacifica*. Recent. D, Dorsal view. ×75. E, Ventral view. ×75. F, Clavate, *Hastigerinopsis digitiformis*. Recent. Side view. ×45. G, Tubulospinate, *Hantkenina alabamensis*. Eocene. Side view. ×41.3. H, Semicircular. *Renulina opercularis*. Eocene. Side view. ×33.8. I, J; Cyclical, *Discospirina tenuissima*. Recent. I, Side view. ×15. J, Peripheral view. ×15. (A–J Reproduced from Cushman JA. *Foraminifera, Their Classification and Economic Use*. Cambridge: Harvard University Press, 1948.)

of calcium carbonate form within the membrane and coalesce to form larger crystal units. The calcium and carbonate ions used in calcification are extracted from sea water.

A different process forms the porcelaneous imperforate walls of miliolid foraminifera. In *Calcituba polymorpha* and *Spiroloculina hyalina*, calcium carbonate crystals are formed in the cytoplasm and transported to the forming chamber wall. A mass of pseudopodia and associated cytoplasm develops outside the test aperture, and the wall is formed of secreted organic material and preformed crystals that are released from vacuoles. No cytoplasmic template is formed.

The benthic foraminifera *Patellina corrugata* (Fig.

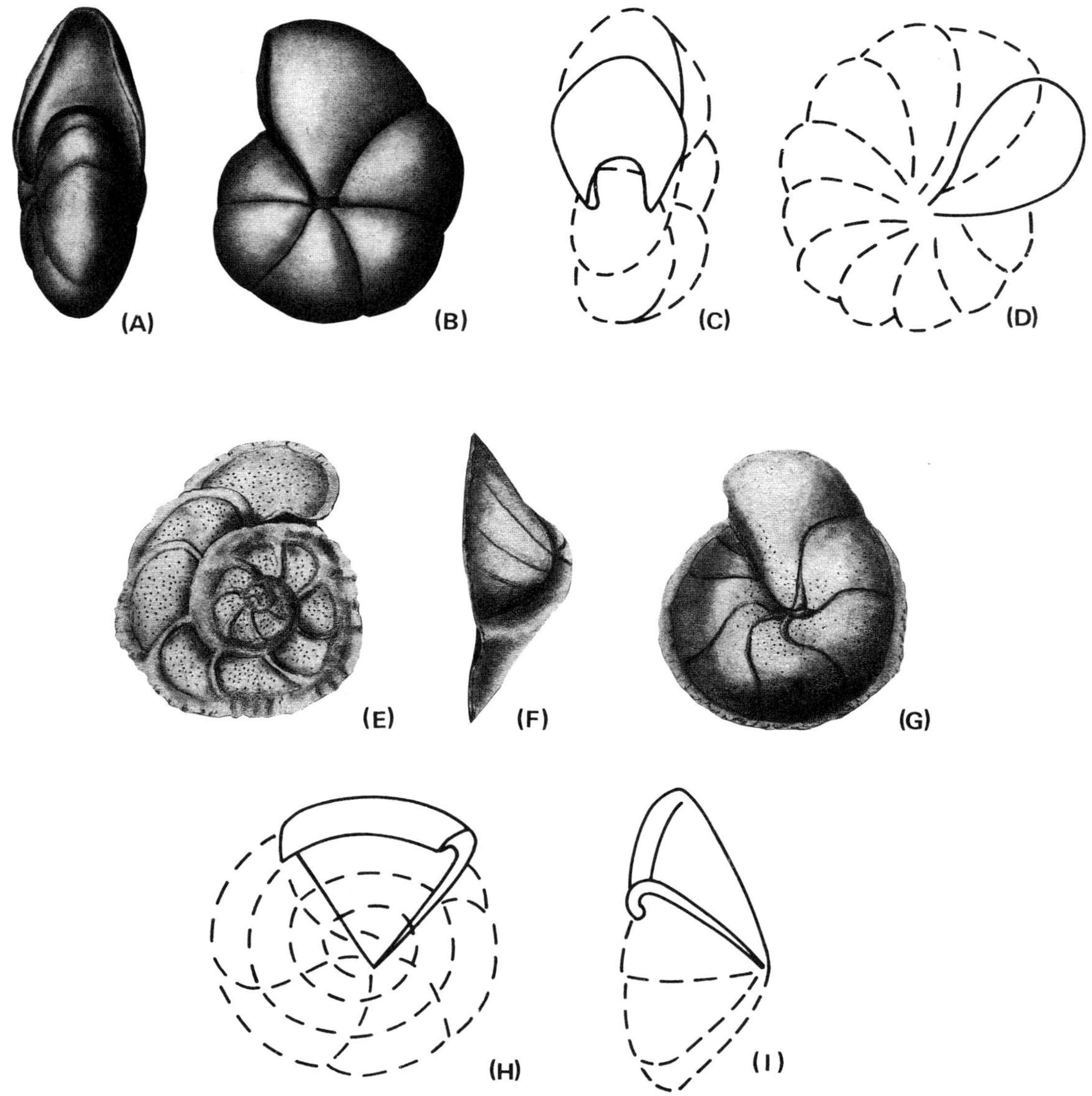

Figure 12.10 Examples of chamber shape. A–D; Lunate, *Pullenia salisburyi*. Pliocene. A, Apertural view. ×45. B, Side view. ×45. C, Diagrammatic apertural view with lunate last chamber emphasized. D, Diagrammatic side view with lunate last chamber emphasized. E–G; Cuneate, *Cibicides refulgens*. Recent. E, Dorsal view. ×30. F, Peripheral view. ×30. G, Ventral view. ×30. H, Diagrammatic dorsal view with cuneate last chamber emphasized. I, Diagrammatic peripheral view with cuneate last chamber emphasized. (A, B, E, F, G Reproduced from Cushman JA. *Foraminifera, Their Classification and Economic Use*. Cambridge: Harvard University Press, 1948.)

12.1G, H, I), belonging to the suborder Spirillinina, is composed of a single calcite crystal, and has yet another pattern of chamber formation. Growth of the test is restricted to the periphery of the chamber wall. Growth cysts are not formed, nor is an organic template produced.

Chamber formation in nonspinose planktic foraminifera appears to be similar to that in calcareous hyaline benthic foraminifera. Cytoplasm bulges from the aperture and assumes the shape of a new chamber. Calcification starts with the secretion of two calcite layers on either side of a primary organic membrane. Calcareous plaques increase in size and number until they coalesce to form a bilamellar wall.

Functional morphology

Foraminiferal **functional morphology** is still in its infancy and consists of suggestions, inferences, and speculations on the function of the whole test, parts of the test, and of the cytoplasm. Among other functions, the test

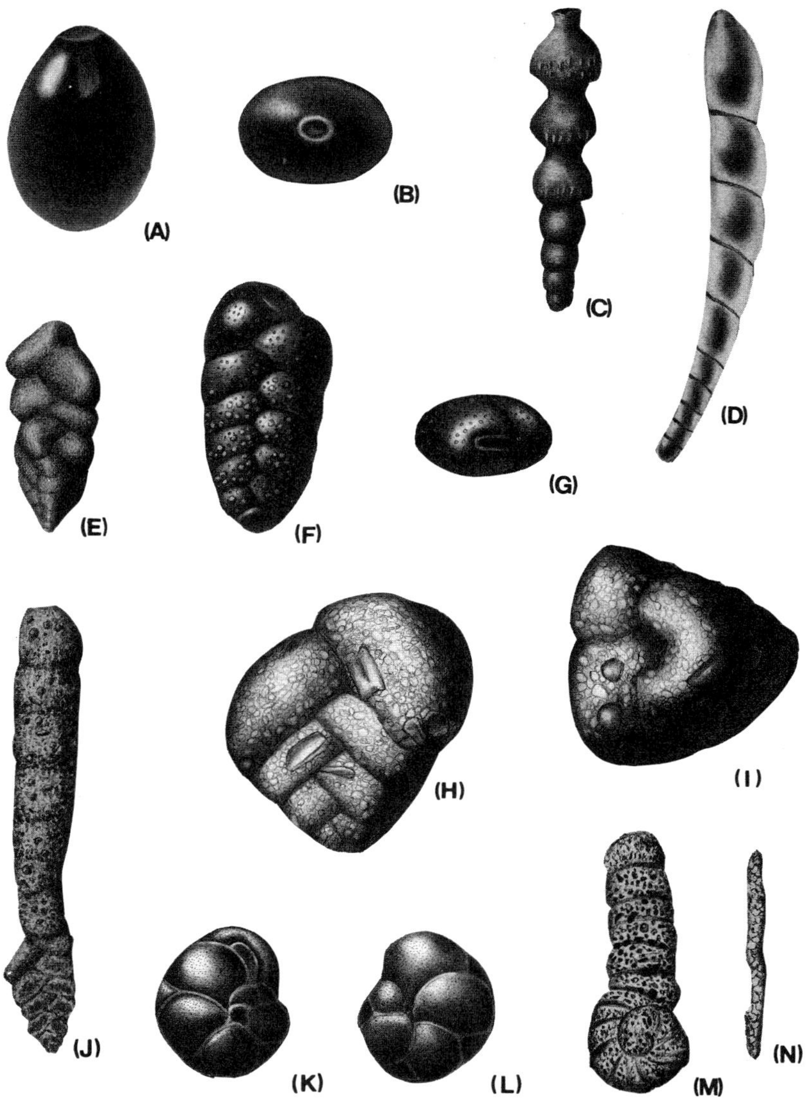

Figure 12.11 Examples of chamber arrangement. A, B; Unilocular, *Fissurina goreaui*. Recent. A, Side view. ×210. B, Apertural view. ×210. C, Uniserial rectilinear, *Nodogenerina lepidula*. Eocene. Side view. ×63. D, Uniserial arcuate, *Dentalina communis*. Eocene. Side view. ×30. E, Triserial to biserial, *Gaudryina triangularis*. Miocene. Side view. ×29. F, G; Biserial, *Bolivina doniezi*. Recent. F, Side view. ×210. G, Apertural view. ×210. H, I; Triserial, *Clavulina tricarinata*. Recent. H, Side view. ×168. I, Apertural view. ×168. J, Biserial to uniserial, *Haimasiella wintoni*. Cretaceous. Side view. ×32. K, L; Biserial enrolled, *Cassidulina barbara*. Pleistocene. K, Side view showing aperture. ×103.6. L, Opposite side. ×103.6. M, N; Planispiral to uniserial, *Ammomarginulina foliacea*. Recent. M, Side view. ×28. N, Edge view. ×28. (A, B, F, G, H, I Reproduced from Buzas MA, Smith RK, Beem KA. *Smithsonian Contrib Paleobiol* 1977; 31:1–139. C, D, E, J, M, N Reproduced from Cushman JA. *Foraminifera, Their Classification and Economic Use*. Cambridge: Harvard University Press, 1948. K, L Reproduced from Buzas MA. *Smithsonian Misc Collections* 1965; 145:1–30.)

probably reduces physical and chemical stress, enhances feeding in particular habitats, enables reproduction, aids in locomotion, and maintains negative buoyancy.

The reticulopodia of foraminifera are used in several life processes: locomotion, collection or capture of prey, digestion, elimination of waste materials, shell construction, and maintenance. However, what are the architectural characteristics of the cytoplasm that allow such a diversity of functions? The Antarctic shallow water benthic foraminifera *Astrammina rara* has cytoplasmic structures that permit capture and retention of struggling metazoans such as crustaceans and juvenile echinoids. The considerable tensile and elastic strength of the delicate reticulopodia is provided by an elaborate **microtubule cytoskeleton**. The microtubules associated with the external surface of the larger reticulopodia are very nu-

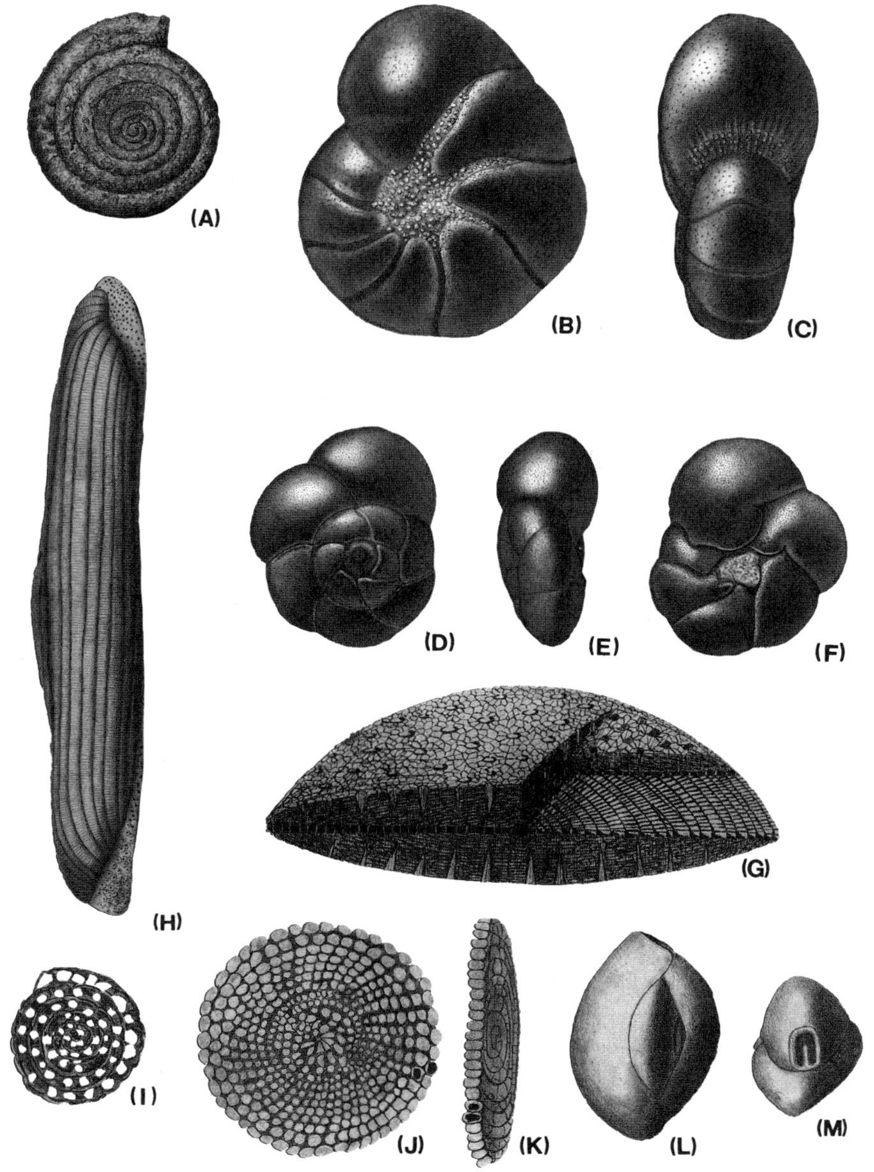

Figure 12.12 Examples of chamber arrangement. A, Planispiral evolute, *Ammodiscus semiconstrictus*. Pennsylvanian. Side view. ×75.9. B, C; Planispiral involute, *Haynesina germanica*. Recent. B, Side view. ×134.6. C, Apertural view. ×134.6. D–F; Trochospiral, *Ammonia tepida*. Recent. D, Dorsal view. ×100.1. E, Peripheral view. ×100.1. F, Ventral view. ×100.1. G, Annular complex, *Discocyclina*. Eocene. ×10.4. H, I; Planispiral fusiform, *Alveolinella quoyi*. Recent. H, Side view. ×10.4. I, Median section. ×10.4. J, K; Annular discoid, *Sorites orbiculus*. Recent. J, Side view. ×24.2. K, Peripheral view. ×24.2. L, M; Milioline, *Quinqueloculina seminula*. L, Side view. ×34.5. M, Apertural view. ×34.5. (A, G–M Reproduced from Cushman JA. *Foraminifera, Their Classification and Economic Use*. Cambridge: Harvard University Press, 1948. B, C Reproduced from Buzas MA, Culver SJ, Isham LB. *J Paleontol* 1985; 59:1075–90. D–F Reproduced from Todd R. *US Natl Mus Bull* 1965; 161:1–139.)

merous and coil around one another to form cables, presumably increasing tensile strength.

The skeleton and cytoplasm have a functional relationship. Supplementary apertures and umbilical and apertural tubercles in the hyaline calcareous benthic foraminifera *Haynesina germanica* (Fig. 12.12B, C) are associated with cytoplasm extrusion. Cytoplasm is extruded from primary and supplementary apertures and spreads to form an extramural layer from which pseudopodia extend. Particulate matter collected by the pseudopodia is transferred to the apertural region by pseudopodial and **extramural** cytoplasmic sheet flow. The tubercles, then, may sieve and disaggregate particulate matter prior to ingestion into the chamber. This

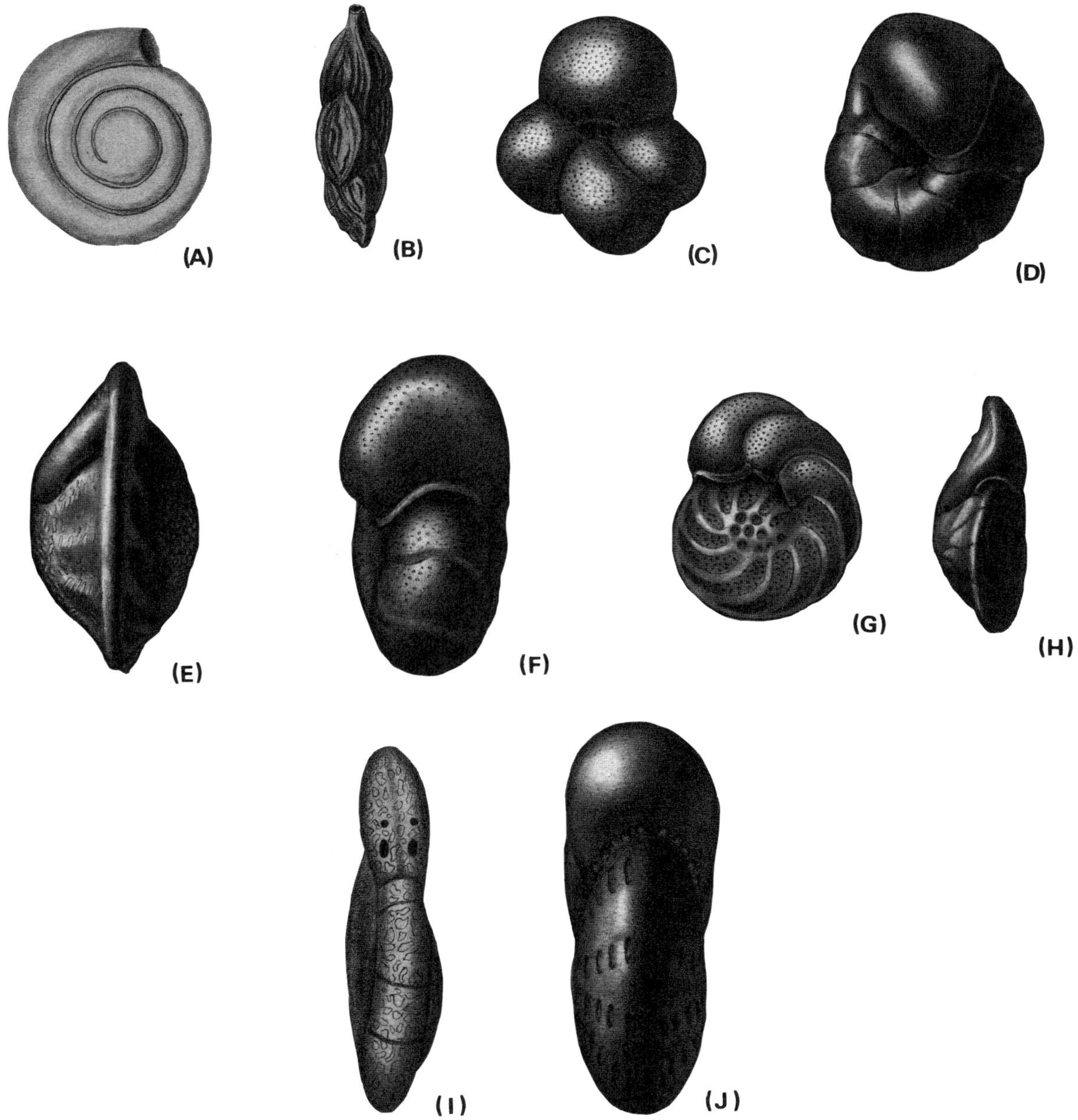

Figure 12.13 Examples of apertural location. A, Open end of tube, *Cyclogyra planorbis*. Recent. Side view. ×88. B, Terminal, *Trifarina angulosa*. Recent. Side view. ×40. C, Umbilical, *Globigerina falconensis*. Recent. Ventral view. ×100. D, Extraumbilical-umbilical, *Eponides* sp. Recent. Ventral view. ×240. E, Extraumbilical, *Nuttalides rugosus*. Recent. Peripheral view. ×145. F, Extraumbilical-peripheral, *Anomalinoides glabrata*. Recent. Peripheral view. ×210. G, H; Peripheral-spiral, *Cibicides wuellersdorfi*. Recent. G, Dorsal view. ×45. H, Peripheral view. ×45. I, Areal, *Entizia tetrastomella*. Recent. Peripheral view. About ×120. J, Interiomarginal, *Elphidium williamsoni*. Recent. Peripheral view. ×125. (A, B, I Reproduced from Cushman JA. *Foraminifera, Their Classification and Economic Use*. Cambridge: Harvard University Press, 1948. C Reproduced from Cole WS, Todd R, Johnson CG. *Bull Amer Paleontol* 1960; 41:77–112. D, E Reproduced from Buzas MA, Smith RK, Beem KA. *Smithsonian Contrib Paleobiol* 1977; 31:1–139. E, G, H Reproduced from Todd R. *US Natl Mus Bull* 1965; 161:1–139. J Reproduced from Buzas MA, Culver SJ, Isham LB. *J Paleontol* 1985; 59:1075–90.)

explains the alignment of tubercles in rows leading towards an aperture (Fig. 12.12C). The rows parallel flow direction in the extramural cytoplasmic sheet.

The general mode of life of foraminifera is commonly reflected in overall test morphology. Benthic foraminifera can attach themselves to a substrate by three mechanisms, clinging by pseudopodia, attached by an organic membrane, and cemented by a calcite layer. Unusual test modifications can result from cementation. *Webbinella concava* exhibits a globular test and probable clinging

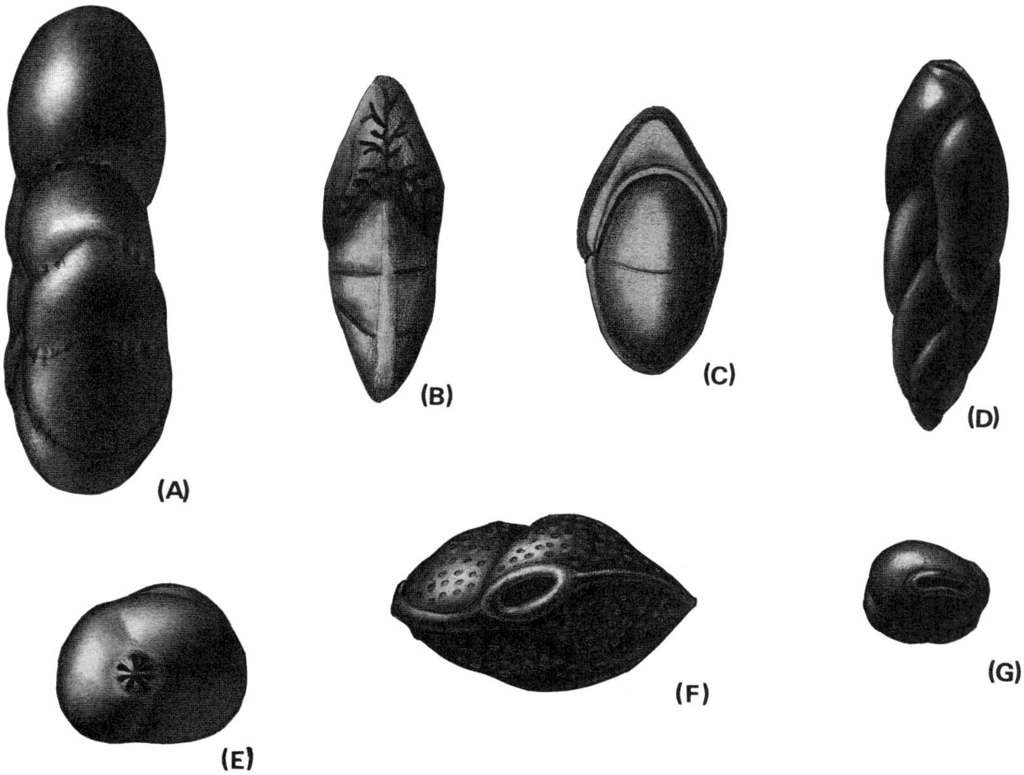

Figure 12.14 Examples of apertural shape. A, Cribrate, *Elphidium subarcticum*. Recent. Peripheral view. ×95. B, Dendritic, *Dendritina arbuscula*. Miocene. Peripheral view. About ×30. C, Slit or arch, *Pullenia compressiuscula*. Recent. Peripheral view. ×50. D, G; Loop, *Fursenkoina pontoni*. Recent. D, Side view. ×120. G, Apertural view. ×120. E, Radiate, *Globulina glacialis*. Pleistocene. Apertural view. ×93. F, Circular or elliptical, *Sigmavirgulina tortuosa*. Recent. Apertural view. ×240. (A Reproduced from Buzas MA, Culver SJ, Isham LB. *J Paleontol* 1985; 59:1075–90. B, C Reproduced from Cushman JA. *Foraminifera, Their Classification and Economic Use*. Cambridge: Harvard University Press, 1948. D, F, G Reproduced from Buzas MA, Smith RK, Beem KA. *Smithsonian Contrib Paleobiol* 1977; 31:1–139. E Reproduced from Buzas MA. *Smithsonian Misc Collections* 1965; 145:1–30.)

Figure 12.15 Example of secondary apertures. A, Interiomarginal (primary aperture lateromarginal), *Hoeglundina elegans*. Recent. Ventral view. ×54. B, Sutural (primary interiomarginal aperture not present in adult), *Candeina nitida*. Recent. Side view. ×50. C, Areal (primary aperture extraumbilical-umbilical), *Poroeponides lateralis*. Recent. Ventral view. ×93. (A Reproduced from Todd R. *US Natl Mus Bull* 1965; 161:1–139. B Reproduced from Cushman JA. *Foraminifera, Their Classification and Economic Use*. Cambridge: Harvard University Press, 1948. C Reproduced from Buzas MA. *Smithsonian Misc Collections* 1965; 149:1–89.)

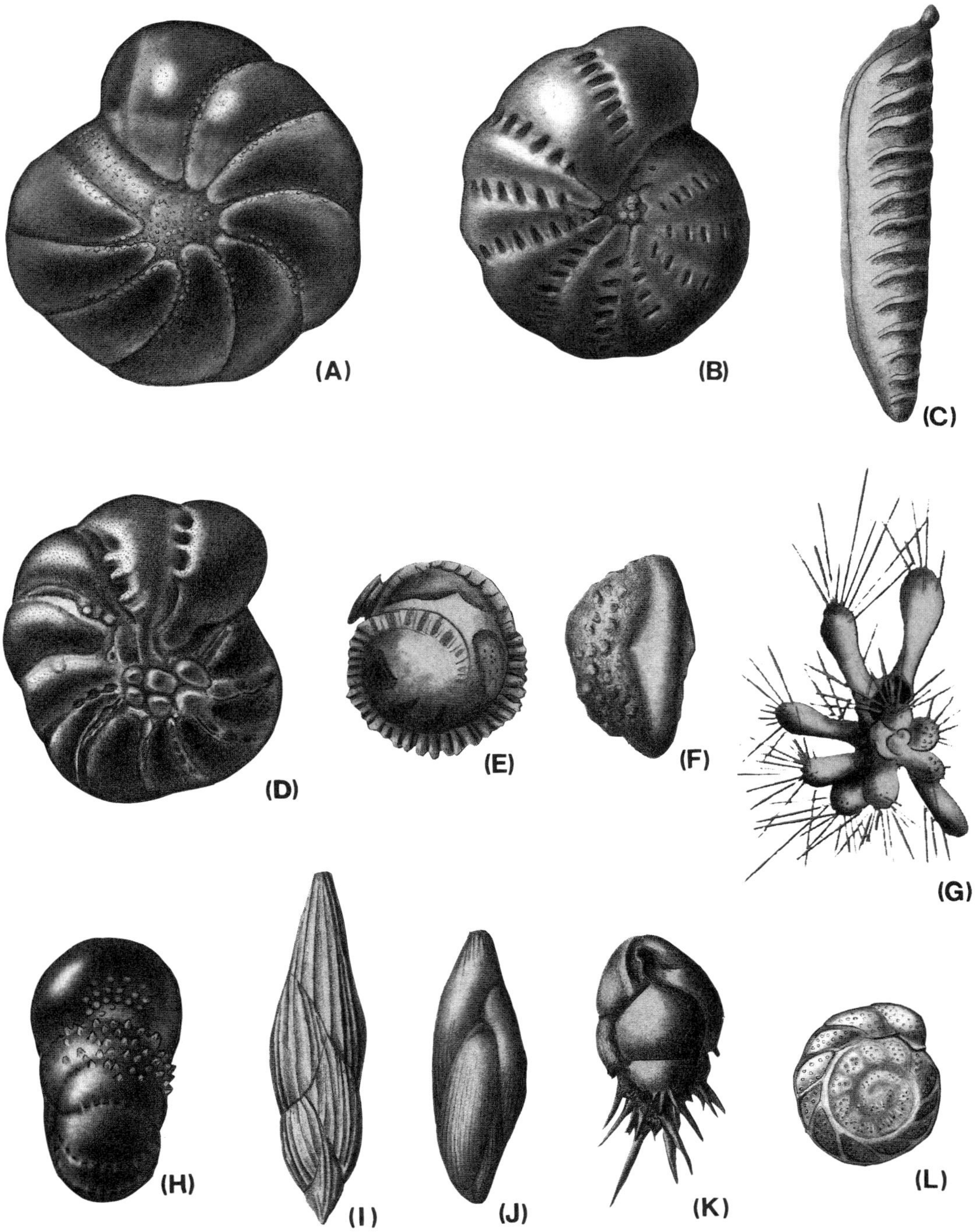

Figure 12.16 Examples of surface sculpture. A, Umbilical, sutural and apertural granulation, *Haynesina orbiculare*. Recent. Side view. ×100.8. B, Sutural bridges, *Elphidium williamsoni*. Recent. Side view. ×100.8. C, Raised sutures, *Vaginulina legumen*. Recent. Side view. About ×10.1. D, Umbilical bosses, *Elphidium excavatum*. Recent. Side view. ×126. E, Keel, *Siphonina tubulosa*. Recent. Dorsal view. ×58.8. F, Tuberculate, *Pegidia dubia*. Recent. Peripheral view. ×35.3. G, Spinose, *Hastigerinopsis digitiformis*. Recent. Side view. ×50. H, Apertural spines, *Elphidium norvangi*. Recent. Peripheral view. ×252. I, Costate, *Pseudopolymorphina rutila*. Oligocene. Side view. ×67. J, Striate, *Guttulina pulchella*. Recent. Side view. ×46. K, Acicular spines, *Bulimina aculeata*. Recent. Side view. ×50. L, Limbate (thickened) sutures, *Cibicides robertsonianus*. Recent. Dorsal view. ×46. (A, B Reproduced from Buzas MA, Culver SJ, Isham LB. *J Paleontol* 1985; 59:1075−90. C, E, F, G, I, J, K Reproduced from Cushman JA. *Foraminifera, Their Classification and Economic Use*. Cambridge: Harvard University Press, 1948. D Reproduced Buzas MA. *Smithsonian Misc Collections* 1965; 149:1−89. H Reproduced from Buzas MA, Smith RK, Beem KA. *Smithsonian Contrib Paleobiol* 1977; 31:1−139.)

Table 12.1 Diagnoses of the order Foraminiferida and its suborders

Order Foraminiferida (Eichwald 1830)

Cytoplasmic body enclosed in test or shell of one or more interconnected chambers; wall may be homogeneous, or of similar or unlike layers or laminae, may be imperforate or finely to coarsely perforate, basically proteinaceous, but may have agglutinated particles, or may deposit the mineral calcite, aragonite, or rarely silica, on the organic base; calcareous wall may be porcelaneous, microgranular, or hyaline and optically or ultrastructurally radiate or granular; canal or stolon systems of varied complexity may be present; commonly test has one or more main apertures through which pseudopodia protrude. Sexual and asexual generations alternate or one generation may be suppressed; gametes biflagellate, triflagellate, or ameboid. Free-living or rarely parasitic; benthic and attached or motile, or pelagic in marine to brackish water, rare in fresh water. Cambrian to Holocene

Suborder Allogromiina (Loeblich & Tappan 1961)

Test unilocular or may tend to become multilocular; test wall membranaceous or proteinaceous, may have ferruginous encrustations or small quantity of agglutinated particles. Upper Cambrian to Holocene. Figure 12.1A

Suborder Textulariina (Delage & Hérouard 1896)

Test agglutinated, foreign particles held in organic or mineralized groundmass. Lower Cambrian to Holocene. Figures 12.1B, 12.7A, B, K, 12.11E, H, I, J, M, N, 12.12A, 12.13I

Suborder Fusulinina (Wedekind 1937)

Test wall of homogeneously microgranular calcite, of tightly packed equidimensional subangular crystals, a few micrometers in diameter. Advanced forms with wall differentiated into two or more layers. Lower Silurian to Upper Permian. Figures 12.1C, D, 12.9C

Suborder Involutinina (Hohenegger & Piller 1977)

Proloculus followed by enrolled tubular second chamber; wall calcareous, perforate, radiate, originally aragonitic but commonly recrystallized to homogeneous microgranular structure, with lamellar thickenings or pillar-like structures in umbilical region of one or both sides. Lower Permian to Upper Cretaceous, Holocene. Figure 12.1E, F

Suborder Spirillinina (Hohenegger & Piller 1975)

Coiling planispiral to high trochospiral; proloculus followed by enrolled tubular undivided chamber, or with few chambers per whorl, chambers may be secondarily subdivided; wall of calcite, optically a single crystal or few to a mosaic of crystals; a-axis preferred orientation along axis of coiling and c-axis parallel to umbilical surface (*Patellina*, see Fig. 12.1G, H, I); may have pseudopores or micropores filled with organic matter and closed by sieve plates, wall formed by marginal accretion, not by calcification of an organic template produced by pseudopodia. Upper Triassic to Holocene. Figures 12.1G, H, I, 12.7C, D

Suborder Carterinina (Loeblich & Tappan 1981)

Test attached, trochospiral, early chambers semicircular, later ones becoming crescentic and finally irregular or spreading in the adult, with undulating surface, earliest chambers simple, later ones may have radial secondary septula resulting from infolding of the wall; wall with organic inner lining and outer layer of rod-like or fusiform secreted spicules, each one crystallographically a single elongated crystal of low magnesium calcite, in a groundmass of small spicules loosely held in an organic matrix. Eocene, Holocene. Figure 12.1J

Suborder Miliolina (Delage & Hérouard 1896)

Test of porcelaneous high magnesium calcite, commonly with organic lining and may have added adventitious material; true pores may occur in protoconch of some, generally imperforate in postembryonic state, but may have pseudopores; may have flexostyle or spiral passage between proloculus and later chambers. Carboniferous to Holocene. Figures 12.1K, L, 12.8H, I, 12.9H, I, J, 12.12J, K, L, M, 12.13A, 12.14B

Suborder Silicoloculinina (Resig, Lowenstam, Echols & Weiner 1980)

Wall imperforate, of secreted opaline silica. Upper Miocene to Holocene. Figure 12.1M

Suborder Lagenina (Delage & Hérouard 1896)

Wall of monolamellar, optically and ultrastructurally radiate calcite, with crystal c-axes perpendicular to surface; crystal units enveloped by organic membranes; primitive taxa without secondary lamination, more advanced forms secondarily lamellar. Upper Silurian to Lower Devonian; Lower Carboniferous to Holocene. Figures 12.1N, 12.9A, B, 12.11A, B, D, 12.14E, 12.16C, I, J

Suborder Robertinina (Loeblich & Tappan 1984)

Test planispirally to trochospirally enrolled; chambers provided with internal partition that attaches near apertural foramen; wall of hyaline perforate, ultrastructurally and optically radiate aragonite (orthorhombic crystal form of calcium carbonate), the hexagonal prisms with c-axis normal to the wall surface and basal pinacoid parallel to the surface, prisms in bundles surrounded by organic sheaths; extremely fine wall perforations may be localized in pore fields. Middle Triassic to Holocene. Figures 12.8A, B, C, 12.1O

Suborder Globigerinina (Delage & Hérouard 1896)

Planktic in habit; test wall of perforate hyaline calcite, optically radiate, preferred crystal orientation with c-axis normal to surface; primary lamination bilamellar, secondary lamination due to addition of shell material during formation of new chamber; surface crust may develop at time of gametogenesis. Middle Jurassic to Holocene. Figures 12.1P, 12.8G, 12.9F, G, 12.13C, 12.15B

Suborder Rotaliina (Delage & Hérouard 1896)

Test multilocular, typically enrolled, but may be reduced to biserial or uniserial, or may be encrusting with proliferated chambers; chambers simple or subdivided by secondary partitions, and median and lateral chambers may be differentiated; wall calcareous, of perforate hyaline lamellar calcite (hexagonal crystal form of calcium carbonate), formed by calcification at each side of an organic membrane, may be optically radial or granular (hyaline oblique), surface smooth, papillate, costate, striate, or cancellate; aperture simple or with internal toothplate, entosolenian tube or hemicylindrical structure; internal canal systems or stolon systems may occur. Triassic to Holocene. Figures 12.1Q, 12.7E, F, G, H, I, J, L, M, N, 12.8D, E, F, J, K, L, 12.9D, E, 12.10A, B, E, F, G, 12.11C, F, G, K, L, 12.12B, C, D, E, F, G, 12.13B, D, E, F, G, H, J, 12.14A, C, D, F, G, 12.15C, 12.16A, B, D, E, F, H, K, L

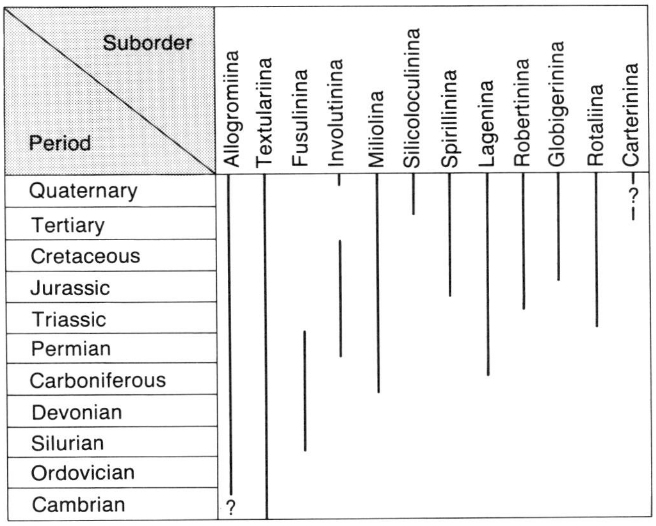

Figure 12.17 Geologic range of foraminiferal suborders.

mode of attachment followed by a firmly cemented phase in its life history. It encrusts through the growth of a wide calcitic skirt (Fig. 12.21A) which is flattened on the basal side, thus affording a firm attachment to the underlying hard substrate. A similar adaptation to high-energy environments is achieved rather differently by *Vasiglobulina reticulata*, which encrusts by extending many **acicular** calcite pillars to the substrate (Fig. 12.21B, C). The tips of the pillars spread and overlap to form a thin calcite layer which is in turn cemented to the substrate.

Test strength, morphology, and habitat of shallow-water **free-living** benthic foraminifera are correlated. Test strength increases with size and with the physical turbulance of the environment. Individuals from high-energy environments have stronger tests than similar-sized individuals from low-energy habitats. The morphology of the strongest species maximizes the area

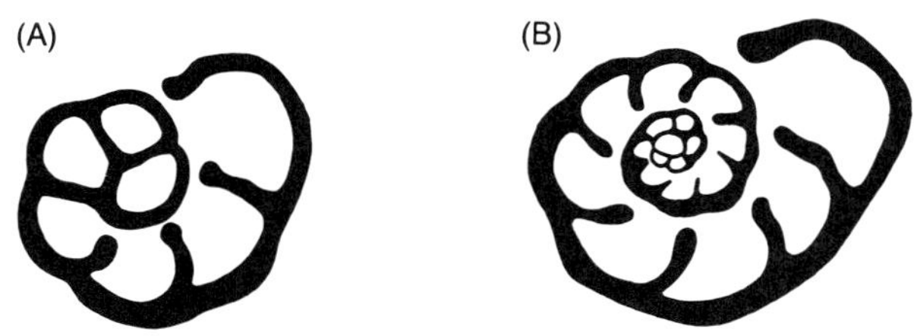

Figure 12.18 Equatorial sections of early part of test of *Planorbulina mediterranensis*. A, Megalospheric test. B, Microspheric test. (Drawn from Le Calvez J. *Archi Zool Experimentale Generale* 1938; 80: 165–333.)

of the test over which a compressive force is distributed, and regions of possible stress concentrations are reinforced or eliminated. Hence, morphologic characters such as thickness of the test wall and overall shape of the test appear to affect strength of the test more than mode of coiling or composition of the test. The strongest species are often biconvex to globular with thickening of the test wall at sutures between chambers and at buttressing plugs in the umbilical region.

Nutrition

Foraminifera are abundant in all marine environments and they exhibit a wide variety of trophic mechanisms (Table 12.2). Few species, however, have been observed feeding under controlled conditions, and those that have were herbivores. Foraminiferal feeding mechanisms are related to test morphology.

Foraminifera that take up dissolved organic matter from their environment should have a large cytoplasmic surface area. *Notodendrodes antarctikos* (Fig. 12.22) from McMurdo Sound, Antarctica, for example, has a branching habit and takes up most of the dissolved organic matter through its root system.

Passive herbivores should be common in food-rich environments. Many attached foraminifera, whose test morphology corresponds to the substrate, may be passive herbivores. Other species (e.g., *Rosalina globularis*) may move in search of diatoms when this source of food becomes scarce. Most active herbivores have trochoid, lenticular, or flattened tests.

Carnivorous foraminifera, both active and passive, have test morphologies similar to herbivores. Pseudopods may be sticky, and the rhizopodal network may be analogous to a spider's web. Small crustaceans and various protozoans are known prey of carnivorous foraminifera. *Spiculosiphon radiata* (Fig. 12.23) incorporates the tests of its foraminiferal prey into its own agglutinated test.

Planktic foraminifera (and probably many benthic species) are omnivorous opportunistic feeders and have been observed to consume autotrophic and heterotrophic protists, metazoans, and organic detritus. The test shapes of benthic omnivores should be similar to herbivores and carnivores as they use similar strategies to capture food.

Foraminifera that extend their pseudopodia into the water column may be suspension feeders. They generally have an erect test with the aperture oriented away from the substrate. Other variations are possible: *Elphidium crispum*, lenticular in shape, can use its pseudopodia to suspend itself between the stipes of coralline algae, much like a spider; *Astrorhiza limnicola* may be an interstitial

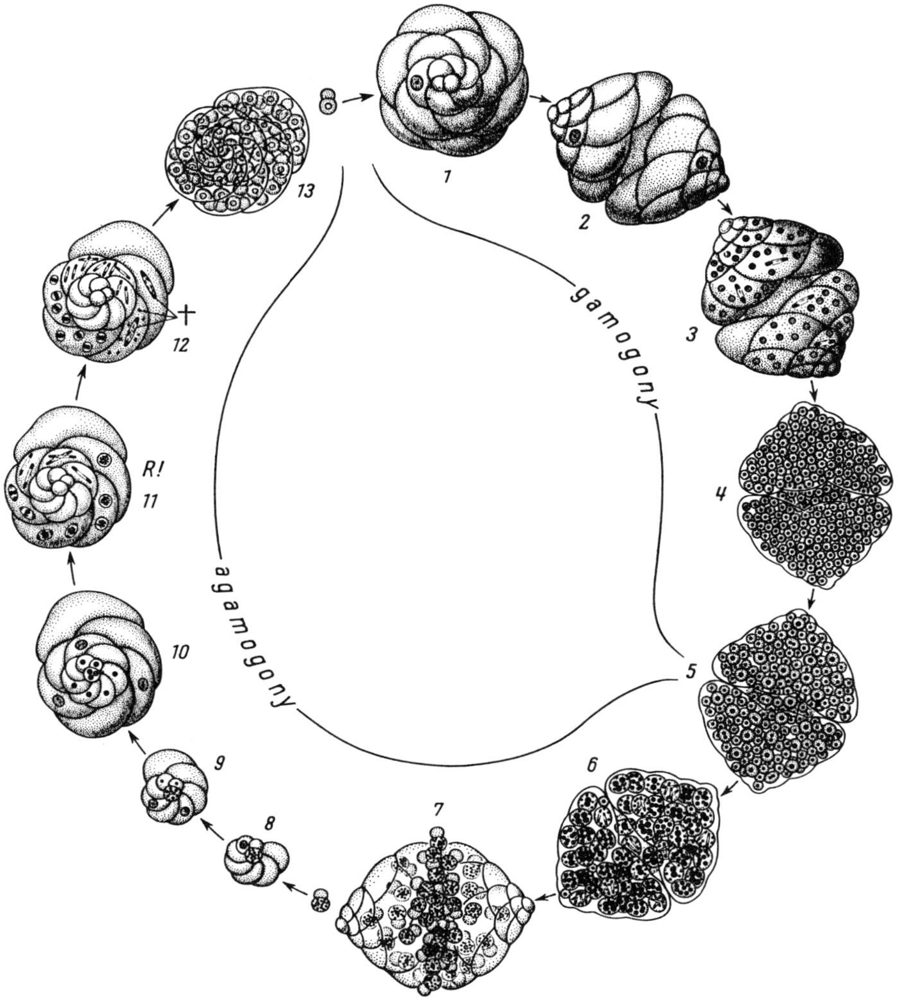

Figure 12.19 Life-cycle of *Glabratella sulcata* Grell. 1, Gamont; 2, mating of two gamonts; 3, gamogeny; 4, gametes; 5, fertilization and zygotes; 6, metagamic divisions; 7, emergence of young agamonts; 8, 9, growth stages of an agamont; 10, adult agamont; 11, first meiotic division (†, R!); 12, second meiotic division (†, R!); 13, formation of the agametes. (Reproduced from Grell KG. *J Foraminiferal Res* 1979; 9:1–13.)

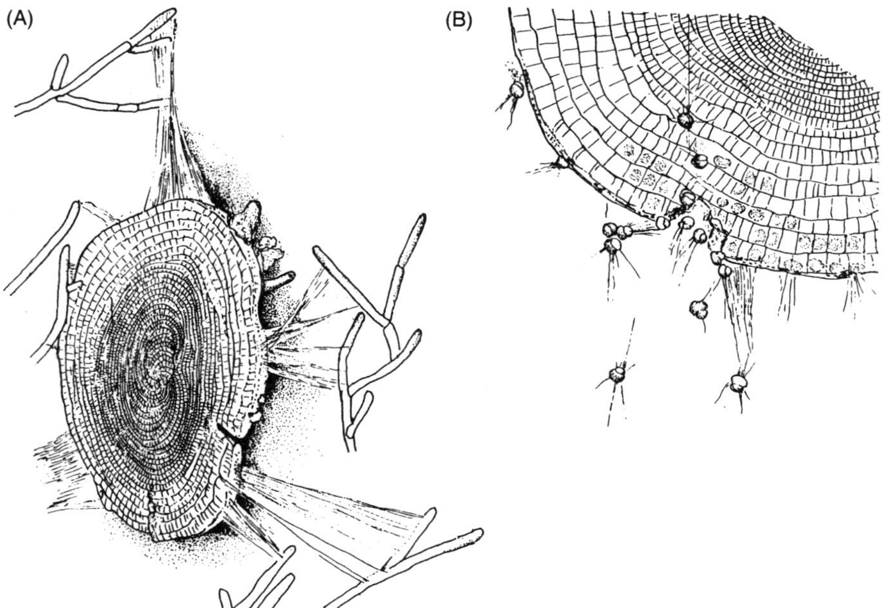

Figure 12.20 A, Adult *Cyclorbiculina* with reproductive chamber attached to the alga *Cladophora* by means of pseudopodial network. B, Portion of same specimen after parts of the test margin were destroyed by the embryos from inside. The first embryos depart from the parental test. Others are still inside but visible through the test wall. (Reproduced from Lutze GF, Wefer G. *J Foraminiferal Res* 1980; 10:251–60.)

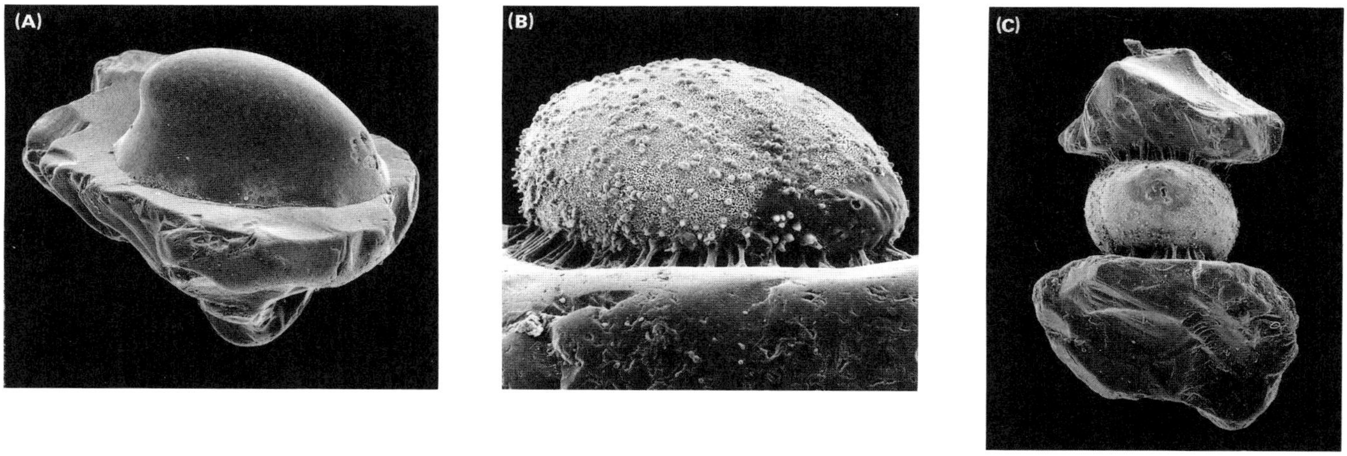

Figure 12.21 A, *Webbinella concava*, oblique lateral view showing cap-shaped test attaching to a quartz grain. Recent. ×74.4. B, *Vasiglobulina reticulata*, oblique lateral view of specimen attached to quartz grain showing smooth lateral mound and two apertural openings at right of specimen. Recent. ×145.7. C, *V. reticulata*, oblique apical view of specimen attached to two quartz grains. Recent. ×52.7. (Reproduced from Poag CW. *Geol Soc Am Bull* 1982; 93:252–68.)

suspension feeder, capturing small crustaceans moving between sediment particles.

Foraminifera that move through fine-grained substrates may be detrital and bacterial scavengers. They commonly have elongate tests (Fig. 12.8J), thus their movement through sediment is facilitated and may result in the production of minute burrows.

Parasitism has been demonstrated for the unilocular benthic foraminifera *Fissurina marginata*. This species captures and digests pseudopodial granules of *Discorbis mediterranensis*. The unilocular genus *Lagena* may also be parasitic.

Cannibalism has been observed in *Glabratella ornatissima*. In this species, zygotes in the closely appressed parent tests (Fig. 12.19) consume each other until only a few individuals are left. Cannibalism has not been noted in mature individuals of *G. ornatissima*.

Table 12.2 Trophic mechanism in foraminifera. (Reproduced from Lipps JH. *Biotic Interactions in Recent and Fossil Benthic Communities*. New York: Plenum Press, 1983:331–76)

Uptake of dissolved organic material	Suspension feeding
	Benthic
	Planktic
	Interstitial
Herbivory	Detrital and bacterial scavenging
Passive	Epifaunal
Active	Infaunal
Carnivory	Parasitism
Passive	Cannibalism
Active	Symbiotism
Omnivory	Ectosymbiotism
Passive	Endosymbiotism
Active	

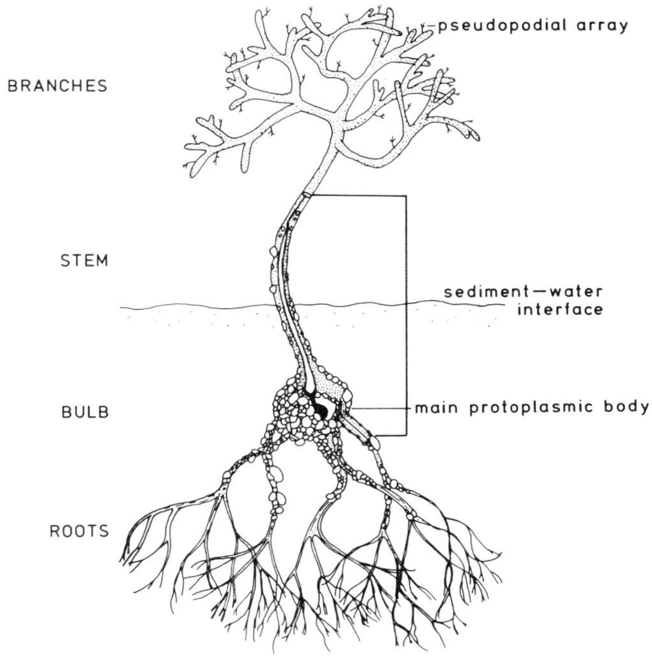

Figure 12.22 *Notodendrodes antarctikos* in life position with root system buried in sediment and branches in the water column. Bracket indicates partial cut-away showing internal organization. Recent. About ×2.35. (Modified from Delaca TE, Lipps JH, Hessler RR. *Zool J Linnean Soc* 1980; 69:205–24.)

Many species of foraminifera that inhabit the photic zone are known to house algal symbionts, usually diatoms, dinoflagellates, or chlorophytes. These foraminifera commonly have large flattened, discoidal (Figs 12.3, 12.9I, J, 12.12G, J, K) or fusiform (Fig. 12.8H, I) tests, sometimes with various kinds of internal partitions to accommodate algal cells. The foraminifera may obtain all or part of their nutritional requirements from their endosymbionts, and test calcification is enhanced. Symbiosis is not restricted to larger benthic foraminifera. Various species of *Elphidium* (e.g., Fig. 12.16B, D), *Nonion*, and *Nonionella* harbor symbionts, as do many planktic foraminifera.

Predation

Foraminifera are key links in marine food chains. Benthic foraminiferal densities may be substantially reduced by macrofaunal predation, either incidentally or selectively. Most incidental consumers of foraminifera are grazers on algae or detritus feeders. Other organisms, particularly some gastropods, are believed to be selective predators of particular species of foraminifera.

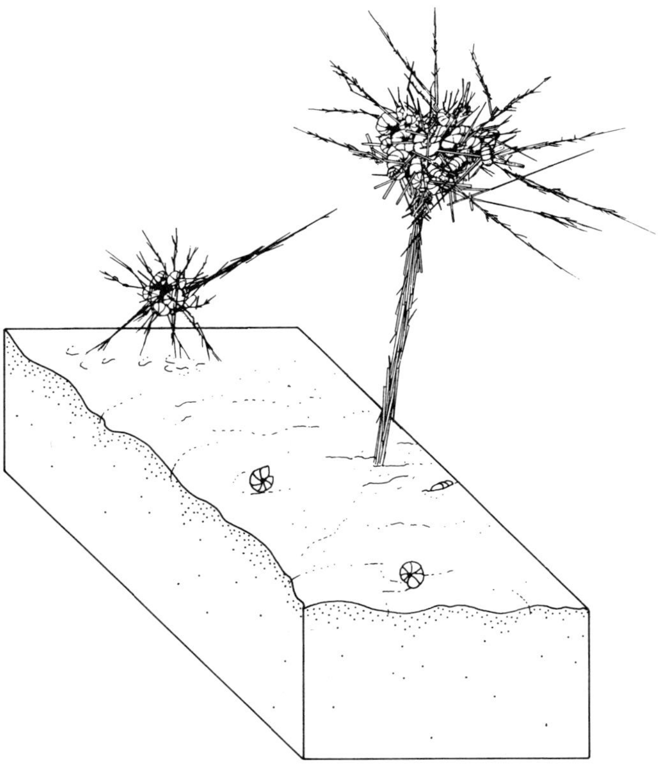

Figure 12.23 Two individuals of *Spiculosiphon radiata* in life position. Specimen on left is moving across the muddy substrate. Specimen on right standing upright on long rod composed of sponge spicules. Recent. About ×8. (Adapted from Christiansen BO. *Astarte Zool Depart Tromsö Mus* 1964; 25:1–8.)

Biogeography

The distribution of foraminifera can be considered at many scales, from a square centimeter to the entire world ocean. Indeed, a third, vertical dimension must also be considered as many benthic foraminifera are infaunal and planktic species exhibit vertical distribution in the world's oceans. The distributions of foraminifera are relatively easy to document but the reasons for those distributions are very complex. A variety of interrelated biotic and abiotic factors must control distributions, but the changing levels of importance of these factors from place to place have proven difficult to unravel. Temperature, salinity, calcium carbonate availability, pH, dissolved oxygen, turbidity, trace elements, illumination, substrate, hydrostatic pressure, currents and tides, and the biotic factors of food availability, predation, and competition are important factors in determining the distribution of foraminifera.

Foraminiferal biogeography can be considered at three scales: small (cm to m), medium (m to km), and large (100s–1000s of km). Centimeter-scale horizontal distributions of benthic foraminifera indicate a distinctly inhomogeneous pattern. This patchiness may result from asexual reproduction or from aggregation around a food source. Knowledge of small-scale patchiness is necessary not only for an understanding of spatial competition, reproduction and behavior, and homogeneity of environment, but also to enable us to estimate the degree of confidence for data from single sample points (e.g., cores, grab samples).

Many benthic foraminifera are epifaunal but true infaunal life-style involving reproduction and feeding beneath the sediment–water interface has been recognized. Benthic foraminifera can live to at least 16 cm depth, but their presence beyond depths of 10 cm or so is likely due to macrofaunal bioturbation. *Quinqueloculina impressa* is able to escape a burial depth of at least 4 cm. Deeper burial prevents escape presumably due to overburden pressures. In fine-grained sediment, escaping foraminifera can produce preservable burrows that are morphologically very similar to structures recognized in X-radiographs of cores taken from 150–3500 m in the Mediterranean.

Benthic foraminiferal biofacies (m- to km-scale distributions) of shelf environments from around the world, recognized by qualitative or quantitative means, are related to various environmental variables (e.g., temperature, salinity, dissolved oxygen). For example, the presence of three faunas in the Tampa-Sarasota Bay area of Florida were related to the influx of freshwater and open-gulf water into the bay; the presence of a miliolid-dominant biofacies in Laguna Madre, Texas, was attributed to hypersalinity and a sandy substrate; foraminiferal assemblages were related to substrate variations in the relatively stable temperature and salinity conditions of the southern Irish Sea.

Although benthic foraminiferal distributions have been related to environmental variables since the 1930s, a causal relationship has rarely been demonstrated, probably because of the complex nature of that relationship. For example, within a temperate, estuarine biofacies, foraminiferal densities over a 12-month sampling period were controlled by a set of several environmental variables. None of the variables, however, was significant individually. A similar study in Discovery Bay, Jamaica, showed that abiotic environmental variables were not significant in controlling foraminiferal densities. In tropical habitats biotic factors may regulate densities of species. This obvious complexity of ecologic relationships between foraminifera and their environment explains the difficulty in truly understanding medium-scale foraminiferal distributions. Even so, along whole continental margins, generally continuous biofacies are recognizable that can be related to depth. These modern depth zones form the basis for **paleobathymetric** interpretation of samples obtained from coastal plain and continental margin wells throughout the world.

Large-scale faunal units defined by species assemblages are biogeographic provinces. They are related to water mass distributions both on the continental margin and in the deep ocean. Provincial boundaries are rarely sharp, and considerable overlap may occur resulting in geographically extensive transition zones.

Nine planktic foraminiferal provinces (Fig. 12.24) are recognized worldwide (Arctic, subarctic, transitional, north subtropical, tropical, south subtropical, transitional, subantarctic, and Antarctic), although specific composition is generally the same for corresponding provinces in adjacent hemispheres. Provincial and water mass distributions are correlated, but vast areas of mixed faunas (Fig. 12.25) result at the margins of oceanic gyres.

The planktic foraminiferal assemblage characterizing the ocean floor does not necessarily correspond with the population of surface waters. Planktic foraminiferal species are markedly stratified with depth in the water column. Bottom sediment assemblages are composites of these stratified populations. Three depth habitats are discernable, shallow (0–50 m), intermediate (50–100 m), and deep (>100 m). Thin-walled, symbiont-bearing, spherical forms characterize near-surface water, while thicker-walled, keeled, symbiont-free forms become more abundant with depth. Depth stratification, physical and seasonal mixing of faunas, and differential dissolution all result in a loss of ecologic information in ocean-floor sediment samples.

Benthic foraminiferal provinces are recognized on continental margins and in the deep ocean. The continental margin around the Americas is perhaps the best example. Foraminiferal provinces on the Atlantic continental margin and in the Gulf of Mexico are related to major water masses. Both latitudinal boundaries on the shelf and depth-relatable boundaries offshore result in seven provinces on the Atlantic continental margin (Fig. 12.26) and four in the Gulf of Mexico (Fig. 12.27). On the Pacific margin of North America, the near-shore, turbulent zone foraminifera are distributed in three faunal provinces between the Strait of Juan de Fuca, Washington, and Cabo San Lucas, Baja California, and are related to coastal water masses.

The South American continental shelf is subdivided into four foraminiferal provinces: West Indian and Argentine on the east coast and Chilean and Panamanian on the west coast (Fig. 12.28). Several subprovinces are also related to water mass distributions in the south Atlantic and south Pacific.

At bathyal and abyssal depths large-scale distribution of benthic foraminifera can be related to water masses despite the relative paucity of data over such a vast area. Faunal provinces, with gradational boundaries, succeed each other both with depth and laterally at similar depths.

Two examples illustrate water mass–faunal correlations in the deep ocean. In the North Atlantic, factor analysis defines three modern assemblages of benthic foraminifera (Fig. 12.29): an *Epistominella exigua* assemblage is associated with Arctic Bottom Water, an *Osangularia umbonifera* assemblage with Antarctic Bottom Water, and a *Hoeglundina* assemblage with North Atlantic Deep Water. These assemblages had different latitudinal and depth distributions over the past 120 000 years due to their migration following changes in circulation of deep water masses.

In the western South Atlantic, North Atlantic Deep Water and Antarctic Bottom Water can be discriminated on the basis of the benthic foraminiferal assemblages in core-top samples. *Osangularia umbonifera* dominates samples underlying Antarctic Bottom Water, whereas assemblages underlying North Atlantic Deep Water are characterized by a more diverse, evenly distributed fauna. Although data are available for nine physical–chemical

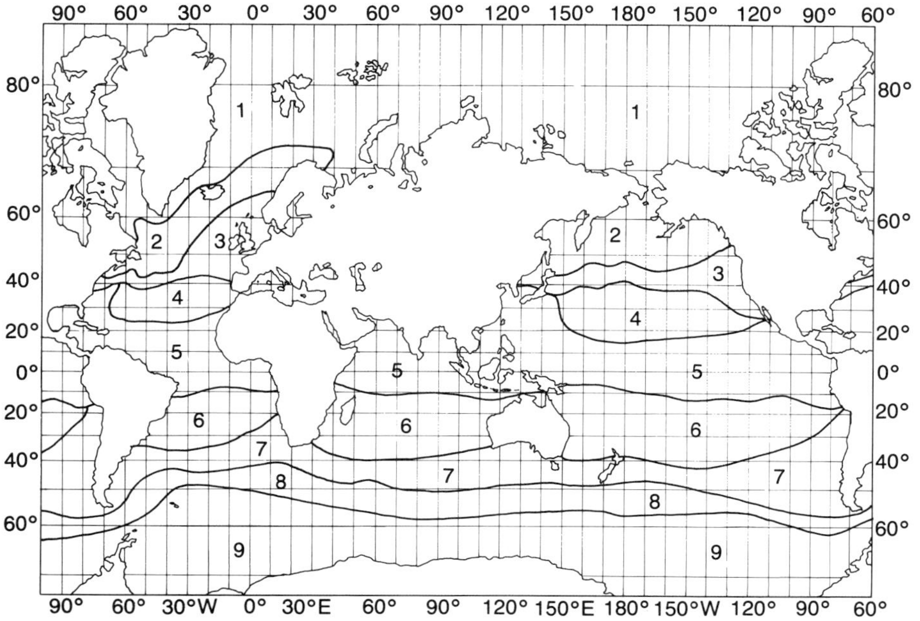

Figure 12.24 Modern planktic foraminiferal provinces. 1, Arctic; 2, subarctic; 3, transitional; 4, subtropical; 5, tropical; 6, subtropical; 7, transitional; 8, subantarctic; 9, Antarctic. (Modified from Bé AWH, Tolderlund DS. *The Micropalaeontology of Oceans*. Cambridge: Cambridge University Press, 1971: 105–49.)

hydrographic variables, a single property controlling the distribution of assemblages cannot be determined. A complex of interrelated environmental variables controls foraminiferal densities and distribution.

Paleobiology

Paleobiologic inferences for foraminifera are scattered throughout the literature, often concern individual species, and are mainly anecdotal in nature. For example, active herbivorous benthic foraminifera have been suggested to have mostly trochoid, lenticular, or flattened tests. Similarly, some trochoid ancestral forms have been suggested to have inhabited clear, open seas and to have given rise to planispiral descendants that inhabited brackish water, coastal or estuarine environments.

C.A. Ross' paleobiologic analysis of test morphology of the extinct Fusulinacea (Fig. 12.1C, D) is an excellent example, however, of what is possible. Based on comparison with the biology, ecology, and shell construction

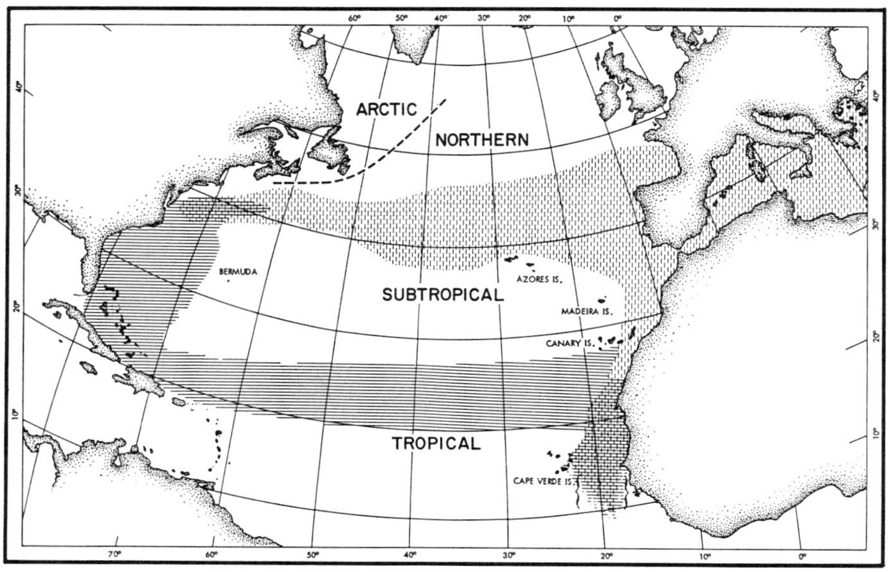

Figure 12.25 Generalized distribution of planktic foraminiferal faunas in the North Atlantic. Horizontal lines indicate a mixture of tropical and subtropical faunas. Vertical lines indicate a mixture of northern and subtropical faunas. Crossed lines indicate a mixture of all three faunas. (Reproduced from Cifelli R, Benier CS. *J Foraminiferal Res* 1976; 6:158–273.)

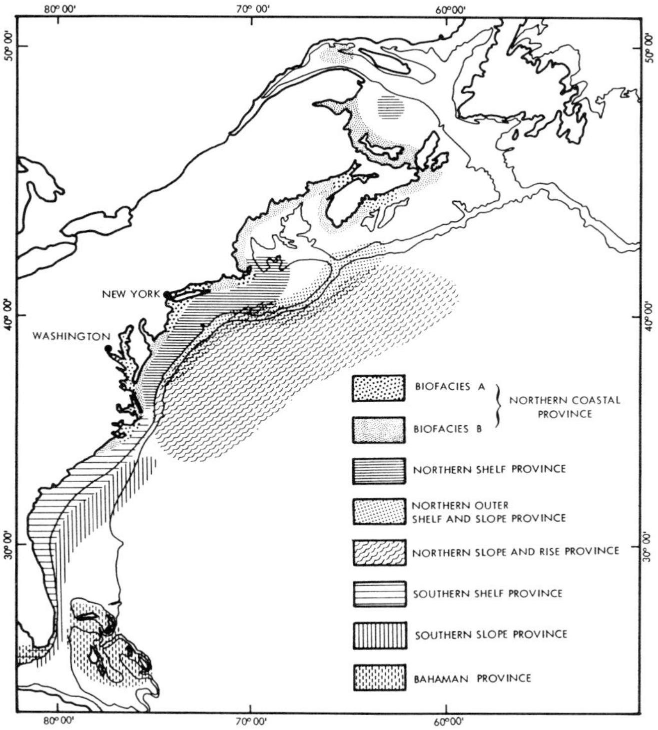

Figure 12.26 Recent benthic foraminiferal provinces on the Atlantic continental margin of North America. (Modified from Culver SJ, Buzas MA. *J Foraminiferal Res* 1981; 11:217–40.)

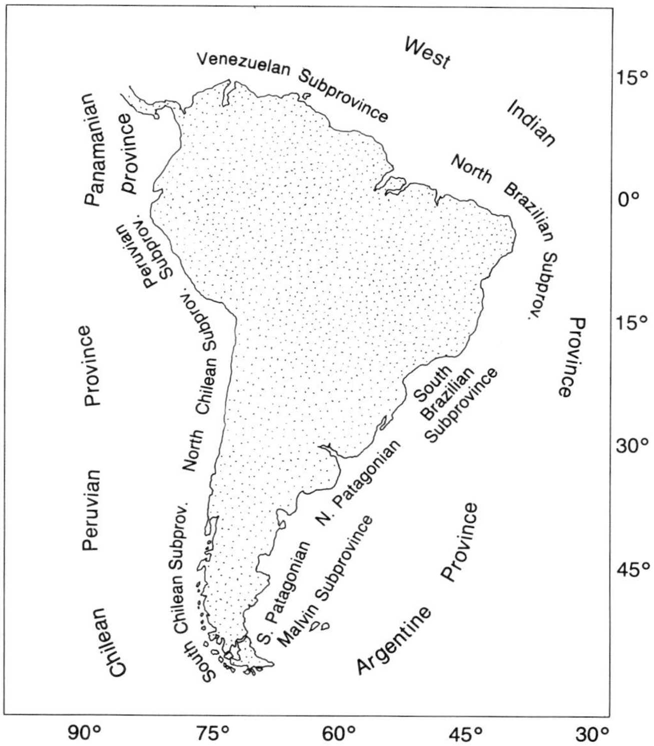

Figure 12.28 Recent benthic foraminiferal provinces of the South American continental shelf. (Modified from Boltovskoy E, Wright R. *Recent Foraminifera*. The Hague: W. Junk b.v., 1976:101.)

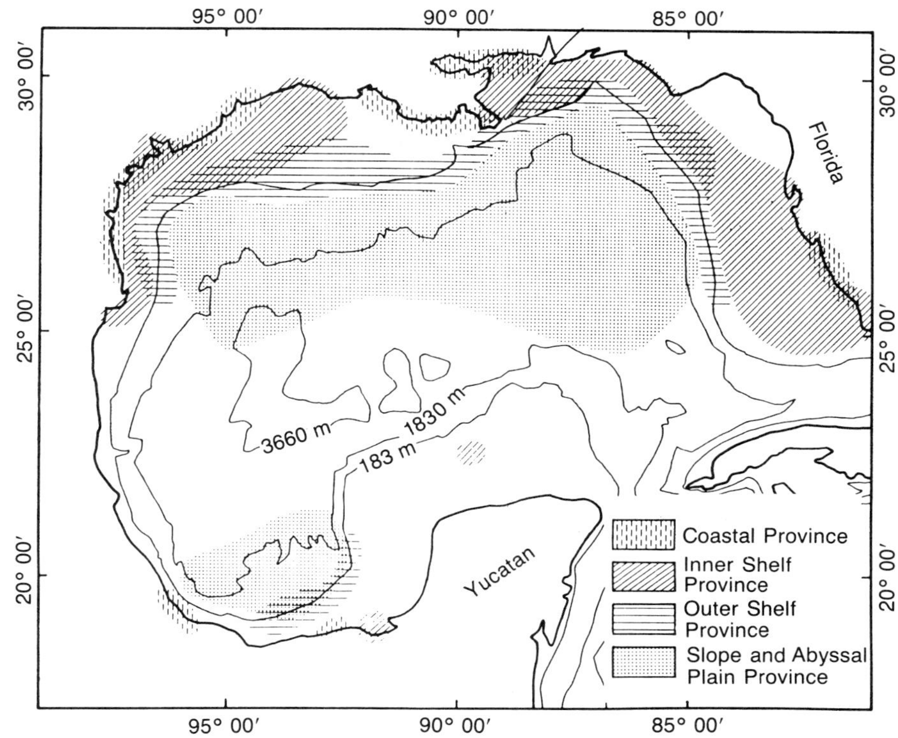

Figure 12.27 Recent benthic foraminiferal provinces in the Gulf of Mexico. (Modified from Culver SJ, Buzas MA. *J Foraminiferal Res* 1983; 13:21–31.)

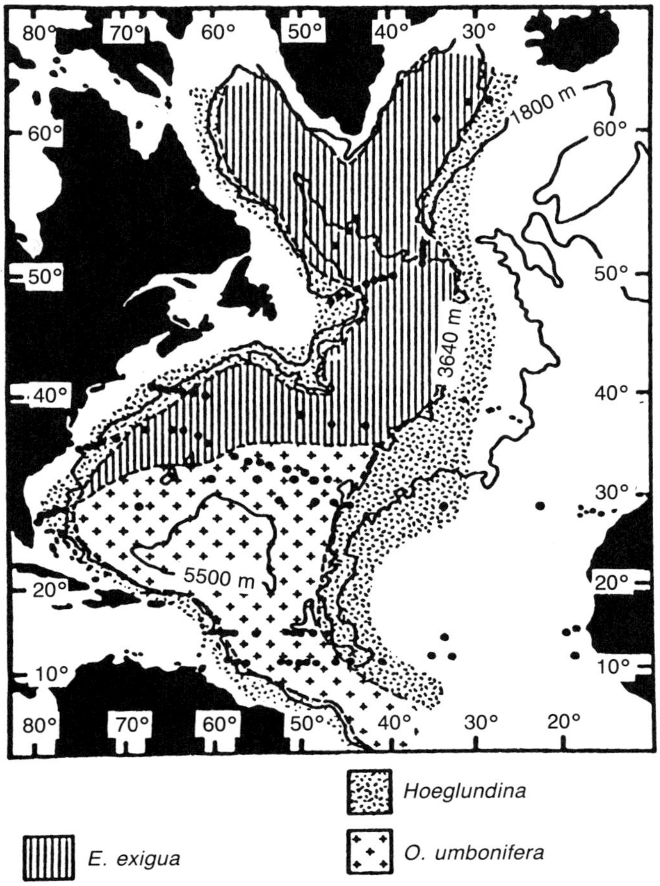

Figure 12.29 Distribution of modern benthic foraminiferal assemblages in the western north Atlantic. (Reproduced from Schnitker D. *Nature* 1974; 248:385–7.)

of modern larger foraminifera (Fig. 12.8H, I), Ross concluded that three species of Permian fusulinaceans reached reproductive maturity at 4 to more than 7 years. Two species from the southwestern USA exhibited two unequal growth seasons per year, whereas another species from northwestern Canada had one nearly uniform season of maximum growth. This pattern suggests that the southwestern USA was located within the solar tropics or in an area where seasonal water currents carried unequal nutrient supply, and northwestern Canada was located outside the solar tropics or perhaps in the path of a nearly uniform seasonal current. The regularity and size of initial chambers indicate that young fusulinaceans may have developed inside the adult test prior to release, a situation similar to modern larger foraminifera.

Delayed maturation and growth to large sizes in modern nonfusulinacean larger foraminifera (e.g., Figs 12.9I, J, 12.12G, H, I, J, K) may occur only under stable environmental conditions where food resources are limited. Algal symbiosis is also advantageous under those conditions in the photic zone. Perhaps, the extinct fusulinaceans also possessed these environmental and biologic characteristics.

Oxygen isotope analysis is a completely different approach to paleobiologic inferences. Depth stratification of Tertiary and Cretaceous planktic foraminifera, similar to that of modern planktic foraminifera, was detected with oxygen isotopic differences. Species with globigerine (globular) morphology (e.g., Fig. 12.13C) occupied shallower depths than species with globorotalid (flattened, keeled) morphology (see Fig. 12.52) since the mid-Cretaceous. Cretaceous biserially arranged species occupied both shallow and deep levels. The factors controlling depth stratification in the ocean probably have not changed over the past 100 My.

Foraminifera are commonly used for paleoenvironmental interpretation, involving paleobiogeography, paleoclimatology, paleoceanography, and paleobathymetry. Planktic and benthic foraminifera have proven equally useful in such studies.

Paleobathymetry

Benthic foraminifera have been used as paleodepth indicators since the 1930s when shelf foraminifera off California were shown to have a consistent zonation related to depth. For years, certain species were thought always to occupy the same depth interval wherever in the world's oceans they were found. Such **isobathyal** species were used to infer paleodepths until the 1970s when foraminiferal regional distributions were shown to be controlled by the distributions of water masses. Because water masses may migrate vertically and laterally through time, the habitat of benthic foraminiferal taxa can also migrate. For example, the depth habitat of *Melonis pompilioides* off California has changed considerably throughout the Quaternary, and the depth distribution of many taxa varies from ocean basin to ocean basin at the present time. Thus depth is merely a way of locating a sample in three-dimensional space, along with altitude, latitude, and longitude and the concept of isobathyal species is no longer strictly adhered to. Benthic foraminifera, however, do show well-developed, easily recognizable, depth-related distributions even in areas of considerable sediment transportation.

The inference of paleodepth based on the known depth distribution of modern foraminiferal taxa is termed the taxonomic approach. Four other methods are useful for assessing paleoenvironments (including paleodepths): trends of species diversity, planktic : benthic ratios, shell-type ratios, and foraminiferal morphology.

In general, species diversity increases offshore from the shoreline to the continental slope where it may stay constant or decline. This trend has been used to reconstruct depositional basins in the Miocene of California. Planktic : benthic ratios exhibit a similar trend and have been applied to paleodepth studies throughout the Cenozoic and into the Cretaceous.

Shell-type (agglutinated, porcelaneous, hyaline) ratios from samples collected in several modern environments define different fields when plotted on a triangular diagram (Fig. 12.30). These fields can then be used to interpret similar triangular plots constructed from fossil data. This approach has been used to reconstruct paleoenvironments in the Paleogene.

The relationship of the morphology of benthic foraminifera to environment is potentially a most powerful tool for paleoenvironmental interpretations. Four biofacies are recognizable in Texas bays using the external test morphology of 45 benthic species considered in six morphologic groups, straight-cylindrical, plano-convex, elongate-flattened, biconvex-keeled, tapered, and rounded-planispiral (Fig. 12.31). This approach to paleoenvironmental analysis has considerable potential as it is independent of species-level taxonomy and may well be applicable as far back as the Cretaceous.

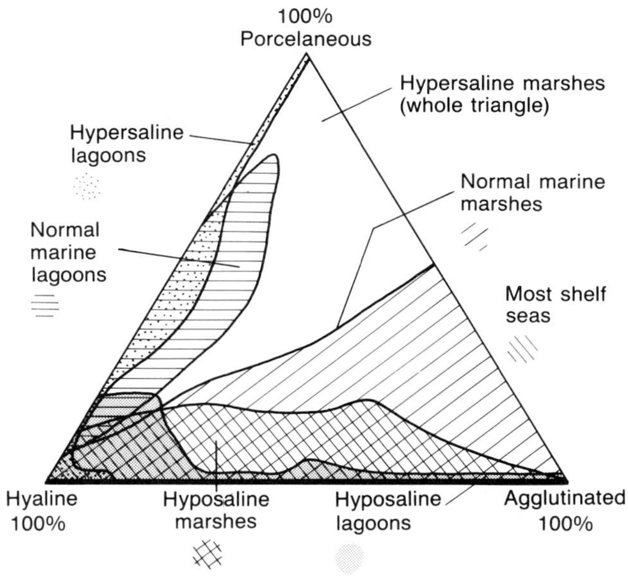

Figure 12.30 Summary triangular plot showing fields defined by foraminiferal shell-type ratios in several modern environments. (Reproduced from Murray JW. *Distribution and Ecology of Living Benthic Foraminiferids*. London: Heinemann Educational Books, 1973:201.)

Paleobiogeography and paleoceanography

The biogeography of modern foraminifera is related to ocean currents and distributions of water masses. Hence, paleobiogeographic patterns of benthic and planktic foraminifera can be used to infer paleoceanography. The three most studied segments of time are the Eocene–Oligocene boundary, the Miocene, and the Quaternary.

At the Eocene–Oligocene boundary, benthic foraminiferal enrichment in ^{18}O indicates cooling of the deep sea and build-up of Antarctic ice. Planktic foraminifera show greatest ^{18}O enrichment in high latitudes as a result of cooling of surface water. A previous oceanic cooling of 2–4°C took place at the Middle to Late Eocene boundary, and coincident major changes occurred in the benthic biota. The following period of cooling at the Eocene–Oligocene boundary did not result in major faunal changes as the surviving species were environmentally tolerant. The observed faunal data do not indicate a mass extinction event but a response to changing paleoceanographic conditions. A major extinction event had occurred earlier just prior to the Paleocene–Eocene boundary (Fig. 12.32).

The planktic faunas also reflect a stepwise general cooling trend from the middle Eocene to Oligocene. Five intervals of major faunal changes in response to oceanic cooling are recognized, the Middle Eocene (44–43 Ma), the Middle–Late Eocene boundary (41–40 Ma), the Late Eocene (39–38 Ma), the Eocene–Oligocene boundary (37–36 Ma), and the Late Oligocene (31–29 Ma). Four of these faunal changes were rapid and were characterized by extinction and origination of planktic species. The exception is the Eocene–Oligocene boundary where cool-water species increased in abundance but no major extinctions took place. These faunal patterns and the presence of hiatuses in deep-sea cores suggest that vigorous bottom-water circulation began in the middle Eocene, well before the development of the cold, deep water of the psychrosphere at the Eocene–Oligocene boundary.

Miocene paleoceanographic changes also indicate a stepwise general cooling trend. Conditions similar to those in the modern ocean can be traced to the early Middle Miocene (12–17 Ma). Prior to that time, the temperature of the bottom water was much warmer (Fig. 12.33). Evidence from the southwestern Pacific Ocean indicates that during this transition interval the Antarctic icecap expanded, oxygen isotope values increased, and the introduction of cold, more oxygenated waters into the ocean basins led to a turnover in benthic foraminifera. Species extinctions and originations occurred and benthic faunas became more bathymetrically restricted as deep-ocean water became increasingly stratified (Fig. 12.34).

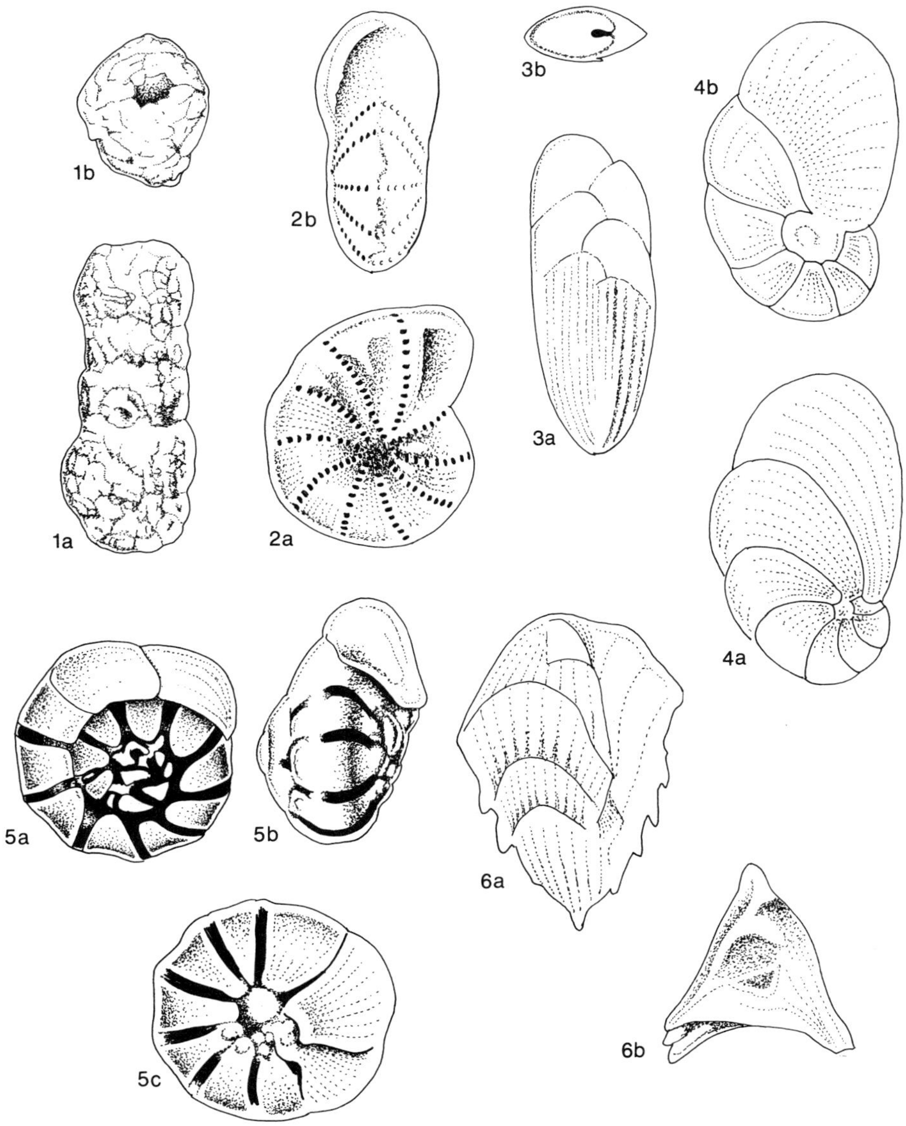

Figure 12.31 Examples of six morphologic groups of benthic foraminifera. 1, Straight-cylindrical; 2, rounded-planispiral; 3, elongate-flattened; 4, biconvex-keeled; 5, plano-convex; 6, tapered. (Redrawn from Severin KP. *Marine Micropaleontol* 1983; 8:65–76.)

Further glacial expansion occurred in the Late Miocene (5–9 Ma) causing a second period of benthic faunal changes as shifts in species distributions occurred. By Late Miocene time, benthic faunas were similar to modern faunas.

Planktic foraminiferal faunas from the equatorial Pacific indicate an earlier major oceanographic event in the Early Miocene. A hiatus in deep-sea cores dated at 22.2 Ma is coincident with the extinction of Oligocene faunas and the origination of Miocene species. This event correlates with a cooling trend, as indicated by oxygen isotope data. Drake Passage may have opened by about 22–25 Ma allowing deep, circum-Antarctic circulation and the introduction of cool Antarctic Bottom Water into the Pacific basin, and possibly also into the Atlantic and Indian Oceans.

Three benthic faunal assemblages are associated with three abyssal water masses in the modern western north Atlantic (Fig. 12.29). The distribution of foraminiferal assemblages during the last glacial maximum (ca. 18 000 b.p.) differed (Fig. 12.35), possibly as a result of cessation of Arctic Bottom Water flow and a modified supply of Antarctic Water. Foraminiferal assemblage distributions during the last interglacial (ca. 120 000 b.p.) were very similar to today, thus indicating similar oceanographic conditions (Fig. 12.36).

Planktic foraminifera from deep-sea cores are probably the principal source of paleoceanographic data for recon-

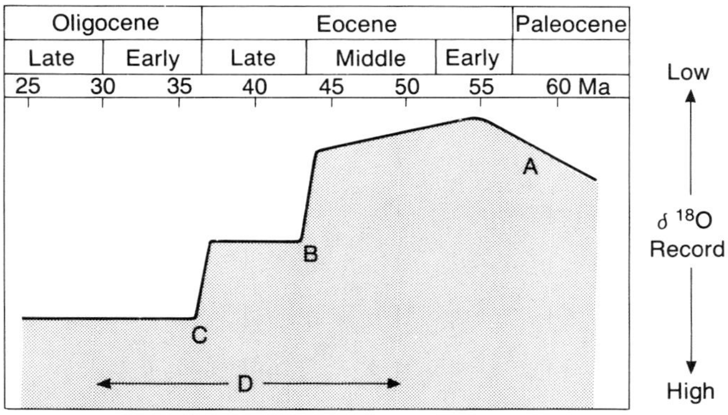

Figure 12.32 Faunal events and the oxygen isotope record from the Paleocene to the Oligocene. A, Major extinction event just prior to Paleocene–Eocene boundary. B, Major reduction in abundance of *Nuttallides truempyi*. C, Faunal assemblage change with no major extinction at Eocene–Oligocene boundary. D, Gradual faunal turnover. (Modified from Corliss BH, Keigwin LD Jr. *Mesozoic and Cenozoic Oceans, Geodynamic Series 15*. American Geophysical Union, 1986:101–18.)

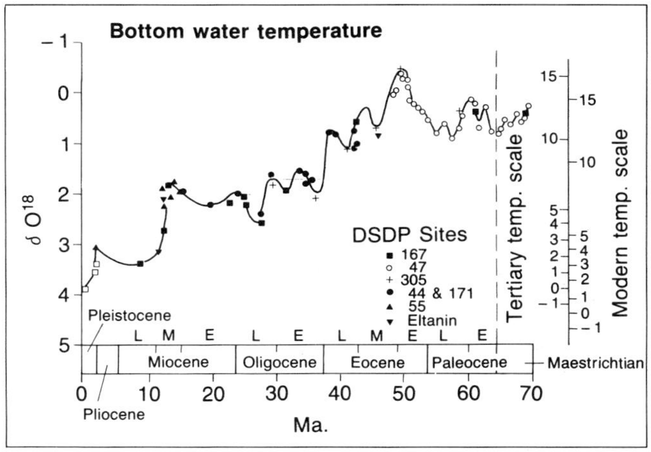

Figure 12.33 Tertiary bottom-water temperature curve for the tropical Pacific. Temperature maxima of the Early Paleocene and Early Eocene are followed by generally declining temperatures throughout the rest of the Tertiary, except for a brief period in the Middle Miocene. (Reproduced from Douglas RG, Woodruff F. *The Oceanic Lithosphere, The Sea*. Vol. 7. New York: John Wiley, 1981:1233–1327.)

structing Quaternary climates and oceanography. Horizontal and vertical distributions of foraminifera together with isotopic analyses of their tests have provided the main body of data, while quantitative techniques for paleoclimatic analysis have resulted in a detailed view of the latest Quaternary.

The **CLIMAP** project made estimates of Pacific sea-surface temperatures for the last glacial maximum (18 000 b.p), using planktic foraminifera, radiolaria, coccoliths, and diatoms. Estimates of temperature for both August and February during glacial times showed marked cooling in the subarctic and equatorial regions in comparison with modern conditions (Fig. 12.37). The areas with greatest cooling also experienced the greatest change in seasonal contrast. Within the center of subtropical gyres, however, temperatures were as warm as or warmer than those of the modern sea surface. The reconstructions provide a broad framework into which more regional oceanographic studies can be placed.

Biostratigraphy

Foraminifera are of some biostratigraphic utility starting in the early Phanerozoic. The Paleozoic agglutinated

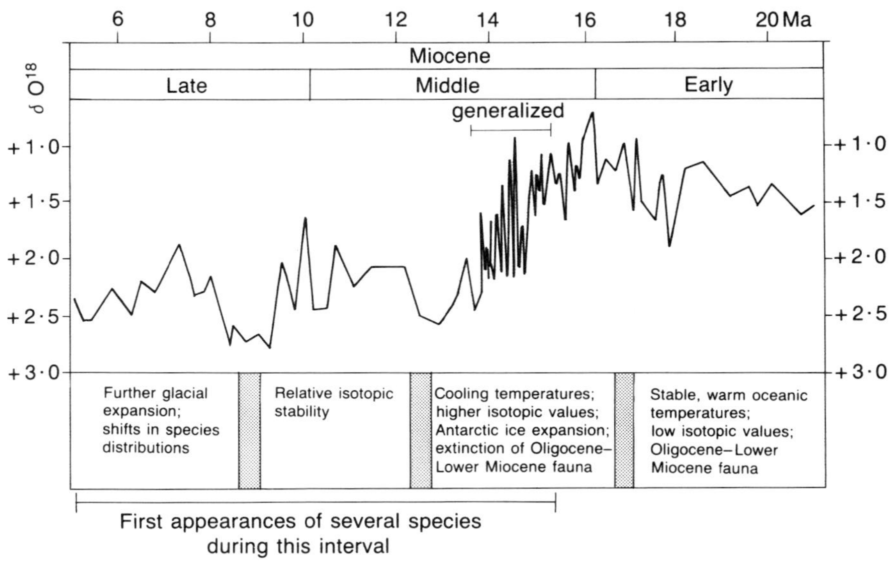

Figure 12.34 Summary of faunal and ocean climate events and the oxygen isotopic record for the Miocene of the Ontong-Java Plateau (southwestern Pacific Ocean). (Adapted from Woodruff F, Douglas RG. *Marine Micropaleontol* 1981; 6: 617–32.)

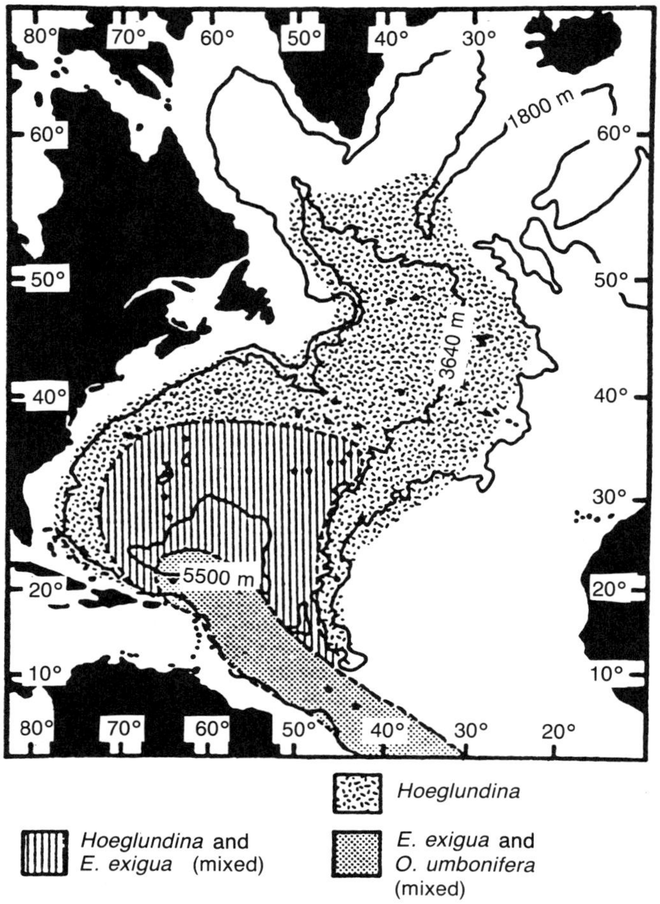

Figure 12.35 Distribution of benthic foraminiferal assemblages in the north Atlantic under glacial conditions (17 000 b.p.). (Reproduced from Schnitker D. *Nature* 1974; 248:385–7.)

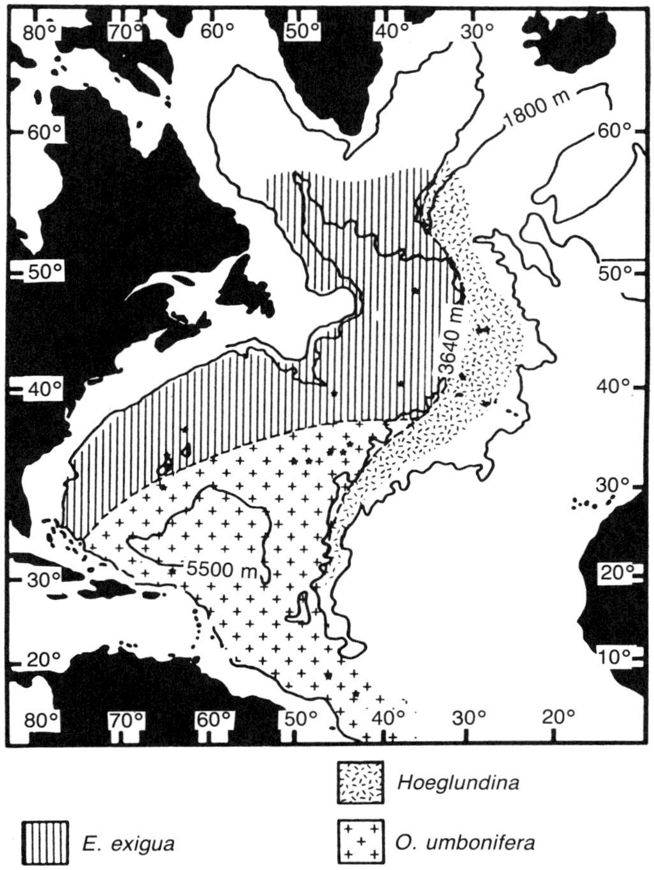

Figure 12.36 Distribution of benthic foraminiferal assemblages in the north Atlantic under interglacial conditions (120 000 b.p.). (Reproduced from Schnitker D. *Nature* 1974; 248:385–7.)

foraminifera of North America may be used for coarse biostratigraphic subdivision of silty shales and shaly sandstones (Fig. 12.38), but not until the Carboniferous are foraminifera of great use for biostratigraphy.

Larger foraminifera, several millimeters to several centimeters in size, arose several times from different smaller benthic ancestors and became particularly abundant and biostratigraphically useful (e.g., Fig. 12.39) in Upper Oligocene–Lower Miocene, Eocene, Cretaceous, and Upper Carboniferous to Permian strata. The larger and morphologically complex fusulinaceans are used to produce detailed Upper Carboniferous to Permian zonations (Fig. 12.40). The microgranular fusulinaceans are typical of stable, shallow, carbonate shelves and can be used to establish correlations of stratigraphic intervals of 3–5 My duration. Although fusulinaceans were provincial, 16 zones can be recognized worldwide for the Upper Carboniferous and the Permian. Their evolutionary antecedents, small, morphologically simple, microgranular benthic foraminifera, are used to subdivide the Lower Carboniferous.

Smaller benthic foraminifera are important in biostratigraphic subdivisions of Cretaceous and Cenozoic continental shelf and epeiric sea strata. Biostratigraphic zonations of the entire Tertiary and much of the Cretaceous of California and the Tertiary of the northern Gulf of Mexico have been developed, chiefly for use in the exploration for petroleum (Fig. 12.41).

As petroleum exploration moved offshore to outer neritic and upper bathyal depths, planktic foraminifera became of increased importance even in such restricted basins as the Gulf of Mexico. Planktic foraminiferal biostratigraphy provides a geochronology not only for petroleum exploration but also for the evolution of the entire oceanic realm from the Cretaceous to the Recent.

Planktic foraminifera are of biostratigraphic use at several taxonomic levels. For example, the families Hedbergellidae, Rotaliporidae, and Schackoinidae appear and dominate in the Lower Cretaceous, whereas the

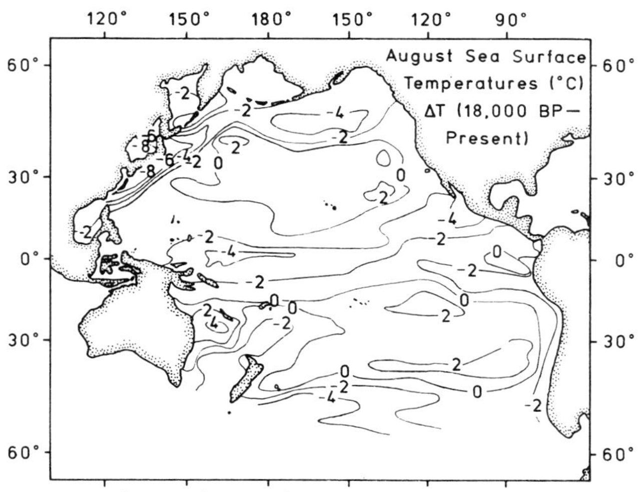

Figure 12.37 Temperature difference in the Pacific (°C between August 18 000 b.p. and average modern August sea-surface temperatures). (Modified from Moore TC, Burkle LH, Geitzenauer K *et al. Marine Micropaleontol* 1980; 5:215–47.)

Upper Cretaceous is characterized by the dominance of the Globotruncanidae and the Heterohelicidae. At the generic level, for example, *Leupoldina* and *Schackoina* appear near the Barremian–Aptian boundary, while *Rotalipora* appears near the Albian–Cenomanian boundary. Indeed, 118 genera of planktic foraminifera can be used for biostratigraphic subdivision of the Cretaceous through to the Recent. At the species level, much more detailed zonations are possible. The 55 My from the Hauterivian to the Maastrichtian is subdivided into 28 zones. The zones are based on the first and last occurrences of **marker species**, some from well-established evolutionary lineages (Fig. 12.42). In the Cenozoic, planktic foraminifera are now used to define 42 zones worldwide in tropical and subtropical regions. These zones are designated with a P for Paleogene or an N for Neogene plus a number for each zone from the oldest to the youngest (Figs 12.43, 12.44). In temperate and cooler regions, these zones cannot be recognized and separate zonations have been defined.

Evolutionary history

Foraminifera have had two major periods of evolution, one in the Devonian and the other in the Triassic to Early Jurassic. Major taxonomic diversification occurred during these periods and has resulted in evolutionary innovations in test construction and morphology. Superimposed upon these major trends is a myriad of faunal changes resulting from varying patterns of species originations and extinctions commencing in the earliest Cambrian.

The oldest fossil foraminifera appear to be the simple, cylindrical, agglutinated tubes present in the earliest Cambrian (Fig. 12.45). Although they have also been interpreted as worm tubes, they closely resemble true foraminifera as they are indistinguishable from the modern benthic genus *Bathysiphon* (Fig. 12.7A). Recently, straight and coiled tubular agglutinated foraminifera have been reported from Lower and Middle Cambrian strata of Africa, the USA, and the USSR. Thus, shelled protozoans seem to have appeared at the same time as shelled metazoans. The subsequent radiation of both groups was probably related to the increasing trophic interaction within these early Phanerozoic communities.

Agglutinated foraminifera became more abundant in the Ordovician, and several new, generally tubular genera appeared. True multichambered, agglutinated, foraminiferal tests may have arisen in the Devonian. By the Carboniferous they were abundant and included biserial, triserial, and trochospirally coiled forms.

Many new genera of multichambered agglutinated foraminifera arose in the Late Jurassic and Cretaceous. Labyrinthic wall structure (Fig. 12.4), a character of possible polyphyletic origin, and partitioned forms became abundant. Many of these genera died out by the end of the Cretaceous, but several new labyrinthic forms appeared periodically throughout the Cenozoic.

Certain Upper Cretaceous to Paleogene agglutinated foraminifera occur commonly in **flysch** deposits around the world associated with the pelitic interval of turbidite sequences. Some elements of these faunas have restricted geologic ranges and may have local and regional stratigraphic utility from the Maastrichtian (Upper Cretaceous) to the Oligocene.

The fusulinaceans of the late Carboniferous and Permian descended from the endothyroids, microgranular-walled forms with simpler morphology, towards the end of the early Carboniferous. Major evolutionary trends of the fusulinaceans that have stratigraphic significance include a general increase in size combined with increased septal folding leading to formation of **chamberlets**; development of an **alveolar wall** with loss of the inner layer of a three-layered wall; replacement of a single, large, septal foramen by numerous foramina (Fig. 12.2).

Two other suborders of benthic foraminifera, the Miliolina and the Lagenina, arose in the Carboniferous although their ancestry is not clear. The miliolines have been considered descendants of calcareous or agglutinated

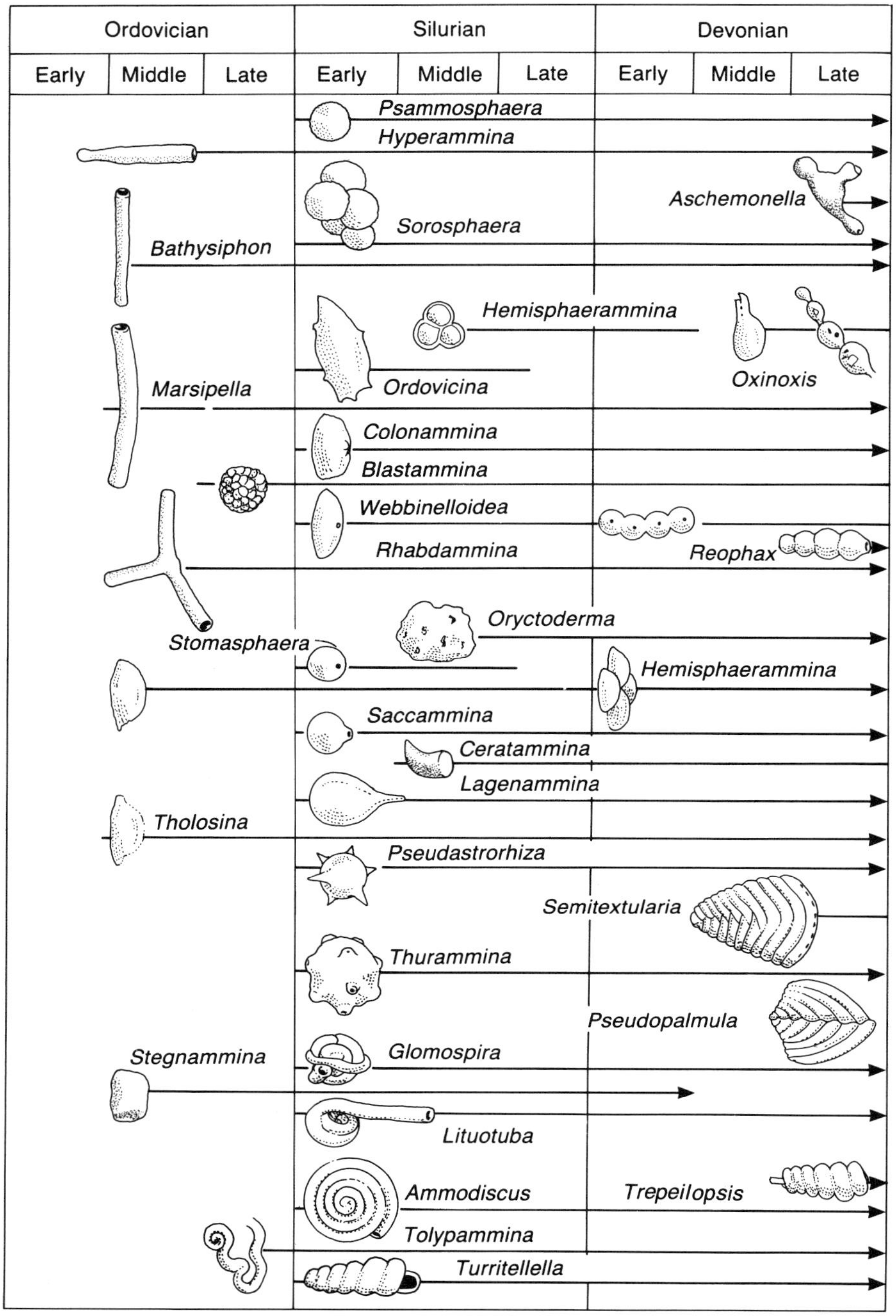

Figure 12.38 Ranges of agglutinated foraminiferal genera in the Paleozoic of the USA showing their biostratigraphic value based on first and last appearances. Arrow indicates range above the Devonian. (Modified from Conkin JE, Conkin BM. *Stratigraphic Micropaleontology of Atlantic Basin and Borderlands*. Amsterdam: Elsevier, 1977:49–59.)

spiral forms. More recently, the miliolines were suggested to have arisen directly from an allogromiid-like ancestor. This follows the discovery that a modern allogromiid species develops one or more chambers similar to those of some simple miliolines. Smaller miliolines range throughout the Mesozoic and Cenozoic, but the larger ones evolved rapidly resulting in short-ranged, biostratigraphically useful forms particularly in the Paleocene and Eocene (Fig. 12.39). These Paleogene alveoline faunas are widespread throughout the Mediterranean region and extended as far north as southern England during the peak Eocene transgression.

Other larger milioline lineages have been shown to exhibit both punctuated and gradual evolutionary changes. These lineages, together with unrelated rotaline larger foraminiferal lineages (Fig. 12.46), developed mainly through gradual unidirectional trends. The first species in some lineages, however, differ fundamentally from their

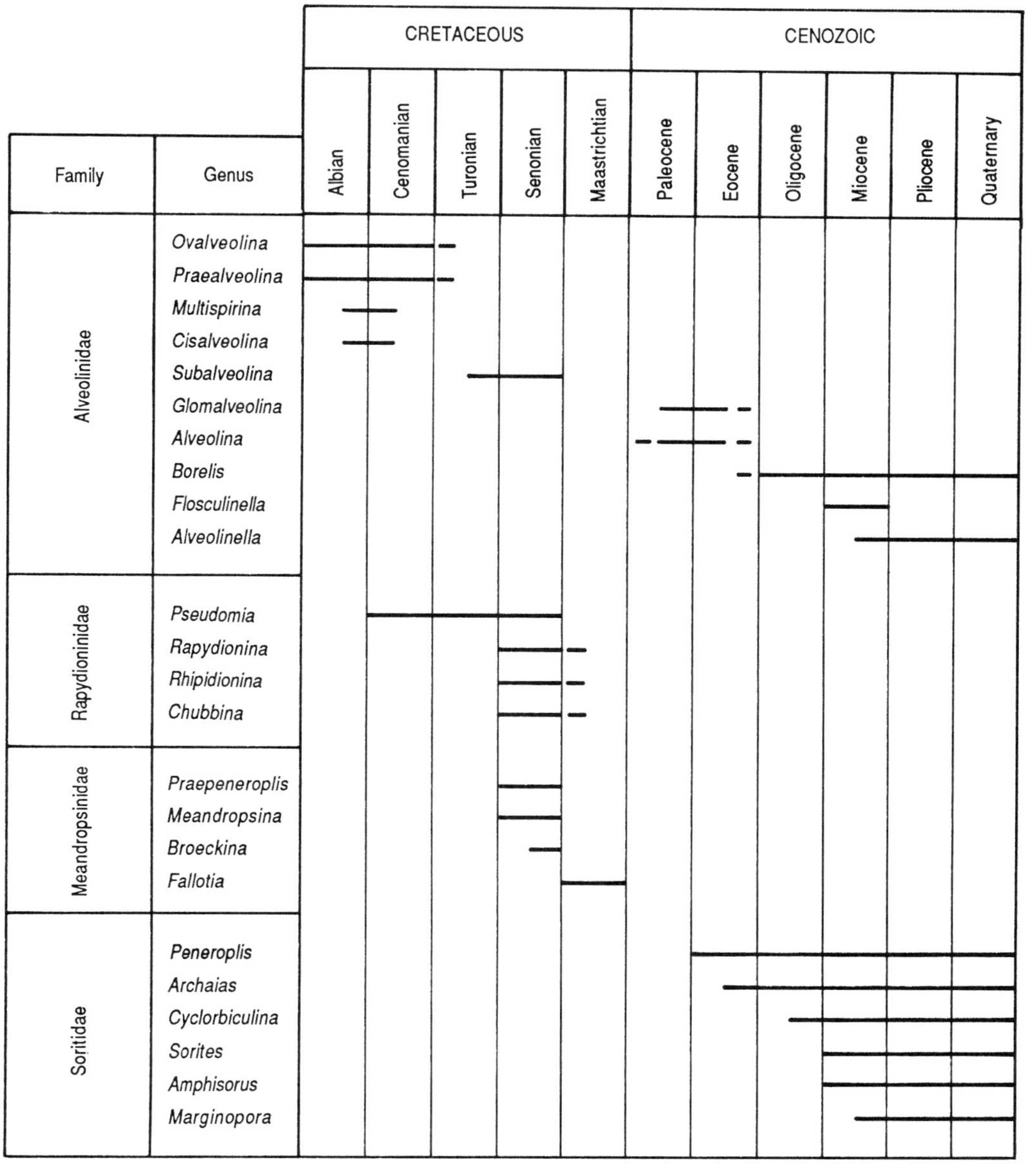

Figure 12.39 Stratigraphic ranges of larger miliolid foraminifera in the Cretaceous and Cenozoic. (Modified from Haynes JR. *Foraminifera*. New York: John Wiley, 1981:160.)

ancestors in possessing new, rapidly developed morphologic characters. Tectonically induced eustatic changes were the most likely cause of worldwide changes in the composition of Cenozoic, larger foraminiferal faunas.

Following Permian extinctions, particularly of the fusulinaceans, smaller benthic foraminifera underwent an evolutionary radiation. Representatives of the suborders Rotaliina, Involutinina, Spirillinina, and Robertinina first appeared in the Triassic. All benthic foraminiferal suborders, except the Carterinina, which has only one genus with a low preservation potential, had originated by this time (Fig. 12.17).

Benthic foraminifera evolved to occupy many habitats, from marshes to the open shelf, from carbonate reefs to deep neritic basins. Increasingly in the Mesozoic, benthic foraminifera are recorded from deeper-water rocks until, by the Late Cretaceous, they are recorded from rocks of abyssal origin.

Rapidly evolving lineages of the Lagenina are useful for local zonations of Jurassic strata. Regional correlations, however, are rarely possible due to facies control of these faunas. In the Upper Cretaceous rapidly evolving species in the *Bolivinoides* lineage similarly are useful in correlation. This zonal scheme has been applied worldwide.

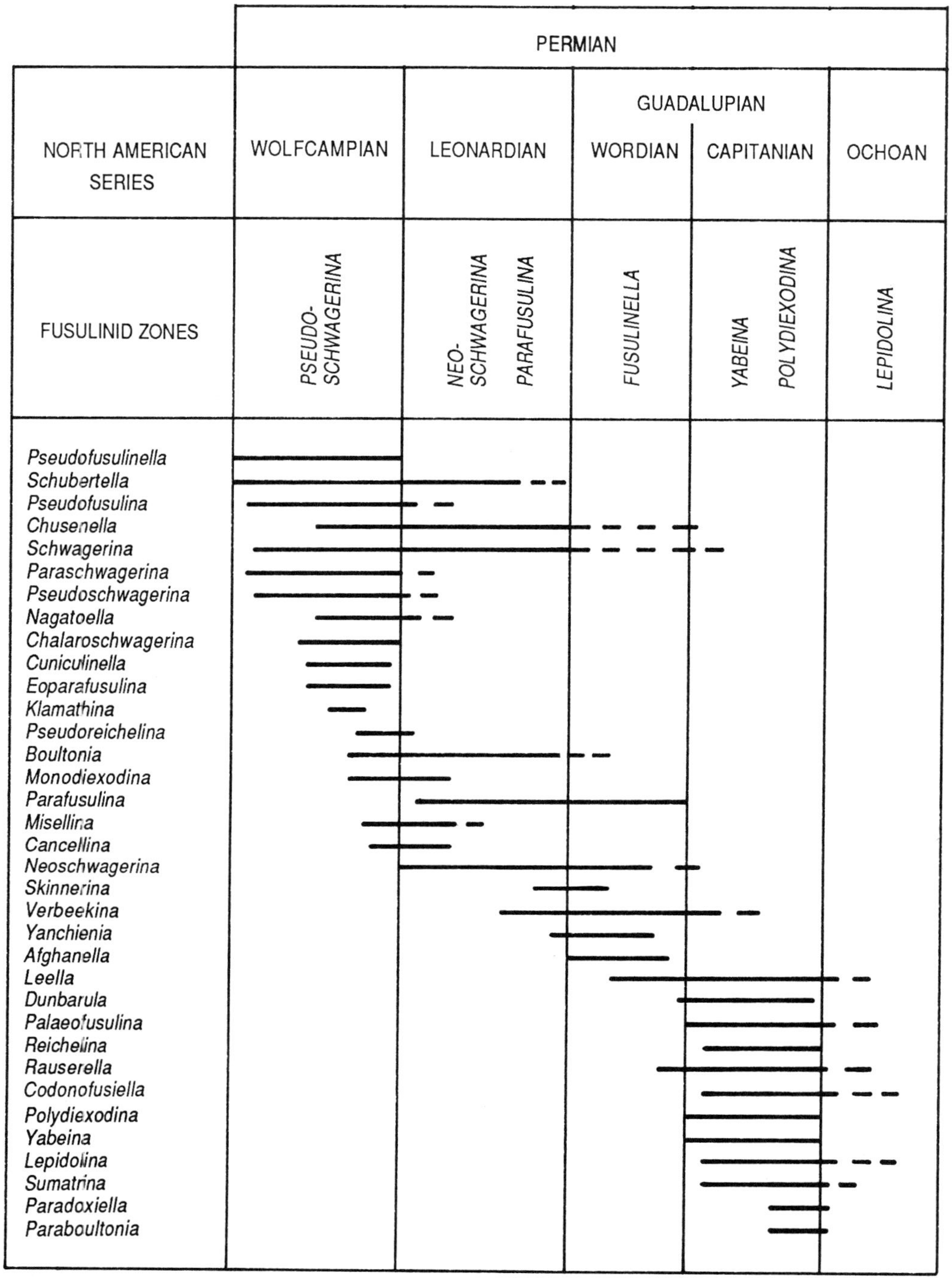

Figure 12.40 Fusulinid zonation in the Permian of the USA. (Modified from Douglass RC. *Concepts and Methods of Biostratigraphy*. Stroudsburg: Dowden, Hutchinson and Ross, 1977:463–81.)

The nature of the benthic foraminiferal faunal changes at the Cretaceous–Tertiary boundary is in some doubt. Neither shallow nor deep-dwelling benthic foraminifera, however, appear to have undergone **mass extinction**. Assemblages on either side of the boundary are similar at the familial and generic level in shallow-water environments and many Cretaceous species range into the Paleocene. Deeper-water benthic assemblages are even more similar. In Trinidad, about 80% of the Cretaceous species range into the Paleocene, whereas all of the Paleocene deep-water species described from the Atlantic Ocean have been recorded from the Cretaceous. A

Series	Formation	Foraminiferida
Pleistocene	Houston	*Candeina nitida* *Globorotalia menardii* (sinistral) *Trimosina denticulata (Trimosina "A")* *Hyalinea balthica* *Globigerina pachyderma* *Globorotalia truncatulinoides*
Pliocene	Citronelle	*Globorotalia menardii miocenica* *Globorotalia inflata* *Globorotalia crassaforma* *Buliminella basispinata (Buliminella "1")*
Miocene	"Fleming"	*Rangia (Miorangia) microjohnsoni* (Bivalvia) *Lenticulina vaughani (Robulus "E")* *Bigenerina floridana (Bigenerina "A")* *Spiroplectammina barrowi (Textularia "L")* *Globorotalia mayeri* *Bigenerina nodosaria directa (Bigenerina "2")* *Textularia stapperi (Textularia "W")* *Globorotalia fohsi fohsi* *Bigenerina humblei* *Uvigerina lirettensis* *Lenticulina (Robulus) cristi (Cristellaria "I")* *Globorotalia fohsi barisanensis* *Hanzawaia carstensi opima* *Amphistegina chipolensis (Amphistegina "B")* *Lenticulina (Robulus) macomberi (Robulus "43")* *Nummulites* sp. (=*Operculinoides* sp.) *Lenticulina (Robulus) chambersi* *Marginulina ascensionensis* *Siphonina davisi* *Lenticulina jeffersonensis* *Siphogenerina fredsmithi*

Figure 12.41 Gulf coast stratigraphic correlation chart for the Neogene, based primarily on smaller benthic foraminifera. (Modified from Skinner HC, Steinkraus WE. *Gulf Coast Stratigraphic Correlation Methods with an Atlas and Catalogue of Principal Index Foraminiferida*. New Orleans: Louisiana Heritage Press, 1972:81–211.)

broader survey of Maastrichtian to early Paleocene deep-water benthic species indicates that some replacement of species is evident across the boundary but, at the generic level, late Cretaceous faunas are nearly identical to those of the early Tertiary.

Although benthic foraminiferal faunas across the Cretaceous–Tertiary boundary have yet to be compared in detail, the mass extinction evident in planktic foraminiferal faunas is much less marked in shallow benthic faunas and did not occur among deeper benthic assemblages.

Deep-water benthic faunas exhibit three major episodes of change during the Cenozoic: in the latest Paleocene, at the Eocene–Oligocene boundary, and in the middle Miocene. The benthic faunas in the latest Paleocene underwent massive extinctions at the generic and specific levels. No major oceanographic changes have been documented for this time. Hence, these benthic faunal changes, the greatest in the Cenozoic, remain unexplained.

Deep-sea benthic faunal changes leading up to the development of the cold, deep-water **psychrosphere** near the Eocene–Oligocene boundary exhibit a stepwise pattern, probably as a result of a general cooling trend of the deep ocean. A major extinction event is not evident at the Eocene–Oligocene boundary. Instead, extinctions and originations of species take place at discrete intervals throughout the Middle Eocene to Early Oligocene (Fig. 12.47).

Middle Miocene (12–17 Ma) cooling of deep-water masses also led to faunal changes in deep-sea benthic foraminifera. Species extinctions and originations occurred, while further cooling in the Late Miocene (5–9 Ma) resulted in less drastic shifts of species distribution and abundance.

Continental margin benthic foraminifera evolved with as much complexity as deeper-water species (Table 12.3). Although modern continental margin faunas are the result of species orginations throughout the Cenozoic and even the Cretaceous, the Miocene is the most important interval for origin of neritic faunas. Only 7% of the modern North American Atlantic shelf benthic species existed in the Oligocene. In comparison, 24% of the deeper-dwelling continental margin species are recorded in Oligocene strata. Further, the evolutionary durations of northern shelf species from Cape Hatteras to Newfoundland are shorter than those of southern shelf species fram Cape Hatteras to Florida. These data (Table 12.3) suggest that evolutionary rates of benthic foraminifera from the continental margin are greatest in the more

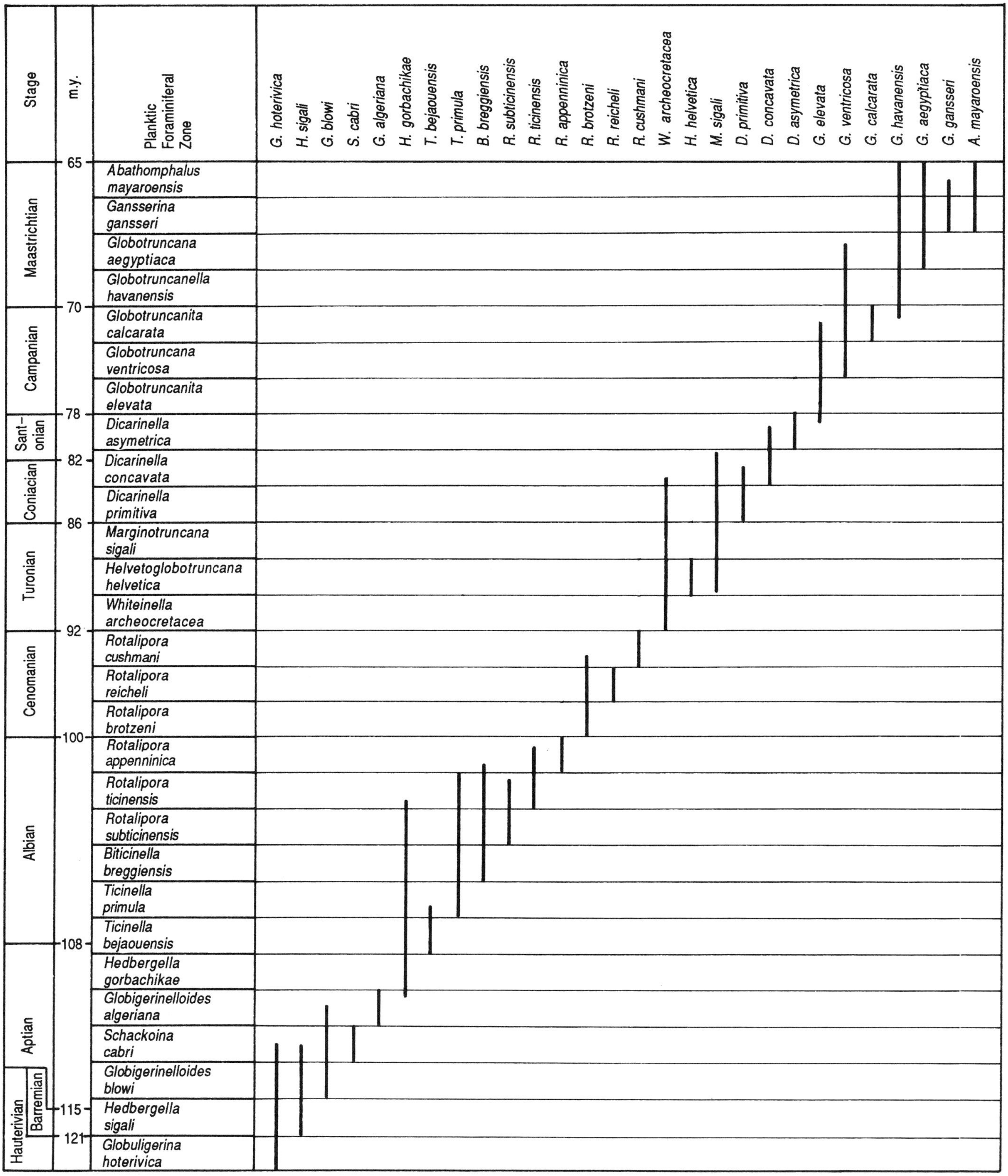

Figure 12.42 Planktic foraminiferal zones and ranges of zonal markers for the Cretaceous (Hauterivian to Maastrichtian). (Modified from Caron M. *Plankton Stratigraphy*. Cambridge: Cambridge University Press, 1985:17–86.)

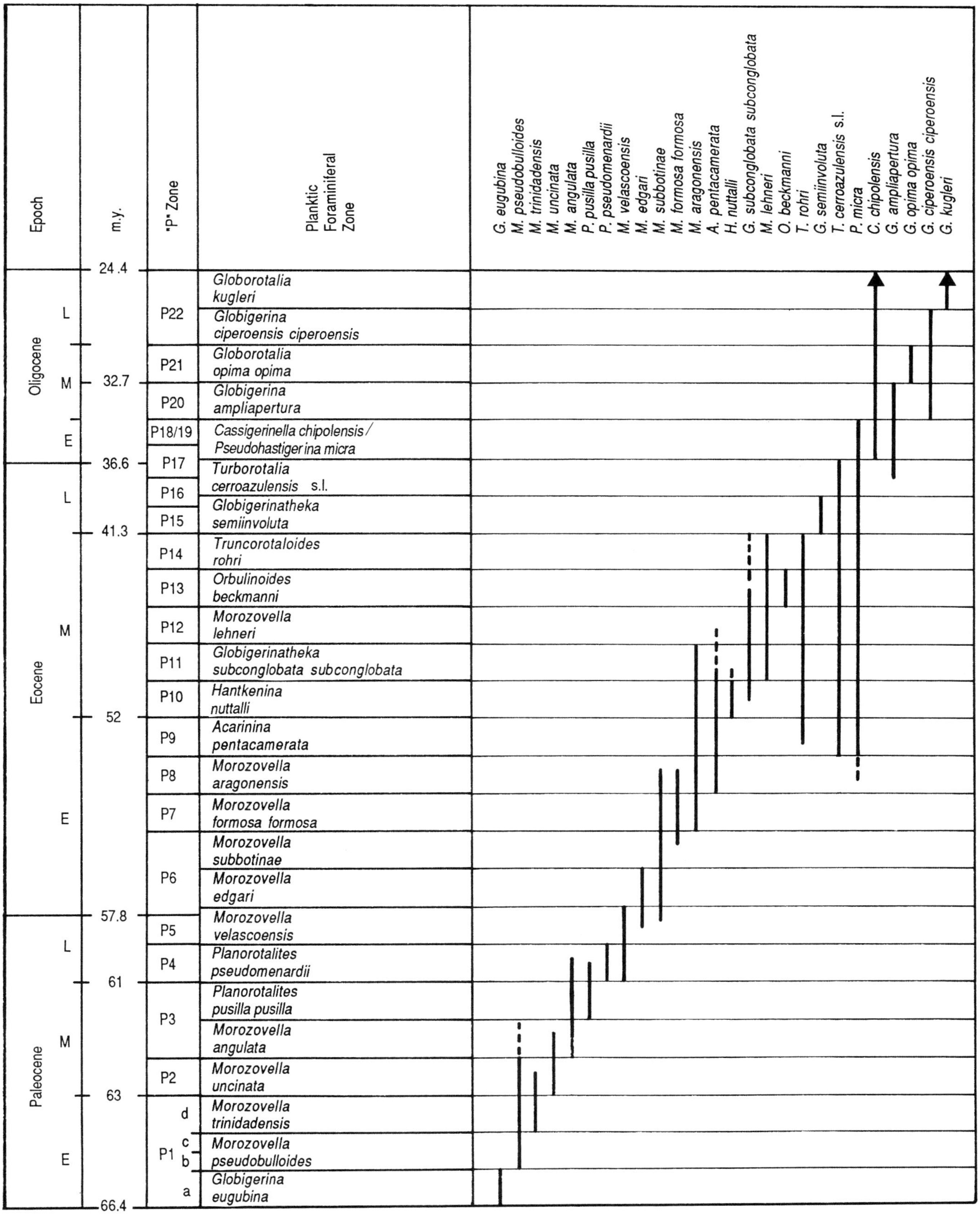

Figure 12.43 Planktic foraminiferal zones and ranges of zonal markers from the Paleocene to Oligocene. (Modified from Toumarkine M, Luterbacher H, Bolli HM, Saunders JB. *Plankton Stratigraphy*. Cambridge: Cambridge University Press, 1985:87–262.)

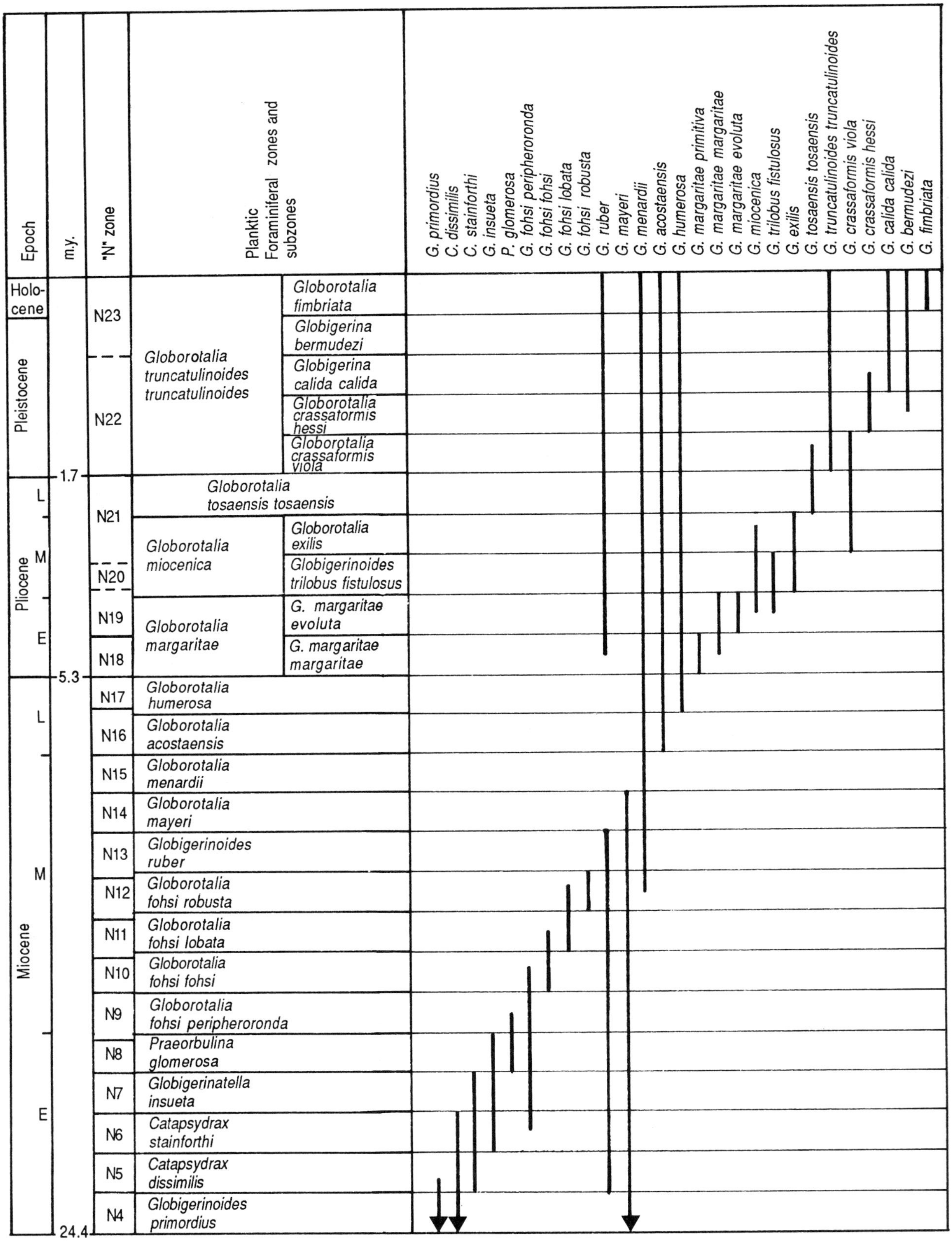

Figure 12.44 Planktic foraminiferal zones and ranges of zonal markers from the Miocene to Holocene. (Modified from Bolli HM, Saunders JB. *Plankton Stratigraphy*. Cambridge: Cambridge University Press, 1985:155–262.)

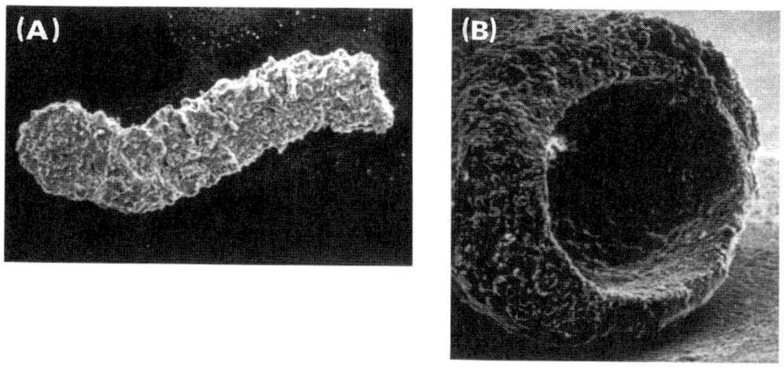

Figure 12.45 A, B *Platysolenites*. Lower Cambrian, Russia. A, Side view of compressed specimen. About ×74. B, Apertural view. About ×148. (Courtesy of J.H. Lipps, University of California, Berkeley.)

variable environments. Foraminifera from less variable subtropical shelf environments evolved less rapidly, while species inhabiting the more stable bathyal and abyssal environments evolved even more slowly.

Planktic foraminifera first appeared in Middle Jurassic strata of the northern margin of the Tethys and the epicontinental basins of Europe. They probably were **meroplanktic** and may have evolved from a group of Triassic to Early Jurassic, trochospiral, benthic calcareous, hyaline foraminifera, the Oberhauserellidae, by acquisition of subglobular chambers and a planktic mode of life late in ontogeny. By the Hauterivian, planktic foramini-

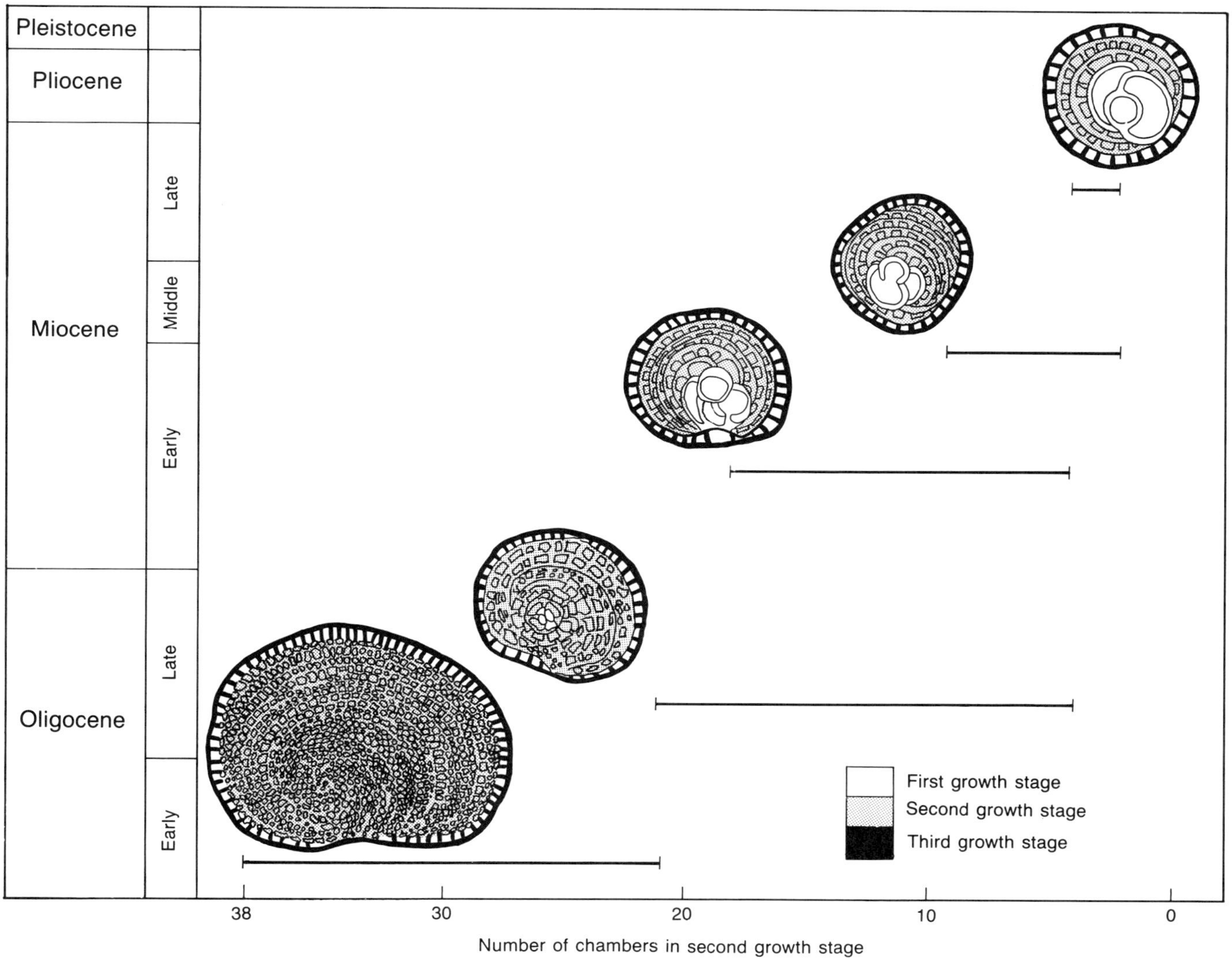

Figure 12.46 Evolution of the second growth stage of the larger rotaline *Cycloclypeus*. The long initial coil (stippled second growth stage) of the Oligocene species is reduced through time. Thus evolution was unidirectional although the rate of evolution was not constant. Only the first annular chamber of the third growth stage is illustrated. Horizontal lines indicate range in number of chambers in the second growth stage. (Modified from Adams CG. *Evolution, Time and Space: the Emergence of the Biosphere*. London: Academic Press, 1983:255–89.)

fera had become widespread throughout the low and midlatitudes of the world and had probably become **holoplanktic**.

The evolution of Mesozoic planktic foraminifera, including changing patterns of population density, species diversity, and test morphology, is related to fluctuations in the oceanic environment corresponding to eustatic cycles. A possible relationship between environmental stress and speciation tactics is summarized in Figure 12.48.

Three principal periods of diversification and evolutionary radiation characterize the fossil record of planktic

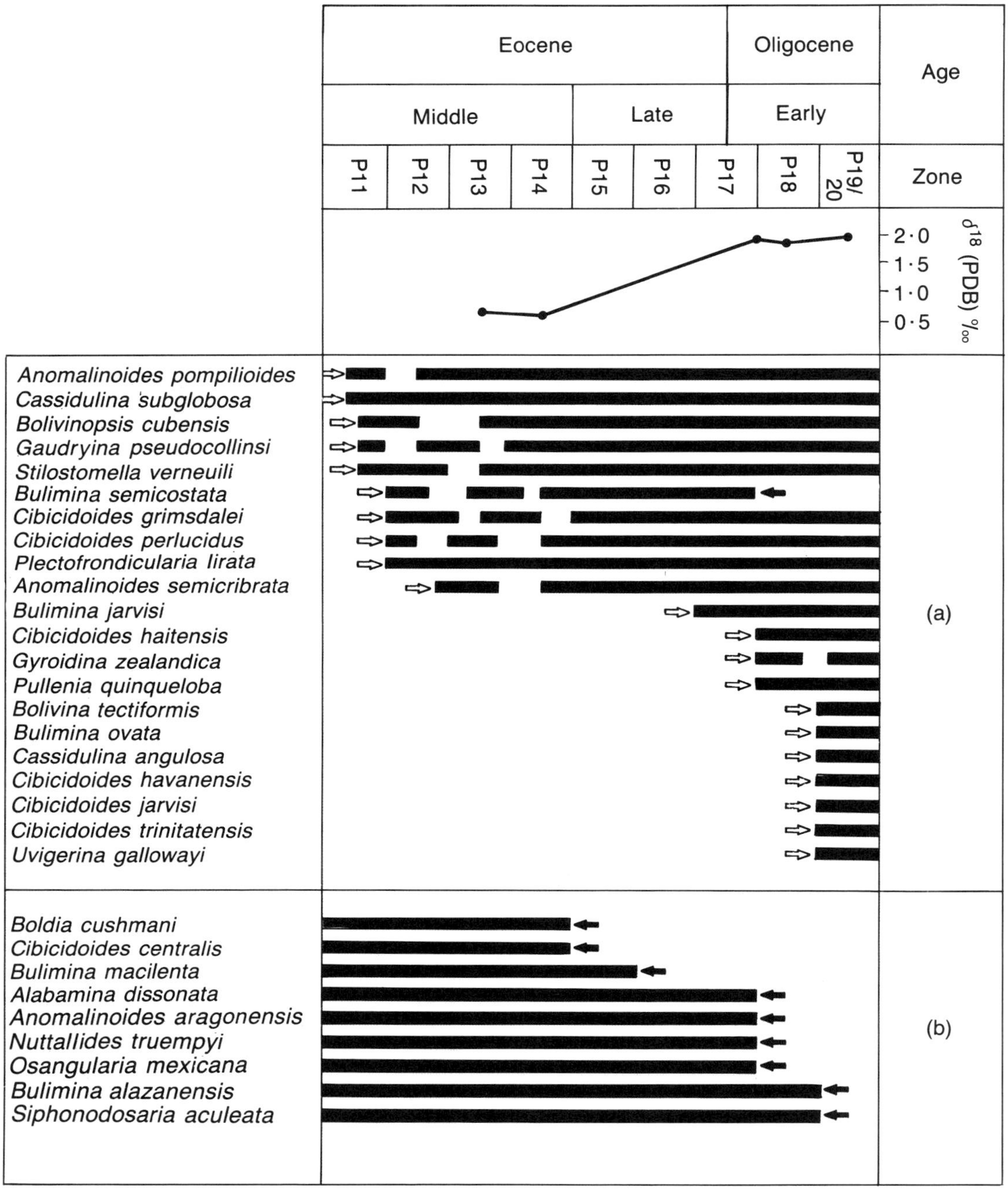

Figure 12.47 Benthic foraminiferal biostratigraphic data for the north central Pacific. a, Species exhibiting first appearances (open arrows) during the Middle Eocene to Early Oligocene. b, Species exhibiting last occurrences (closed arrows) during the Middle Eocene to Early Oligocene. (Modified from Corliss BH. *Marine Micropaleontol* 1981; 6:367–84.)

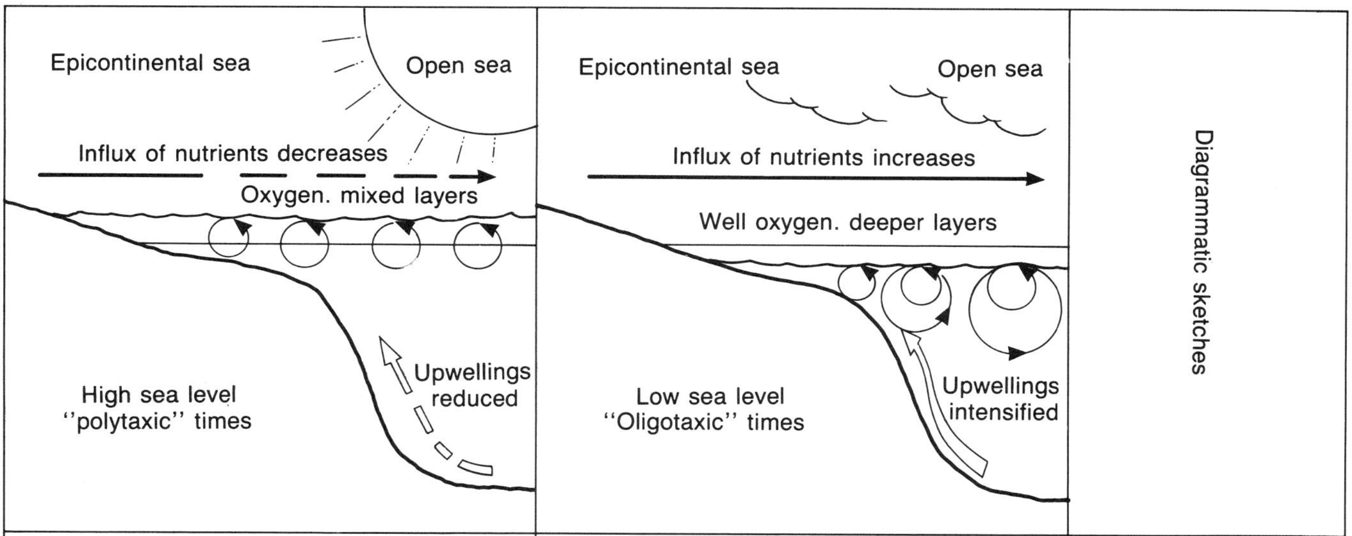

Rise of sea level	Drop of sea level	Eustatic oscillations	Oceanic control
Warmer/stable	Colder/unstable	Temp./climate	Oceanic control
Sluggish oceans	Highly convective oceans	Currents	Oceanic control
Less contrasted	Sharper latitudinally and vertically	Temp. gradients	Oceanic control
Well depth stratified	High mixing	Water masses	Oceanic control
Lower	High	Net nutrient supply	Oceanic control
Warmer, stable, less fertile = mature	Cooler, unstable, fertile = low maturity	Ecosystem	
Total biomass efficient → High	Large wasting → Lower	Gross primary production	Biologic accommodation of planktic foraminifers
High assimilation *efficiency*	Large *reproductive* effort	Speciation tactics	Biologic accommodation of planktic foraminifers
K-selection = speciation into multiple niches of deeper waters	*r*-selection = favoring rapid population growth within upper oceanic layers	Speciation tactics	Biologic accommodation of planktic foraminifers
Numerous species; fewer individuals life cycle longer; greater size and complex morphotypes (keeled species)	Fewer species; numerous offspring; primitive, cosmopolitan, small sized morphotypes (Globigerina-like species)	Speciation tactics	Biologic accommodation of planktic foraminifers
Maximum diversity ↓ Complex communities, depth stratified	Low diversity ↓ Blooms of "opportunist" species	Rhythm of evolution	Biologic accommodation of planktic foraminifers
Periods of stability e.g. Late Cretaceous	Periods of (re)colonization eg Middle Jurassic, Early Cretaceous, Early Paleocene	Geologic record	

Figure 12.48 Environmental stress and speciation tactics among early planktic foraminifera. (Redrawn from Caron M, Homewood P. *Marine Micropaleontol* 1983; 7:453–62.)

Table 12.3 Average species duration (My) of benthic foraminifera on the North American Atlantic continental margin. (Reproduced from Buzas MA, Culver SJ. *Science* 1984; 225:829–30)

	Species (No.)	Duration (My)
Entire Atlantic continental margin	131	20
Less than 200 m	75	16
More than 200 m	33	26
At all depths	23	25
Less than 200 m (Florida to Newfoundland)	24	18
Less than 200 m (Florida to Cape Hatteras)	30	20
Less than 200 m (Cape Hatteras to Newfoundland)	21	7

foraminifera: the mid-Cretaceous (Aptian to Turonian), the Paleocene to Middle Eocene, and the Early Miocene (Fig. 12.49). Following the initiation of each phase, morphologic radiation produced forms that were **homeomorphic** of taxa in other phases of evolutionary radiation.

Following the Middle Cretaceous radiation, keeled trochospiral planktic foraminifera (e.g., *Globotruncana*) became abundant and diverse. Several evolutionary lineages are valuable in biostratigraphy (Fig. 12.42). The

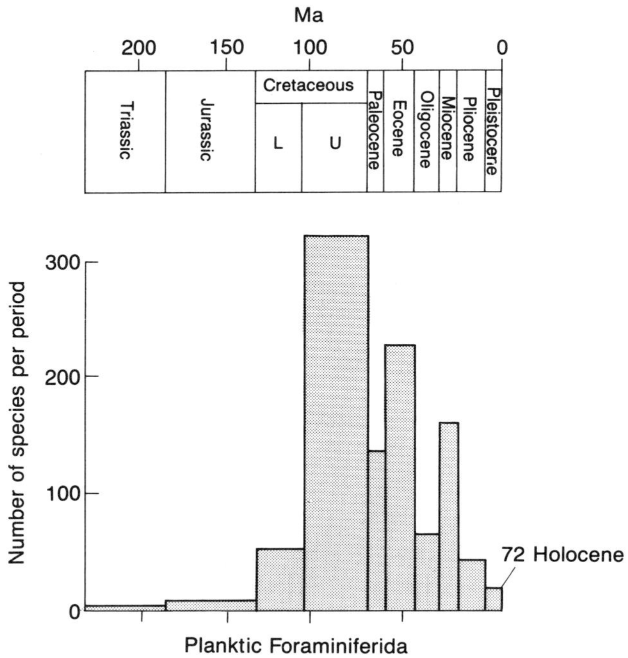

Figure 12.49 Species diversity in planktic foraminifera. The Early Paleocene assemblage has far fewer species than is suggested by the figure. (Redrawn from Tappan H, Loeblich AR Jr. *Earth-Sci Rev* 1973; 9:207–40.)

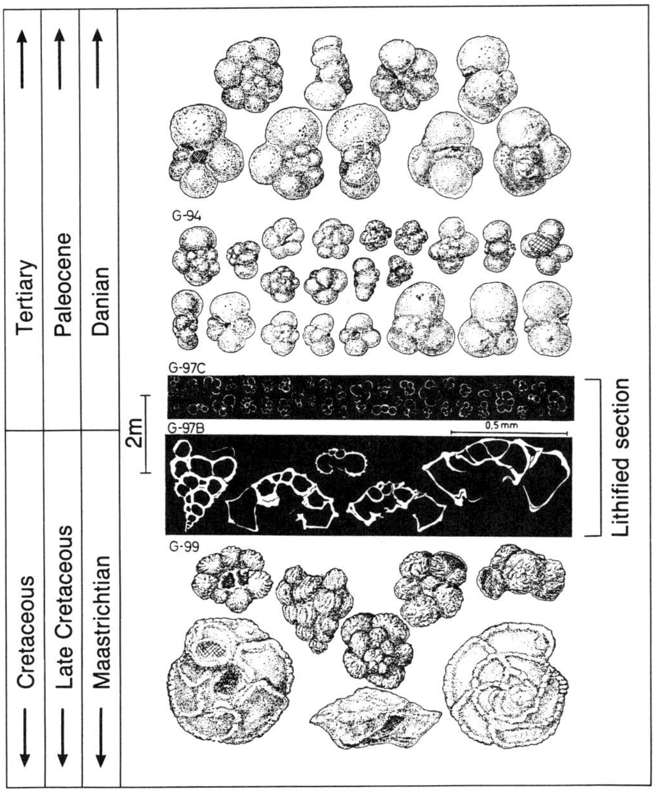

Figure 12.50 Planktic foraminiferal changes across the Cretaceous–Tertiary boundary at Gubbio, Italy. Large ornate globotruncanids become extinct and are succeeded by an early Danian fauna composed of small, unkeeled globigerinids. (Modified from Luterbacher H, Premoli Silva I. *Riv Ital Paleontol* 1964; 70:67–128.)

Cretaceous planktic foraminifera suddenly became extinct at the Cretaceous–Tertiary boundary. Late Maastrichtian assemblages, dominated by large, keeled globotruncanids, are vastly different from those of the early Danian which consist of small, unkeeled, trochospiral, and biserial forms (Fig. 12.50). According to some workers, all but one species (*Guembelitria cretacea*) became extinct at the Cretaceous–Tertiary boundary. More recent work in Texas and Tunisia incorporates microstratigraphic sampling covering a period of some 200 000 years which spans the boundary. At this scale elements of planktic faunas appear to have undergone extinction before, at, and after the Cretaceous–Tertiary boundary. Ten or more "Cretaceous" planktic species range into the Tertiary (Fig. 12.51).

Although some researchers do not agree with this interpretation, the Tertiary planktic faunas are now generally considered to have descended from just two, small, planktic species that survived the Cretaceous–Tertiary extinction event. Despite intense study of the boundary

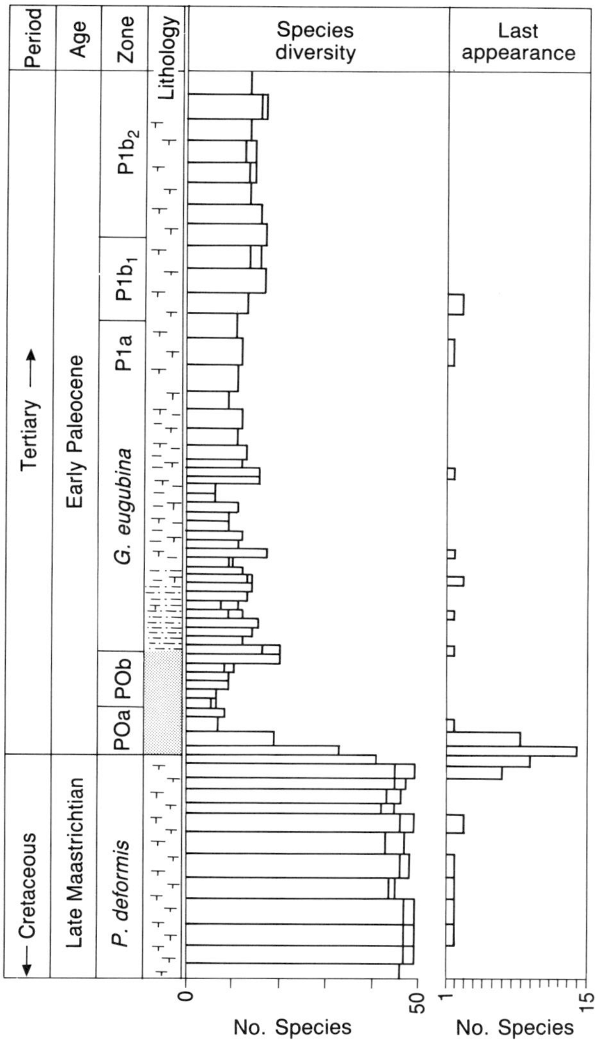

Figure 12.51 Species diversity and last appearance of planktic foraminiferal species in Late Maastrichtian to Early Paleocene sediments at El Kef, Tunisia. (Modified from Keller G. *Marine Micropaleontol* 1988; 13:239–63.)

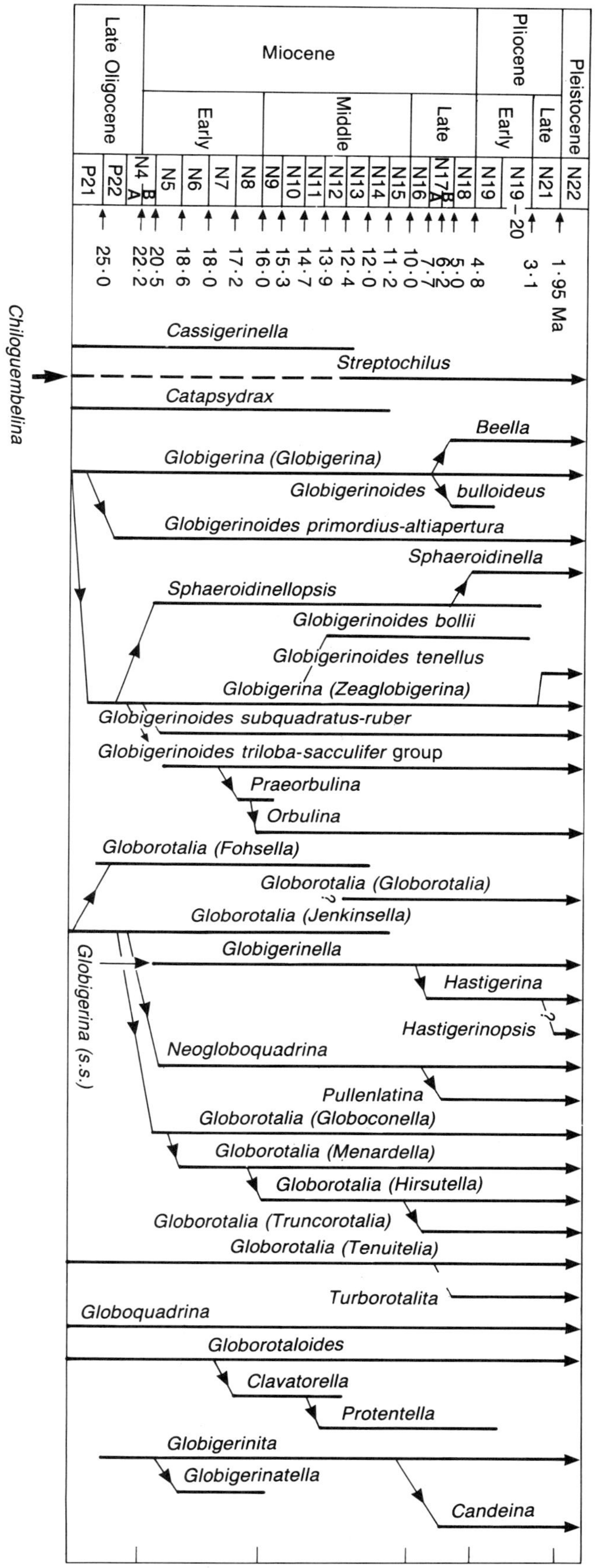

Figure 12.52 (*Right*) Lineages of Neogene planktic foraminiferal genera and subgenera. (Reproduced from Kennett JP, Srinivasan MS. *Neogene Planktonic Foraminifera: A Phylogenetic Atlas.* Stroudsburg: Hutchinson Ross, 1983:10.)

sequence, however, we are still far from understanding the nature of the faunal changes at the Cretaceous–Tertiary boundary, let alone the cause of the catastrophic extinctions.

Planktic foraminiferal evolutionary lineages are well documented for the Paleogene and, in particular, for the Neogene (Fig. 12.52). These lineages form the basis for the high-resolution biostratigraphic zonation of the Cenozoic into some 42 zones. More recently, planktic foraminiferal data have been applied to studies of the mode and tempo of evolution. The data suggest that no simple model of evolution can be applied to planktic lineages. On the contrary, the few Neogene lineages subjected to detailed morphometric study show considerable variation in evolutionary pattern. To complicate matters further such patterns may be interpreted in different ways by different authors. The *Globorotalia conoidea–Globorotalia inflata* lineage shows **phyletic gradualism** according to some workers, but to others a punctuational pattern. These two views may result from observations of the lineage in different water masses. The gradual changes may have occurred in cool, subtropical water masses, whereas evolutionary branching occurred in warmer waters.

Morphometric analysis of the Early Pliocene *G. crassaformis* lineage suggests a gradual transition from *G. cibaoensis* to *G. crassaformis* although the rate of morphologic change increased. Similar analysis of the Late Neogene *G. tumida* lineage indicates periods of relative morphologic stasis interrupted by episodes of relatively rapid evolutionary change. The term **punctuated gradualism** was introduced to describe this pattern (Fig. 12.53).

Not all species and genera of planktic foraminifera have known ancestors. For example, *Hantkenina* (Fig. 12.9G), appearing abruptly at the base of the Middle Eocene, is probably a descendent of *Pseudohastigerina*, although no intermediaries are known.

Both phyletic gradualism and **punctuated equilibrium** have been demonstrated to characterize the evolutionary history of different planktic foraminifera. The morphometric analyses of various lineages, however, indicate that patterns of evolution in planktic foraminifera are

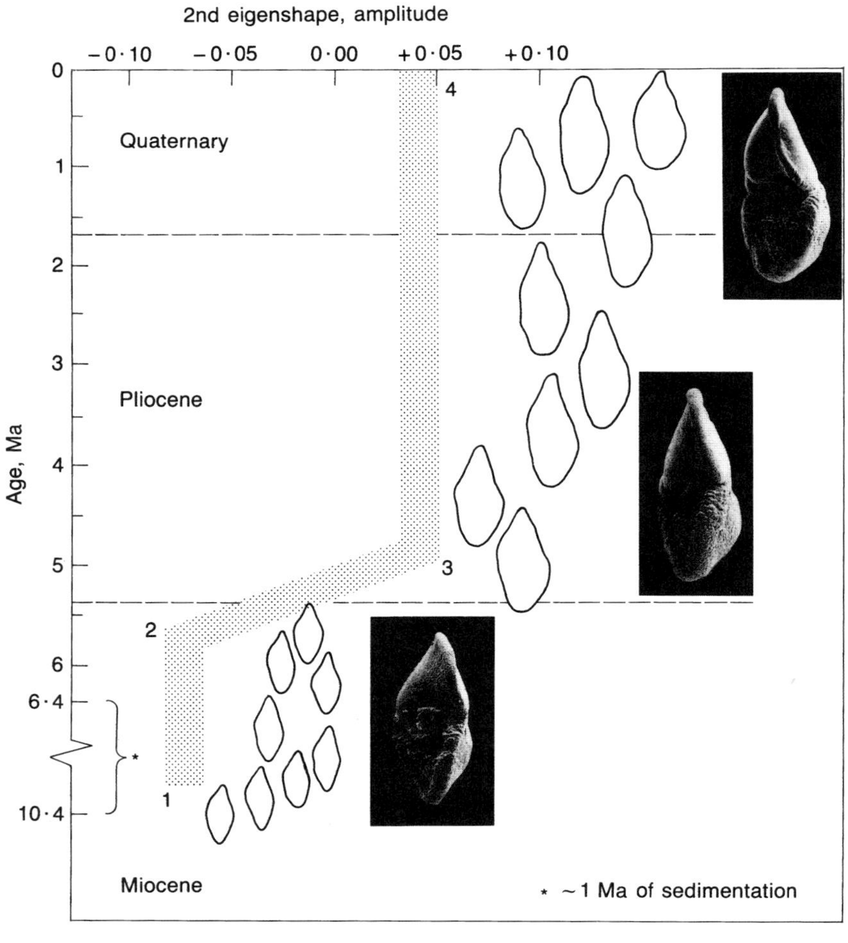

Figure 12.53 Variation in shape (2nd eigenshape) through time of the *Globorotalia tumida* lineage at Deep Sea Drilling Project Site 214, Indian Ocean. Stippled pattern indicates general trend of shape change. Variations of peripheral view outlines indicated for the latest Miocene (*Globorotalia plesiotumida*), Pliocene and Quaternary (*G. tumida*). Period of relative morphologic stasis (1–2) followed by rapid evolutionary change (over 0.6 Ma) (2–3) followed by renewed morphologic stasis (3–4). Scanning electron micrographs are of peripheral views of *G. plesiotumida* (latest Miocene, ×70), *G. tumida* (earliest Pliocene, ×55), *G. tumida* (Late Pleistocene, ×42.9). (Modified from Malmgren BA, Berggren WA, Lohmann GP. *Paleobiology* 1983; 9:377–89.)

much more complex than this simple dichotomous scenario.

Supplementary reading

Bolli HM, Saunders JB, Perch-Nielsen K. *Plankton Stratigraphy.* Cambridge: Cambridge University Press, 1985:1032. The most up-to-date text on the biostratigraphic use of planktic microfossils.

Boltovskoy E, Wright R. *Recent Foraminifera.* The Hague: W. Junk, b.v., 1976:515. Comprehensive text covering all aspects of modern benthic and planktic foraminifera.

Broadhead TW, ed. *Foraminifera: Notes for a Short Course.* University of Tennessee Department of Geological Sciences, Studies in Geology 6, 1982; 219. Brief reviews of Paleozoic through Recent foraminifera.

Cushman JA. *Foraminifera, Their Classification and Economic Use.* Cambridge, MA: Harvard University Press, 1948:605. General text on foraminifera written by the "father" of North American foraminiferal studies.

Douglass RC. *The Development of Fusulinid Biostratigraphy.* In: Kauffman EG, Hazel JE, eds. *Concepts and Methods of Biostratigraphy.* Stroudsburg, PA: Dowden, Hutchison and Ross, 1977:463–81. Summary of the use of fusulinids in biostratigraphy.

Hallock P. Why are larger Foraminifera large? *Paleobiology* 1985; 11:195–208. A discussion of the coevolution of algal symbiosis and growth to large size in stable, food-poor environments.

Haynes JR. *Foraminifera.* London: Macmillan, 1981:433. The most up-to-date general text, with emphasis on biostratigraphic utility, on foraminifera.

Hedley RH, Adams CG, eds. *Foraminifera.* Vols 1–3. New York: Academic Press, 1974:276, 1976:265, 1978:290. Review articles on various aspects of foraminiferal research.

Hemleben C, Spindler M, Anderson OR. *Modern Planktonic Foraminifera.* Berlin: Springer-Verlag, 1989:363. Comprehensive text on living planktic foraminifera.

Kennett JP, Srinivasan S. *Neogene Planktonic Foraminifera: A Phylogenetic Atlas.* Stroudsburg, PA: Hutchinson, Ross, 1983:265. Well-illustrated guide to evolutionary relationships of Neogene planktic foraminifera.

Lipps JH. Biotic interactions in benthic foraminifera. In: Tevesz MJS, McCall PL, eds. *Biotic Interactions in Recent and Fossil Benthic Communities.* New York: Plenum Press, 1983:331–76. A review of various aspects of benthic foraminiferal biology.

Lipps JH, Berger WH, Buzas MA, Douglas RG, Ross CA. *Foraminiferal Ecology and Paleoecology. SEPM Short Course No. 6.* Houston: Society of Economic Paleontologists and Mineralogists, 1979:198. Reviews on benthic and planktic foraminifera by several experts in each field.

Loeblich AR Jr, Tappan H. *Treatise on Invertebrate Paleontology, C: Protista 2, Sarcodina chiefly "Thecamoebians" and Foraminiferida.* Vols 1 and 2. Lawrence, KS: University of Kansas Press, 1964:900. Classification, description, and illustration of foraminiferal genera.

Loeblich AR Jr, Tappan H. *Foraminiferal Genera and their Classification.* Vols 1 and 2. New York: Van Nostrand Reinhold, 1988:970. Lavishly illustrated basic reference on foraminiferal genera; an update of the 1964 *Treatise.*

Ross CA, Haman D, eds. *Timing and Depositional History of Eustatic Sequences: Constraints on Seismic Sequences.* Cushman Foundation for Foraminiferal Research Special Publication 24, 1987:228. Case studies by various authors on micropaleontologic constraints on sequence stratigraphy.

Tappan H, Loeblich AR Jr. Foraminiferal evolution, diversification, and extinction. *J Paleontol* 1988; 62:695–714. An explanation of the patterns of foraminiferal evolution.

Radiolaria

13

Richard E. Casey

Introduction

Radiolaria are marine, **holoplanktic** protozoans belonging to the superclass Actinopodea of the subphylum Sarcodina. The term radiolaria itself is now used informally. In the fossil record, only the class Polycystinea (or **polycystine**) encompassing the orders Spumellar (**spumellarians**) and Nassellar (or **nassellarians**), which possess solid opaline skeletal structures, and the order Phaeodarea (or **phaeodarians**), which possess hollow skeletal structures of an admixture of silica and organic matter, are preserved. Radiolarian skeletons range in size from about 50 to 200 μm. The class Acantharea (or **acantharians**) are common in nearshore waters and are sometimes involved in plankton blooms, but they are never preserved as fossils. Herein, radiolaria refers to both members of the Polycystina and the Phaeodara.

The polycystine radiolarians have the longest geologic range (Cambrian to Holocene), the widest biogeography (pole to pole, surface to abyss), and the most diverse taxonomy of the well-preserved microzooplankton. They are used extensively in biostratigraphy, in paleoceanography, and in studies on the tempo and mode of evolution. Phaeodarians may be used to infer various water and geologic conditions.

History of study

The history of radiolarian research can be separated into six periods: an early recognition of the group in both living and fossil forms; work by Ernst Haeckel; early work on living forms; a resurgence of radiolarian work by William Riedel; radiolarian work associated with the **Deep Sea Drilling Project** (DSDP) and the **Ocean Drilling Program**; and the current diverse modern work.

In the early 1800s living and fossil radiolarians were first described. Ehrenberg developed the first radiolarian classification by the mid 1800s and called the group polycystina. A few years later Müller coined the term radiolaria. At the end of the nineteenth century Hertwig developed a classification based on their unique feature, the central capsule, and Haeckel published his extensive work on radiolaria including a new classification of radiolarians and a monographic report in the **HMS *Challenger*** expedition. His report, although out of date, was the principal radiolarian reference for almost 100 years. It notably described the radiolaria collected from the plankton and sediments during the *Challenger* expedition. It also contained a classification, material on the biology and zoogeography of radiolarians, and a complete account of all the species known to that time. Although Haeckel was the German champion of Darwin and an outstanding evolutionist in his own right, his radiolarian classification was very artificial and not at all natural or phylogenetic, perhaps partly because of the enormity of his task. Haeckel stated that he expected to have finished the *Challenger* report within a few years, that it took 10, and that a really comprehensive treatment could not be done in a lifetime. In the early 1900s radiolarian research concerned the zoogeography of living forms collected from major oceanographic expeditions. In the mid 1900s W.R. Riedel began modern radiolarian work in the USA by developing a Tertiary tropical radiolarian biostratigraphy and teaching, either directly or indirectly, most of the radiolarian students of that time in the USA and some from Europe. A similar development occurred in Russia at about the same time with Reshetnjak, Petrushevskaya, Lipman, Strelkov, and Zhamoida leading that effort. Radiolarians became useful with the development of deep-sea drilling in the late 1960s. Riedel's tropical biostratigraphic zonation that had been developed using conventional deep-sea coring techniques was just in time to aid the DSDP, and radiolarian specialists were part of the scientific parties of almost every DSDP cruise. Some of Riedel's students ventured into other aspects of radiolarian research, such as the distribution of living radiolarians in plankton and recent sediments, and into early attempts at paleoceanographic interpretations. J. Hays used radiolarians to trace the Antarctic Convergence and paleotemperatures through time and to develop a Neogene

Antarctic biostratigraphy that could be related to the paleomagnetic scale. The late 1960s to late 1980s was an exciting time with much being done to improve and correlate radiolarian biostratigraphies to magnetostratigraphies and other biostratigraphies and to use radiolarians as paleoceanographic indicators.

Perhaps some of the most important advances during this time were those of Roger Anderson and his work on living radiolarians. Anderson collected and maintained radiolarians in the laboratory and studied the living organisms. In 1983 Anderson published much of his work and a review of other radiolarian work in a book titled *Radiolaria*. In the late 1900s Mesozoic and Paleozoic radiolarians were studied in Europe, Russia, Japan, and the USA. In the western USA, Emile Pessagno and members of the United States Geological Survey sorted out suspect terranes using pre-Cenozoic radiolarians. In the summer of 1988 about 80 radiolarian specialists met in Germany for Interrad I (The First International Radiolarian Conference). Although those 80 probably are only a tenth of those involved in radiolarian research throughout the world, they represented more than 20 countries with specialists actively involved in radiolarian research.

Morphology and systematics

Morphology

Polycystine radiolarians are usually divided into spumellaria, with skeletons based on a spherical symmetry (Figs 13.1A, B, 13.2), and nassellaria, with skeletons with a nonspherical, usually bilateral, symmetry (Figs 13.1C, D, 13.3, 13.4). Nassellarians in turn are divided into *cyrtida* with general conical skeletons (Fig. 13.1C) and *spyroida* with skeletons possessing a usually distinct, D-shaped sagittal ring (Fig. 13.1D). The simplest spumellarian or nassellarian skeleton is a *spicule*: spumellarian spicules (Fig. 13.2A) can usually be distinguished from nassellarian spicules by being round, whereas the nassellarian spicules are usually facetted and possess a distinct organization of spines (Fig. 13.3A). Both spumellarian and nassellarian spicules can be distinguished from most sponge spicules because the radiolarian spicules are solid and do not possess the central space common to many sponge spicules. Spines are protuberances from more complicated radiolarian skeletons (Fig. 13.2B). The internal shells are referred to as *medullary shells*, and the outmost shell of a multishelled or single-shelled form is referred to as the *cortical shell* (Fig. 13.2B). *Bars* or

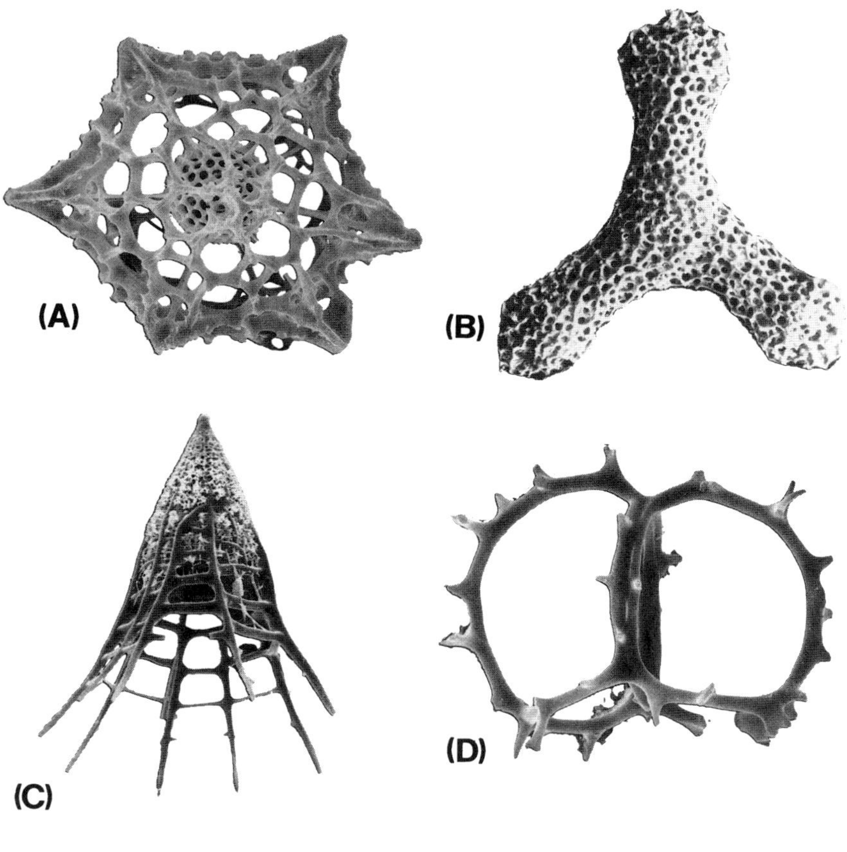

Figure 13.1 Polycystine radiolarians of the subgroups spumellaria (A, B) and nassellaria (C, D).

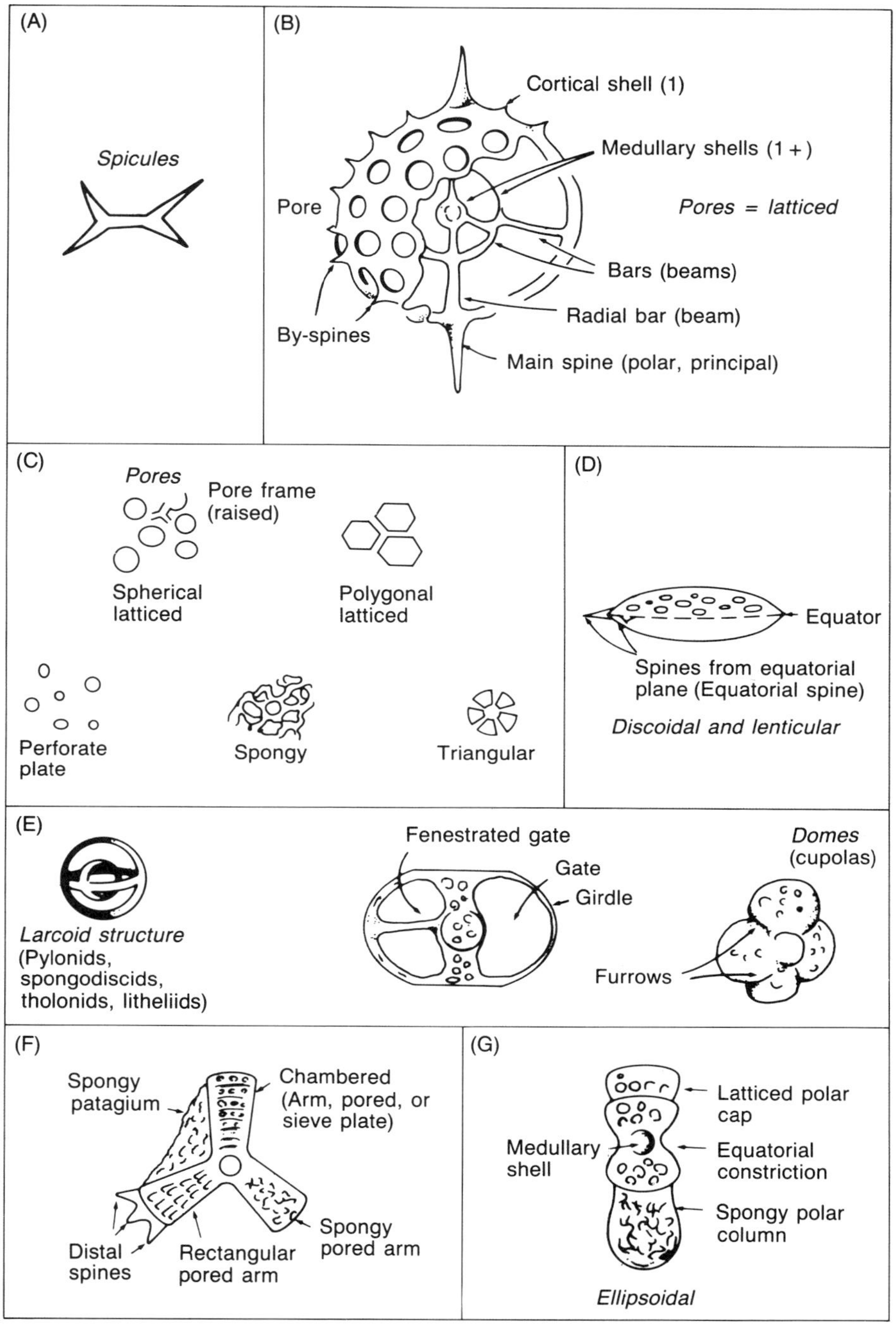

Figure 13.2 Spumellarian skeletal morphology. The letters are referred to in the text.

beams refer to the connections between shells and structures (Fig. 13.2B). Main (or polar or principal) spines are external projections supported by inner radial bars, whereas those spines not supported by bars are called by-spines (Fig. 13.2B). Pores may take on many different forms and arrangements (Fig. 13.2C) and may differ between shells of a multishelled form. The shape and arrangement of pores, ornamentation (such as spines around pore edges), and the amount of surface area taken up by either pores or the inner-pore space may be of taxonomic importance. The general shapes of spumellarian skeletons range from spherical (commonly domed or spiraled) to discoidal to various extensions from that spherical or discoidal plan such as ellipsoidal to armed (Fig. 13.2D). A very complicated coiled structure with gates (holes) and girdles (latticed structure) is referred to as *larcoid* (Fig. 13.2E). The *patagium* is usually a spongy structure that connects arms and appears to develop as the spumellarian ages (Fig. 13.2F). The arms themselves may be chambered or spongy, and may possess such

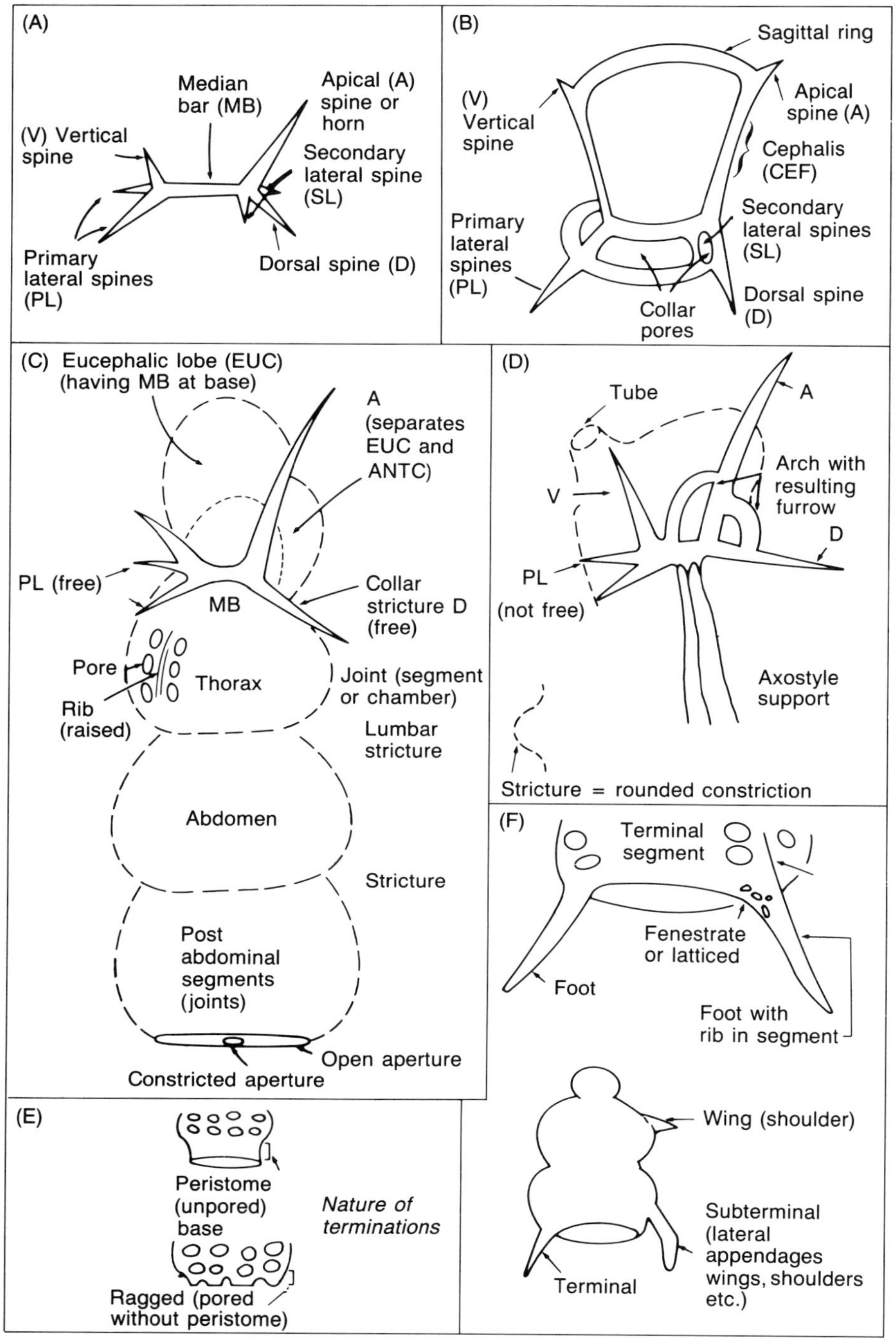

Figure 13.3 Nassellarian skeletal morphology. The letters are referred to in the text.

features as distal spines (Fig. 13.2F). Some ellipsoidal forms may have an equatorial constriction (Fig. 13.2G). The distal portions of some ellipsoidal forms may possess latticed polar caps, spongy polar columns, or both (Fig. 13.2G).

Nassellarian skeletal terminology (Fig. 13.3) is even more complex than spumellarian terminology. Here, as in the spumellarians, the simplest forms are spines (Fig. 13.3A). A skeletal form referred to as a sagittal ring or D-ring has an extension of the vertical and apical spines and their connection by a bar (Fig. 13.3B). These spines appear to be very orderly in arrangement. Other spines on the median bar are given specific names (Fig. 13.3A, B, C, D). In fact some classifications of radiolarians place considerable emphasis on the median bar and these spines, and ascribe not only systematic but also phylogenetic importance to these features (compare median bars and spines on Fig. 13.3A, B, C, D).

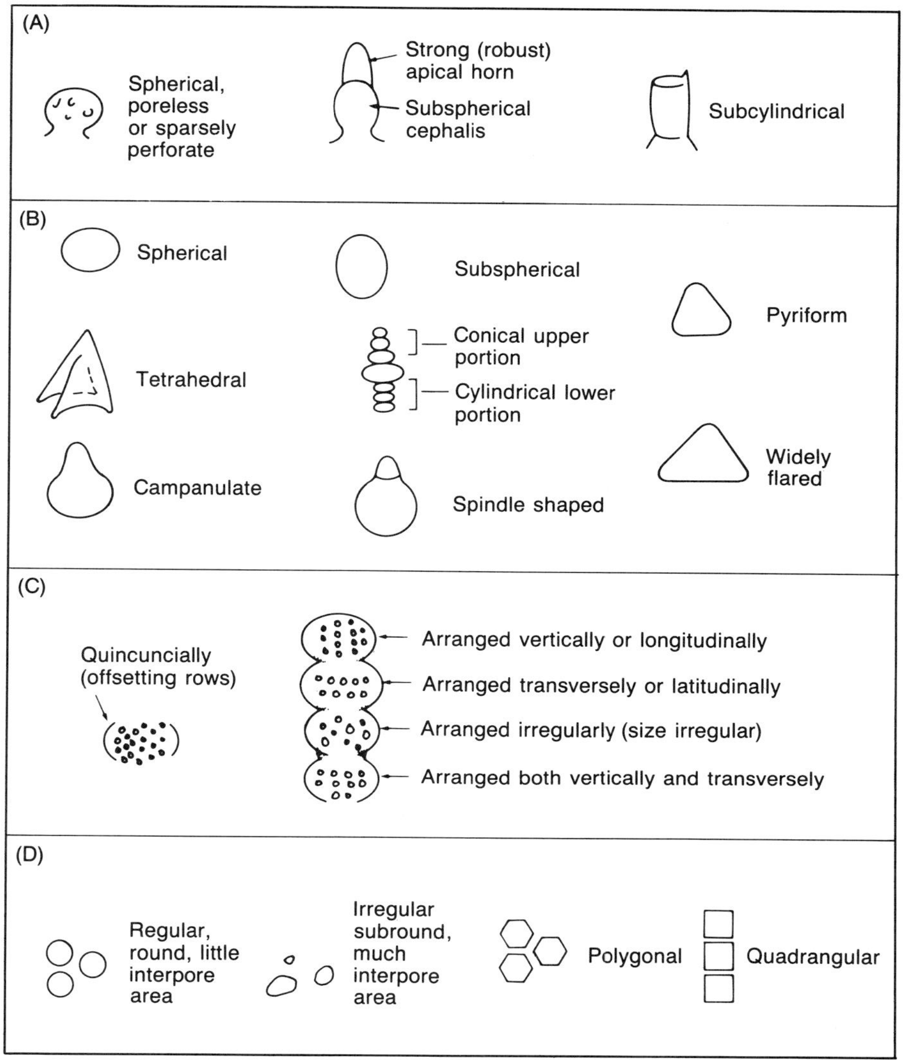

Figure 13.4 Nassellarian skeletal morphology continued. The letters are referred to in the text.

This arrangement of median bar and associated spines is sometimes referred to as the collar structure, which is the bar arrangement within the collar stricture (or collar constriction labeled a collar stricture on Fig. 13.3C) for it rests between the **cephalis** (head) and **thorax** (the second joint or chamber). In fact the spines from the median bar may divide the cephalis into smaller units such as the eucephalic and antecephalic lobes, or they may be associated with other cephalic or collar features such as the eucephalic and antecephalic lobes, or tubes (Fig. 13.3D). The median bar may possess downward projections referred to as axostyle supports (Fig. 13.3D). The joint or segment following the thorax is the **abdomen** and the two are separated by the lumbar stricture (Fig. 13.3C). All remaining segments are referred to as post-abdominal, and the terminal segments may possess either an open or constricted aperture (Fig. 13.3C). This terminal segment may also possess a peristome (basal ring) that may be ragged or poreless (Fig. 13.3E). Subterminal appendages may be referred to as wings or shoulders depending on which segment they are attached to; terminal appendages are referred to as feet (Fig. 13.3F). The cephalis may be spherical or subspherical or elongate, may possess processes (horns, spines, etc.) and may be perforate (pored) or poreless (Fig. 13.4A). The segments or the entire shell may have shapes that are important taxonomically (Fig. 13.4B). The pores on the segments may be arranged in a number of different fashions, commonly of taxonomic importance (Fig. 13.4C). The pores themselves may be round with little interpore area or irregular with much interpore area, or quadrangular (Fig. 13.4D).

Systematics

Although various classifications of radiolaria had been

proposed since Ehrenberg in 1847, modern classifications have only recently developed, chiefly by paleontologists. These classifications are based primarily on the skeletal morphology of Recent and Cenozoic forms and are biased towards characters that preserve in fossils. Mesozoic and Paleozoic radiolaria differ from Cenozoic ones and separate classifications have been established for these, but they have not been integrated together. The higher radiolarian taxa in this chapter therefore are referred to informally (Fig. 13.5). These informal categories are closest to families and are similar to others in common use now.

Today most radiolarian workers use a combination of Riedel's and Petrushevskaya's classifications. In fact most use the combined scheme of Nigrini and Moore's and Nigrini and Lombari's works, which are used in catalogs of modern and Miocene radiolarians, respectively. Table 13.1 gives the combined classification of Nigrini and Moore. These were primarily devised as aids to the Climate Long Range Investigation Mapping and Planning (**CLIMAP**) program and the Cenozoic Paleoceanography (**CENOP**) program, which are discussed in the section "Paleobiology" below. However, these classifications are based primarily on the skeletons of Recent and Cenozoic radiolarians and were conceived by micropaleontologists, which presents at least two major biases; one toward the Cenozoic and another toward fossilized characteristics. Because most fossil radiolarian workers are Cenozoic, Mesozoic, or Paleozoic specialists, little effort has been expended on combining the systematic developments within each of the eras. The higher radiolarian taxa in this chapter, therefore, will be referred to on an informal basis. Figure 13.5 gives the informal higher taxonomic units used in this chapter; these units are close to family level. This informal classification is similar to others currently in use.

Table 13.1 Combined polycystine radiolarian classification scheme. (Reproduced from Nigrini C, Moore TC. *A Guide to Modern Radiolaria*. Cushman Foundation for Foraminiferal Research, Special Publication 16, 1979; 342)

Subclass Radiolaria (Müller 1858)
Order Polycystina (Ehrenberg 1838) emend. (Riedel 1967)
Suborder Spumellaria (Ehrenberg 1875)
Family Collosphaeridae (Müller 1858)
Family Actinommidae (Haeckel 1862) emend. (Riedel 1967)
Family Sponguridae (Haeckel 1862) emend. (Petrushevskaya 1975)
Family Phacodiscidae (Haeckel 1881)
Family Spongodiscidae (Haeckel 1862) emend. (Riedel 1967)
Family Pyloniidae (Haeckel 1881)
Family Litheliidae (Haeckel 1862)
Suborder Nassellaria (Ehrenberg 1875)
Family Plagoniidae (Haeckel 1881) emend. (Riedel 1967)
Family Trissocyclidae (Haeckel 1881) emend. (Goll 1968) = Acanthodesmiidae (Haeckel 1862 in Riedel 1971)
Family Carpocaniidae (Haeckel 1881) emend. (Riedel 1967)
Family Theoperidae (Haeckel 1881) emend. (Riedel 1967)
Family Pterocoryidae (Haeckel 1881) emend. (Riedel 1967)
Family Artostrobiidae (Riedel 1967) emend. (Foreman 1973)
Family Cannobotryidae (Haeckel 1881) emend. (Riedel 1967)

Biology

Radiolarians are holoplanktic marine protists. The main radiolarian subgroups, the polycystines and phaeodarians, are paleontologically important for they are preserved in the fossil record. The acantharian radiolarians, although of interest biologically, never leave a fossil record.

The living individual

Living radiolarians are not well known biologically because they are difficult to maintain under laboratory conditions. Although most radiolarians with skeletons exist as single, individual cells, some, such as the collosphaerids, are colonies centimeters long. In some of the skeletonless colonial forms the maximum length may be several meters. The central capsule is a unique feature of radiolarian soft-part anatomy. It divides the endoplasm (intracapsulum) and ectoplasm (extracapsulum), separates the radiolaria from other organisms, and is used to divide the radiolaria into major subgroups. One or many nuclei are found within the endoplasm of the central capsule as are a number of other subcellular structures.

In the ectoplasm of many spumellarians the morphology of the skeleton is completely concealed by alveoli or bubble-like fluid or gas vesicles. This froth can be so extensive as to make the radiolaria visible to the naked human eye and has allowed them to be collected by scuba divers. In these radiolarians the function of the skeleton, if any, apparently is internal and not external. However, in many spumellarians and most nassellarians, the skeletal morphology is reflected in the outline of the living individual and it might have a hydrodynamic or feeding function. Many deep-living forms have elongated

POLYCYSTINES (skeleton of opaline silica)

Spumellarians (sphere)
Polycystines that are generally spherical or based on sphere.

1

Beloids (needle)
Skeleton of scattered spicules (1) that are usually circular in outline. Beloids can be distinguished from sponge spicules because beloids do not possess a central canal as sponges do. Range, Paleozoic to Recent; environment, shallow cosmopolitan.

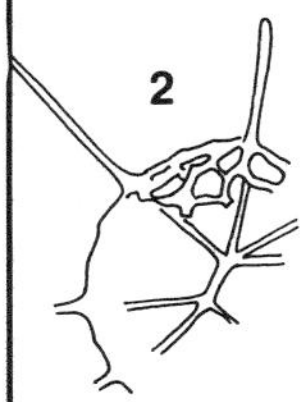

2

Entactinids (ray spine within)
Skeleton spherical and ellipsoidal with a latticed wall structure, distinguished from other groups by bars running to the center of the skeleton (2). Range, Cambrian to Permian (biostratigraphically important); environment, warm-water sphere.

3

Orosphaerids (mountainous sphere)
A large spherical or cup-shaped skeleton with a coarse, polygonal lattice and commonly club-shaped spines (3), distinguished by large size of complete specimens (a few mm) but in sediments usually find only pieces of lattice or whole spines (some Paleozoic forms resemble orosphaerids, some paleoactinommids and especially the rotasphaerids). Range, Eocene to Recent (could be biostratigraphically important); environment, probably all deep sea or cold-water sphere, occur in red clays as one of the few preserved fossils, possible biostratigraphic indicators in red clays.

4

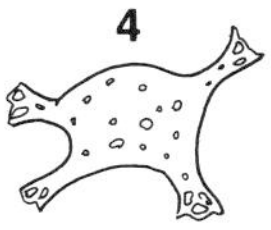

5

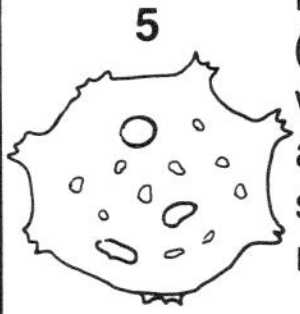

Collosphaerids (neck sphere)
Single spheres with usually more interpore area than pore area, many times with internal or external tubular or spindly external projections, can be distinguished from similar forms by the above characteristics and usually frail (not robust nature) (4, 5). Range, Lower Miocene (or Late Oligocene) to Recent (biostratigraphically important); environment, warm-water sphere, dominate in oligotrophic anticyclonic gyres, colonial and possessing symbiotic algae (zooxanthellae). A few more robust forms appear to be cold shelfal tolerant.

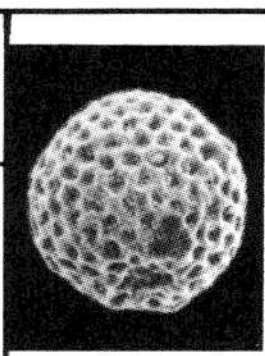

Actinommids (ray spined)
A catchall polyphyletic group containing generally spherical forms (some rough subgroups below). Range of "group," Paleozoic?, Triassic to Recent.

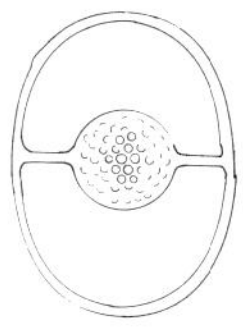

6

Saturnalins (like Saturn)
Spherical latticed skeleton possessing a circular outer ring connected to a latticed or sometimes spongy shell (6). Range, Triassic to Recent forms with ornamented outer ring appearing to be restricted to the Mesozoic; environment, warm-water sphere.

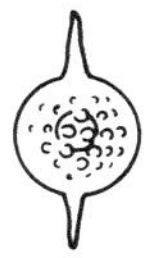

7

Stylosphaerins (pillar sphere)
Latticed shells with polar spines, one or two (the saturnalins are probably a subgroup of this group) (7). Range, Paleozoic? to Recent; environment, cosmopolitan.

8

Cubosphaerins (six-spined sphere, in cube pattern)
Latticed shells with polar and equatorial spines (six spines) (8) (tripospherids, 3 spines; staurosphaerids, 4 spines; pentasphaerids, 5 spines). Range, Paleozoic to Recent; environment, cosmospolitan.

9

Astrosphaerins (star sphere)
Latticed shells with numerous spines (9). Range, Paleozoic to Recent; environment, cosmopolitan.

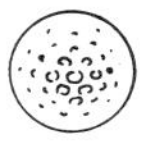

10

Cenodiscids (empty disk)
Lattice shell (single) without obvious spination, can usually be distinguished from collosphaerids by having less interpore area, thicker skeleton, and no tubes (10). Range, Cambrian to Recent; environment, wide range of environments but great abundance in sediment samples usually means the sample reflects cold surface waters or has undergone considerable dissolution.

11

Liospherids (layered sphere)
Lattice shell (multiple) without obvious spines (11). Range, Paleozoic to Recent; environment, warm-water sphere, symbiotic algae, shelfal tolerant.

Figure 13.5 Informal taxonomic units, descriptions, and characteristics of those units, used in this chapter. The informal taxonomic units are in bold or italics followed by the English meaning of that usually Latin name in parentheses, followed by a description of the group, its geologic range, and the environment it lives in or is inferred to have lived in.

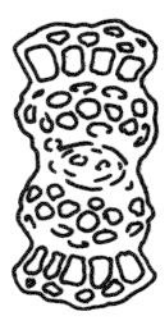

12

Artiscids (loaf of bread)
Elliptical outer (cortical) latticed shell many times with an equatorial constriction and caps and / or spongy columns (12). Range, Oligocene to Recent (biostratigraphically important); environment, warm-water sphere, symbiotic algae, shelfal tolerant.

13

Phacodiscids (lens disk)
Discoidal, biconvex, latticed outer shell (13). Range, Mesozoic to Recent, environment, warm-water sphere.

14

Coccodiscids (kernal disk)
Lens-shaped with a latticed center and spongy chambered girdle or arms (14) (orbiculiformids and maybe cavaspongidae included here). Range, Mesozoic to Oligocene; environment, warm-water sphere.

Spongodiscids (spongy disk)
Polyphyletic group of generally discoidal and spongy nature. Group range, Devonian to Recent (subgroups below).

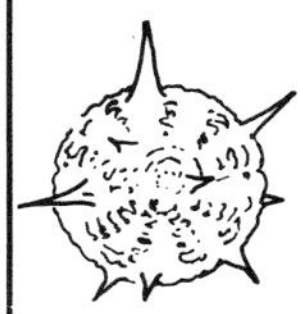

15

Spongotrochins (spongy wheel)
Spongy lens (sometimes with evidence of "rings") and many times with large numerous spines (15). Range, Mesozoic (maybe Paleozoic) to Recent (includes orbicuformids?); environment, commonly cold-water sphere, good indicators of upwelling.

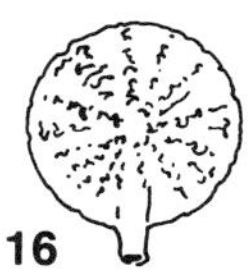

16

Spongopylins (spongy with tube)
Spongy lens without spines but with a pylome (16). Range, Mesozoic to Recent, environment, cold-water sphere.

17

18

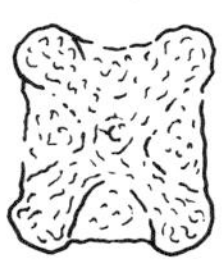

19

Spongasterins (spongy stomach)
Spongy disk with margins faceted and radiating (17, 18, 19). Range, Paleocene to Recent (Cretaceous?) (biostratigraphically important). Environment, warm-water sphere, tolerant of nearshore conditions (indicator of such), with symbiotic algae.

20

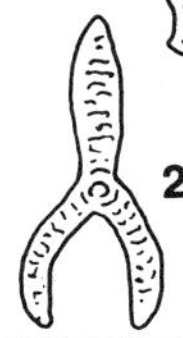

21

Dictyocorins (netted head)
Spongy disk with three arms (chambered or spongy) (20, 21). Range, Mesozoic to Recent; environment, mainly warm-water sphere, common, (dictyocorins) (20) not euchitonids (21) in eutrophic environments (indicative of such), some with symbiotic zooxanthellae (there are cool and warm morphotypes).

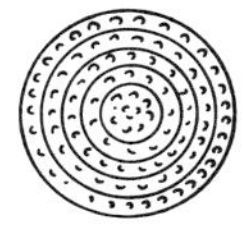

22

Stylodictins or *Porodiscins* (chambered or pored net or disk)
Disk with rings of lattice, may be a subgroup of spongodiscids (22). Range, Permo–Triassic to Recent; environment, appear to be mainly subsurface and cold-water forms, high latitude and upwelling.

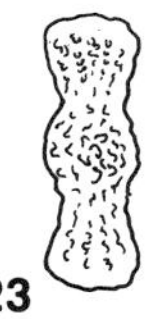

23

Spongurins (spongy)
Spongy cylinder (amphirhopalins and prunobrachids included here) (23). Range, Mesozoic to Recent; environment, cosmopolitan, cold and warm morphotypes.

24

Phaseliformins (pouch form)
Subellipsoidal spongy (24). Range, Late Cretaceous; environment, ?

25

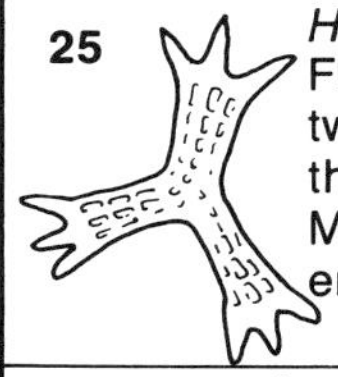

Hagiastrids (holy star)
Flat "disk" with spongy-rectangular mesh and two to four (usually three) radial arms (large) that may possess spines (25). Range, Mesozoic (biostratigraphically important); environment, warm-water sphere.

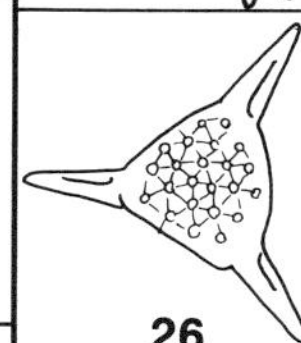

26

Pseudoaulophacids (false tent lens)
Lens-like or more commonly triangular spongy and triangular mesh, usually with a few prominent marginal spines (26). Range, Mesozoic; environment,?

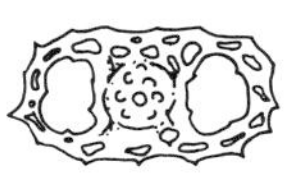

27

Pylonids (gateways)
Shell an ellipsoid of girdles with gates (holes) (27). Range, Eocene to Recent; environment, mainly warm-water sphere.

28

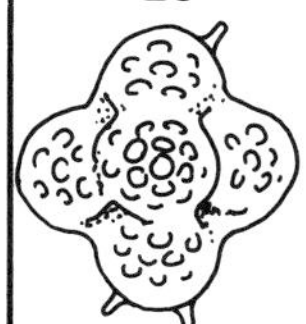

Tholonids (domes)
Outer (cortical) shell elliptical with bulb-like extensions of the shell and some spines (28). Range, Miocene to Recent; environment, deep and/or cold-water sphere (in intermediate waters).

29

Lithelids (sphered)
Coiled latticed shell (29) (includes larcospirins and sphaeropylins). Range, Carboniferous to Recent, environment; cosmopolitan with cold water morphotype tightly coiled and warm water sphere morphotype loosely (more open) coiled.

30

Paleoactinommids (old spiney one)
All the Paleozoic Spumellaria with single latticed shell or two or more concentric shells and lacking internal spicular system of entactinids and characteristic lattice structure of rotasphaerids.

Nassellarians (nose)
Polycystines that are generally cone-shaped.

Plagonids (snares)
Polyphyletic group of single forms, spicule to two joints.

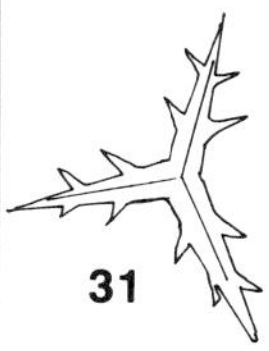

31

Plagiacanthins (thorn snare)
Thorax and sometimes walls of cephalis reduced, sometimes to the extreme of only spicules remaining; these "spicules" can be distinguished from beloids in that the plagiacanthins usually possess faceted spines and the beloids round spines (31). Range, Miocene to Recent (but "spines" in Mesozoic and Paleozoic); environment, cosmopolitan in surface and subsurface.

32

Lophophaenins (open neck)
Cephalis and thorax about equally developed, usually two lobes (or three) in cephalis; apical, dorsal and primary spines usually obvious and extending beyond lattice shell (32). Range, Cretaceous to Recent; environment, cosmopolitan, mainly shallow but also deep members (the antarcticins appear to be good shallow Antarctic forms, and so far are the only radiolarians recognized from bottom water in the northern hemisphere).

33

Sethoperins (sieve bearer)
Cephalis and six latticed plates extending from cephalis, or modified as a "basket"-like thorax (33). Range, Mesozoic to Recent; environment, warm-water sphere.

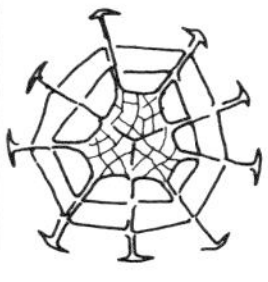

34

Sethophormins (sieve basket)
Cephalis large, hemispherical with a flattened umbrella-shaped thorax (34). Range, Cretaceous to Recent; environment, cosmopolitan but many in cold-water sphere.

35

Cyrtentactins (cage with spine within)
Looks like "cross" between entactinid (spumellarian) and lophophaenin (nassellarian), may be earliest nassellarian (35) (pylenonemids included here). Range, Devonian to Carboniferous; environment, ?

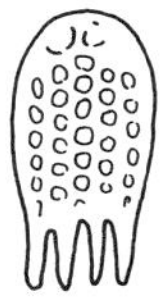

36

Carpocanids (fruit basket)
Cephalis small and recessed into the thorax. Thorax usually elongate and usually with longitudinally arranged pores (36) (williriedellids of the Mesozoic included here, popofskyellids probably here also). Range, Mesozoic to Recent; environment, warm-water sphere mainly.

Theoperids (devine pouch)
Small spherical, usually poreless cephalis, with one or more postcephalic chambers, usually only one small apical spine on top of cephalis.

37

Eucyrtidins (true cage)
Usually more than one postcephalic chamber (joint), and pores not quadrangular (37) (dictyomitrids, neosciadiocapsids, parvacingulids, spongocapsulids, syringocapsids, ultranaporaids, xitids. Range, Mesozoic to Recent (biostratigraphically important); environment, cosmopolitan but mainly a warm-water sphere.

38

Theocalyptrins (god-like sheath)
Helmet-shaped robust forms cold water (38), nonrobust warm-water sphere. Range, Mesozoic to Recent; environment, mainly cold-water sphere.

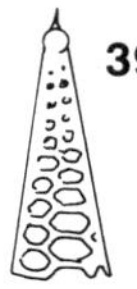

39

40

Plectopyramins (woven pyramid)
Usually only one postcephalic joint and this joint cone-shaped and usually with longitudinally arranged quadrangular pores (39, 40). Range, Mesozoic to Recent; environment, cosmopolitan but mainly deep or cold-water sphere (enhanced under high productivity and upwelling regions) divergences.

41

Amphipyndacins (double)
Small spherical, usually poreless cephalis with several postcephalic joints; cephalis divided into two vertical segments by transverse internal ledge (probably an eucyrtidin) (41). Range, Cretaceous to Early Tertiary (biostratigraphically important); environment, warm-water sphere mainly (a few living "similar" forms are deep sea).

42

Pterocorids or *Pterocorythids* (winged helmet)
Large elongate cephalis with one or more postcephalic chambers; cephalis usually consists of a main lobe (eucephalic) or two lateral lobes separated by furrows; the cephalis is pored and usually bears a large vertical spine (apical) that emerges from the side of the cephalis (42). Range, Late Cretaceous to Recent (biostratigraphically important); environment, warm-water sphere.

43

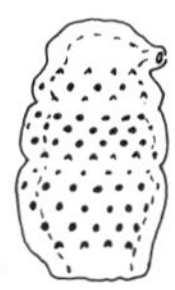

44

Artostrobids (loaf top)
Cephalis of eucephalic lobe, other lobes, and sometimes tubes; many postcephalic joints usually with latitudinally arranged pores; spines may sometimes be seen protruding from the collar structure into the postcephalic chamber inner area. Range, Cretaceous to Recent (may be good for cosmopolitan biostratigraphy); environment, cosmopolitan, warm water (43) and cold water (and deep water) (44), morphotypes can be distinguished (robust artostrobids appear to be upwelling enhanced).

45

Cannobotryids (tube cluster)
Cephalis of eucephalic lobe, other lobes which may extend as tubes; usually only one or two postcephalic joints; pores commonly appear to be randomly arranged (45). Range, Eocene (Cretaceous) to Recent; environment, cosmopolitan but more diverse in warm-water sphere; cold (deep) and warm-water morphotypes can be distinguished.

Acanthodesmiids (thorn bundle)
D-shaped ring or latticed, bilobed chamber with internal D-shaped sagittal ring.

46

47

Acanthodesmins
D-shaped ring (46) or modification of such, or sagittal ring (D-ring) surrounded by two latticed lateral lobes (47); apical and basal poles not obvious. Range, Eocene to Recent; environment, mainly warm-water sphere, usually shallow and apparently with symbiotic algae.

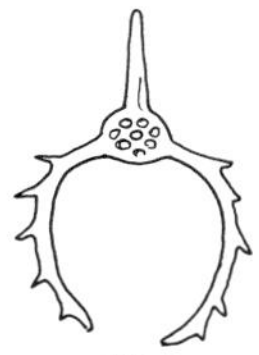

48

Triospyrins (three wraps)
Sagittal ring (D-ring) surrounded by two latticed lateral lobes, apical and basal poles obvious; commonly with basal feet and apical spines (sometimes very long) (48). Range, Paleocene to Recent (biostratigraphically important); environment, mainly warm-water sphere.

RADIOLARIA INCERTAE SEDIS (probably spumellarian-like)

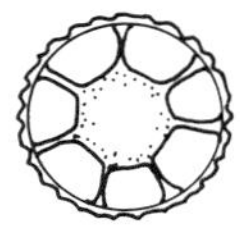

49

Rotaformids (wheel form)
Lens-shaped, central area enclosing what appears to be a nassellarian cephalis, this central area is connected to an outer ring by radial bars (49). Range, Cretaceous (biostratigraphically important); environment, ?

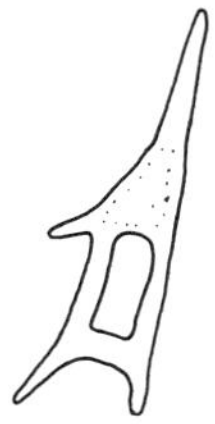

50

Albaillellids (slim ring)
Skeleton consisting of three spines in a plane forming a triangle. Rib-like elements may protrude from the spines giving the skeleton a jointed appearance, may be oldest nassellarian (50). Range, Silurian? to Permian (maybe Triassic) (biostratigraphically important); environment, ? (open ocean, perhaps deep living).

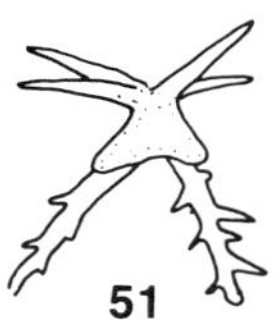

51

Palaeoscenids (old tent)
Skeleton consisting of four diverging basal spines with rib-like elements protruding, topped by two or four "apical" spines (51). Range, Devonian to Carboniferous (biostratigraphically important); environment, ?

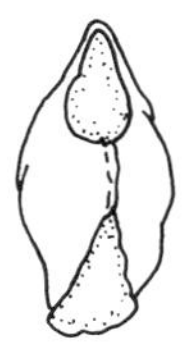

52

Corythoecids (God's helmet)
Tubular or conical, imperforate, siliceous shells, with one or two longitudinal ribs in opposite sides of the shell wall; one bilobed lamellar wing extends laterally from shell wall between the two ribs (52). Range, Late Paleozoic; environment, cold-water sphere, open ocean, deep?

53
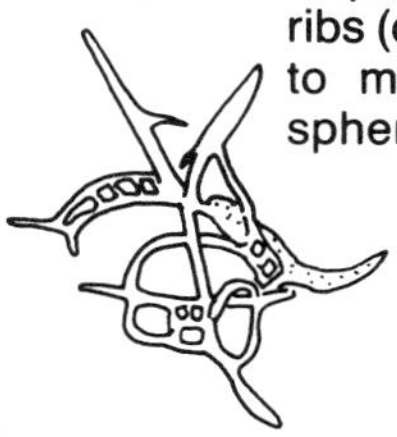

Ceratoikisids (thorny bow)
Cephalis-like structure with a main arch and ribs (down) and thorns (up) (53). Range, Early to mid-Paleozoic; environment, cold-water sphere, open ocean, deep?

54
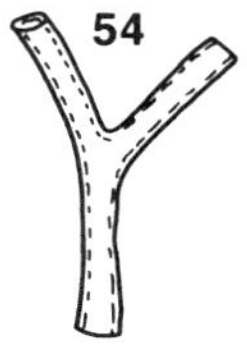

55
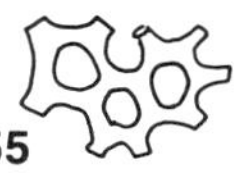

Phaeodarians (skeleton of mixture of opaline silica and organic matter)
Skeletons rarely preserved, and when preserved usually only as spines (54) or lattice (55). Range, Cretaceous to Recent (perhaps Permian or older). Environment, usually subsurface to deeper living forms. Presence in sediment indicates an environment of deposition low in oxygen, also useful as indices of thermal alteration for basin analysis.

or flattened skeletons, and these radiolarians commonly are attached to the outside of deep-sea **marine snow**, so an obvious functional morphology here may be attachment to enable digesting portions of the marine snow or associated bacteria. The skeletons probably function in many ways because they are so diverse. One main function may simply be ballast.

Axopoda or stiffened ray-like pseudopods radiate from the periphery of living radiolarians (Fig. 13.6A). Another structure that may radiate from the periphery of some spumellaria is an axoflagellum that is composed of many axopods and appears to originate from a specialized region in the central capsule wall (Fig. 13.6B). Within the ectoplasm of many spumellarians and nassellarians are symbiotic algae, all of which appear to be dinoflagellates (Fig. 13.6C, D). In a living colonial radiolarian these **symbionts** disperse in the light of day and contract back to the central capsules at night, probably via the rhizopodial network of the radiolarian.

Radiolarians reproduce by **fission** and possibly sexually by the release of tiny flagellated cells, called swarmers. Swarmers are formed from the **central capsule** after the ectoplasm has been discarded. The central capsule sinks hundred of meters deeper than where the adults live, forms a bulge which fragments, releasing the swarmers. However, there is no evidence that the swarmers ever come together to form new individuals, and so their role in sexual reproduction remains inferential only.

Most radiolarians have life-spans of a few weeks. In the laboratory radiolarian cultures may form swarmers within about 2 weeks. Field experiments using the flux of radiolarian skeletons into natural sediment traps (anoxic basins) and manmade sediment traps, compared to the standing crops of those same species in the overlying water, indicate life-spans from a few days to a few months for various species of radiolarians. The average, shallow-living radiolarian probably lives for about 2 weeks.

Formation of the skeleton

Most radiolarians possess skeletons. The polycystines have skeletons of opaline silica, the phaeodarians have siliceous skeletons containing much organic material and hence are not well preserved as fossils, and the acantharians

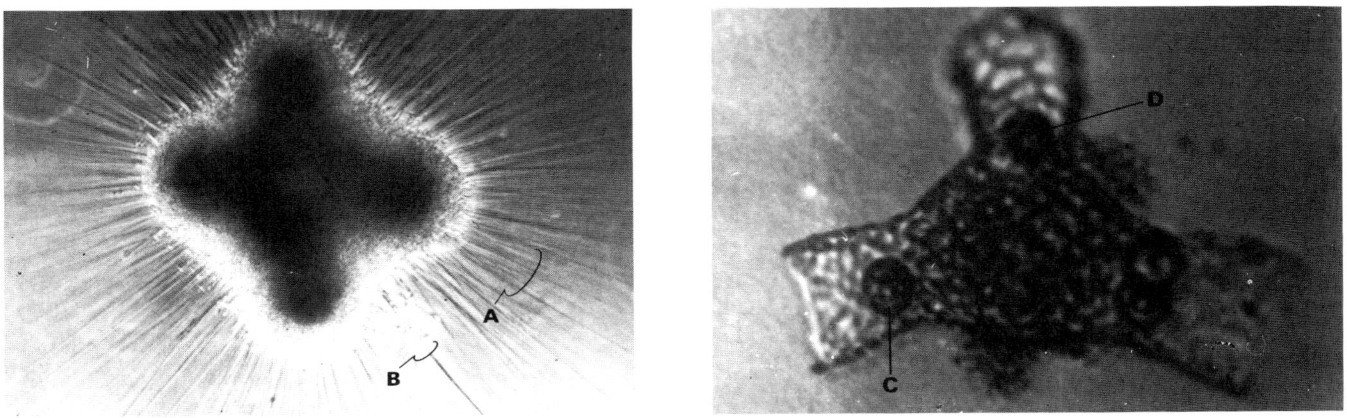

Figure 13.6 Soft-part anatomy of living radiolarians. The spongasterin exhibits fine axopods protruding from its periphery (A) and an axoflagellum (B). The collosphaerid contains symbiotic **zooxanthellae** (C, D). (Spongasterin from a slide courtesy of O.R. Anderson.)

have skeletons of strontium sulfate which do not preserve as fossils. Radiolarians are also able to repair damaged skeletons and to grow by adding to their skeletons. Skeletons from plankton tows and sediment samples, however, do not show a gradation from small juvenile to larger adult forms, indicating the growth is not by simple addition of skeletal material alone.

The skeleton is deposited within a cytoplasmic sheath. The sheath, in one case (Fig. 13.7), is just long enough to encase the central capsule indicating that the formation of an initial skeleton would probably protect the capsule and other vital organelles. The sheaths appear to be perforated (Fig. 13.7A).

Ecology

Radiolarians are **bacterivores**, **nannoherbivores**, **herbivores**, **omnivores**, **symbiotrophes**, **detritivores**, and maybe **osmotrophes**. The different trophic types live in different places in the oceans. The larger phaeodarian radiolarians live deep and are believed to be detritivores and bacterivores, as are the deep-living polycystines (Fig. 13.8). These deep-water forms are also more abundant underneath highly productive regions because their food comes directly (for detritivores) or indirectly (for bacterivores) from these productive waters above. The warm, shallow waters are dominated by symbiotrophes, herbivores, and omnivores. Herbivores are generally restricted to the upper 200 m of the column; symbiotrophes dominate the **oligotrophic** subtropical anticyclonic gyres and warm shelf and neritic-influenced regions; and both detritivores and bacterivores dominate the shallow subsurface, high-latitude regions. In these shallow waters, at depths where high concentrations of acantharian radiolarians occurred, polycystine radiolarians were absent; where shelled tintinnids, phaeodarians, planktic foraminifera, and pteropods were most abundant, the standing crops of both nassellarian and spumellarian radiolarians decreased. Polycystine radiolarian densities and diversities decline in the presence of high concentrations of diatoms in high-latitude, shallow waters and in the eastern boundary currents. Perhaps radiolarians decline in abundance due to direct competition for food sources, or are themselves a food source where tintinnids, phaeodarians, planktic foraminifera, and pteropods are abundant. Possibly radiolarians decline when competing with diatoms for dissolved silica, and perhaps blooming acantharians release an extrametabolite that inhibits polycystine radiolarians.

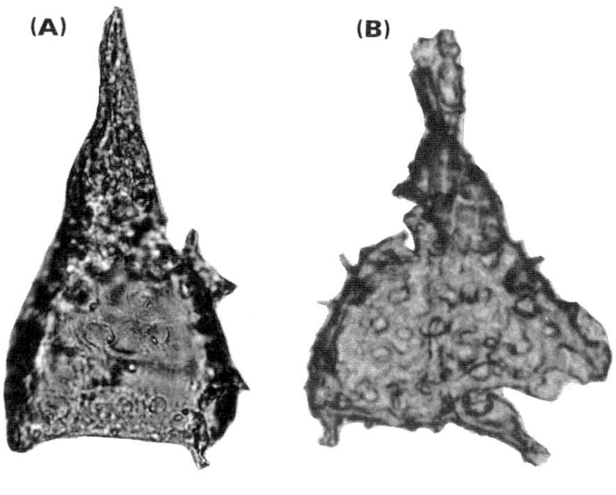

Figure 13.7 Suspected cytoplasmic sheath of a nassellarian. This suspected sheath was picked wet from a plankton tow sample. As the specimen dried the originally poreless sheath exhibited pores (A) that became more apparent as drying continued (B).

Distribution

Polycystine radiolarians are geographically the greatest ranging (pole to pole, surface to abyss) and taxonomically the most diverse of the well-preserved marine planktic protozoa. There are about 400 to 500 relatively common species living in the oceans. Approximately 200 species live in shallow (0–200 m), warm, central tropical waters; 40–50 in high-latitude (poleward of the subtropical or polar convergences), shallow (0–400 m), cool waters; 150–200 in water deeper than 200 m; and about 40–50 species are **eurybathyal**. Many of the deep-dwelling species appear to live in shallower water in high-latitude, cool seas, and submerge below the warmer tropical water. Most radiolarian species are also vertically stratified in the water column. Radiolarian zoogeography, like that of other plankton, is related to general oceanic circulation and water mass patterns (Fig. 13.9). Provinces with co-occurring radiolarian species correspond to water masses of similar characteristics. Where the water characteristics change, the radiolarian species also change. Thus, the chief radiolarian zoogeographic boundaries correspond to major current convergences in the subtropical and tropical regions and extensions of them at depth (Figs 13.10, 13.11). Most of the truly warm-water taxa are restricted to equatorward regions between the subtropical convergences, and most of the cold-water varieties of the warm-water taxa occur between the subtropical and polar convergences. In the modern ocean, about eight shallow-water, polycystine provinces and seven deep-water provinces are present (Figs 13.9–13.11).

The 15 provinces are characterized by different radio-

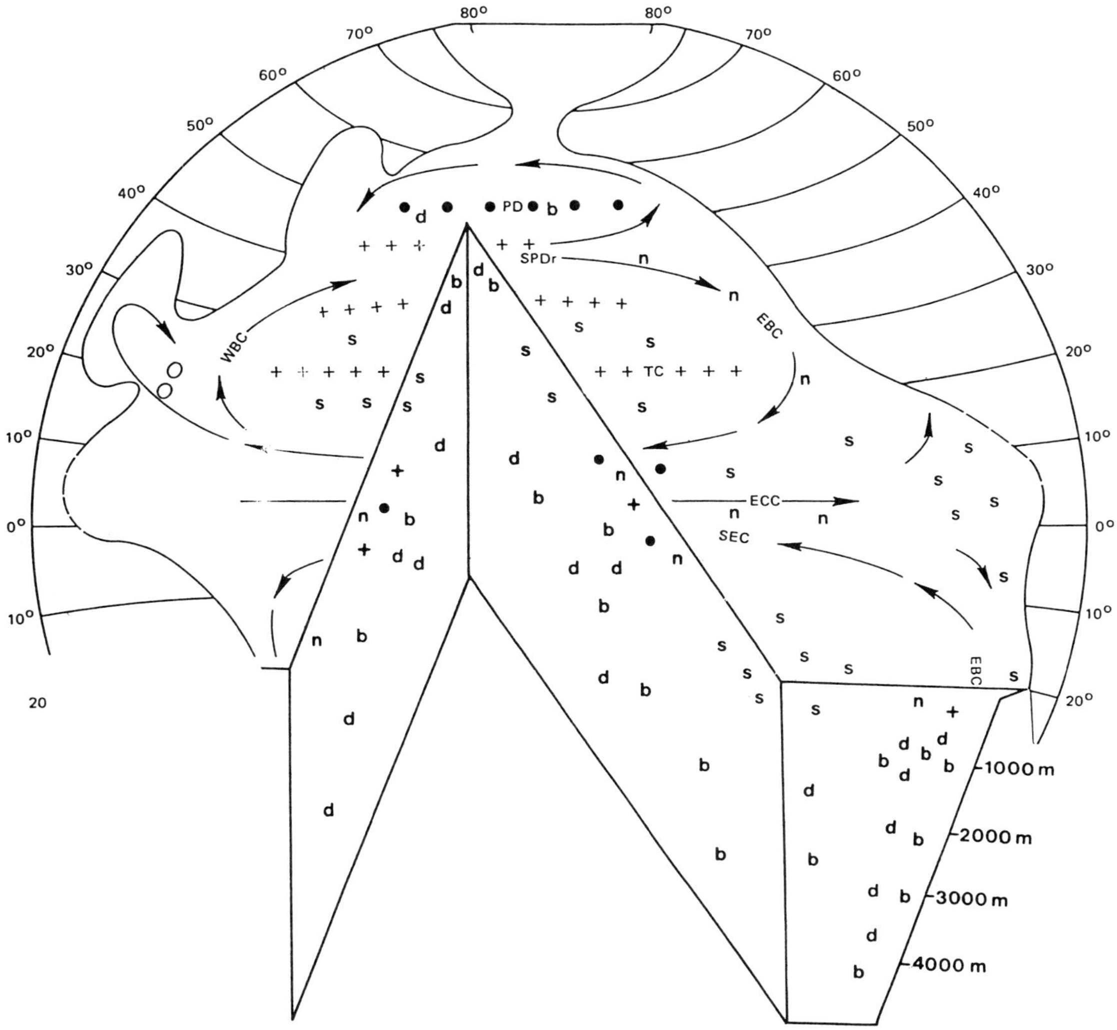

Figure 13.8 Radiolarian feeding behaviors and distributions depicted on an ideal ocean. The letters refer to the presence of the four basic feeding behaviors referred to in the text: n, nannoherbivore or herbivore; s, symbiotrophe; b, bacterivore; d, detritivore. The density of the packing of letters represents the relative density of the individuals (the relative standing crop) of each group. (Modified from Casey RE, Gust L, Leavesley A *et al. Trans Gulf Coast Assoc Geol Soc* 1979; 29:216–22.)

larians (Table 13.2). The subpolar cyclonic gyres contain relatively few species with few endemic forms in the northern gyre and a moderate number in the southern gyre. The Transition water mass contains a moderately diverse fauna dominated by cold-water varieties of warm-water species and few endemic forms. The subtropical, anticyclonic gyre province exhibits the highest radiolarian diversities (hundreds of species), densities (thousands per cubic meter), and endemism of any of the provinces. This province is dominated by radiolarians hosting symbiotic algae, and this association may be the reason they dominate these regions. Some evidence suggests that these symbiotic forms may vertically migrate to take advantage of both warm sunlit waters and the **nutricline** below. The equatorial province is very similar to the subtropical anticyclonic gyre province; however, it does not exhibit as high a diversity, density, or degree of endemism. The eastern tropical province is dominated by symbiont-bearing radiolarians like the subtropical anticyclonic gyre, but unlike the gyre, there are few endemics and only low to moderate diversities and densities.

The deep-water provinces appear mainly to be related to the water masses found at depth (Figs 13.10, 13.11). The Central water masses are mainly formed by winter cooling of western boundary current waters in the shallow Transition region; these waters then dive below tropical waters at the subtropical convergences. They are inhabited by a radiolarian fauna of moderate diversity, low density, and a moderate number of endemic species. The Polar–Circumpolar waters appear to result from the

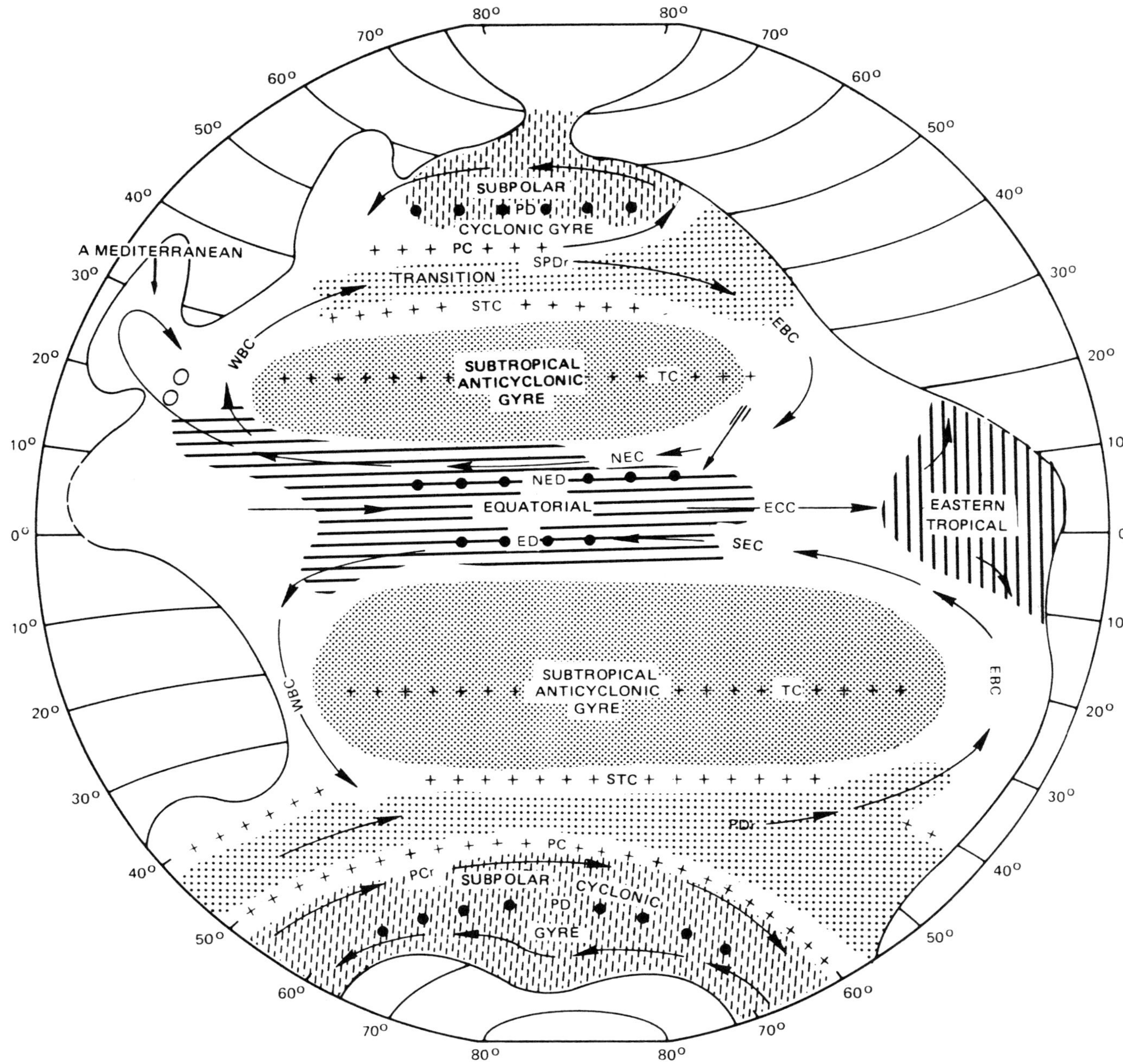

Figure 13.9 The eight major shallow-water radiolarian environments (provinces) of a modern ocean. EBC, eastern boundary current; ED, equatorial divergence; NEC, north equatorial current; NED, north equatorial divergence; PC, polar convergence; PCr, polar current; PD, polar divergence; PDr, polar drift; SEC, south equatorial current; STC, subtropical convergence; TC, tropical convergence; WBC, western boundary current; ●, water upwelling; +, water downwelling; ⇐, winds at surface of ocean; ←, currents at surface of ocean. (Modified from Casey RE. *Palaeogeogr Palaeoclimatol Palaeoecol* 1989; 74:15–22.)

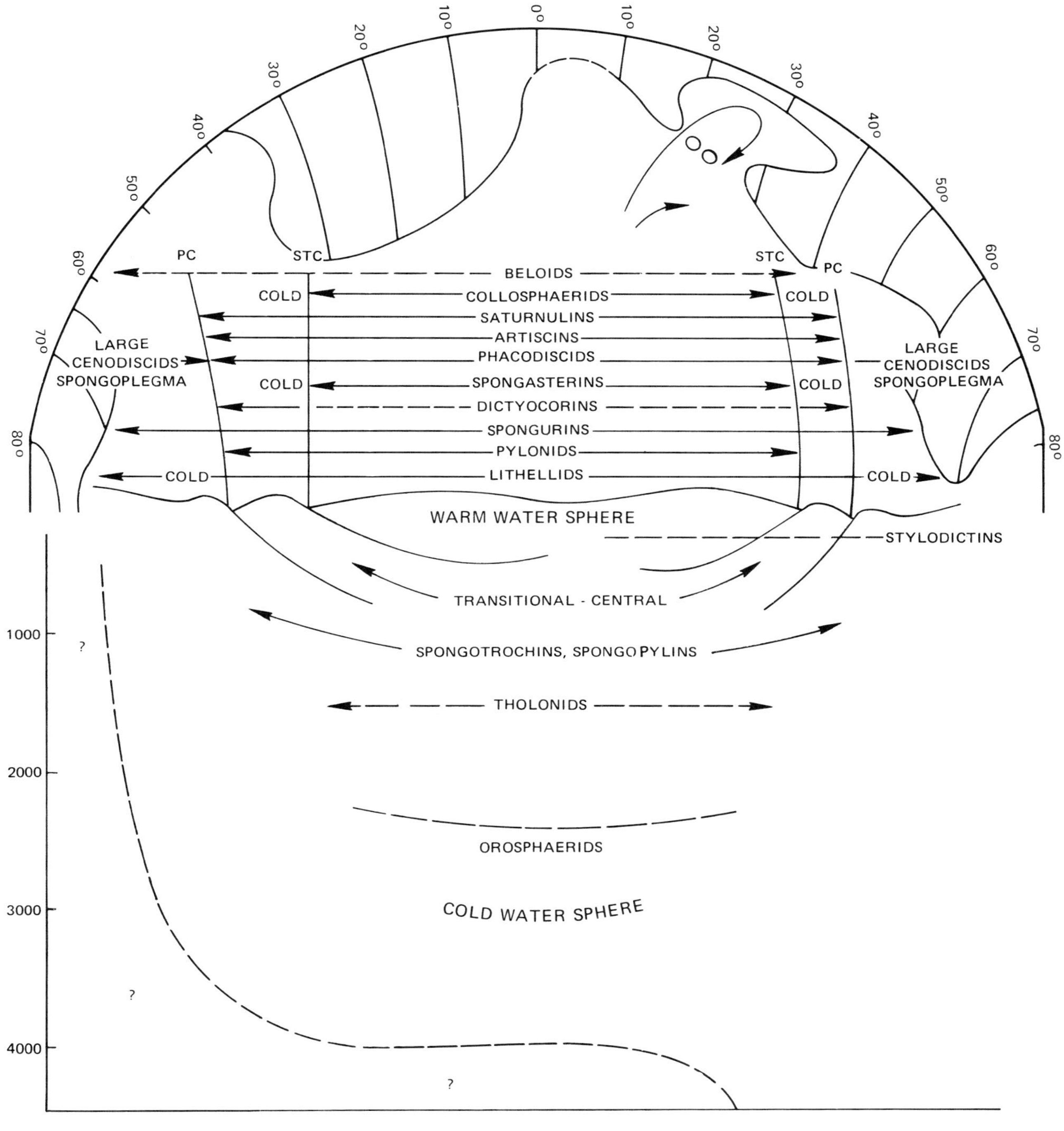

Figure 13.10 Dominant geographic (top) and depth (bottom; depth in m on left) distributions of spumellarian radiolarians in an ocean. (Reproduced from Casey RE, Spaw JM, Kunze FR. *Trans Gulf Coast Assoc Geol Soc* 1982; 32:319–32.)

mixing of north Atlantic Deep and Bottom Water with Antarctic water masses. The diversities, densities, and degree of endemism of the radiolarian fauna in these waters is probably low. Deep or Common waters refer to the waters between the intermediate and usually Polar Bottom waters, and in part include north Atlantic Deep and Bottom Water and Pacific Common Water. The radiolarian diversity and density of these waters is low (densities of a few to hundreds per cubic meter); however, the degree of endemism is moderate. The only living form found in the Polar or Antarctic Bottom Water to date is a type related to an antarctisin, and the radiolarian

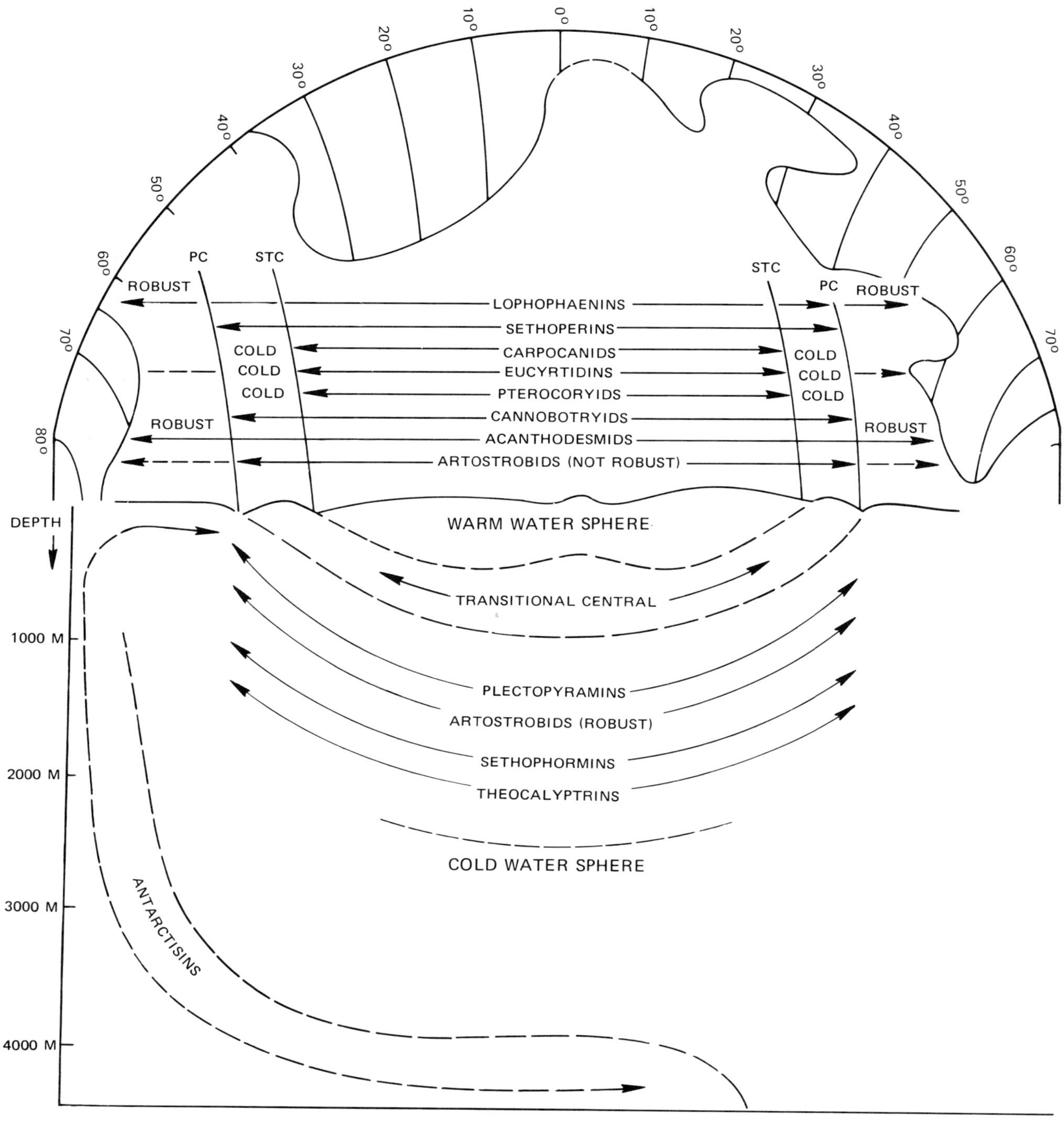

Figure 13.11 Dominant geographic (top) and depth (bottom; depth in m on left) distributions of nassellarian radiolarians in an ocean. (Reproduced from Casey RE, Spaw JM, Kunze FR. *Trans Gulf Coast Assoc Geol Soc* 1982; 32:319–22.)

diversity, density, and degree of endemism appears to be very low.

Radiolarian taphonomy

The physical and chemical processes in the water column control the kinds of shelled microplankton that are present in the overlying waters, their living distributions, production, dissolution processes in the water column, and preservation at the sediment–water interface, and sediments. Around a south polar continent, terrigenous sediments from continental glaciers and grounded ice shelves are transported northward but are usually held poleward of the polar divergence resulting in a dominance

Table 13.2 Oceanographic and radiolarian characteristics of the major radiolarian provinces (provinces shown in Figs 13.10 and 13.11). (Some biologic and oceanographic information taken from McGowan JA. *The Biology of the Oceanic Pacific*. Corvollis: Oregon State University Press, 1974:19–28, and Reid JL, Briton E, Fleminger A, Venrich EL, McGowan JA. *Advances in Oceanography*. New York: Plenum Publications, 1978:65–130)

Zone hypothetical ocean	Primary productivity	Degree stratified (layering)	Leakage (loss to other provinces)	Neritic influence	Radiolarian characteristics				
					Diversity	Density	% Endemics	Dominant	Province age
Subpolar cyclonic gyre									
Northern	High	Low	Great	Great	Low	Low	Few if any	Cenodiscids (d), robust lithellids (d), lophophaenids (d), and stylodictins (d) + an intermediate component	Pliocene?
Southern	High	Low	Great	Moderate	Low	Low	Moderate	Same as in northern with stylodictins replaced by antarctisins (d)	Pliocene
Transition									
Northern and southern	Moderate	Low	Great	Little	Moderate	Moderate	Few	Cold-water morphotypes of warm taxa + a central component	Disappears and then reappears
Subtropical anticyclonic gyre									
Northern and southern	Low	High	Very low	None	Very high	Very high	High	Symbiont-bearing and warm-water sphere such as collosphaerids (s), artiscins (s), spongasters (s)	Mid-Miocene
Equatorial	High to moderate	High to low	Moderate	Little	High	High	Moderate	Similar to gyre but not as many symbiont bearers	Mid-Miocene
Eastern tropical	High	Complex	Moderate	Moderate to great	Low to moderate	Low to moderate	Few	Symbiont bearing	Pliocene
Central									
Northern and southern (northern sometimes of greater salinity)	—	—	Very low	None	Moderate	Low	Moderate	Eucyrtidins, pterocoryids, spongopylins (d), robust artostrobids (d)	Mid-Miocene
Intermediate									
Northern and southern (northern or southern sometimes of greater salinity)	—	—	Very low	None	Moderate	Low	Moderate	Spongodiscids (d), spongotrochins (d), tholonids (d), plectopyramids (d), artostrobids (robust) (d), theocalyptrins (d)	Mid-Miocene
Polar circumpolar	—	—	Moderate	None	Probably low	Probably low	Probably low	?	
Deep or Common	—	—	Moderate	None	Low	Low	Moderate	Orosphaerids (d), plectopyramids (d)	
Polar Bottom	—	—	Moderate	Little	Very low	Very low	Very low	Antarctisins ? (d)	

(d), detritovore; (s), symbiotrophe.

of glacial marine sediments underlying these areas (Fig. 13.12). Radiolarians preserved in the sediments south of the polar divergence are usually robust subsurface southern ocean or intermediate or deep-living forms (also robust). The radiolarian content of these sediments is usually much less than the diatom component unless dissolution has removed the diatoms and left the more robust radiolarians.

Between the polar divergence and polar convergence the **biogenous** counterpart of the sediments are dominated by diatoms. The radiolarians in this region are similar to the radiolarians south of the divergence. The polar convergence controls the northward extension of diatom-dominated siliceous oozes except for regions of the eastern boundary currents, where at times in the past siliceous **oozes** have been deposited. Nutrient- and silica-rich waters **upwelled** at the polar divergence support a large standing crop of

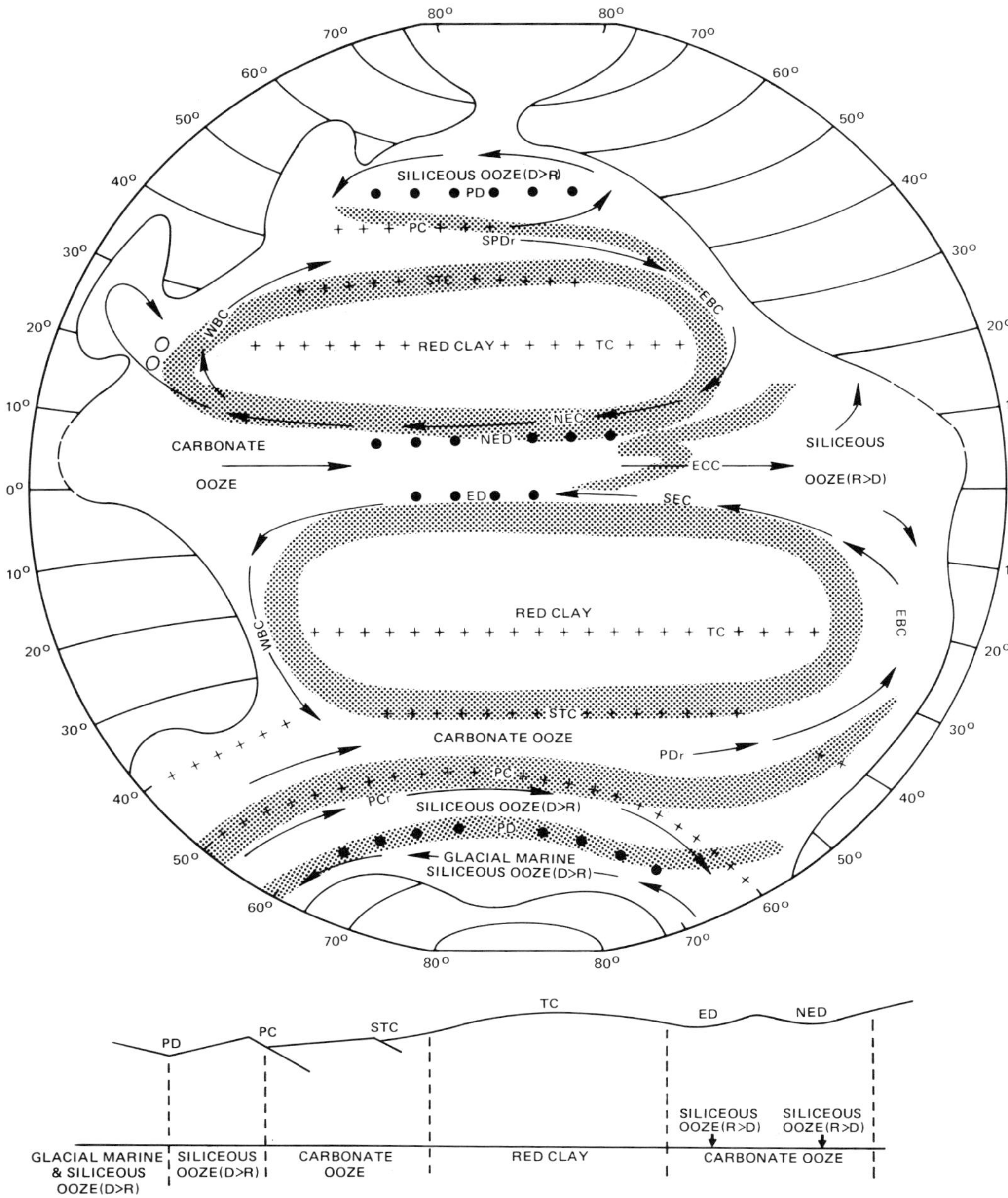

Figure 13.12 Dominant sediment distributions in an ocean.

diatoms until these waters and some of their endemic radiolarians dive below the photic zone at the polar convergence. The warmer waters north of the polar convergence support shelled microplankton and nannoplankton assemblages dominated by planktic foraminifera and coccolithophorids which produce dominantly calcareous oozes between the polar and subtropical convergence. In shallow waters north of the southern subtropical convergence the same calcareous plankton dominate the living microplankton but their standing crops drop an order of magnitude (in some regions). This is due to negligible upwelling in the southern subtropical anticyclonic gyre and a low rate of flux of calcareous skeletons. Dissolution during descent, or shortly after reaching the sediment–water interface, results in sediments usually devoid of biogenous sediments except for orosphaerid radiolarians and **ichthyoliths**. Rates of sedimentation under these gyres are in the order of millimeters per 1000 years. The dominant sediments are red or brown clays carried by winds and currents from the continents, or are authigenic sediments derived from precipitation from the water column or reprecipitated from dissolved biogenous, terrigenous, or volcanogenic sediments. In many areas under the gyres, biogenous sediments are more or less evenly distributed in these upper 5 cm which corresponds to the average depth of bioturbation. Therefore, the final destruction of the biogenous part of these sediments may be by dissolution in the gut of detritivores or by stripping some protective coating such as a fecal pellet or metallic covering that may protect the siliceous skeletons from dissolution.

Biogenous sediments again dominate under the equatorial current systems (south equatorial, equatorial, and north equatorial currents) where primary productivity increases because of an increase in nutrients through local upwelling, and the eastern boundary currents advecting nutrients that were carried from higher latitudes or upwelled within the boundary current systems. The region is mainly dominated by calcareous oozes with siliceous components (mainly radiolarian) increasing at times to form radiolarian oozes under the eastern tropical, eastern equatorial, and northern equatorial divergence regions. Apparently, under the regions of highest productivity in the equatorial region enough soft-part organic material gets to the bottom, probably by fecal pellet sedimentation, that the oxidation of this organic material produces conditions acidic enough to dissolve some of the calcareous material and enhance the radiolarian component.

The northern hemisphere sedimentation is essentially a mirror image of the southern hemisphere, with two major exceptions. Dissolution of calcareous oozes is controlled to a large degree by the presence of southern polar (Antarctic) Bottom Water. These waters bathe all of the southern hemisphere deep oceans and aid in removing the calcareous oozes from under the southern anticyclonic subtropical gyre; however, the rates of sedimentation of biogenous oozes under the equatorial current system exceed the dissolution rate caused by south polar water. Under the northern subtropical anticyclonic gyre; again the south polar waters remove the calcareous biogenous component but only as far into the north as they bathe the bottom; poleward calcareous components are preserved. Another exception is in the sedimentation under the subpolar cyclonic gyre. No glacial marine sediments are deposited in this region due to the absence of glacial and sea ice resulting in siliceous oozes (diatom dominated) throughout the gyre.

In comparing the dominant ideal deep-sea sediment types (Fig. 13.12) with the actual distributions of deep-sea sediments in the present-day world ocean some obvious differences can be seen. The distributions of glacial marine, siliceous, and calcareous sediments in the real southern ocean is very similar to that of the ideal ocean and can be explained in the same manner. The distributions of brown (red) clay, authigenic sediments, and calcareous oozes of the real ocean under the subtropical anticyclonic gyres differ greatly from the ideal ocean. The main reason for this is that the ideal ocean is an **isobathal** ocean of abyssal depths, the real ocean is not. In general the calcareous oozes of the subtropical anticyclonic gyres follow the strike of and drape the ocean highs (midoceanic ridges or rises and shoaling regions). This is because these regions are shallower than the depth where calcareous skeletons are dissolved as rapidly as they are deposited (the **calcite compensation depth** or **CCD**). The equatorial radiolarian oozes are found to any degree only in the equatorial Pacific where significant upwelling, primary productivity, and fecal pellet sedimentation to the bottom occur. Although most shallow-living radiolarians that are preserved reach the bottom in fecal pellets, the deep-living radiolarians, probably just as numerous as the shallow-living forms, mainly get to the bottom as settling of individual tests.

Paleobiology

The paleobiology of radiolarians has been chiefly inferred from their **paleobiogeography**. Polycystine radiolarians are more useful in paleobiology than phaeodarians because they range in age from Cambrian to Recent rather than Cretaceous to Recent, they are most widely distributed, they are most diverse, and their biology is well documented. Paleobiogeography of radiolarians is determined in two ways provided that a short time interval is used and samples are widely distributed over the area. One way is to compare the distribution of extant species in the plankton to the distribution of extinct species, and another is to compare the distributions of both taxa in the sedimentary record. An extinct species that corresponds in its distribution to a living species is then assumed to have the same distribution and similar life requirements. Thus, species from each modern province can be used as analogs for species as old as Miocene. The paleobiology of older species that have no modern analogs can be inferred by comparing the distributions of increasingly older species back through time.

Ancient feeding behavior may also be inferred from these paleobiogeographies. Forms that are reconstructed to be intermediate- or deep-water forms are most likely to have been detritivores or bacterivores. Forms that are more abundant under paleoupwelling regions and that resemble extant deep forms such as elongated nassellarians or spongy disk-shaped spumellarians probably also fed on detritus and bacteria. Species that dominated the paleosubtropical gyres probably harbored symbionts.

Radiolarians as paleoceanographic indicators

The paleobiogeography of radiolaria provides a basis for inferring the **paleoceanography** of regions and entire oceans. Radiolarians may be used to determine ancient water temperatures, circulation patterns, productive regions, ocean depths, upwelling, and the age of water masses.

Paleotemperatures of the oceans have been determined by comparing radiolarian species in modern sediments to the overlying surface water temperatures and then extrapolating these temperatures to similar radiolarians in older sedimentary rocks in the same area. In the Antarctic oceans, comparison of subantarctic, Antarctic, and other assemblages in the sediment to those in older rocks showed that the Antarctic waters have cooled since Late Tertiary time. Radiolarians have also been used to aid paleotemperature reconstructions for entire oceans at specific times (Fig. 13.13).

Paleocirculation patterns can also be inferred from ancient radiolarian distributions by comparing the modern to the fossil assemblages. Reconstructions of past currents have been done for ocean basins like the North Pacific, single current systems, and events within current

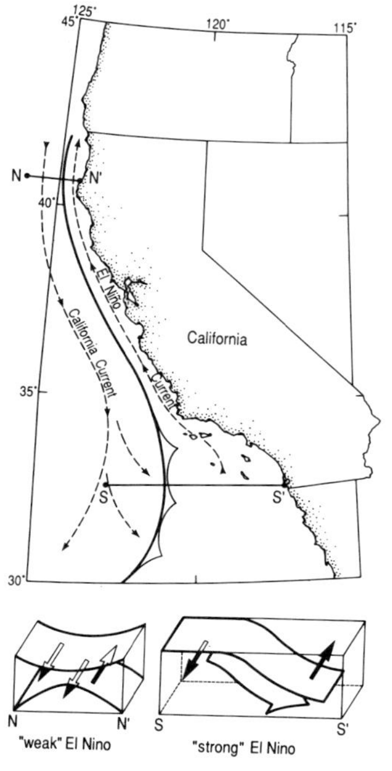

Figure 13.14 Model of El Niño circulation conditions for the California Current. (Modified from Casey RE, Weinheimer AL, Nelson CO., *Aspects Climate Variabil Pacific West Americas* 1989; 55:85–92.)

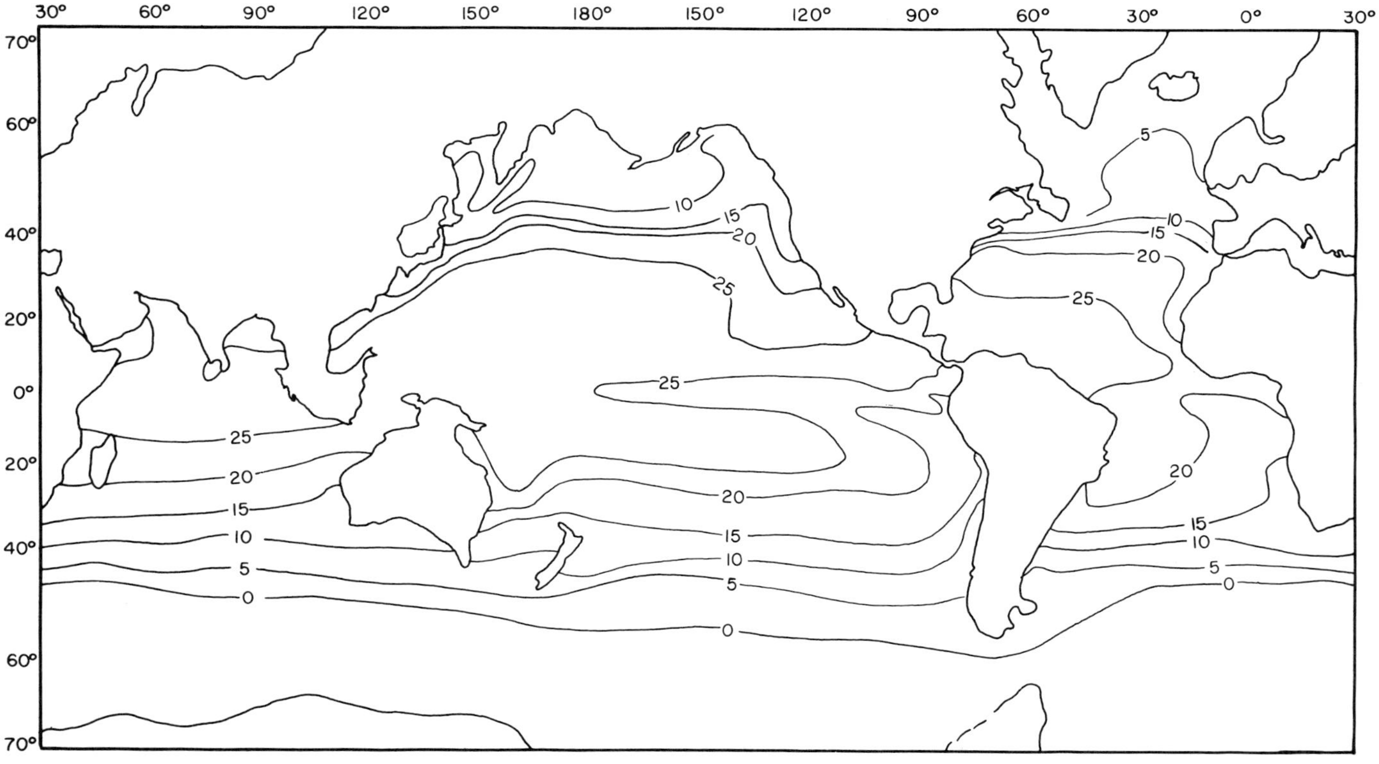

Figure 13.13 Sea-surface temperatures for northern hemisphere summer (August) 18 000 years ago. (Modified from Project CLIMAP project members. *Science* 1976; 191:1131–7.)

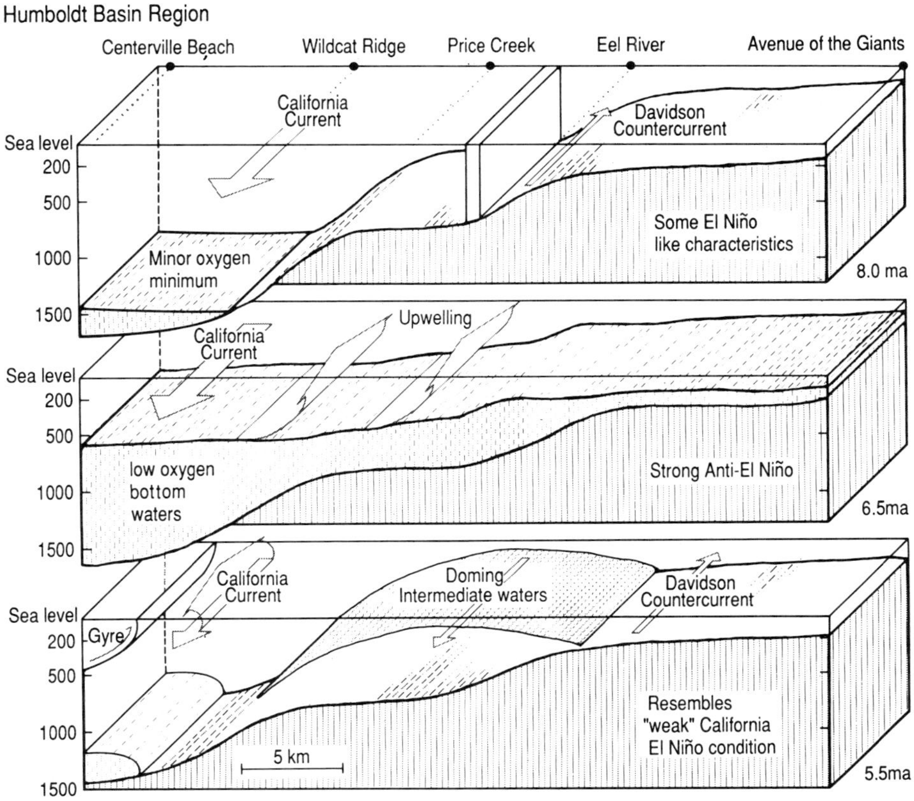

Figure 13.15 Paleoceanographic reconstructions of the Humboldt Basin region (northern California) at 8, 6.5, and 5.5 Ma. (Modified from Casey RE, Weinheimer AL, Nelson CO. *Aspects Climate Variabil Pacific West Americas* 1989; 55:85–92.)

systems like El Niño events in the California Current system (Fig. 13.14), and in a portion of a single system (Fig. 13.15). They have also served well in the interpretation of modern seasonal circulation patterns along the modern Texas coast which encourages their use in paleocirculation reconstructions (Fig. 13.16).

Because some radiolarians are endemic to specific oceanographic water masses, the evolution of those forms may be used as a conservative estimate of the time of initiation of those water masses. This method was employed in estimating the province ages in Table 13.2.

Radiolarians may also be useful in estimating ancient ocean depths, although foraminifera are more commonly used. Usually an increase in the number of radiolarians per gram of sediment has been used to indicate increased paleodepth. The number of radiolarians in Recent sediments from the Continental Borderland off southern California (Fig. 13.17) was very low for shelf sediments (usually less than 50/g) and increased fairly consistently with increased depth (to about 800/g at about 1300 m). In an area of known upwelling, a peak in radiolarian number occurred at shelf break depth when compared to shallower and deeper depths. This peak probably results from increased mortality of deep-living forms upwelled onto the shelf. In a shallow-water section, such a peak in radiolarian number or peak in deep-living forms may be used as an index of shelf break depth.

Radiolarians extracted from Mesozoic sediments have been used to develop a neritic versus deep water index. Mesozoic deep-sea rocks, for example, exhibit the highest indices and neritic sections the lowest (Fig. 13.18). However, radiolarian-rich sediments deposited prior to the evolution of diatoms in the Cretaceous or perhaps Jurassic, may well have been deposited commonly in shallow water in prediatom times. In fact, high concentrations of radiolarians are found in some Norwegian fjord sediments today.

Spumellarian radiolarians dominate shelf waters, whereas nassellarian radiolarians are more common to open-ocean waters. Therefore, the ratio of spumellarians to nassellarians has been used to interpret near-shore versus offshore conditions and to suggest fluctuations in sea level. Low radiolarian diversity may also indicate near-shore, and high radiolarian diversity indicate open-ocean environments even for Paleozoic sediments.

Radiolarians from plankton tows may distinguish upwelling conditions. The appearance of radiolarians usually is restricted to depths deeper than 200 m in the upper waters off the California coast and these may reflect upwelling conditions. The presence of certain radiolarian

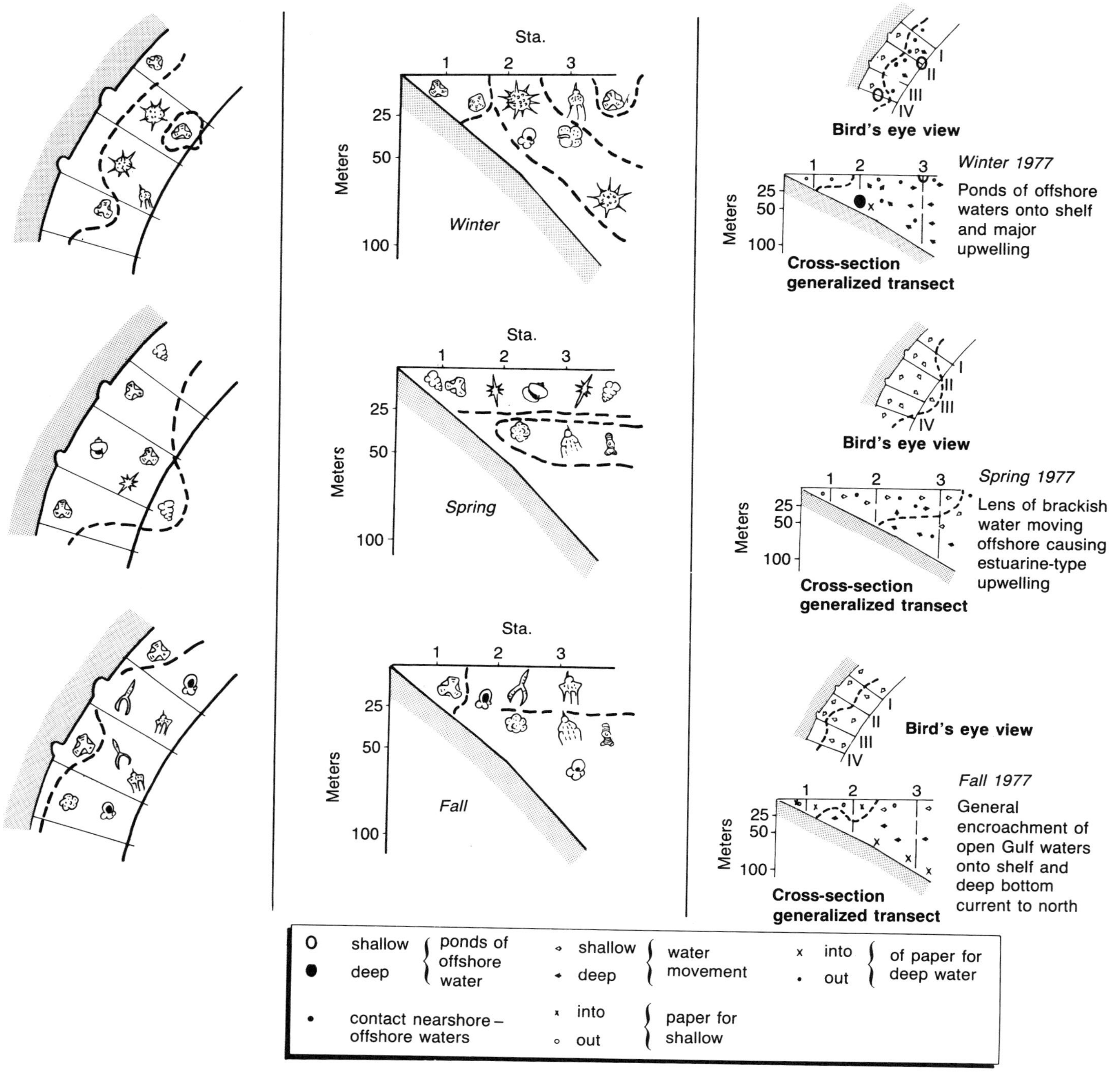

Figure 13.16 General oceanographic trends for 1977 along the south Texas coast derived from interpretation of radiolarian and other shelled microplankton data. (Reproduced from Leavesley A, Bauer M, McMillen KJ, Casey RE. *Trans Gulf Coast Assoc Geol Soc* 1978; 28:229–38.)

species in the shelf waters off the south Texas coast indicates upwelling conditions because these species are normally found at depths greater than shelf depths (Fig. 13.16).

Unusually high accumulation rates of biogenic silica in general have been used to suggest regions of high fertility. More specifically the accumulation and good state of preservation of radiolarians in sediments is associated with regions of high productivity. In comparing **eutrophic** southern California borderland and more oligotrophic Gulf of Mexico and Caribbean radiolarian assemblages, the greater percentage and diversity of known deep-water polycystine radiolarians in the southern California samples result from the more intensive upwelling con-

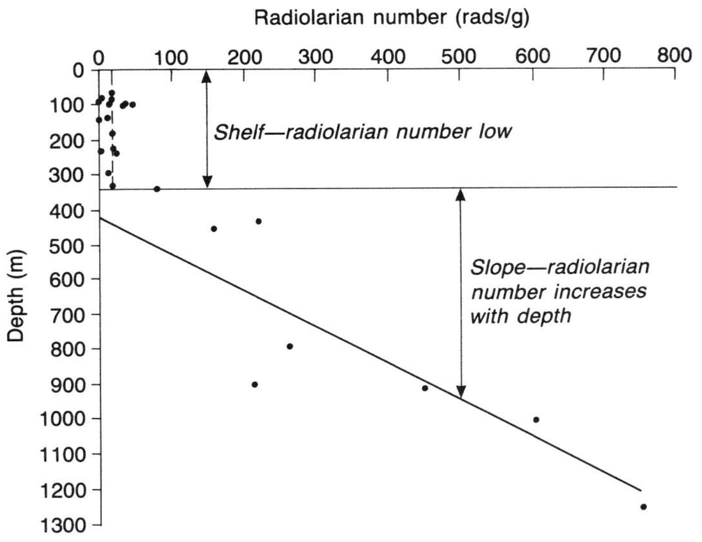

Figure 13.17 Radiolarian number (number/g of sediment) vs. depth for 24 surface sediment samples from the Southern California Continental Borderland. (Reproduced from Cleveland MN, Casey RE. *Pacific Section, Society of Economic Paleontologists and Mineralogists, Book 45,* 1986:21–30.)

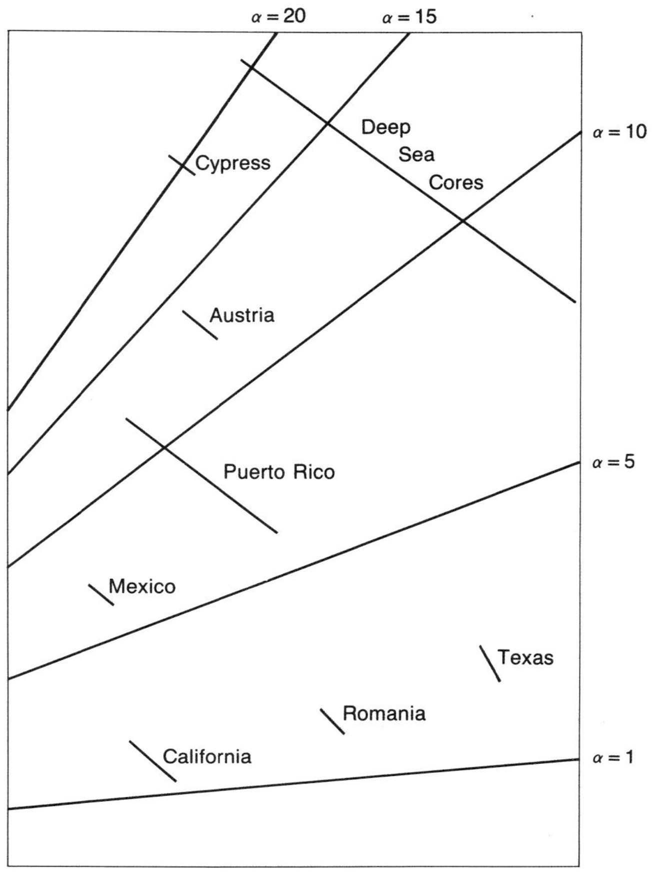

Figure 13.18 A plot of the diversities (Fisher Alpha Diversity Index) of different Mesozoic radiolarian assemblages. Deep Sea Drilling Project cores exhibit the highest diversities because they are the deepest and most "open ocean" and the Californian and Romanian sections the lowest because they are neritic. (Modified from Empson-Morin KM. *Micropaleontology* 1984; 30:87–115.)

ditions off California which caused an enhancement of deep-water radiolarian diversity and number in the sediments. On a larger geographic scale, intermediate and deep-water radiolarians are enhanced in Recent sediments underlying regions of upwelling: the polar cyclonic gyres, eastern boundary currents, and oceanic convergences and divergences (Fig. 13.19).

Low oxygen conditions on the sea floor can be inferred by the presence of abundant phaeodarian radiolarians. Phaeodarian skeletons contain so much organic material that they are usually only preserved where oxygen concentrations are so low that oxidation of the organic material cannot take place. The ratio of the total surface area of phaeodarian skeletons preserved to the dry weight of sediment provides an index of oxygen concentration. In the Neogene of northern California, a basin floor with anoxic conditions, the main oxygen minimum zone impingement on the continental slope, and a shallower oxygen minimum zone on the shelf break could be determined (Fig. 13.15).

Biostratigraphy

Radiolaria have been used in biostratigraphy for all Phanerozoic sedimentary rocks. Paleozoic and Mesozoic rocks are not well studied biostratigraphically and zonations apply chiefly to local regions. These rocks generally require difficult extraction, processing, and evaluation techniques that have impeded the development of detailed biostratigraphy for the Paleozoic and Mesozoic. Radiolaria have been utilized, however, with much success for biostratigraphy in the Cenozoic.

Cenozoic radiolarian biostratigraphy is best developed for the tropical regions where species diversity and evolutionary rates are high. The currently used tropical biostratigraphic zonation contains 26 zones ranging in age from the Late Paleocene to the Recent, and a recent revision recognized 29 zones that conform to the usage of the International Stratigraphic Guide (Fig. 13.20). These zones can be recognized generally through the tropical regions of the entire world.

Higher latitude radiolarian biostratigraphy, as with most planktic protists, is less well defined and useful only in local areas. For the most part, these zonations are based mainly on shallow-water species with localized distributions. Deep-water radiolaria provide a basis for worldwide biostratigraphy because they are so widely

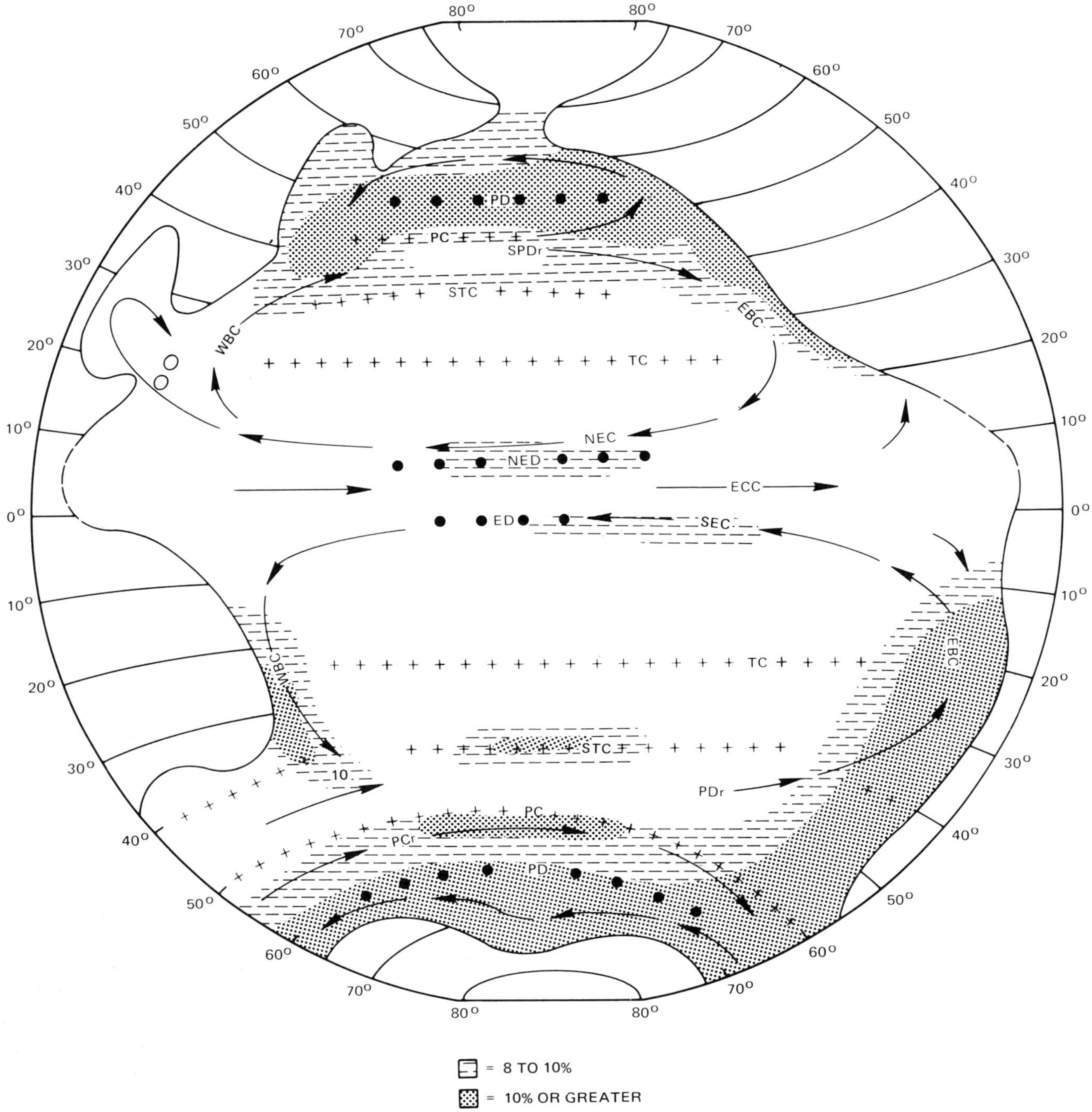

Figure 13.19 The percentage of intermediate- to deep-water forms of entire radiolarian fauna in well-preserved sediments in a diagrammatic ocean. The higher percentages are found underlying upwelling regions. (Reproduced from Casey RE, Spaw JM, Kunze FR. *Trans Gulf Coast Assoc Geol Soc* 1982; 32:319–32.)

distributed. However, the zonations based on the fewer deep-water species are broad and consist of only six zones for the Neogene (Fig. 13.21). These zones may be used to correlate between more detailed biostratigraphies used regionally.

The Cenozoic radiolarian zones and first and last occurrences of species, particularly in the tropical regions, have been related to well-dated magnetostratigraphy (Figs 13.22–13.24). The dates associated with the magnetostratigraphy allow the determination of radiolarian events to within a fraction of a million years.

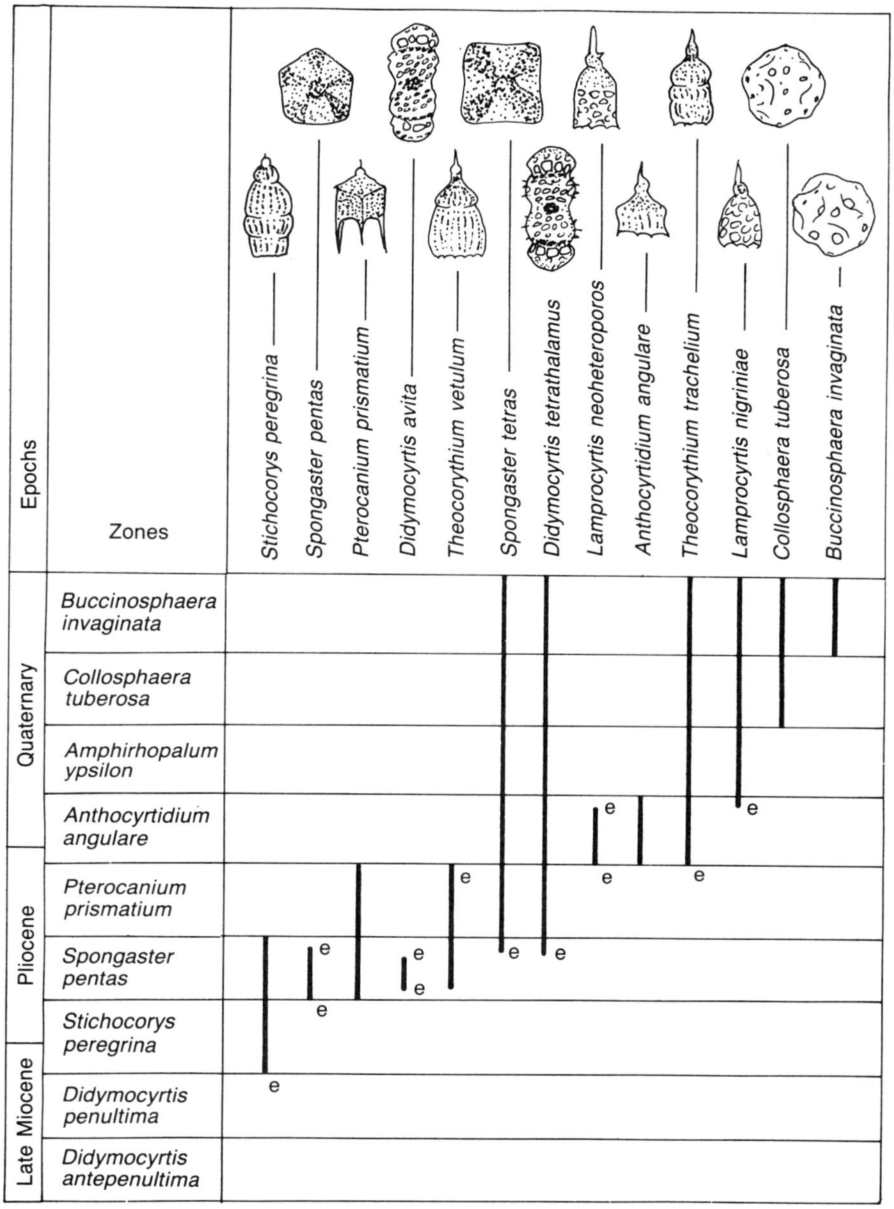

Figure 13.20 A portion of the most recent Cenozoic tropical radiolarian zonation. (Modified from Sanfilippo A, Westberg-Smith J, Riedel WR. *Plankton Stratigraphy*. Cambridge: Cambridge University Press, 1985:631–712.)

Evolutionary history

Macroevolution

Radiolaria first appeared in the Cambrian and underwent a number of changes in morphology and environment throughout the rest of the Phanerozoic (Figs 13.25–13.28).

Simple paleoactinommid and other radiolarians are present in Cambrian rocks. The oldest, well-preserved, and carefully documented radiolarians are from the Upper Cambrian and Lower Ordovician. These radiolarians exhibit the three basic radiolarian skeletal types: the unshelled spicule, closed sphere, and cone. These fossils were apparently deposited in a comparatively shallow environment, although some may indeed have been deeper forms that invaded these shallower environments during times of open ocean intrusion. These Paleozoic radiolarians can thus be inferred to have preferred deep or shallow water (Fig. 13.25). The cold-water and warm-water radiolarians are distinguishable in the Cambrian and a deep radiolarian fauna had developed by at least Silurian times (Fig. 13.25).

Possibly the initiation of the mid Paleozoic cold water (Fig. 13.25 no.1) set the stage for the evolution of the first deep, cold-water forms (albaillellids and, perhaps, ceratoikisids). This Late Paleozoic cold water may have resulted in a major radiolarian reorganization of deep, cold-water radiolarians (with extinctions of albaillellids

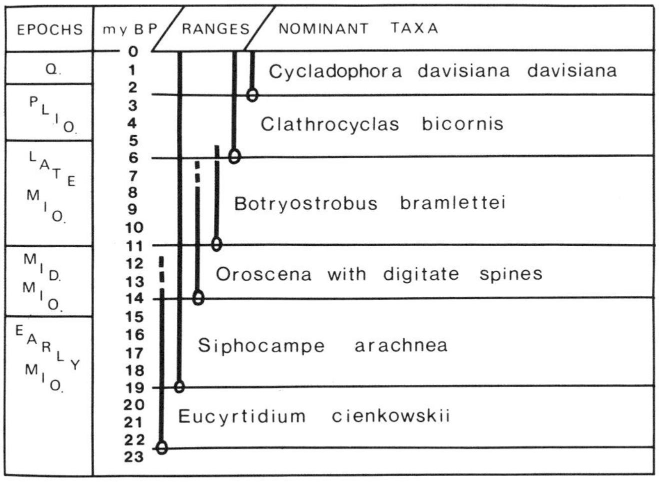

Figure 13.21 The ranges of the six nominant taxa used to erect the Neogene deep-living Radiolarian Zonation (NdRZ) are related to one another and to geologic time. (Reproduced from Casey RE, Wigley CR. American Association of Petroleum Geologists, Abstracts with Programs. New Orleans, 1985.)

and popofskyellids), and the evolution of deep, cold-water, spongodiscid-like forms and corythoecids. Permian radiolarian faunas are reasonably diverse, at least for the Early Permian. However, the Late Paleozoic and Paleozoic–Mesozoic transition represents a major radiolarian reorganization with the extinctions of albaillellids, rotasphaerids, entactinids, and paleoactinommids (Fig. 13.25 no.2). This reorganization may have been caused by the reduction in the number of surface current and water mass environments as a result of the suturing of the continents and eutrophication due to Permian glaciation.

The first unquestionable nassellarians appear in the Triassic. Approximately half of the extant groups (Fig. 13.25) evolved during the Mesozoic, and there are obvious generic affinities with radiolarians of the Late Cretaceous. Characteristic Mesozoic (especially Late Mesozoic) forms are amphipyndacids, dictyomitrids, and hagiastrids (Fig. 13.27).

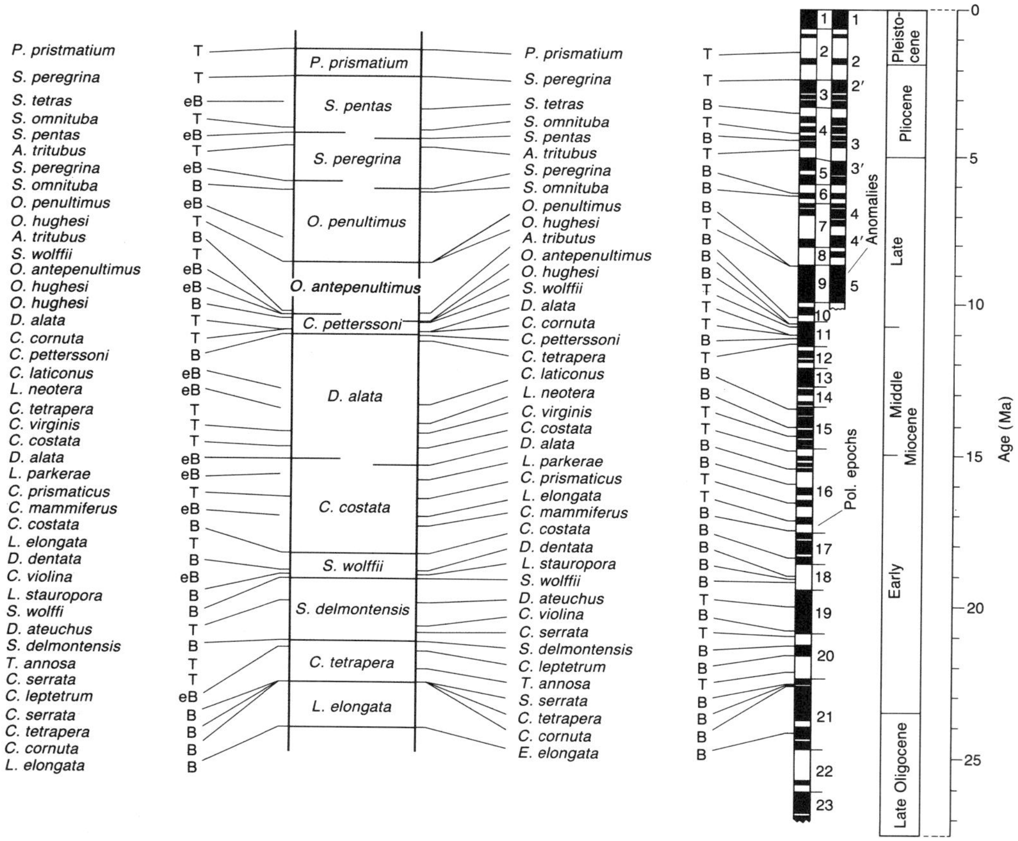

Figure 13.22 A comparison of the tropical radiolarian zonation with the paleomagnetic time-scale. B and T indicate the bottom and top of the stratigraphic range of the species. eB indicates the first evolutionary appearance of the species. (From Theyer F, Mato CY, Hammond SR. *Marine Micropaleontology* 1978; 3:377–95.)

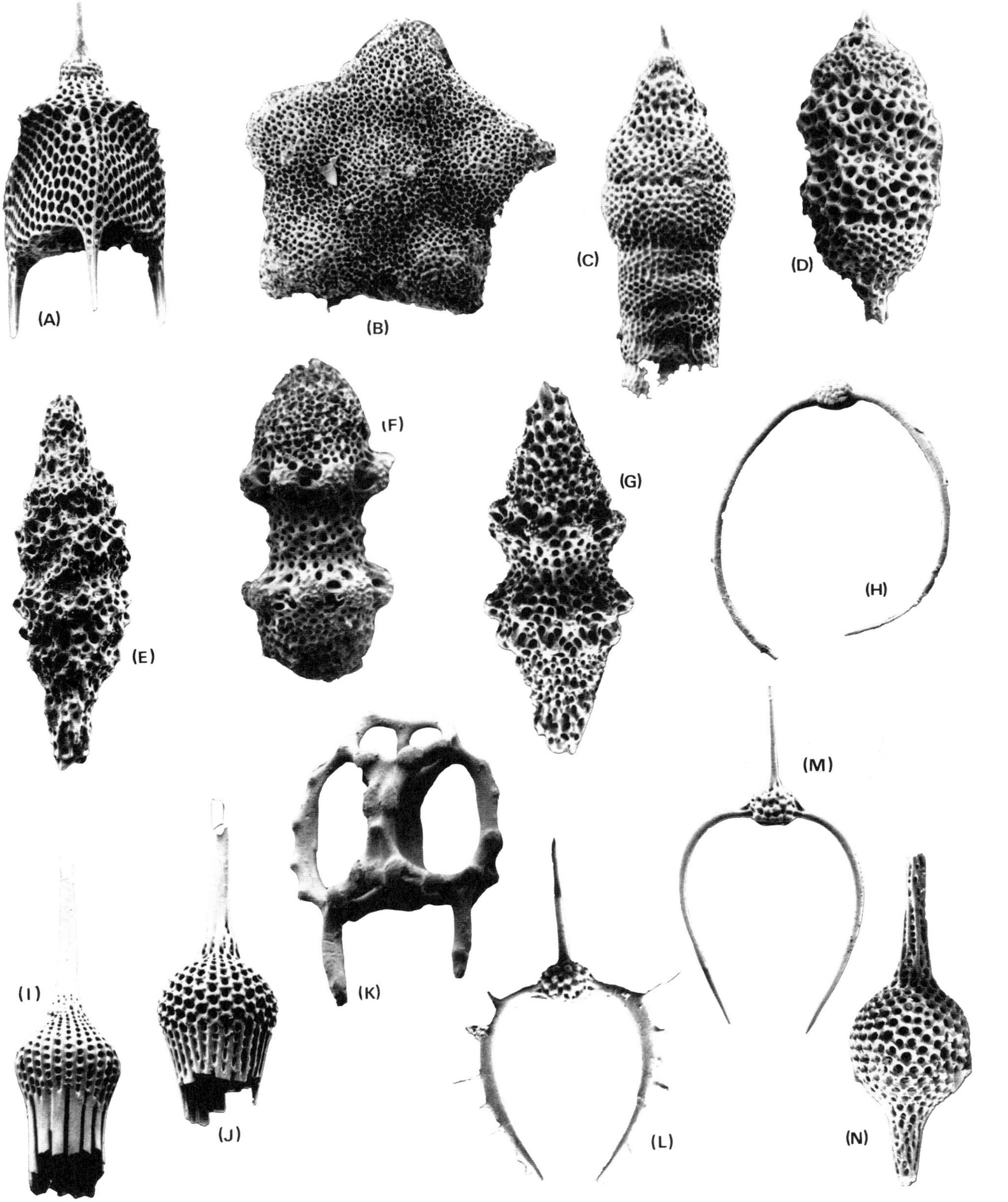

Figure 13.23 Radiolarians used in tropical radiolarian zonation. The illustrations and names can be matched to the events in Figure 13.22. The radiolarians on this figure are: A, *Pterocanium prismatium*; B, *Spongaster pentas*; C, *Stichocorys peregrina*; D, *Ommatartus penultimus*; E, *Ommatartus antepenultimus*; F, *Cannartus petterssoni*; G, *Cannartus laticonus*; H, *Dorcadospyris alata*; I, *Calocycletta costata*; J, *Calocycletta virginis*; K, *Giraffospyris toxaria*; L, *Dorcadospyris dentata*; M, *Dorcadospyris forcipata*; N, *Cannartus prismaticus*. (Reproduced from Theyer F, Mato CY, Hammond SR. *Marine Micropaleontology* 1978; 3:377–95.)

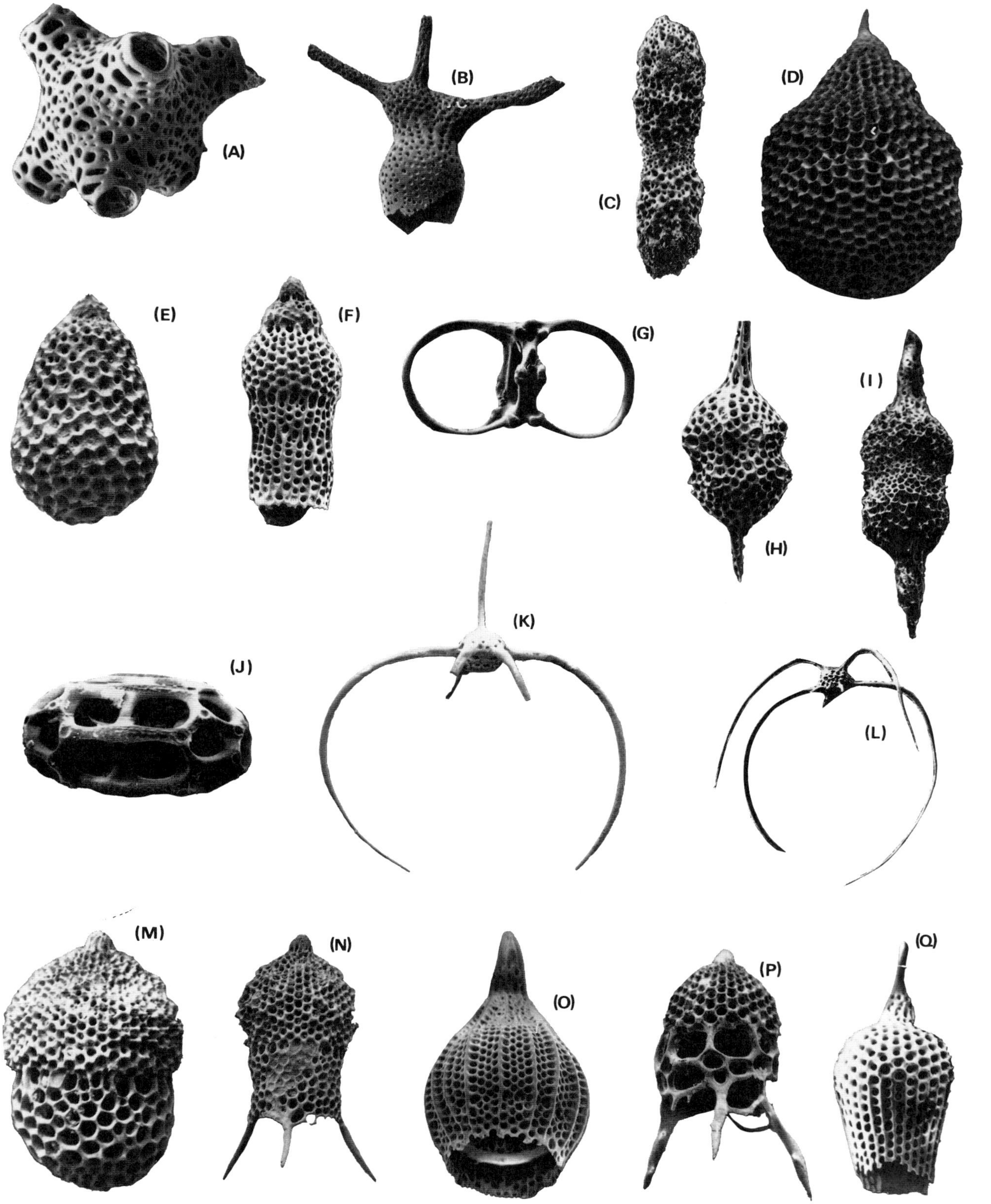

Figure 13.24 Radiolarians used in tropical radiolarian zonation; continued from Figure 13.23. The illustrations and names can be matched to the events in Figure 13.22. The radiolarians in this figure are: A, *Solenosphaera omnituba*; B, *Acrobotrys tritubus*; C, *Ommatartus hughesi*; D, *Cyrtocapsella cornuta*; E, *Cyrtocapsella tetrapera*; F, *Stichocorys wolffii*; G, *Acrocubus octopylus*; H, *Cannartus violina*; I, *Cannartus tubarius*; J, *Tympanidium binoctonum*; K, *Dorcadospyris papilio*; L, *Dorcadospyris riedeli*; M, *Cyclampterium leptetrum*; N, *Cyclampterium leptetrum*; O, *Theocyrtis annosa*; P, *Cyclampterium pegetrum*; Q, *Calocycletta robusta*. (Reproduced from Theyer F, Mato CY, Hammond SR. *Marine Micropaleontology* 1978; 3:377–85.)

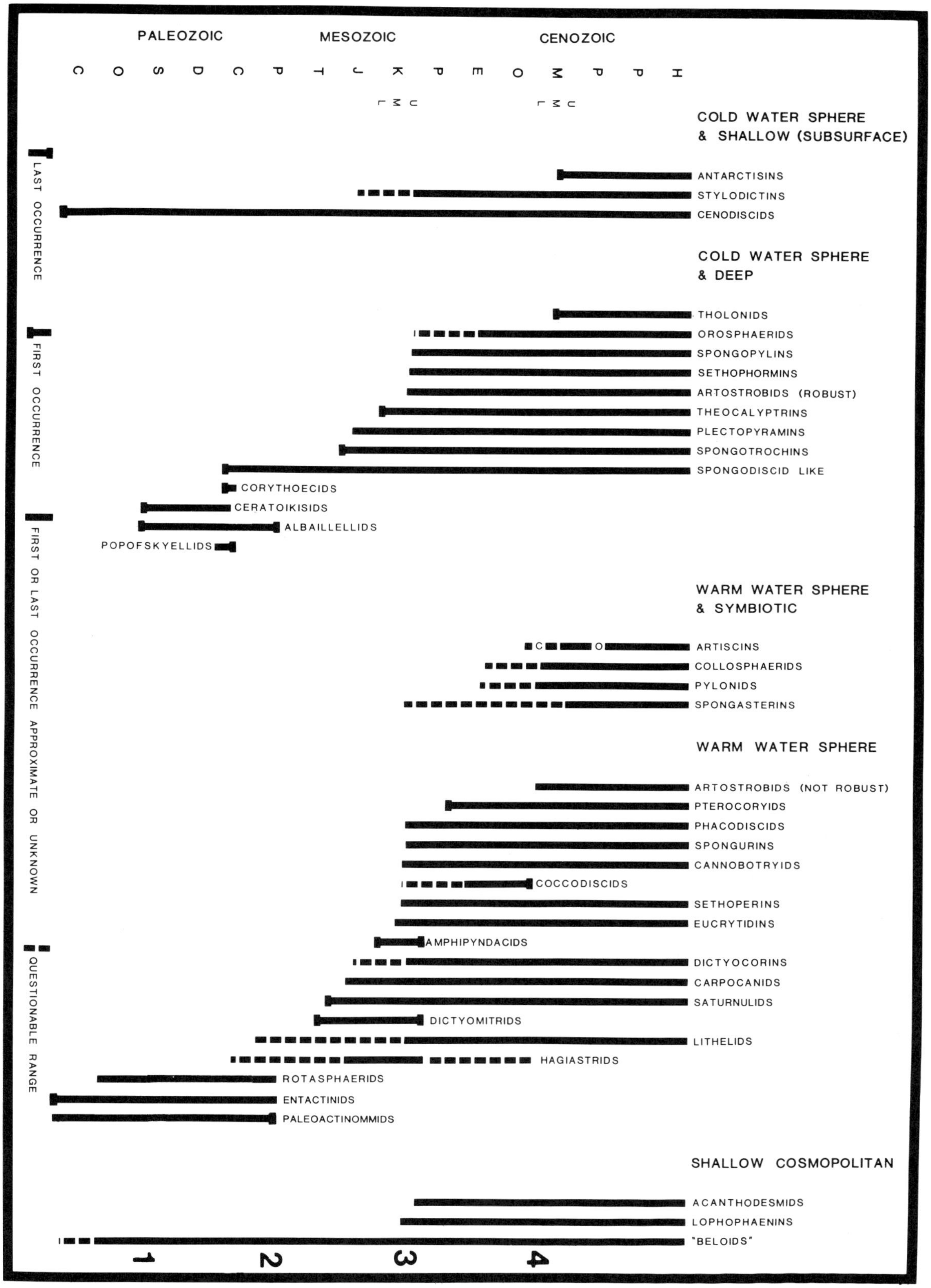

Figure 13.25 Geologic ranges of radiolarian "family level" groups, arranged by environmental group. The numbers are referred to in the text. (Modified from Casey RE, Wigley CR, Perez-Guzman AM. *Paleobiology* 1983; 9:363–76.)

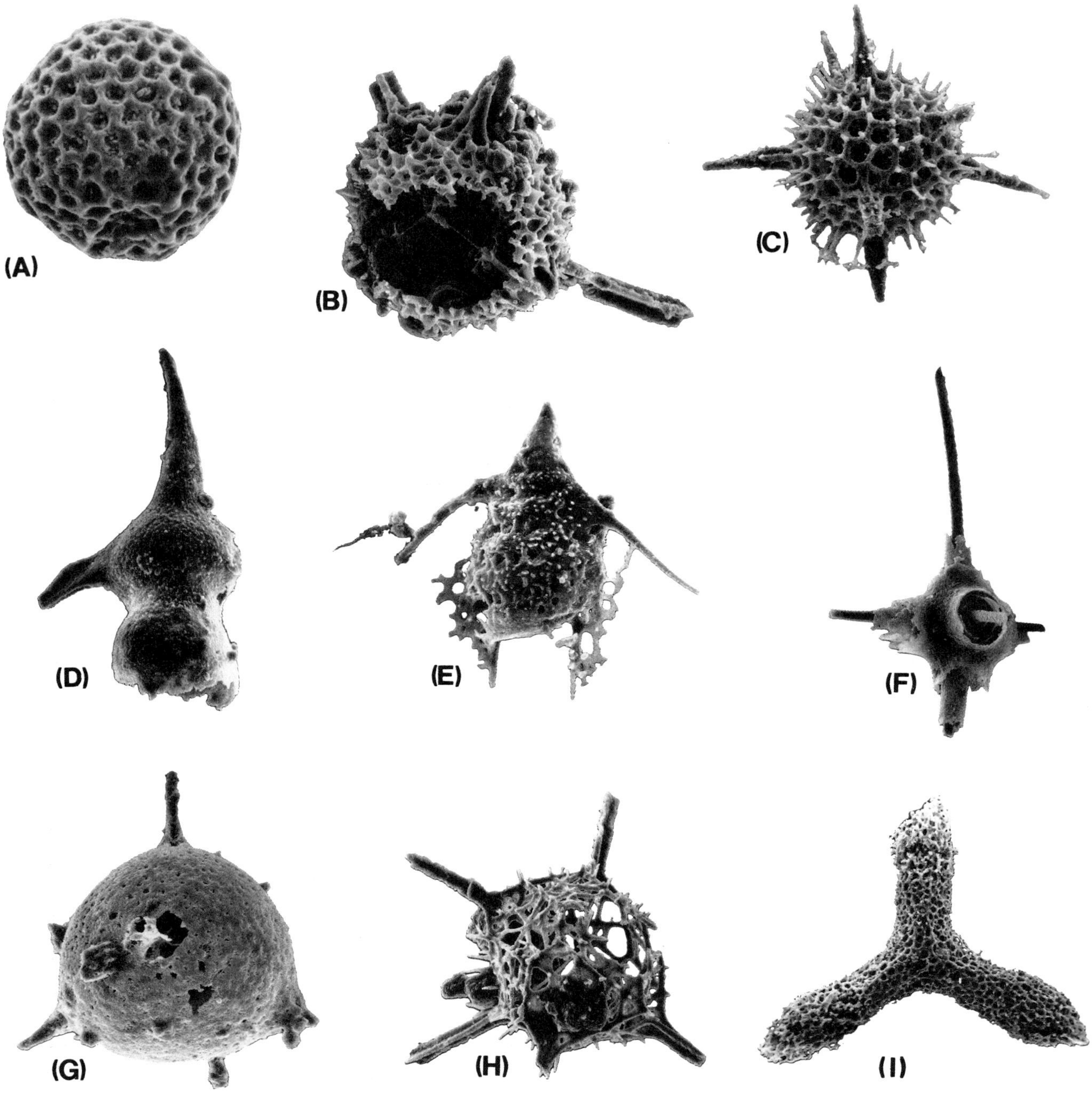

Figure 13.26 Paleozoic radiolarians. The general "cenodiscid" from (A is a Mesozoic example) is found from Cambrian to Recent; however, the characteristic junction of the bars in an entactinid can be seen in a broken specimen (B). A complete entactinid can be seen as specimen C. Early "nassellarians," the albaillellids, are illustrated as D and E with E exhibiting joints and pores characteristic of later nassellarians. Many of the skeletal characteristics of post-Paleozoic radiolarians occur in the Permian radiolarians shown in F–I; F possesses hollow spines similar to recent Phaeodarians, G possesses a pore field with small pores and lots of pore area similar to Cenozoic collosphaerids, H exhibits squarish pores similar to Cenozoic orosphaerids, and I exhibits a spongy wall structure developed by many post-Paleozoic groups. (A courtesy of J.H. Lipps, University of California, Berkeley. Permian specimens courtesy of W. Cornell, University of Texas, El Paso. Scanning electron photography courtesy of J. Straw and Chevron Oil Co.)

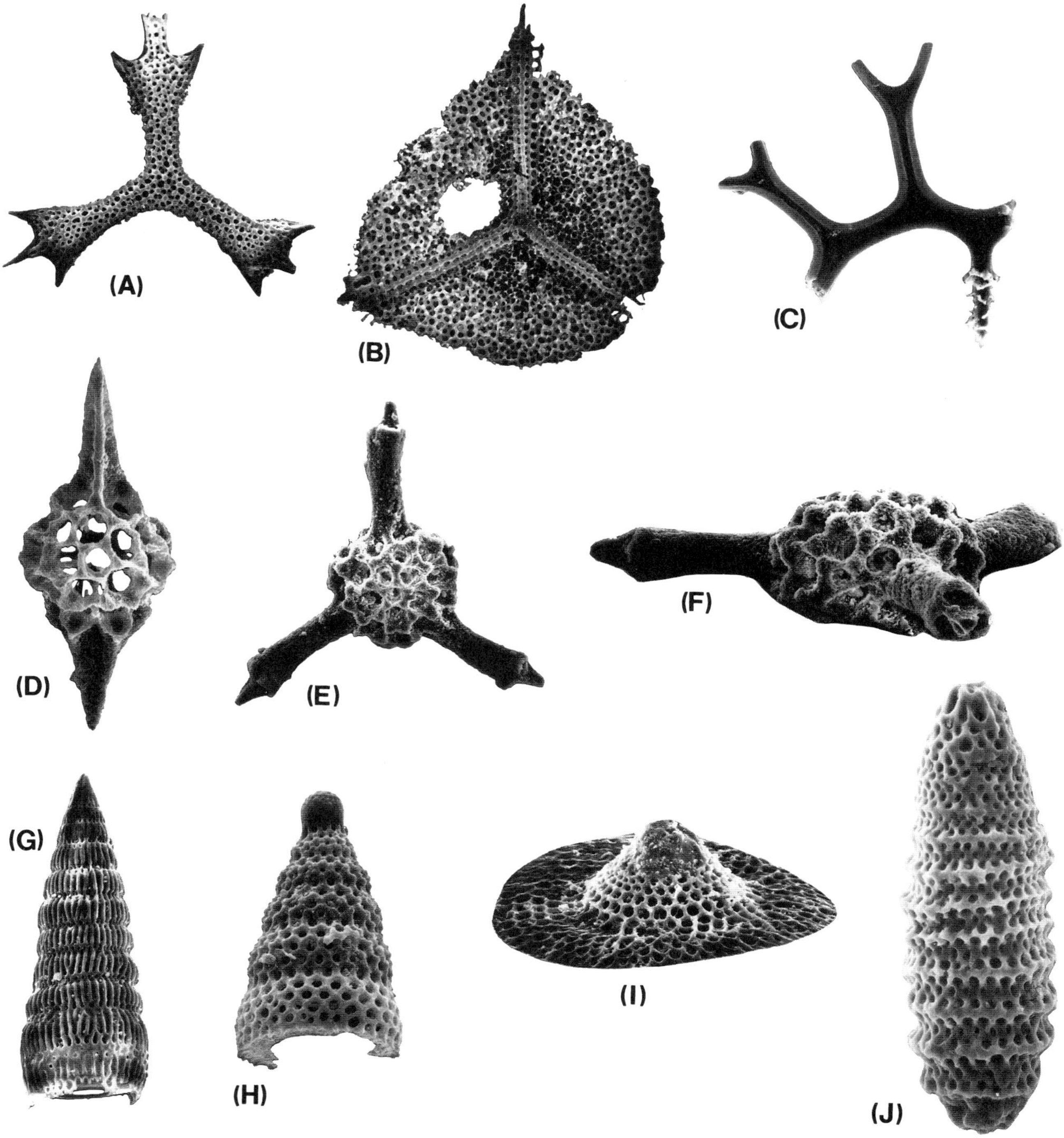

Figure 13.27 Mesozoic radiolarians. Hagiastrids are illustrated as specimens A and B, the B specimen exhibits a patagium. A portion of a saturnulid is shown as specimen C; even this small piece gives away its Mesozoic affinity by being robust and ornate. A stylosphaerid is shown as D, and a three-spined form with hollow spines (seen in side view F) characteristic of the Mesozoic is shown as E and F. Nassellarians characteristic of the Mesozoic include dictyomitrids (G) and amphipyndacids (H). Mesozoic ancestors of dominantly Cenozoic forms include the early theocalyptrin (I) and dictyomitrid-artostrobid-like form (J). (All photographs, except G, courtesy of D. Wagner and Chevron Oil Co.)

Figure 13.28 Cenozoic radiolarians. Specimens A and B are eucyrtidins and specimens C and D are pterocorids. Specimens E and F represent the symbiot-bearing groups typical of the Cenozoic, the collosphaerids and spongasterids respectively. Specimen G is a theocalyptrin probably evolved from a Mesozoic ancestor such as shown as Figure 13.27I. Specimen I is tholonid and specimens J and K are spines of orsphaerids; both tholonids and orosphaerids are deep-living groups. Specimens L and M are phaeodarians, specimen L is a spine, and specimen M is an entire skeleton.

Although most other planktic fossil groups undergo a drastic taxonomic reduction at the Mesozoic–Cenozoic transition, the major radiolarian faunal reorganization occurred within the Early Cenozoic. The radiolarians differ from these other microfossil groups in being both shallow and deep living, where the other groups are mainly shallow. This may be the reason for the different behavior of the radiolarians at the Mesozoic–Cenozoic transition. In exceptionally well-preserved Paleocene radiolarian assemblages, amphipynacids and dictyomitrids can be found. Hagiastrids, so characteristic of the Mesozoic, are commonly found in early Paleogene and to some extent in later Paleogene sediments (Fig. 13.25). Therefore, apparently the major radiolarian events during the Late Mesozoic, Mesozoic–Cenozoic transition, and shortly thereafter, are not mass extinctions of Mesozoic groups, but the evolution of Cenozoic groups (however, extinctions at the species level during this time may have been very significant).

Many Cenozoic groups evolved at this time (Fig. 13.25 no.3). An environmental pressure related to this rediversification in radiolarians might have been the rediversification of planktic foraminiferans and other planktic groups, after the terminal Cretaceous extinctions within those groups, or related to deep-living groups within the radiolaria.

From the Cretaceous on, radiolarians and planktic foraminifera were both nannoherbivores and symbiont bearers. The evolution of diatoms and their rise to dominance in the Cretaceous may have had a major impact on radiolarian evolution. Apparently with the development of diatoms, a competition for dissolved silica resulted that radiolarians had not previously experienced. The average weight of radiolarian skeletons decreased during the Cenozoic as diatoms flourished. Although shallow, warm-water radiolarians exhibited a loss of skeletal silica throughout the Cenozoic, deep, cold-water forms did not, indicating that only those radiolarians associated with diatoms were affected.

The last major radiolarian reorganization occurred at the Paleogene–Neogene transition (Fig. 13.25 no.4). A number of groups either evolved or increased in dominance and diversity at this transition so that they are considered Neogene groups. The evolution of the cold-water groups in the mid Miocene was considered to be, at least in part, a response to the initiation of the Neogene water mass regime at that time, and the establishment of the new intermediate and circumpolar water masses that they invaded. The warm-water, symbiont-bearing groups evolved as the Neogene water mass regime with oligotrophic subtropical gyres developed. This development of eutrophic and oligotrophic provinces may have been aided by the evolution of a significant biomass of symbiotic radiolarians which sequestered enough carbon to lower the greenhouse effect. Similar oligotrophic environments were probably available and occupied by other radiolarians in the Paleogene; however, the Neogene reorganization of water masses eliminated these forms.

Speciation and microevolution

The geographic occurrences of species of the *Lamprocyrtis* and *Stichocorys* lineages in plankton and in Recent and fossil sediments indicate both **allopatric** and **sympatric** speciation. Speciation by branching (allopatric speciation), with the ancestor coexisting in time, occurred for a significant time when the new species occupied an environment different than its ancestor. Speciation by replacement (possibly **sympatric**) with the brief ancestor–descendant co-occurrence, occurred when the new species remained in the ancestral species' geographic environment. These long overlaps in the sedimentary record might suggest gradualism and the brief overlaps suggest punctuation; however, both appear to occur almost instantaneously (except perhaps for the cases of allopatry described) when the paleogeographies, especially the water column paleogeographies, are reconstructed. Some significant amount of time is apparently required for an allopatrically evolved species to inhabit its maximum geographic range, but very little time is required for a sympatrically evolved species to inhabit its maximum geographic range.

Spongaster pentas (the pentagonal form, Figs 13.5 no.19, 13.18, 13.23B) is an important biostratigraphic indicator (Fig. 13.27) whose first and last occurrences (evolution and extinction) have been dated in the equatorial Pacific paleomagnetically as 4.6 and 3.4 Ma, respectively. However, it is living in the Gulf of Mexico and is perhaps a relict left when the Isthmus of Panama separated the previously connected equatorial Atlantic and Pacific Oceans. *S. pentas* is also living in, and apparently restricted to, the eastern tropical Pacific (Fig. 13.9). In the north Pacific, *S. pentas* exhibits a broad distribution at about 5 Ma. At about 4 Ma its range began to contract to the tropical regions and ultimately to the eastern tropical Pacific where it occurs in both Quaternary sediments and the plankton. *S. tetras tetras* (Fig. 13.5 no.17; the quadrangular form) evolved from *S. pentas* at approximately 3.4 Ma and *S. tetras irregularis* (similar to *S. tetras tetras* except that it is rectangular rather than squarish in outline, Fig. 13.5 no.18) evolved from *S. tetras tetras* sometime later than 3.4 Ma. Both evolutionary events may have resulted from a progressive decline of the parent *S. pentas* stock in middle and high latitudes due to climatic deterioration.

Radiolarian zoogeography and the physical oceanography are important in radiolarian evolution. Cenozoic radiolarian microevolution and paleophysical oceanography is related to the number of potential radiolarian provinces, the differences between tropical surface and bottom water paleotemperatures, the relative radiolarian species diversity, relative diversity and abundance of deep-living species, and relative paleoproductivity derived in the equatorial region (Fig. 13.29). A positive correlation is apparent for all of these factors since the latest Oligocene to the present (Figs 13.29, 13.30B, C, D, E). As the number of radiolarian provinces or the temperature difference between surface and bottom waters increases, so do the relative total radiolarian and deep-water species diversities, relative deep-water species abundance, and relative paleoproductivity. The number of potential radiolarian provinces since the latest Oligocene has almost doubled. This has led to significant increases (approximately double) in the relative radiolarian diversities of both shallow- and deep-water radiolarian species. The easiest explanation for this is that the increase in oceanographic differences allowed more ecologic opportunities for radiolarians, and they have then allopatrically evolved into new species.

Specific oceanographic events may well have caused the development of these new provinces (water masses) in the Neogene. At about 22 Ma the Antarctic Convergence may have been formed, and coupled to that formation was the initiation of a new water mass, the Antarctic Intermediate waters (the dashed lines on Fig. 13.30C). At about 6 Ma the development of a northern intermediate water added one more deep- and one more shallow-water mass (Fig. 13.30E). Three more water masses were added when glaciation in the northern hemisphere (about 3 Ma) resulted in more north Atlantic Deep and Bottom Water and a development of enough mixing to form Circumpolar Water, a secondary (mixing) water mass. The last Pacific water mass to be formed was the eastern Tropical Pacific Water which was formed at about 3.5 Ma related to the uplift of the Isthmus of Panama.

Thus eight or nine new provinces have been added in the Neogene resulting in a total of 15, or about twice as

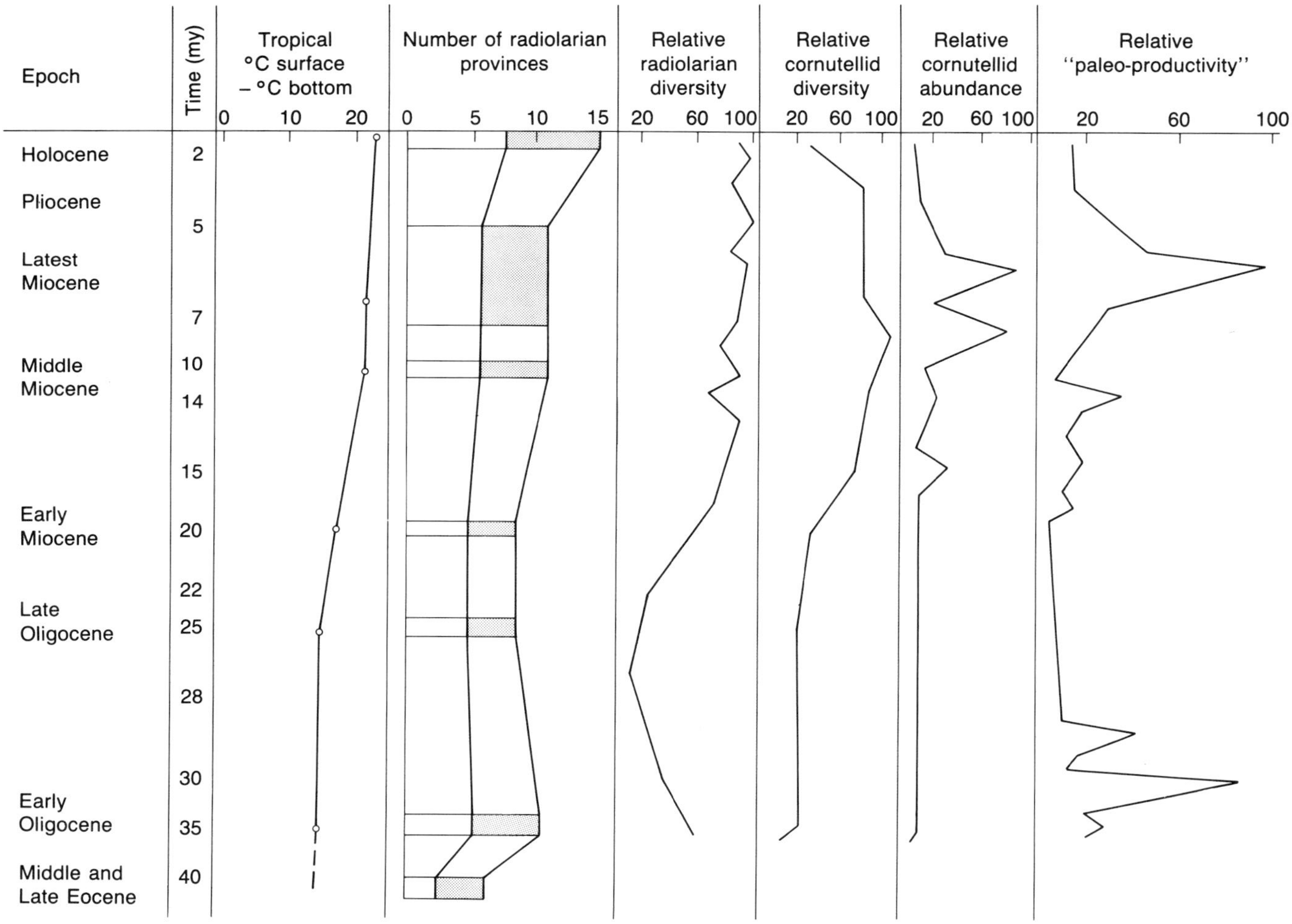

Figure 13.29 A comparison of the number of radiolarian provinces through the Cenozoic with the relative radiolarian species diversity, relative cornutellid (a deep-living radiolarian) "diversity" and relative abundance, and relative paleoproductivity. (Modified from Casey RE, Weinheimer AL, Nelson CO. *Bull Marine Sci* 1990; 47:221–32.)

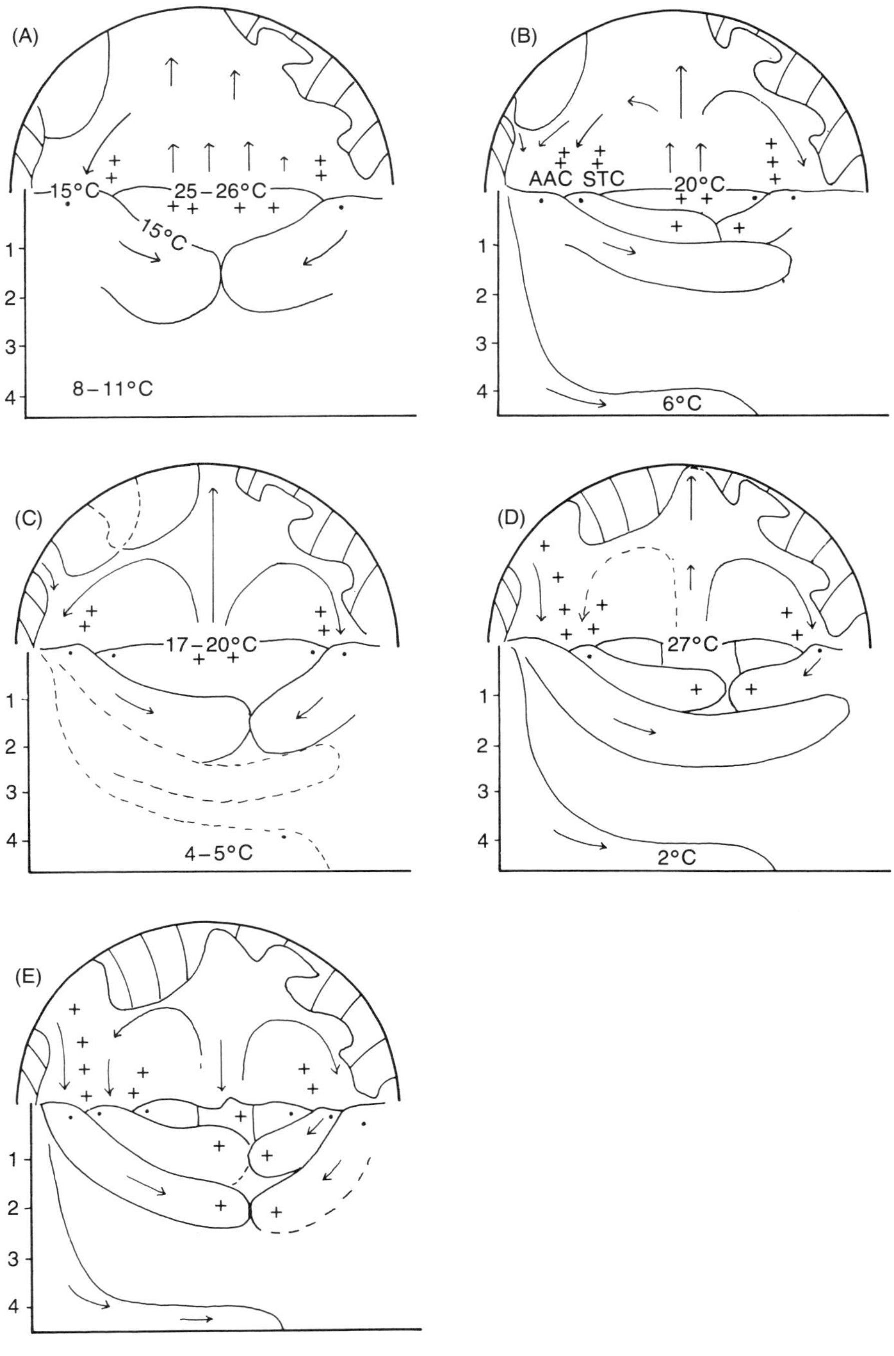

Figure 13.30 Geographic (N–S; tops of figures) and depth (bottoms; depth in km) reconstructions of the Pacific during: A, Middle Ecocene (44–46 Ma) and Late Eocene (40–42 Ma); B, Early Oligocene (34–36 Ma); C, Late Oligocene (24–26 Ma) and Early Miocene (18–20 Ma); D, Middle Miocene (10–12 Ma); E, Late Miocene (75 Ma) and Recent. All drawings illustrate the cross-sectional water masses with solid lines, and on the surface of the ocean the convergences with + and divergences with +. In the cross-sectional part of the drawings x refers to water movement into the page and • refers to water movements out of the page. The two drawings of Australia in C illustrate the movement of Australia northward from the 24–26 to the 18–20 Ma time slice. (Reproduced from Casey RE, Weinheimer AL, Nelson CO. *Bull Marine Sci* 1990; 47:221–32.)

many as in the latest Oligocene. The evolution and expansion of symbiotic radiolarian biomass may have resulted in a lowering of greenhouse conditions which results in the formation of these water masses that become new radiolarian provinces. These new provinces have allowed for increases in shallow radiolarian diversities (three or four more provinces) and deep radiolarian diversities (five more provinces).

Supplementary reading

Anderson OR. *Radiolaria*. New York: Springer-Verlag, 1983:355. An excellent survey of the work to date on the biology of radiolarians, with a good review of other aspects of radiolarian research.

Bolli HM, Saunders JM, Perch-Nielsen K, eds. *Plankton Stratigraphy*. Cambridge: Cambridge University Press, 1985:1032. Contains up-to-date Paleozoic, Mesozoic, and Cenozoic radiolarian biostratigraphic zonations.

Campbell AS. Radiolaria. In: Moore RC, ed. *Treatise on Invertebrate Paleontology, Part D, Protista 3*. Boulder and Lawrence: Geological Society of America and University of Kansas Press, 1954:D11–D163. Good illustrations and readily available.

Casey RE, Barron JA, eds. *Siliceous Microfossil and Microplankton Studies of the Monterey Formation and Modern Analogs*. Pacific Section, Society of Economic Paleontologists and Mineralogists, Special Publication 45, 1986:147. Contains a variety of studies exhibiting the geologic and paleoceanographic uses of radiolarians.

Cline RM, Hayes JD, eds. *Investigations of Late Quaternary Paleoceanography and Paleoclimatology*. Geological Society of America Memoir 145. Boulder: Geological Society of America, 1976:465. Use of radiolarians and other organisms to reconstruct paleoceanography and paleoclimatology.

Funnell BM, Riedel WR, eds. *The Micropalaeontology of Oceans*. Cambridge: Cambridge University Press, 1971:828. Contains some of the modern radiolarian classifications and some radiolarian ecologic and biostratigraphic work.

Haeckel E. Report on the radiolaria collected by HMS *Challenger* during the years 1873–1876. *Rept Voyage Challenger, Zoology 18* 1887; 1803. Contains excellent illustrations of radiolarians.

Kennett JP. *The Miocene Ocean: Paleoceanography and Biogeography*. Geological Society of America Memoir 163. Boulder: Geological Society of America, 1985:337. Reconstructions of paleoceanography.

Kling SL. Radiolaria. In: Haq BU, Boersma A, eds. *Introduction to Marine Micropaleontology*. New York: Elsevier, 1978:203–44. A survey of radiolaria and radiolarian work up to that time.

Nigrini C, Lombari G. *A Guide to Miocene Radiolaria*. Cushman Foundation for Foraminiferal Research, Special Publication 22, 1984:422. A survey of some common Miocene radiolarians. For each species a figure or a photograph (or photographs or drawing) is given as are descriptions, dimensions, remarks, and sometimes geologic and geographic distributional information.

Nigrini C, Moore TC Jr. *A Guide to Modern Radiolaria*. Cushman Foundation for Foraminiferal Research, Special Publication 16, 1979:342. A survey of some of the most common living radiolarians. For each species a figure is given as are descriptions, dimensions, remarks, and sometimes distributional information.

Tintinnids

14

Helen Tappan

Introduction

The tintinnids, almost the only ciliate protozoans to have a fossil record, are an important component of the present-day microscopic zooplankton. The highly differentiated protozoans of the subphylum Ciliophora are probably the most advanced as well as the most homogeneous protozoan group. The cell is covered by many tiny hair-like cilia that are used in feeding and locomotion of the protozoan. About 14% of the estimated 7200 species of living ciliates are tintinnids, a group characterized by having an external cup-shaped or tubular organic covering or lorica that may be quite firm and resistant and may also be covered partly or completely by attached quartz particles or skeletal fragments of other microscopic organisms, such as the diatoms or coccolithophorids upon which they feed. In the present seas, the abundant tintinnids are a major food source for the larger zooplankton.

Rare occurrences have been reported of soft-bodied ciliates, other than tintinnids, preserved in chert, amber, and bitumenous shale, but even these few records are not universally accepted. They are stated to range from about 60 to 280 μm in diameter, none yet is thoroughly described and illustrated, and the geologic age of many is also uncertain.

The probable fossil tintinnids or calpionellids have been described from thin sections of Jurassic and Lower Cretaceous pelagic limestones. They occur in the Alpine and Mediterranean areas of Europe and correlative strata in northern Africa, Cuba, and Mexico. A relationship with modern tintinnids was first suggested in the mid 1930s, but since then similar fossils thought to be tintinnids have been recognized from the Paleozoic, elsewhere in the Mesozoic, and from Tertiary strata in the USA and Europe. Because of their small size, as compared to other microfossils such as foraminifers and radiolarians, they may have been overlooked even when present.

History of study

The ciliates were originally termed *Infusoria*, because of their common appearance in infusions of dried grasses. Such ciliates as the freshwater *Paramecium* excysted as water became available, as did associated flagellates and amebans. Later morphologic emphasis led to proposal of the name Ciliata for the infusorians, thus contrasting this group with the Mastigophora (with one or more flagella for locomotion) and the Rhizopodea (with impermanent pseudopodia).

Living tintinnids have been known for about two centuries, the earliest described genus being *Tintinnus*, proposed by Schrank in 1803. In contrast to the nonloricate freshwater ciliates, many early studies of tintinnids were based on plankton samples obtained by oceanographic expeditions from the various seas, and descriptions emphasized the nature of the lorica. More detailed studies of the living cell, using the light microscope, were done in the 1920s and 1930s, but many features were only clarified after the scanning electron microscope was used to examine freeze-dried specimens.

Fossil calpionellids were described from thin sections of Upper Jurassic and Lower Cretaceous limestones during the 1930s and 1940s and suggested to represent loricae of oceanic tintinnids; their known geologic range has gradually been extended since then. Neritic tintinnids have also been described from the mid-Cretaceous of Colorado, the Eocene and Oligocene of Europe, and the American Gulf Coast and Caribbean.

Morphology and systematics

Ciliates are distinguished from other protozoans in having cilia or compound ciliary organelles during at least some stage of their life-cycle. In addition, a functional mouth or cytostome is provided with a well-defined buccal cavity related to the ciliary organelles. Internally, the cells have two kinds of nuclei, smaller micronuclei with numerous recognizable chromosomes, and larger irregularly shaped macronuclei. The ciliate subclass Spirotrichia, which includes the tintinnids, has differentiated cilia with conspicuous compound ciliary structures in the buccal (mouth) region and a reduced cover of cilia over the remainder of the cell body (Figs 14.1–14.3).

Living tintinnids range in length from 10 to about 1000 μm, commonly being between 100 and 300 μm. The cell is conical or bell shaped and enclosed within a **lorica** that may be 10 times as large (Figs 14.4, 14.5). Truncate and concave at the oral end, the cell body tapers to a stalk-like **peduncle** at its base, by which it is attached to the interior wall of the lorica near the base of the cavity or at one side in the lower third. The peduncle may taper gradually to an adhesive tip or branch and anastomose at the attachment. As the protozoan appears to turn freely within the lorica, the attachment may be temporary and may change in position. By extending or contracting the peduncle as well as the main body, the tintinnid may protrude from the lorica or be entirely retracted within it.

The mouth or cytostome lies in a deep preoral cavity and is surrounded by a sinistrally spiraling (counterclockwise) whorl of from 16 to 24 long narrow compound

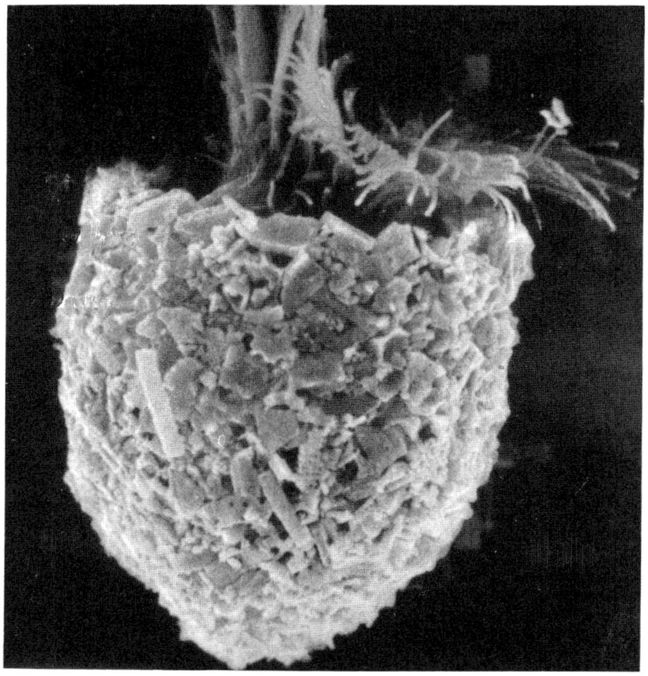

Figure 14.1 *Tintinnopsis parva* (Merkle). Scanning electron micrograph (SEM) of cultured specimen, fixed and dried by critical point method, showing agglutinated cap-like lorica, with ciliary membranelles of the protozoan protruding from the aperture. Note that the fringed edge of the membranelles continues as individual cilia. Approximately ×2100. (Reproduced from Gold K. *J Protozool* 1979; 26:415–9.)

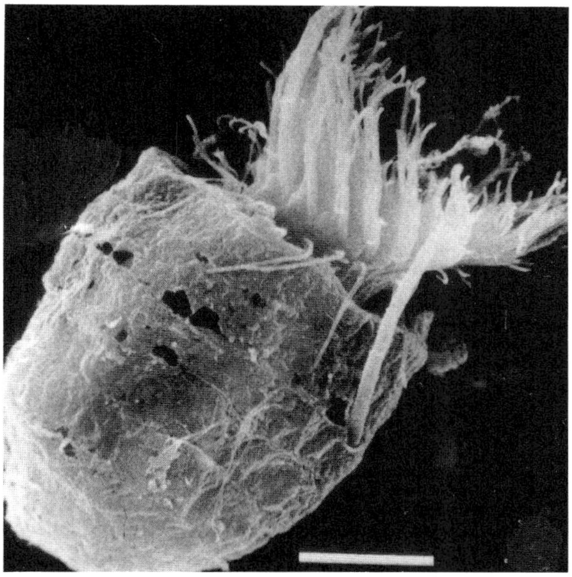

Figure 14.2 *Tintinnopsis parva* (Merkle). SEM of cultured specimen, fixed and dried by critical point method. This specimen was grown in the absence of particulate material that could be used for agglutination, hence the lorica is entirely organic, although well formed. The ciliary membranelles protruding from the lorica opening show their oblique attachment. Scale bar, 10 μm. (Reproduced from Gold K. *J Protozool* 1979; 26:415–9.)

Figure 14.3 *Tintinnopsis parva* (Merkle). SEM of cultured specimen, fixed and dried by critical point method. This specimen was prepared shortly after asexual reproduction, and being the anterior individual or proter it left the original lorica and was free swimming for a time before producing a new lorica. Note the sparse short cilia on the body, in comparison to the longer cilia of the complex ciliary membranelles around the cytostome. Scale bar, 5 μm. (Reproduced from Gold K. *J Protozool* 1979; 26:415–9.)

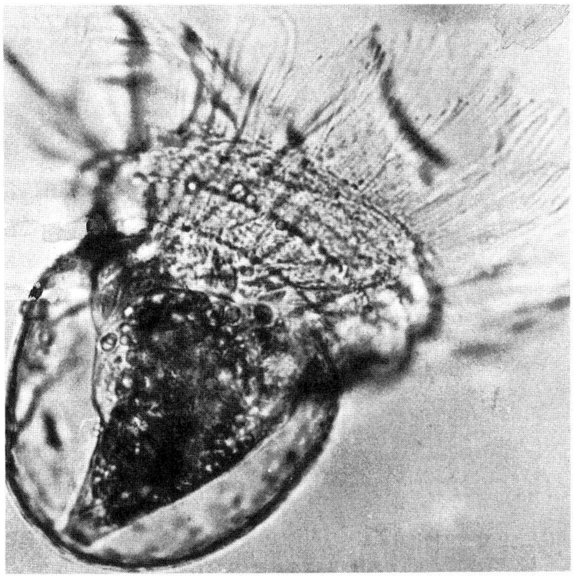

Figure 14.4 *Petalotricha ampulla* (Fol). Mediterranean Sea. Side view, in transmitted light, showing conical body attached at the base of the cup-like lorica, and with the adoral membranelles extended from the opening of the lorica. Approximately ×450. (Reproduced from Laval-Peuto M. *Protistologica* 1972; 8:369–86.)

ciliary plates, known as **membranelles**. The ciliary membranelles fasten obliquely and are deeply fringed inward into the individual cilia that function in locomotion and feeding.

The transparent, colorless, and homogeneous cytoplasm is covered with a fine **pellicle** (Fig. 14.6). Food vacuoles are abundant and waste such as the skeletal elements of prey organisms is extruded from the cell through a cytopyge on the ventral surface of the body, just above the peduncle.

The lorica

The external covering or **lorica** of the tintinnid may have the form of an inverted cap, helmet, cup, goblet, acorn, horn, or tube. Most are radially symmetrical, but some are laterally compressed. The protozoan body may protrude from the oral end of the lorica, which usually is closed at the opposite end but rarely is open at both. Cultures of living species have shown that the size and proportions of the lorica may change with age or the

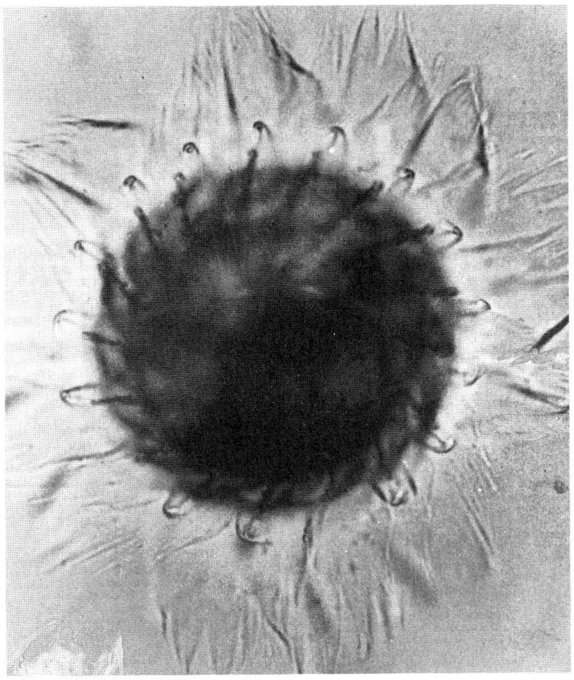

Figure 14.5 *Petalotricha ampulla* (Fol). Mediterranean Sea. View from above, showing the adoral membranelles spreading out around the cytostome. Approximately ×450. (Reproduced from Laval-Peuto M. *Protistologica* 1972; 8:369–86.)

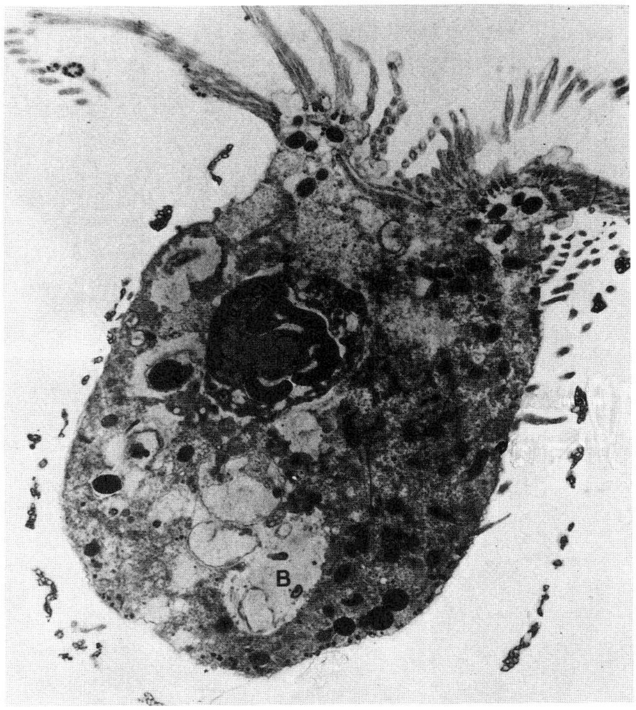

Figure 14.6 *Tintinnopsis parva* (Merkle). Cultivated specimen, longitudinal section in transmission electron microscopy, showing adoral membranelles at top, sections of some showing the ring of tubules characteristic of all cilia and flagella, and very sparse body cilia; digestive vacuole in center of figure, showing ingested prey. B, Symbiotic bacteria. (Reproduced from Laval-Peuto M, Gold K, Storm ER. *Trans Amer Microscop Soc* 1979; 98:204–12.)

physiologic state of the individual but that the oral diameter of the lorica remains constant.

The wall of the lorica is organic, although the exact composition is not known. It ranges from clear, hyaline, and transparent to thick and opaque, it may be simple and undifferentiated in structure or have distinct layers, and the consistency varies from gelatinous or membranous to firm. Some have agglutinated quartz grains or food debris such as coccoliths and diatom frustules at the exterior of the lorica. The agglutinated particles may be sparse or may be tightly packed and firmly cemented (Fig. 14.7), resulting in a yellowish or brownish color. Whether the lorica is wholly organic or agglutinated appears to be genetically controlled, although agglutinating species may use whatever material is locally available. A single plankton haul may contain species with a wide variation in lorica type, whether wholly agglutinated (Codonellopsidae: Fig. 14.8; Codonellidae: Fig. 14.9), having an agglutinated bowl and hyaline organic collar (Codonellopsidae: Fig. 14.10; Dictyocystidae: Fig. 14.11), or wholly organic and hyaline (Tintinnidae, Xystonellidae: Fig. 14.12).

Tintinnids having compound walls may have two layers separated by an open space, or the intervening space may be filled with closely packed debris such as coccoliths or have a honeycomb-like reticular structure. The lorica surface may be smooth or textured, with ribs, plications,

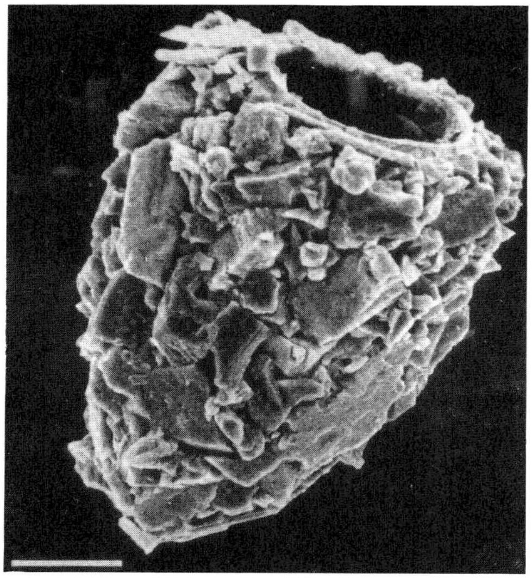

Figure 14.8 *Stenosemella oliva* (Meunier). Atlantic, Woods Hole, Massachusetts. Empty lorica, viewed in SEM, showing agglutinated lorica, and rimmed aperture. Scale bar, 8 μm. (Reproduced from Gold K, Morales EA. *Biol Bull* 1976; 150:377–92.)

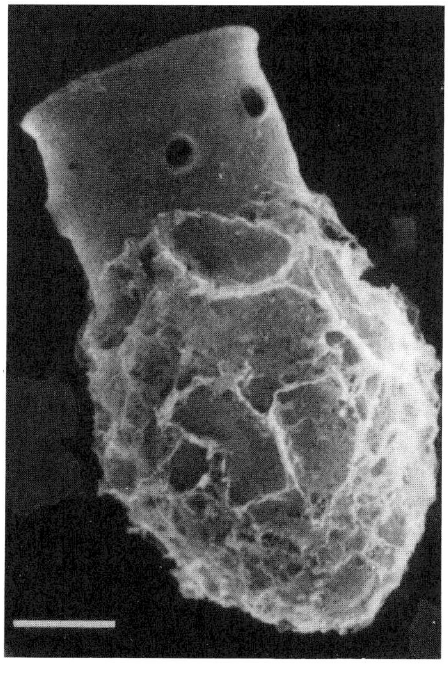

Figure 14.7 *Codonellopsis americana* (Kofoid & Campbell). Enewetak Atoll, Pacific. Empty lorica, viewed in SEM, showing smooth, hyaline organic neck with perforations; the agglutinated lorica bowl has been treated with hydrochloric acid to remove calcareous particles, and shows only the remaining organic material that formerly cemented the grains to the wall. Scale bar, 8 μm. (Reproduced from Gold K, Morales EA. *J Protozool* 1977; 24:580–7.)

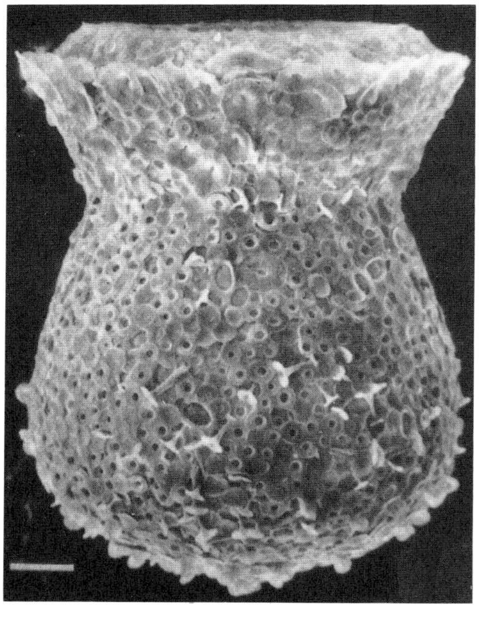

Figure 14.9 *Codonaria oceanica* (Brandt). Enewetak Atoll, Pacific. Empty lorica in SEM, showing distinctive neck, flaring collar, and suboral cone above the collar. Lorica bowl densely encrusted with coccoliths of *Discosphaera, Rhabdosphaera, Syracosphaera, Umbilicosphaera, Umbellosphaera,* and *Emiliania*, probably from the cells that were ingested as food. Scale bar, 9.5 μm. (Reproduced from Gold K, Morales EA. *J Protozool* 1977; 24:580–7.)

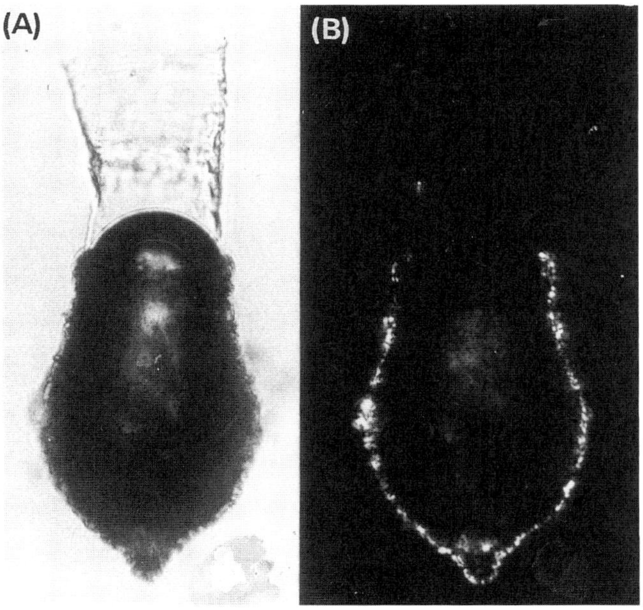

Figure 14.10 A and B, *Codonellopsis* sp. Plankton sample from Gulf of California, Mexico. A, Specimen viewed in light microscope, showing thick agglutinated wall of the lorica bowl, and thin hyaline wall of the elongate neck; air bubble in the lorica appears dark in the photograph. B, Same specimens viewed in cross polarized light, showing calcareous coccoliths in the agglutinated bowl that appear light-colored, but absence of any calcareous material on the rock. A and B ×370. (Reproduced from Tappan H, Loeblich AR Jr. *J Paleontol* 1968; 42:1378–94.)

fluting, shelves, and perforations (Fig. 14.13). The most distinctive modifications occur in the oral region, and the collar may be differently ornamented than the bowl of the lorica. The oral region may have a smooth to denticulate differentiated rim, or may have a distinct collar. The

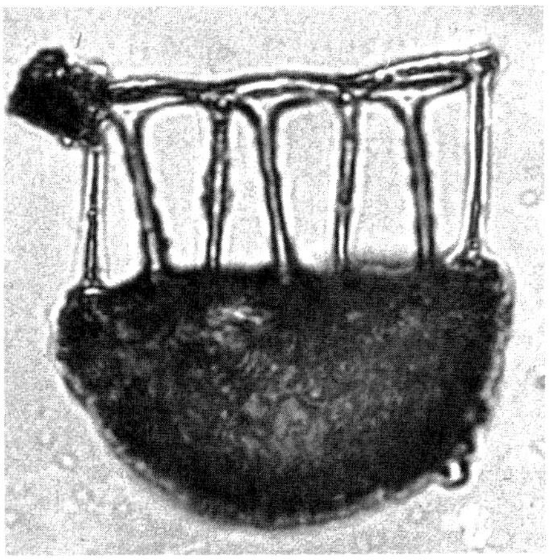

Figure 14.11 *Dictyocysta* sp. Eastern Pacific, off San Diego, California. Viewed in transmitted light, showing roughly finished bowl exterior, probably agglutinated, and transparent wholly organic arched frames of the neck. ×770. (Reproduced from Tappan H, Loeblich AR Jr. *J Paleontol* 1968; 42:1378–94.)

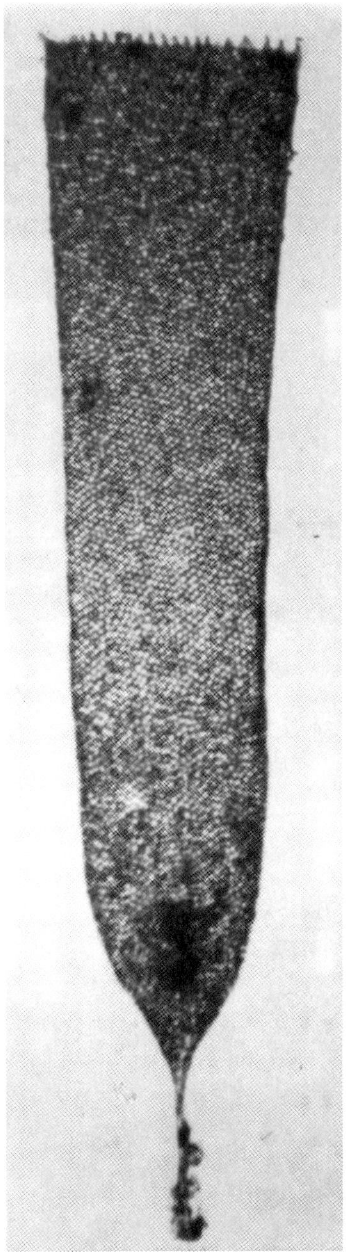

Figure 14.12 *Parafavella gigantea* (Brandt). Western Atlantic, New York Bight. Air-dried lorica of unusually large species, showing elongate and reticulate lorica bowl, denticulate rim, and aboral horn. Length before drying was 463 μm. (Reproduced from Gold K, Morales EA. *Trans Amer Microscop Soc* 1975; 94:142–5.)

collar may be single or double, short, or elongate, and may have large window-like openings (as in Fig. 14.11). The collar may restrict the opening or be tubular or broadly flaring. An internal shelf or diaphragm at the junction of the bowl and collar may constrict the opening. Other modifications may occur at the basal extremity, that may terminate in a short stump or an elongate horn

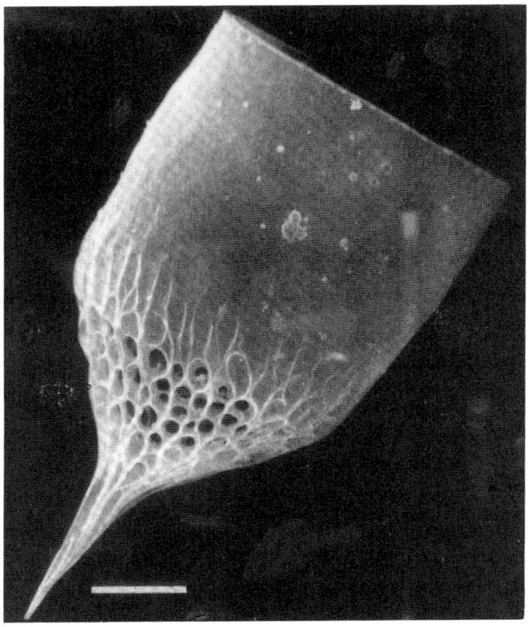

Figure 14.13 *Epiplocylis pacifica* (Kofoid & Campbell). Enewetak Atoll, Pacific. Empty lorica in SEM, showing hyaline organic wall, produced aboral horn, and costate and pierced reticulate ornamentation of the base of the bowl. Scale bar, 16 μm. (Reproduced from Gold K, Morales EA. *J Protozool* 1977; 24:580–7.)

Figure 14.14 *Rhabdonella elegans* (Jörgensen). Enewetak Atoll, Pacific. Empty lorica in SEM, showing hyaline organic wall of the slender elongate lorica, produced aboral horn, and costate ornamentation. Scale bar, 14 μm. (Reproduced from Gold K, Morales EA. *J Protozool* 1977; 24:580–7.)

or pedicel (Fig. 14.14). The basal horn may be hollow or solid, and smooth to elaborately ornamented. In other species, the base may be open and the lorica tubular (Fig. 14.15).

An organic lorica is commonest among modern tintinnids, but only a few organic ones have been reported as fossil. In contrast, fossil calpionellids of the Jurassic and Cretaceous, most Paleozoic representatives, and the pseudarcellids of the Tertiary have a calcareous lorica (Fig. 14.16). Some fossil representatives also have agglutinated loricae, some with quartz particles are known from the mid-Cretaceous, and many Jurassic–Cretaceous calpionellids also include numerous coccoliths in their lorica walls (Fig. 14.17).

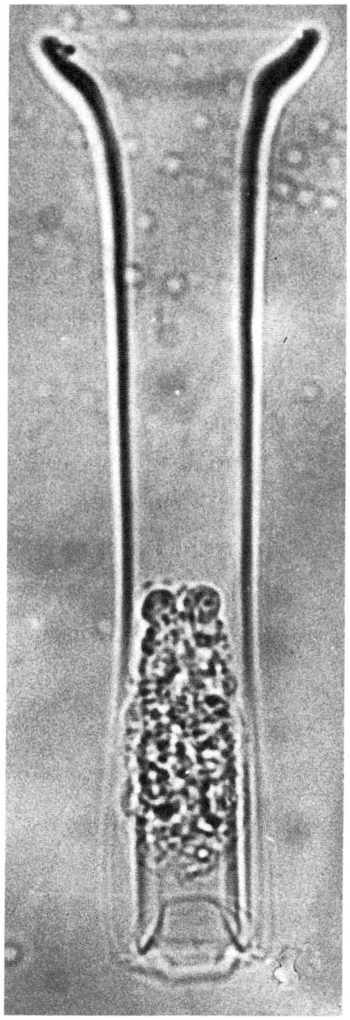

Figure 14.15 *Eutintinnus* sp. eastern Pacific, off Baja California, Mexico. Tubular hyaline organic lorica, open at both ends, but flaring at the apertural end. Enlarged ×871. The contracted preserved protozoan is visible near the base. (Reproduced from Tappan H, Loeblich AR Jr. *J Paleontol* 1968; 42:1378–94.)

Various suggestions have been proposed to explain the calcareous fossils, as no comparable secreted calcareous lorica is known in living tintinnids. At first, an originally organic lorica was suggested to have been diagenetically replaced by calcite; others suggested that the detrital quartz grains of an agglutinated lorica had been replaced by calcite, as occurred rarely with the opaline silica of radiolaria and sponge spicules. However, the crystalline silica of quartz is much more stable than the opaline silica of the radiolarians, and would not be similarly replaced. In addition, the common association of unchanged agglutinated foraminiferans together with calcareous fossil tintinnids also made replacement appear unlikely. Another suggestion was that the lorica was originally constructed of tightly packed coccoliths (as are those of some modern species) and that the coccoliths were diagenetically recrystallized and cemented in the calcareous pelagic muds of the sea floor. A suggestion that the lorica was originally of aragonite that was later converted to calcite was shown to be unlikely as the calcite loricas were found in association with well-preserved agglutinated foraminifera, and the Eocene species occurred with both calcitic and aragonitic foraminiferal tests that showed no diagenetic alteration.

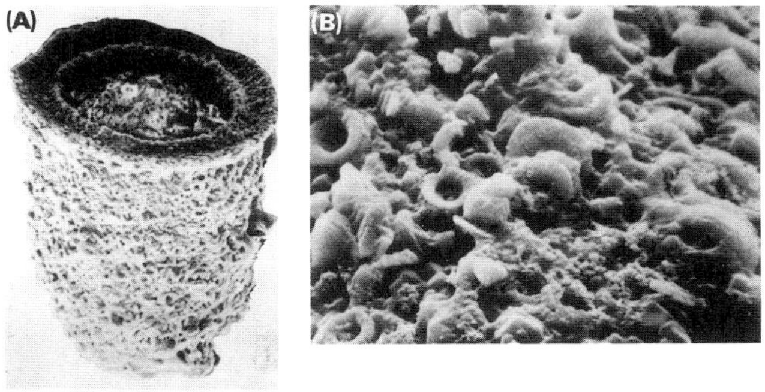

Figure 14.17 *Remaniella* sp. aff. *cadischiana* (Colom). Lower Cretaceous (Berriasian), western Atlantic, off the Bahamas, Deep Sea Drilling Project, Leg 11, Site 100. A, Side view of lorica, showing flaring collar. ×345. B, Enlargement of surface, showing agglutinated coccoliths. ×1518. (Reproduced from Luterbacher HP. *Initial Rep Deep Sea Drill Project* 1972; 11:561–93.)

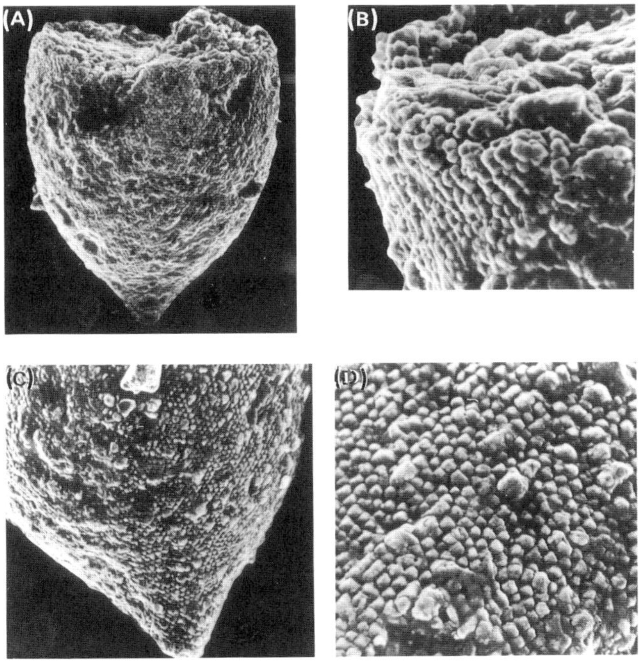

Figure 14.16 *Remaniella murgeanu* (Pop). Lower Cretaceous (Hauterivian), eastern Atlantic, off northwest Africa, Madeira Islands, Deep Sea Drilling Project, Leg 50, Site 416A. A, Side view of lorica, upper part with collar broken, ×637.5. B, Enlargement of rim of lorica, in oblique view, showing small nodular crystallites that appear to form several layers, ×2550. C, Enlargement of apiculate base of the bowl, showing oblique alignment of the crystallites, ×984. D, Enlargement of surface of lorica, showing rows of small nodular crystallites and occasionally interspersed larger ones, ×2460. (Reproduced from Bolli HM. *Initial Rep Deep Sea Drill Project* 1980; 50:525–43.)

Also supporting an originally secreted calcite lorica is the differing optical orientation of the calcite crystals in the apertural region of *Calpionellopsis* and those of the remainder of the lorica. Some calpionellids have a calcitic wall with dark median layer that is probably a remnant of a primary organic one. The Eocene *Remanellina* also has a distinctly layered wall, with smooth inner and outer layers enclosing an alveolar intermediate one (Fig. 14.18), and the associated *Tytthocorys* has a cryptocrystalline bowl surmounted by a collar constructed of distinctly larger calcite crystals.

Thus the calcite of the calpionellid lorica and that of other fossil tintinnids probably was original, and genetically controlled, as was the construction of loricae with organic walls, walls of tightly cemented coccoliths, and those with agglutinated quartz grains held in an organic matrix.

The above described features of the lorica, including the layered wall structure of some, presence of tightly cemented coccoliths or quartz grains agglutinated to an organic matrix in others, and still others of wholly organic composition, as well as the morphologic similarities in general form and size, differentiated bowl and neck structure, an inner shelf at the junction of bowl and neck, and the presence of a basal horn or pedicel, all suggest that these fossils are true tintinnids. The lorica of living tintinnids is constructed in a helical fashion, and the crystals of the wall of *Calpionella alpina* also have been shown to have a helicoid arrangement. In addition, the sutured

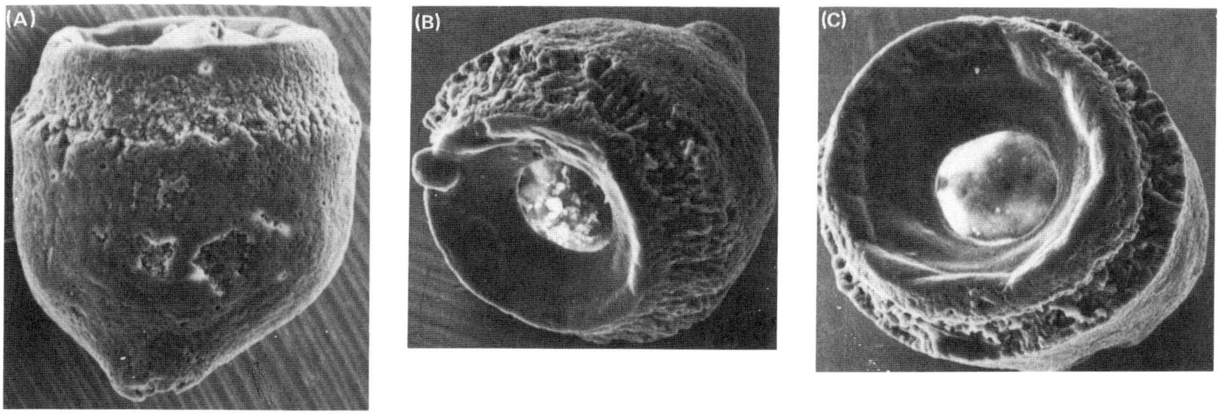

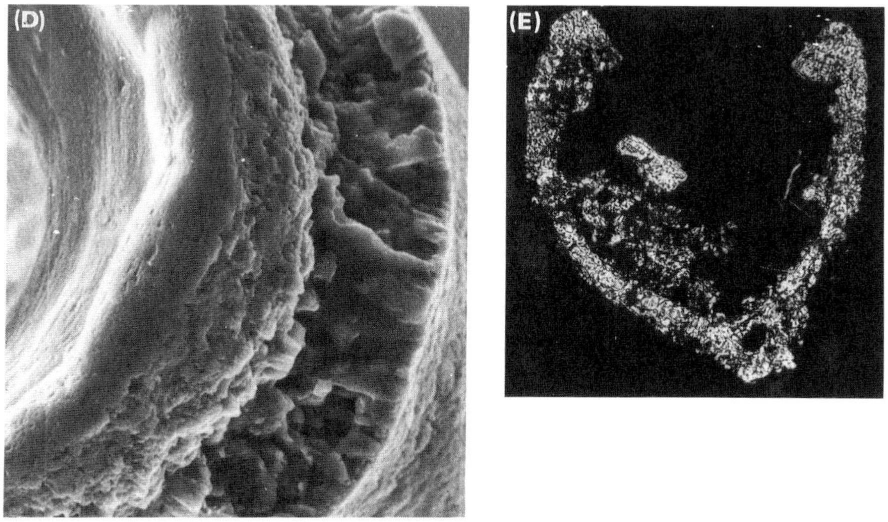

Figure 14.18 *Remanellina eocenica* (Tappan & Loeblich). Upper Eocene, Moodys Branch Formation, Jackson, Mississippi. A, Side view of holotype, showing smooth lorica bowl, high constricted collar, and short, blunt aboral horn. ×270. B, Oblique view of paratype, showing sloping circumoral shelf that constricts the opening. C, Slightly oblique oral view of a paratype, showing somewhat spiraling inner sculpture. A–C all ×270. D, Enlargement of specimen of C showing smooth inner and outer surfaces and alveolar interior. ×800. E, Longitudinal section of paratype, seen in crossed polarized light, showing gradual slope upward of the bowl, the internal circumoral shelf, and aboral horn. ×248. (Reproduced from Tappan H, Loeblich AR Jr. *J Paleontol* 1968; 42:1378–94.)

junction of collar and bowl of the calpionellid fossil *Colomiella*, and the differing optical orientation of the crystals of collar and bowl, strongly suggest a later addition of the collar, just as occurs in lorica construction by the living *Favella* (discussed below under "Construction of the lorica"). Possibly, the composition of the lorica was more varied in the geologic past, those with calcareous secreted lorica evolving during favorable periods of warmer waters but later disappearing in response to oceanographic changes that affected circulation patterns and temperatures.

Classification

In general, the present classification of the tintinnids has changed little over the past 60 years, although a few genera have been added. In 1929 Kofoid and Campbell included 12 families and 51 living genera, whereas in 1977 Corliss listed 13 families and 108 genera (including 30 fossil). Fossil and living taxa have been placed in the same families here, as by Corliss and others, although some place the fossil taxa in a separate superfamily and families, or, emphasizing the differences in composition, place them in separate ciliate orders and even in different protistan subphyla (Table 14.1).

Many species of tintinnids were described from material obtained in plankton samples by oceanographic expeditions. Because disturbances cause the protozoan to abandon its lorica, the plankton samples may contain only empty loricae; hence, many were differentiated by relative proportions of the lorica, or presence or length of necks, collars, and aboral horns. However, these features recently have all been shown to vary in monospecific cultures according to the age and physiologic state of the protozoan. Thus many specific names, and even some described genera, have been shown to be synonymous. Fossil species described from thin sections of limestones also may show considerable variation, depending on whether the section is completely axial, or somewhat oblique (hence failing to intersect the true base), and appearing rounded although in fact the species may be apiculate or even have a basal horn.

Biology

Locomotion and nutrition

Tintinnid locomotion results from the rhythmic pulsation of the oral membranelles. Some show a somewhat jerky backward movement, whereas *Tintinnopsis* normally moves with the oral end foremost. However, under stress any tintinnid may reverse its direction. Whether the direction and character of movement is due solely to the morphology and movement of the membranelles or is modified by shape of the lorica is not known, although the long aboral pedicel of some has been suggested to serve as a counterweight to the expanded organism protruding from the lorica opening.

Tintinnids feed upon other small protists, such as diatoms, dinoflagellates, coccolithophores, silicoflagellates and other chrysomonads, bacteria, radiolaria, and other ciliates. Some of these food organisms have been found in vacuoles within the tintinnid cytoplasm, and their skeletal elements commonly are used in constructing the lorica. The skeletal and other waste material accumulates within the cytoplasm until fission occurs and is then used in construction of a new lorica. The choice of agglutinated material on the lorica (coccoliths or diatom frustules representing certain species) has been suggested to indicate selectivity in food preference, but may only reflect local availability of material.

Because of their abundance, second only to diatoms in the Antarctic plankton, tintinnids form an important link in the oceanic food chain. They are voracious feeders on phytoplankton, and in turn are used as food by larger zooplankton and fish.

Reproduction

Production of unlike gametes in sexual reproduction typically does not occur in the ciliates. Sexual nuclear recombination appears to consist of a process of isogamous conjugation. Two or rarely more individuals join by their oral surfaces, and the micronuclei fuse during the several hours that the protozoans remain united at their peristome margins. Throughout this period, they cannot withdraw into the lorica, but may continue to swim. Such paired lorica have been observed in specimens obtained from plankton tows, and similar paired fossils have been reported from the Upper Jurassic of Greece. Unlike previously reported conjugation of similar individuals, *Tintinnopsis* was observed in culture to produce microgametes by lateral budding from the posterior part of the mother cell. These microgametes differ somewhat in structure from the normal vegetative cell, having short and conspicuous but not very active cilia, and they apparently are not free swimming. Their organelles appear to be used only for attachment to a receptive macrogamete (apparently identical to the normal vegetative cell), to which the microgamete transfers upon contact.

Asexual binary fission is the most common method of reproduction. The new individual first develops as a lateral bud at the rear of the parent cell (Fig. 14.19A, B, C). Membranelles form about the new cytostome on the bud, but before nuclear division commences two separately functioning individuals have developed, connected only by a thin strand of very fluid cytoplasm. Both the micronuclei and macronuclei then divide, the resultant smaller nuclei enlarging rapidly. The posterior daughter cell (or opisthe) inherits the parent lorica to which its peduncle remains attached, and the anterior daughter cell (or proter) leaves, forming a new lorica just before separation or soon thereafter.

Construction of the lorica

As the new peristome forms for the future opisthe during binary fission, a large number of opaque granules of material that later will be used for construction of the lorica arise near the base of the cell and migrate to the cell anterior. Just prior to separation of the daughter cells, the granules accumulate as a dark crescent in the **proter**, centered above the peristome of the **opisthe**.

Studies of cultures show that at least one species, *Favella ehrenbergi*, may construct three different types of lorica that had formerly been regarded as distinct species. All are elongate and cylindrical, taper at the base, and may have an aboral horn. The wall is transparent and homogeneous, with a single layer of alveolar structure (refringence may give this a three-layered appearance in the microscope). The typical or *Favella*-form has a smooth-walled, continuous bowl, and an aboral horn and may have a collar of the same diameter but a spiraling, overlapped structure. The overlapped spiraled wall of the *Coxliella*-form consists of a spirally wound band, and the lorica base may be flattened to apiculate, or rarely open, but lacks a horn (Fig. 14.19D). The *Decipiens*-form has a roughened surface, a short aboral horn, and a spiraled structure, although unlike in the other two forms, adjacent bands do not overlap.

Using the accumulated granules of lorica material it inherited, within a few minutes after separation of the

Table 14.1 Classification of Tintinnida

Subphylum Ciliophora (Doflein 1901)

Class Ciliata (Perty 1852)

Class Polyhymenophora (Jankowski 1967)

Subclass Spirotrichia (Bütschli 1889)

Order Oligotrichida (Bütschli 1887)

Protozoa with elongate body, thickened pellicle, reduced body covering of cilia, and adoral zone of ciliary membranelles

Suborder Tintinnina (Kofoid & Campbell 1929)

Loricate ciliates, with conspicuous oral membranelles. Free-swimming, mainly pelagic but some occur in the neritic plankton, dominantly marine but with some freshwater representatives

Taxon	Characteristics	Example	Occurrence
Superfamily Codonellidea (Kent 1881)	Pronounced ciliary membrane, lorica with spiral elements, may have oral constriction; agglutinated particles		Ordovician, Upper Devonian to Recent
Family Tintinnidiidae (Kofoid & Campbell 1929)	Tubular or flaring lorica, no distinct collar; gelatinous wall, may have agglutinated particles, spiral suboral structure; two living genera	*Tintinnidium* (Kent 1881)	Recent; marine, brackish, fresh water
Family Codonellidae (Kent 1881)	Lorica tubular to cup shaped, may have collar and constricted throat; aboral end apiculate or with horn, wall may be reticulate or agglutinated; 26 living genera (two also reported in Cretaceous, one in Pleistocene), 26 only fossil	*Codonella* (Haeckel 1873); *Dicloeopella* (Eicher 1965); *Tintinnopsella* (Colom 1948); *Tintinnopsis* (Stein 1867); *Tytthocorys* (Tappan & Loeblich 1968)	Ordovician–Upper Devonian, Upper Jurassic–Cretaceous, Eocene, Oligocene, Pleistocene, Recent. Fresh water, marine, neritic, and oceanic
Family Codonellopsidae (Kofoid & Campbell 1929)	Lorica top shaped, base rounded to apiculate; hyaline collar with spiral or annular ridges; bowl reticulate, wall thick or agglutinated; three living genera (one also Pleistocene), 12 fossil	*Calpionella* (Lorenz 1902); *Codonellopsis* (Jörgensen 1924); *Stenosemella* (Jörgensen 1924)	Upper Triassic, Upper Jurassic–Lower Cretaceous, Pleistocene–Recent. All marine, neritic, and oceanic
Family Coxliellidae (Kofoid & Campbell 1929)	Lorica tubular or goblet shaped; spiral lamina; may have aboral horn; wall structure prismatic or agglutinated; seven living genera (one also fossil), three only fossil	*Climacocylis* (Jörgensen 1924); *Rhabdonelloides* (Colom 1939)	Lower Cretaceous, Recent. Marine, neritic, and oceanic

Taxon	Characteristics	Example	Occurrence
Family Epiplocylididae (Kofoid & Campbell 1939)	Lorica acorn shaped, has suboral shelf; aborally blunt, acuminate or with horn; wall thin, hyaline, reticulate; three living genera	*Epiplocylis* (Jörgensen 1924)	Recent. Marine, oceanic
Family Petalotrichidae (Kofoid & Campbell 1929)	Lorica cup shaped, oral rim smooth or denticulate, one or two collars; may be agglutinated; seven living genera	*Petalotricha* (Kent 1881)	Recent. Marine, mostly oceanic, rarely neritic
Family Dictyocystidae (Kent 1881)	Lorica with dense reticulate bowl, agglutinated; collar of hyaline open arches or windows; three living genera	*Dictyocysta* (Ehrenberg 1854)	Recent. All marine, oceanic
Superfamily Tintinnidea (Claparède & Lachmann 1838)	No ciliary membrane; lorica tubular or flaring, wall thin, hyaline, organic, not agglutinated; vertical or transverse ornamentation (not spiral)		Jurassic, Recent
Family Cyttarocylididae (Kofoid & Campbell 1929)	Lorica bell or kettle shaped, collar flared above constriction; wall coarsely reticulate; one living genus	*Cyttarocylis* (Fol 1881)	Recent. Marine, oceanic
Family Ptychocylididae (Kofoid & Campbell 1929)	Lorica bell or kettle shaped, bowl elongate, oral rim denticulate or furrowed; wall three layered, with median alveolar layer; four living genera	*Favella* (Jörgensen 1924); *Ptychocylis* (Brandt 1896)	Recent. Marine, oceanic
Family Rhabdonellidae (Kofoid & Campbell 1929)	Lorica conical to chalice shaped, with broad and grooved oral rim and aboral horn; wall with vertical ribs; four living genera	*Rhabdonella* (Brandt 1906)	Recent. Marine, oceanic
Family Xystonellidae (Kofoid & Campbell 1929)	Lorica elongate, chalice shaped, aboral end narrow, with pedicel; reticulate wall; four living genera, one fossil (not validly described)	*Parafavella* (Kofoid & Campbell 1929); *Xystonella* (Brandt 1906)	Jurassic, Recent. Marine, oceanic
Family Undellidae (Kofoid & Campbell 1929)	Goblet-shaped lorica with suboral ledge or inner collar; wall of three layers; eight living genera	*Amplectella* (Kofoid & Campbell 1929); *Undella* (Daday 1887)	Recent. Marine, oceanic
Family Tintinnidae (Claparède & Lachmann 1858)	Lorica elongate, orally flared; both ends open, or base closed; wall hyaline, organic; 23 living genera	*Salpingella* (Jörgensen 1924); *Tintinnus* (Schrank 1803)	Recent. Marine, oceanic; rare in brackish water

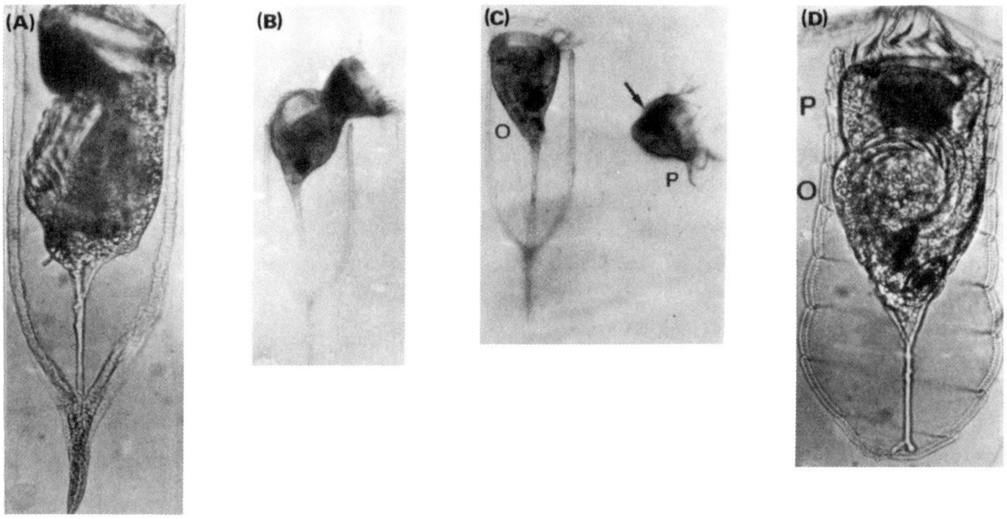

Figure 14.19 *Favella ehrenbergii* (Claparède & Lachmann). Bay of Villefranche-sur-Mer, France. A, Individual with favella-form lorica beginning to divide, and showing attachment by the pedicel; area at center left shows the development of the new opisthe daughter cell peristome of spiraling ciliary membranelles; just above the new peristome is a dark area that is an accumulation of lorica-building granules that will be utilized by the proter daughter cell to form a new lorica. ×221. B, Dividing favella-form. ×102. C, Favella-form immediately after separation, the posterior daughter cell or opisthe (O) remains attached within the retained parent lorica, while the anterior daughter cell or proter (P) becomes free-swimming before building a new lorica; arrow points to the crescent of accumulated lorica building material. ×102. D, Dividing individual with coxliella-form lorica of spiraling band, showing circular ring of the completed new peristome of the opisthe (O) and the dark area of lorica building granules in the anterior daughter cell (P). ×221. (Reproduced from Laval-Pueto M. *Protistologica* 1981; 17:249–72.)

daughter cells the proter constructs both the bowl and horn of a *Favella*-form (rarely a *Decipiens*-form) lorica, regardless of whether the parent cell was of *Favella*- or *Coxliella*-form. The granules are rapidly extruded from a pore beneath the crown of membranelles, becoming alveolar upon contact with sea water. Propelled by action of the body cilia, the granules flow backwards in a ribbon and the base of the bowl is molded around the cell body before solidifying; the horn is added at the end of the backward flow. The anterior part of the lorica is then helically constructed, added from the already formed base toward the anterior.

Although the anterior daughter cell builds a new lorica, the posterior daughter cell retains the original parent lorica, to which it later may add a spiraled collar. If the original lorica is abandoned for some reason during the period between cell divisions, due to injury or trauma, a new *Coxliella*-form lorica is slowly constructed over a period of 3 to 4 hours. A damaged lorica apparently is not repaired.

Spherical, ovoid, or flask-like resting cysts have been described within the lorica of about 20 tintinnid species (Figs 14.20, 14.21). The cyst may have a short neck and is closed by a discoid operculum. Flask-like cysts are thick-walled, hyaline, resistant to acids, and may have a smooth or spiny surface. They appear to be a routine part of the life-cycle, but whether their formation is related to reproduction is unknown. Spherical cysts commonly are thin walled. Protective cysts also develop in cultures in response to a reduced salinity or high temperatures.

Ecology

Tintinnids are dominantly marine and pelagic organisms, living in the upper water levels in all oceans. Many are typically oceanic (holoplanktic) in occurrence, but about 10% of those known occur nearshore and are neritic (meroplanktic). The Tintinnidiidae and Codonellidae commonly are neritic, whereas the Dictyocystidae, Rhabdonellidae, and Xystonellidae are typically oceanic, although most families include representatives from both environments.

Some neritic species (of *Tintinnus, Tintinnopsis, Codonella, Tintinnidium*, and *Leprotintinnus*) occur in brackish water. About 10 species (of the genera *Tintinnopsis, Codonella, Tintinnidium*, and *Coxliella*) are found in freshwater lakes.

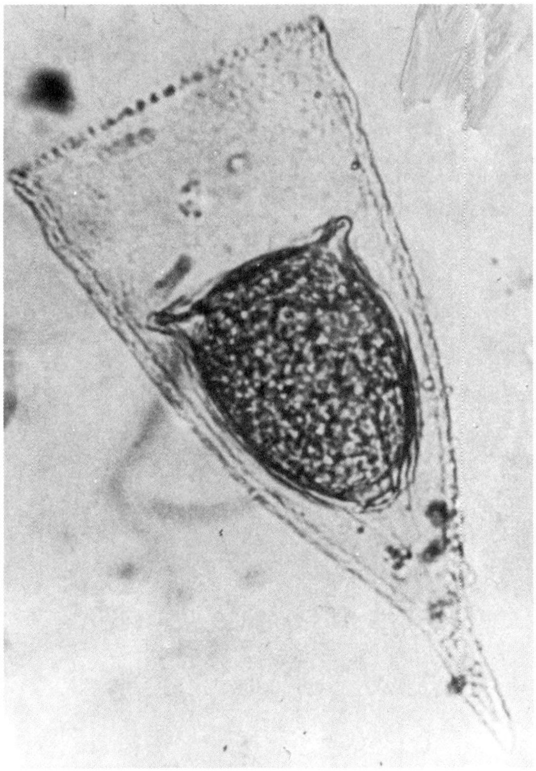

Figure 14.20 *Favella serrata* (Möbius). Mouth of Bristol Channel, England. Hyaline organic lorica with serrate oral rim, and short aboral horn, enclosing a thick-walled flask-shaped cyst that probably serves as a resting stage. ×414. (Reproduced from Reid PC, John AWG. *J Marine Biol Assoc UK* 1978; 58:551–7.)

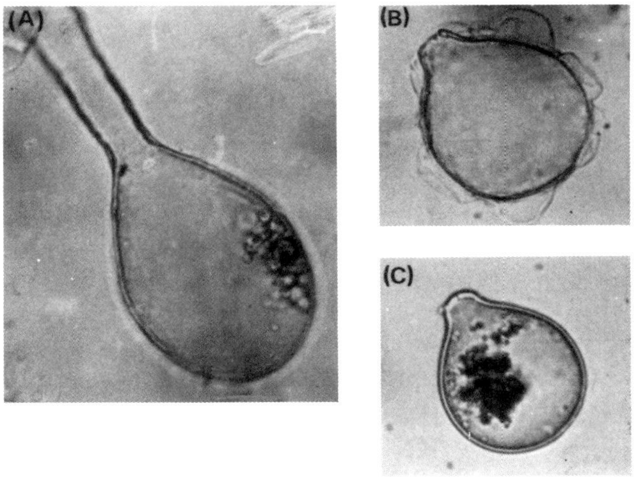

Figure 14.21 Unidentified tintinnid cysts. A, Flask-shaped cyst from Firth of Tay, Scotland. B, Smooth-walled cyst with undulating outer membrane, from Galway, Ireland. C, Smooth-walled cyst without an outer membrane, from Port Gordon, Scotland. All ×465. (Reproduced from Reid PC, John AWG. *J Marine Biol Assoc UK* 1978; 58:551–7.)

Biogeography

In cooler regions of the seas, the tintinnid fauna is not very diverse, but many of the species are relatively large and some may be abundant. As many as 5100 individuals have been reported in a liter of sea water in the Kieler Bucht of Germany in June. In tropical waters, variety is greater but individual species are smaller and less numerous than in higher latitudes.

Some species are seasonal in occurrence and numerous only at particular times of the year. For example, near Antarctica, tintinnid abundance began to increase in January, reached a maximum in late March, then decreased to a minimum by the following November and December. Other species have slightly different winter and summer strains. A winter strain of *Tintinnopsis tubulosa* is characteristic of water temperatures of 3–5°C but actively grows in culture of 10°C and will continue to reproduce up to 15°C. At higher temperatures, they decrease in number and fail to produce a lorica. The morphologically similar summer variety, isolated from waters of 20°C, could not be grown in culture at 10°C, but at 19–22°C continued to grow and produce new loricae for several months, although many cells remained naked. At higher temperatures (24°C) only a tiny proportion of the cells produced a lorica and these were thin walled and commonly aberrant in form; however, the inhibition was reversible if the temperature was reduced.

Paleobiology

Affinities of the microfossils here regarded as tintinnids are not universally agreed upon, particularly those with calcareous lorica (calpionellids). Both the earliest described calpionellid and the Oligocene *Pseudarcella* were originally described as foraminifera, although they differ in being considerably smaller, in having a relatively thick imperforate wall, and a much larger aperture. The Eocene–Oligocene representatives have been compared to arcellid thecamebians, although that group has only a membranous test, and only inhabits freshwater rivers, lakes, bogs, or soils. Both calpionellids and pseudarcellids have even been regarded as so unrelated to the tintinnids because of their calcareous wall that they have been assigned to separate orders (of an unknown subphylum). As far as is known, all modern tintinnids do have either a wholly organic wall or one with agglutinated particles. However, their many morphologic as well as ecologic similarities support a close relationship and indicate that perhaps a greater diversity characterized this group in the geologic past, as has been demonstrated for many other groups of microplankton.

As the same calpionellid species have been reported from correlative Upper Jurassic and Lower Cretaceous rocks of southern Europe, northern Africa, Cuba, and Mexico, calpionellid occurrence has been regarded as showing no provincialism. However, paleogeographic reconstructions (Fig. 14.22) show all these occurrences to have been from the ancient Tethyan province, characterized by relatively warm carbonate-rich subtropical or tropical waters. None are found in contemporaneous deposits of high latitudes. At site 534A of the Deep Sea Drilling Project, east of the Black Plateau, the *Crassicollaria intermedia* and *Calpionella alpina* Zones (see section on "Biostratigraphy" below) are well represented, but no calpionellids were found that represented the *Calpionella elliptica* or *Calpionellopsis* Zones (Berriasian–Early Valanginian), although coccoliths and other microfossils documented the presence of strata of this age. Possibly a cooling trend in the Early Cretaceous eliminated the warm-water calpionellids of the stratigraphically higher zones from this area.

Similarly, the calcareous pseudarcellids of the Late Eocene and Early Oligocene are characteristic of the relatively warm waters of the Paleogene (Fig. 14.23). In contrast, the organic-walled or agglutinated tintinnid specimens from the Triassic of Ireland, Jurassic of Germany, and mid-Cretaceous of Colorado, Wyoming, and South Dakota (Fig. 14.24) all represent more temperate nearshore or even brackish waters. Based on modern species distribution, tintinnids may have been oceanic in distribution throughout their history, but the characteristic delicate organic lorica has rarely been preserved; in contrast, a firm organic or agglutinated lorica constructed of any available materials has been characteristic of the nearshore plankton throughout the Mesozoic and Cenozoic.

At certain periods in the geologic past during which oceanic circulation was dominated by circumequatorial currents and the warm waters were supersaturated in carbonates, tropical offshoots that secreted a calcareous lorica arose from the primitive tintinnid taxa. Bonet proposed the superfamily Calpionellidea for such calcareous Mesozoic taxa. Changes in the patterns of oceanic circulation that resulted in somewhat colder seas led to the disappearance of the calcareous calpionellids, more rapidly in some areas than in others. The process was repeated in the Eocene with the appearance of the pseudarcellids and their mid-Cenozoic disappearance when cooling oceans also led to contemporaneous ex-

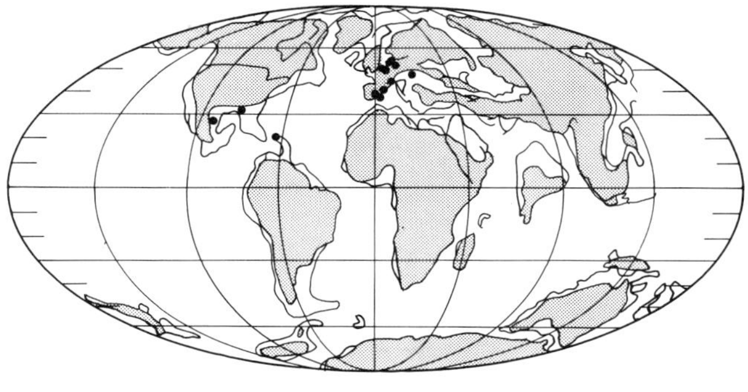

Figure 14.23 Known distribution of tintinnids of the Eocene–Oligocene, plotted on a paleogeographic reconstruction for the Late Eocene. Land areas shown stippled. These neritic tintinnids lived in epicontinental seas, rather than oceanic regions, and had a broader latitudinal range than the Mesozoic calpionellids.

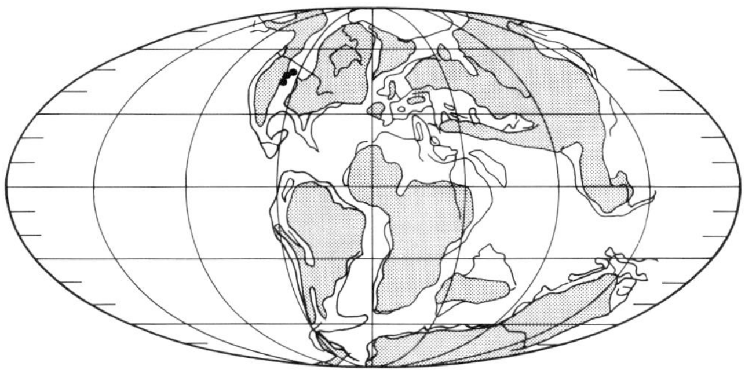

Figure 14.24 Known distribution of tintinnids of the Middle Cretaceous (Late Albian to Early Cenomanian), in the western interior seaway of North America, plotted on a paleogeographic reconstruction for the Cenomanian. Land areas shown stippled. The neritic tintinnids occurred farther north than did the calpionellids, and in a north–south trending inland sea.

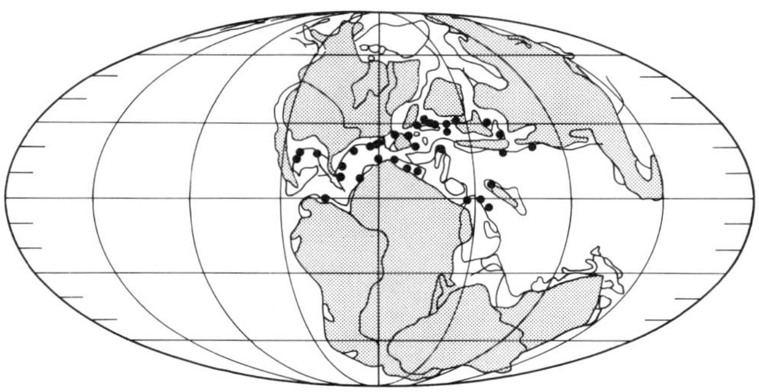

Figure 14.22 Known distribution of calpionellids of the Late Jurassic (Tithonian Stage) to Early Cretaceous (Valanginian Stage), plotted on a paleogeographic reconstruction for the Tithonian. Land areas shown stippled. The oceanic calpionellids are restricted to the Tethyan region, within approximately 40° of the paleoequator.

tinctions of other groups of plankton (such as the specialized planktic foraminifers and rosette-type discoasters of the Paleogene).

Modern tintinnids also occur in fresh water, but fossil freshwater examples are very rare. The living species *Codonella cratera* (Leidy) has been recorded from Pleistocene lake deposits in Sweden and Finland. Whether tintinnids only migrated into fresh waters late in their history or whether insufficient search has yet been made for other examples is uncertain.

Biostratigraphy

Although generally rare as fossils, tintinnids have a fairly long geologic history. Rare calcareous ones have been described from thin sections from the Lower Ordovician (Upper Tremadocian or Lower Arenigian) of Algeria, Upper Silurian (Ludlovian) of Spain, Upper Silurian, Lower Devonian, and Upper Devonian (Fammenian) of the African Sahara, Upper Devonian of France, and Upper Devonian of western Canada.

Tintinnids with organic lorica were illustrated from palynologic preparations of a brackish water facies of Lower Triassic rocks of Ireland (Fig. 14.25), but they may originally have been agglutinated, as the preparation method would have dissolved any calcareous or siliceous material that was originally present. A few tintinnid specimens, with the original organic reticulate wall apparently well preserved, were described from phosphatic Jurassic coprolites from Germany, associated with fragments of ammonites, radiolarians, and rare foraminiferans.

Calcareous tintinnids again arose in the Early Mesozoic, *Praecalpionellopsis* being described from the Upper Triassic of the Yugoslavian Carpathians, and typical calpionellids becoming stratigraphically important for Upper Jurassic to Lower Cretaceous strata of the Tethyan seaway. They are known in thin sections of the deep water pelagic limestone facies of Spain, France, Switzerland, Italy, Germany, Austria, Hungary, Romania, Poland, Czechoslovakia, Bulgaria, Yugoslavia, the USSR, Oman, Turkey, Iraq, Iran, Israel, Tunisia, Algeria, Tibet, deep-sea cores of the eastern and western North Atlantic (Figs 14.26, 14.27), Cuba, Mexico, Venezuela, and Texas, and into the Aptian and Albian of Mexico. Calpionellids commonly are associated with coccoliths, nannoconids, radiolarians, pelagic crinoids, and ammonites.

Some stratigraphically useful calpionellids are widely distributed (Figs 14.28, 14.29). The Upper Jurassic (Tithonian) *Chitinoidella* has a microgranular wall, appearing brownish in transmitted light and white in reflected light. The slightly younger *Praetintinnopsella* has a similar microgranular outer wall layer, but the inner layer is hyaline, like that of typical calpionellids. Typical calpionellids with wholly hyaline calcitic lorica are represented by widely recognized zones in the Jurassic–Cretaceous of the Tethyan region (Table 14.2). The *Crassicollaria* Zone is of Late Tithonian (latest Jurassic) age,

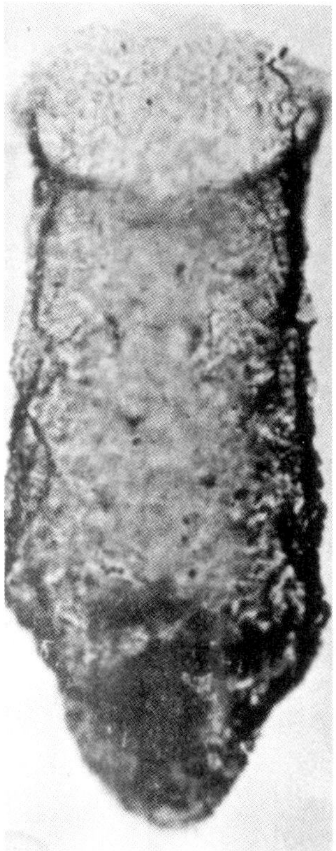

Figure 14.25 Unnamed tintinnid with organic lorica obtained from acid residue, hence possibly originally agglutinated. Lower Triassic of Kingscourt, Ireland. ×926. (Reproduced from Visscher H. *Bull Geol Surv Ireland* 1970; 1:61–4.)

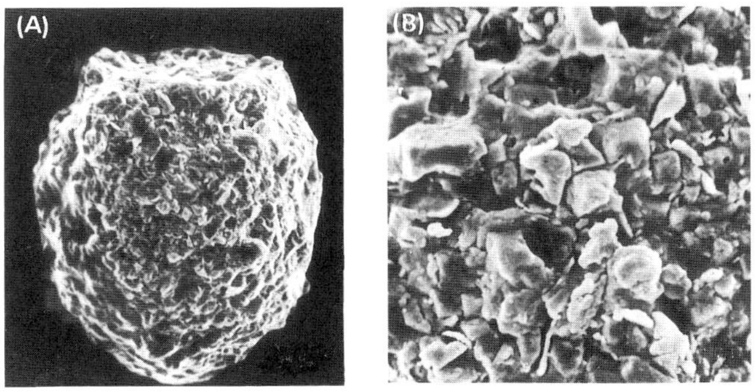

Figure 14.26 *Calpionella alpina* (Lorenz). Lower Cretaceous (Lower Valanginian), eastern Atlantic, off northwest Africa, Madeira Islands, Deep Sea Drilling Project, Leg 50, Site 416A. A, Side view of lorica. ×638. B, Enlargement of surface of same specimen. ×2552. (Reproduced from Bolli HM. *Initial Rep Deep Sea Drill Project* 1980; 50:525–43.)

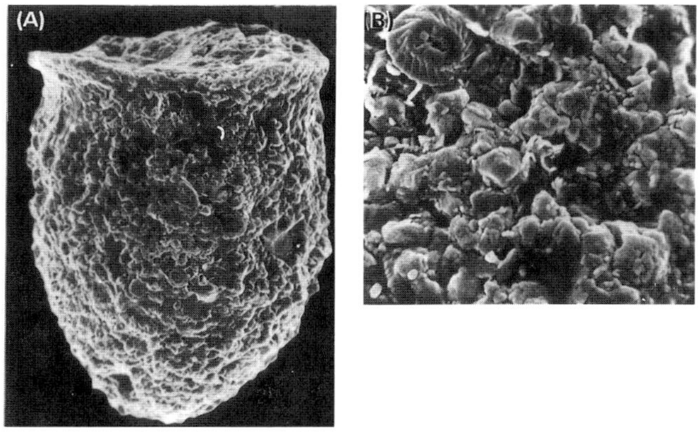

Figure 14.27 *Tintinnopsella carpathica* (Murgeanu and Filipescu). Lower Cretaceous (Valanginian), eastern Atlantic, off northwest Africa, Madeira Islands, Deep Sea Drilling Project, Leg 50, Site 416A. A, Side view of lorica, showing abruptly flaring rim. ×600. B, Enlargement of surface of same specimen, showing attached coccolith of *Watznaueria* at upper left. ×2400. (Reproduced from Bolli HM. *Initial Rep Deep Sea Drill Project* 1980; 50:525–43.)

the succeeding *Calpionella* Zone is of latest Tithonian and Early Berriasian (Early Cretaceous) age, and the *Calpionellopsis* Zone is of Late Berriasian age. Some calpionellid species persisted locally into the *Calpionellites* Zone of the Valanginian, and a few into the Hauterivian before their numbers were sharply reduced. Only *Colomiella* is import-

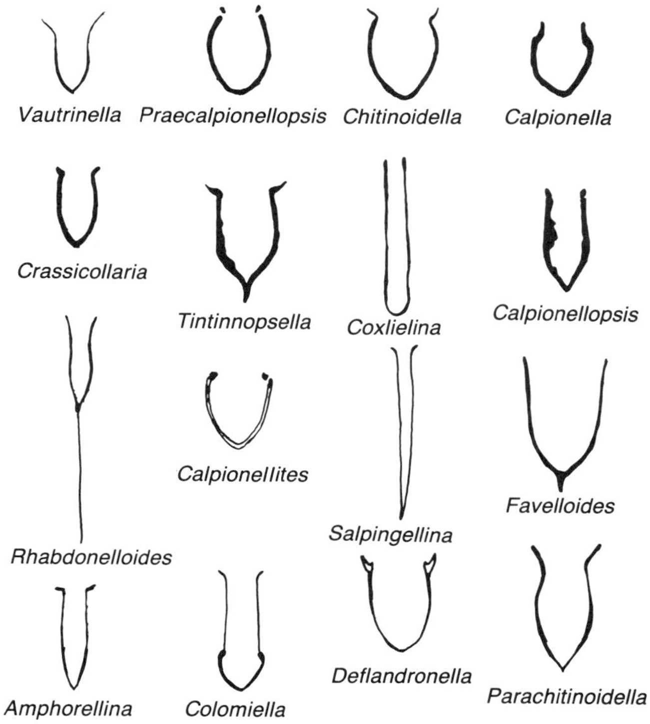

Figure 14.29 Diagrammatic longitudinal sections of some representative calpionellid genera, shown as they commonly have been identified and studied from sections of limestones.

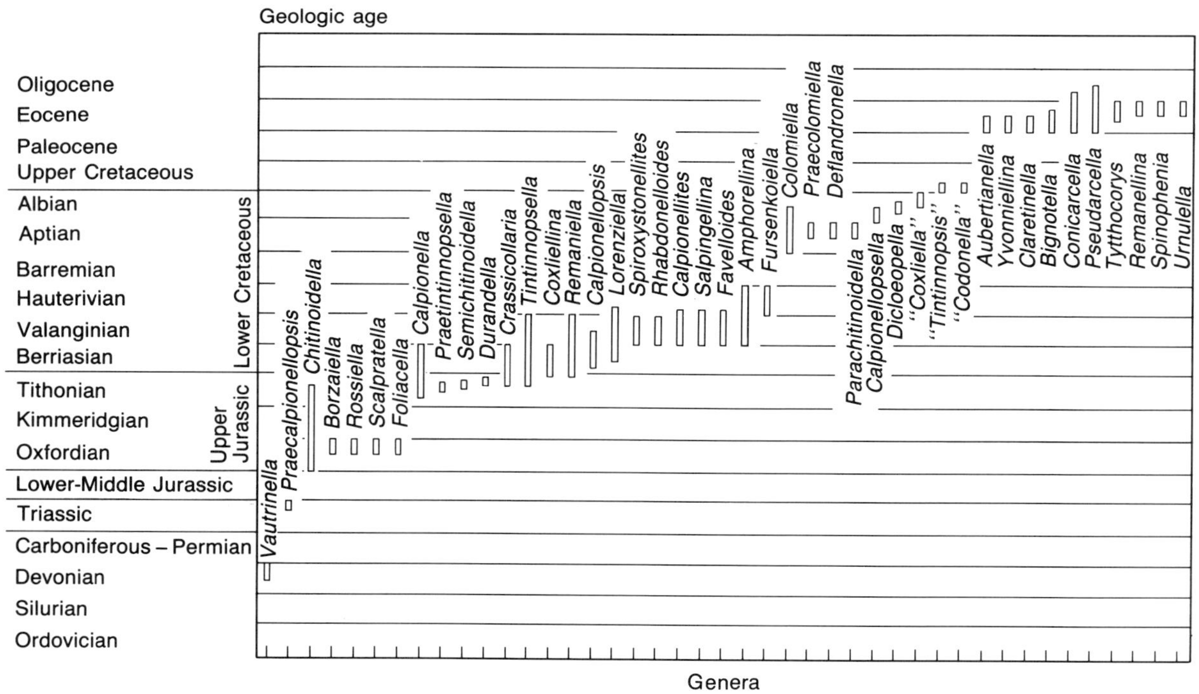

Figure 14.28 Geologic range of fossil tintinnid genera (the Eocene *Aubertianella* is now regarded as synonymous with *Claretinella*). Paleozoic and Mesozoic records of unnamed species are not included. In order to show detailed ranges of the calpionellids, time represented by the geologic periods and stages is not shown to scale.

Table 14.2 Tintinnid Zonation for the Late Jurassic and Early Cretaceous

Period	Stage	Zone	Biostratigraphic event
Lower Cretaceous	Lower Valanginian	*Calpionellites* Zone	
			←1st *Calpionellites darderi*
	Berriasian	*Calpionellopsis* Zone	←1st *Lorenziella hungarica*
			←1st *Calpionellopsis simplex*
		Calpionella Zone	←large *Tintinnopsella carpathica*
Upper Jurassic	Upper Tithonian		←Abundant *Calpionella alpina* (small)
		Crassicollaria Zone	←1st *Calpionella alpina*
			←1st *Tintinnopsella carpathica* (small)
	Middle Tithonian	*Chitinoidella* Zone	

ant in the Barremian, Aptian, and Albian, but because its collar and bowl are separated by a suture and their calcite crystals commonly show different angles of extinction under polarized light, it has been placed by some in a family (Colomiellidae) distinct from the typical calpionellids, although as mentioned above, this probably reflects a later addition of the collar to the bowl, as occurs in living tintinnids. *Colomiella* is known from Mexico, Cuba, southern France, and Tunisia.

Agglutinated tintinnids have been described from the epicontinental seas of the Albian and earliest Cenomanian of the western USA in Colorado, Wyoming, and western South Dakota (Fig. 14.30). These represent a neritic facies and probably somewhat cooler temperatures than those in which the calpionellids dominated.

The Late Cretaceous and Cenozoic record of the tintinnids is sparse, although well-preserved calcareous loricae are known from Lower, Middle, and Upper Eocene strata and rarely from Lower Oligocene beds. Unlike the Jurassic–Cretaceous calpionellids, these pseudarcellids represent a neritic facies, being associated with highly diverse foraminiferal faunas, as well as an abundance of other invertebrates and considerable detrital sediment. Paleogene pseudarcellids have been described from the American Gulf Coast in Mississippi (Fig. 14.31), Mexico, Puerto Rico, Spain, France, Belgium (Fig. 14.32), Germany, Poland, and Hungary.

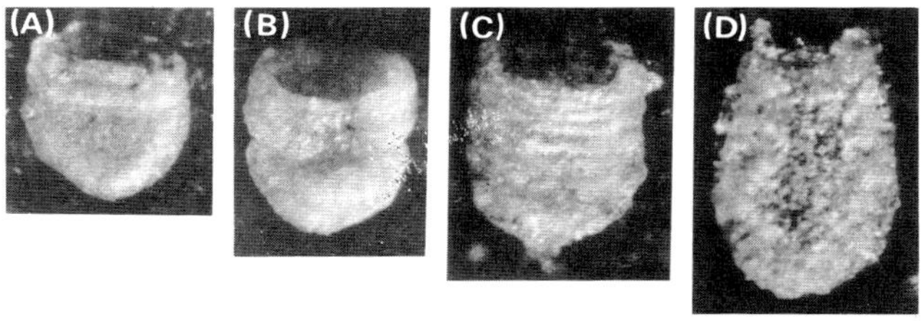

Figure 14.30 A, *Dicloeopella borealis* (Eicher). Lower Cretaceous (Albian) Thermopolis Shale near Mayoworth, central Wyoming. Side view of lorica. ×131. B, *"Codonella" bojiga* (Eicher). Upper Cretaceous (basal Cenomanian) Graneros Shale, central Colorado. Referred to the living genus *Codonella*, but may be generically distinct. Side view of lorica. ×131. C, *"Coxliella" coloradoensis* (Eicher). Upper Cretaceous (basal Cenomanian) Graneros Shale, central Colorado. Referred to the living genus *Coxliella*, but may be generically distinct. Side view. ×131. D, *"Tintinnopsis" parovalis* (Eicher). Upper Cretaceous (basal Cenomanian) Graneros Shale, central Colorado. Referred to the living genus *Tintinnopsis*, but may be generically distinct. Side view. ×131. (A–D, Reproduced from Eicher DL. *Micropaleontology* 1965; 11:449–56.)

Evolutionary history

Because of the strong similarity, even to their ultrastructure, of cilia and flagella of all eukaryotic organisms, the protistan flagellates and ciliates are generally agreed to have arisen from a common flagellate ancestor. The rather compact ciliate group might be assumed to be of much later origin than the more diverse flagellates, but little direct early fossil evidence is available for either group.

The geologic record of the ciliates is almost exclusively that of the tintinnids, known only from their fossilized loricae. If the Ordovician and Silurian records are accepted, the tintinnids have a known geologic history of almost 500 million years, slightly greater than the oldest record of dinoflagellates. The skeletal rhizopods, foraminiferans and radiolarians, occur earlier, but these also probably had a flagellate ancestor.

Within the Tintinnida, the family Tintinnidiidae, with a gelatinous to agglutinated lorica, has been regarded as most primitive. However, the oldest fossils known are those with a calcareous lorica. Whether or not these were originally agglutinated is uncertain, but some Mesozoic taxa do have a typically arenaceous lorica. Fossil examples of a true hyaline organic lorica are known only from an extremely rare and specialized occurrence in the Jurassic. The most cosmopolitan and most varied of the living tintinnids are the Tintinnidae, also generally regarded as most highly evolved, but because of their thin and homogeneous hyaline organic lorica, fossilization would be most unlikely, and their antiquity cannot be determined.

The meagre fossil record of the tintinnids limits their usefulness for discussion of paleobiogeographic patterns and inferences and allows only brief speculation as to the possible iterative appearance of offshoots with a calcareous lorica whenever paleoecologic conditions were favorable. Each evolutionary branch later disappeared when oceanic current patterns and paleotemperatures changed, perhaps resulting in other disruptions of the marine pelagic food chains.

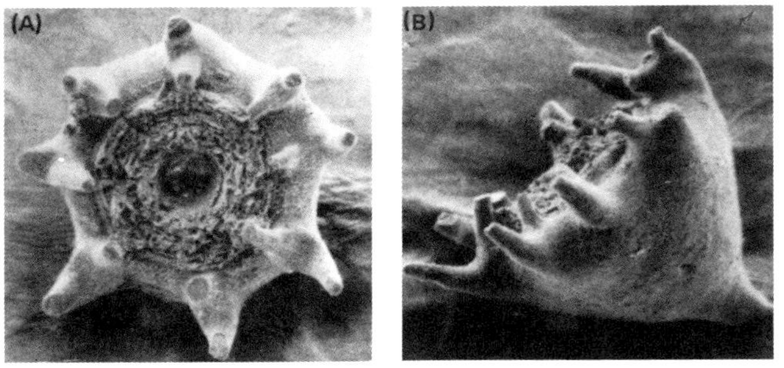

Figure 14.32 *Claretinella keiji* (Szczechura). Lower Eocene (Ypresian), Mont Saint Aubert, Belgium. A, Oral view of holotype, showing spines projecting around the opening, and inner circumoral shelf. B, Side view of holotype, showing bifurcating circumoral spines and aboral horn. Both ×137. (Reproduced from Szczechura J. *Acta Paleontol Polonica* 1979; 24:265–74.)

Supplementary reading

Colom G. Fossil tintinnids: loricated infusoria of the order of the Oligotricha. *J Paleontol* 1948; 22:233–63. Discussion of fossil calpionellids, relating these to the living tintinnids; summary of all known fossil oceanic species, described from thin sections of Jurassic–Cretaceous limestones.

Corliss JO. *The Ciliated Protozoa: Characterization, Classification, and Guide to the Literature*. Oxford: Pergamon Press, 1961:455. General summary of ciliate protozoans, including the tintinnids.

Corliss JO. Annotated assignment of families and genera to the orders and classes currently comprising the Corlissian scheme of higher classification in the phylum Ciliophora. *Trans Amer Microscop Soc* 1977; 96:104–40. Revised classification of ciliate protozoans, including living and fossil tintinnids.

Echols RJ, Fowler GA. Agglutinated tintinnid loricae from some Recent and Late Pleistocene shelf sediments. *Micropaleontology* 1973; 19:431–43. Occurrence of neritic tintinnid loricae in Late Cenozoic sediments, illustrated with SEM.

Eicher DL. Cretaceous tintinnids from the western interior of the United States. *Micropaleontology* 1965; 11:449–56. Description and illustrations of Cretaceous neritic tintinnids with agglutinated loricae.

Gold K, Morales EA. Studies on the sizes, shapes, and the development of the lorica of agglutinated Tintinnida. *Biol Bull* 1976; 150:377–92. Description and illustration in SEM of living neritic species of tintinnids with agglutinated lorica.

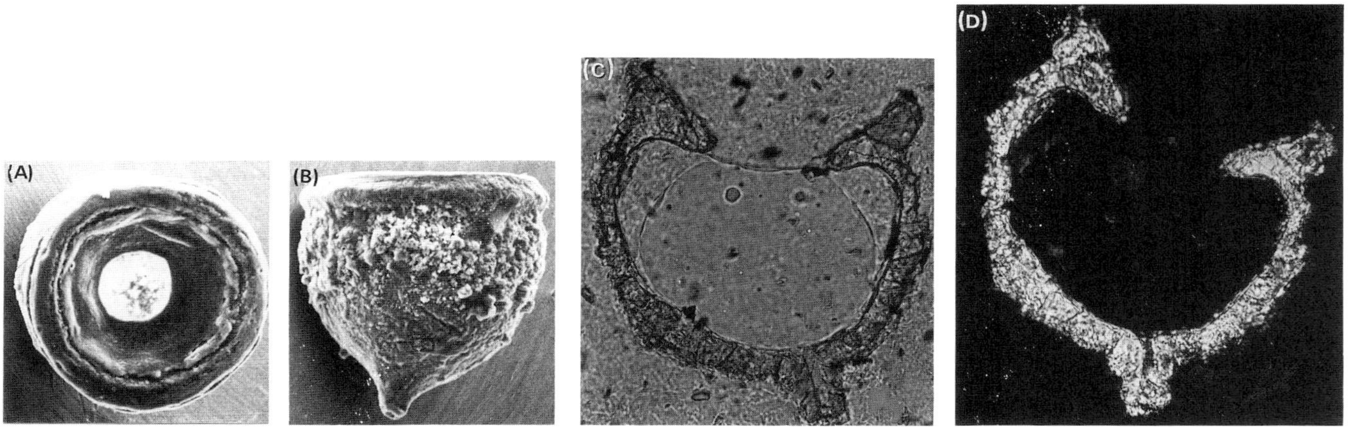

Figure 14.31 *Tytthocorys coronula* (Tappan & Loeblich). Upper Eocene, Moodys Branch Formation, Jackson, Mississippi. A, Oral view, showing smoothly finished surface and inward sloping circumoral shelf. B, Side view of holotype, showing low projecting collar, suboral constriction, and short aboral horn. Both ×173. C, Longitudinal section viewed in transmitted light, showing projecting collar, inward sloping circumoral shelf, and perforated aboral horn. D, Same in cross polarized light, showing the cryptocrystalline calcite of the bowl, and somewhat larger crystals in the neck region. Both sections ×242. (Reproduced from Tappan H, Loeblich AR Jr. *J Paleontol* 1968; 42:1378–94.)

Kofoid CA, Campbell AS. A conspectus of the marine and freshwater Ciliata belonging to the suborder Tintinnoinea, with descriptions of new species principally from the Agassiz Expedition to the eastern tropical Pacific, 1904–1905. *Univ Calif Publ Zool* 1929; 34:1–403. Monographic study of living tintinnid species, based on the lorica.

Kofoid CA, Campbell AS. The Ciliata: the Tintinnoinea. *Bull Mus Comparative Zool Harvard* 1939; 84:1–473. Illustrations and descriptions of known living tintinnids.

Laval-Peuto M. Construction of the lorica in ciliata Tintinnina. *In vivo* study of *Favella ehrenbergii*: variability of the phenotypes during the cycle, biology, statistics, biometry. *Protistologica* 1981; 17:249–72. Life history, reproduction, lorica construction, and variability of a living tintinnid species.

Loeblich AR Jr, Tappan H. Annotated index to genera, subgenera and suprageneric taxa of the ciliate order Tintinnida. *J Protozool* 1968; 15:185–92. Index with complete reference citations for living and fossil tintinnid genera.

Remane J. Untersuchungen zur Systematik und Stratigraphie der Calpionelen in den Jura-Kreide-Grenzschichten des vocontischen Troges. *Palaeontographica* 1964; A123:1–57. Description and occurrence of Jurassic and Cretaceous calpionellid species, regarded as not related to modern tintinnids.

Tappan H, Loeblich AR Jr. Lorica composition of modern and fossil Tintinnida (ciliate protozoa), systematics, geologic distribution and some new Tertiary taxa. *J Paleontol* 1968; 42:1378–94. Discussion of geologic occurrence, and relationship of calpionellids and Tertiary and living tintinnids, including new Tertiary genera and species with calcareous loricae, illustrated in SEM and in thin section.

Appendix: Collection and preparation techniques

The techniques described here are useful, generalized ones that can be used by anyone interested in collecting and observing living and fossil prokaryotes and protists of paleontologic importance. As every situation is unique, these methods may require some modification for successful results. A little experimentation with the procedures described here should give good results for most rocks and microfossils; if not, the suggested reading list will provide additional solutions. Various chemicals are used in these techniques and they all should be regarded as dangerous for one reason or another and handled carefully to avoid contact with them or breathing in their fumes.

Living prokaryotes and protists

Living prokaryotes and protists are collected and prepared for many purposes in paleontology. Almost all of the skeleton-bearing forms are marine and live on the bottom or in the water column. Prokaryotes live everywhere and usually require special techniques for collection and maintenance in the laboratory. Living foraminifera and diatoms are collected in samples obtained from the ocean bottom, whole foraminifera, radiolaria, coccolithophorids, diatoms, dinoflagellates, silicoflagellates, actiniscans, and ebridians are plankton and may be collected from the water column. Diatoms, thecamebians, and dinoflagellates may also be collected from fresh waters using the same techniques described here. Simple methods of collecting living material for the casual investigator or classroom instructor are provided here.

Mat-building prokaryotes are easily collected from special terrestrial and marine environments (see Chapter 5). These mats are usually colorful, and can be sliced, scraped, or cored. For study, the mat samples should be dispersed in the same water from which they were collected, spread on a glass slide, covered with a cover slip, and viewed at high-power magnifications in a compound microscope. Dispersed material may also be placed in an appropriate viewing dish and studied with an inverted microscope.

Collection of planktic protists from the water column is most easily done with a simple plankton net. These nets can be obtained from biologic supply houses in different mesh sizes, different net-opening diameters, and with different brindles. Net-opening diameters and brindle size and type should be selected for ease of handling depending on the size of the vessel and the towing gear. For hand operation from a small boat, a net diameter of 0.5 m or less is easiest. A mesh opening of 10–35 μm will collect foraminifera, diatoms, radiolaria, and sometimes smaller phytoplankton. The nets may become clogged by other zooplankton, diatoms, and debris commonly found in nearshore waters. If zooplankton are abundant, especially gelatinous kinds, then the protists become entangled and sorting of live specimens is difficult. The larger, gelatinous forms must be removed by hand or forceps and the protists isolated by picking or decanting before being placed in a collection jar. Similar treatment may also be necessary in the laboratory later. For skeletons only, the entire sample may be dried upon return to the laboratory and incinerated in a low-temperature furnace to burn off the organic components of the sample.

Benthic foraminifera can usually be found in any marine sediment, except for some sands. They live on and in fine-grained substrata from marshes to the deep sea, among holdfasts and stipes of various nearshore marine plants, and attached to shells, cobbles, reef debris, and other firm substrata. For fine-grained sediments, foraminifera may be seived in the field with sea water over a 180-mesh stainless-steel sieve that will retain most of the foraminifera and allow the sediment to pass through. For foraminifera in plants a 5–10 mm sieve should be inserted above the smaller sieve to catch large plant debris. Gently scrub and rub the plants and their holdfasts over the larger sieve to remove the foraminifera. For attached foraminifera, pieces of the substrata should be collected, kept damp with sea water-saturated paper towels or seaweed, and examined in the laboratory. Benthic diatoms can be similarly collected, except that they live on the surface of sediment or hard substrata. Some diatoms dwell interstitially in sand or in surface films. These can be collected by scraping or coring.

All samples of living organisms should be kept in an insulated container (thermos bottles or food chest) to prevent thermal shock. The living organisms are best observed by diluting the sample with sea water, and examining dispersed material in a petri dish set on a dark background with a dissecting microscope, or from below with an inverted microscope. Samples may also be preserved in solutions of 10% ethanol or buffered formalin. These samples may be examined much later at leisure.

Collecting rock samples for fossil prokaryotes and protists

The collection of rock samples in the field can be of critical importance in the study of fossil prokaryotes and protists. Nearly all marine and freshwater sedimentary rocks contain fossil protists and some contain fossil prokaryotes. However, not all rocks, even within a single outcrop, contain abundant specimens.

Samples may be from single localities or from part of a sequence of samples in a measured section. Four items are especially important in collecting rock samples:

1 The precise geographic location must be recorded with the sample. This record should include latitude and longitude, as well as a written reference to permanent geographic features.

2 The stratigraphic position must be included with the sample. This record usually includes the position (in meters) relative to some integral stratigraphic marker (a distinctive bed, for example). Stratigraphic distance can be determined by measurement with a tape or by use of a Jacobs staff.

3 A sample sufficiently large to complete a particular analysis and to have some remaining for later confirmation or reprocessing is best. The amount collected will be based on the lithology, estimated abundance of microfossils, and the group under study. Generally, fine-grained sedimentary rocks are more likely to have abundant microfossils than coarse-grained rocks. Larger microfossils, like foraminifera, radiolaria, and diatoms, can commonly be seen with a ×10 hand lens in a fresh piece of rock, which will allow an estimate of their abundance. The larger microfossils may require samples as large as 2–5 kg, whereas samples for silicoflagellates and calcareous nannofossils may be as small as a single rock chip. Collect more material than deemed necessary because recollecting a locality later is more costly and less efficient than collecting too much initially.

4 Contamination must not occur, or erroneous or mixed outcomes may result. Before collecting, the outcrop should be scraped or dug to solid, unweathered rock. Rusty-colored rock indicates considerable oxidation, which removes many organic-walled microfossils, and should be avoided. Only pieces from below the surface of the outcrop should be placed in an impervious sample bag (plastic) that can be tightly sealed for the trip back to the laboratory. Indurated limestones, cherts, and other rocks may be broken from outcrops and washed in the laboratory to avoid contamination.

Sample preparation

Fossil prokaryotes and protists come in a wide variety of sizes, compositions, and enclosing rock types. This great variety requires that a large number of techniques be employed to examine the microfossils (Table A.1). These techniques are designed to avoid mechanical damage to specimens so that details of their morphology are maintained for study. However, each rock or sediment sample is rather unique, and no single method can be applied equally to all of them even when recovering the same microfossils from similar rocks. The best results are commonly obtained by experimentation with variations of the general techniques described here.

Embedded specimens

Nearly all kinds of fossil prokaryotes and protists are found in rocks from which they cannot be separated with any technique. They have become part of the rock through processes of cementation or diagenesis. In particular, free specimens are difficult or impossible to obtain from cherts, limestones, and indurated clastic rocks. The only way to examine microfossils in such rocks is to make a thin section of the rock in which the microfossils can be seen in the rock itself.

Thin sections are simply slices of rock thin enough that light can be transmitted, and the internal constituents can be viewed through a microscope. Geologists and petrologists have standard techniques and specifications for thin sections of rocks they wish to study, and these thin sections are adequate for many fossil applications. However, the search for microscopic fossils requires that larger and thicker sections of rock be made for examination in order to assure that enough specimens will be encountered for study or that thicker specimens can be studied properly.

The rock sample must be studied to determine the best

Table A.1 Fossils, lithologies in which they occur, and the best techniques for their study

Fossil groups	Common lithologies	Techniques
Acritarchs, dinoflagellates, prasinophytes, prokaryotes	Fine-grained clastics, carbonates, cherts	Acid maceration and dispersal on a slide; thin sections
Diatoms, radiolaria, silicoflagellates, ebridians, actiniscans	Fine-grained clastics, siliceous shales, cherts	Disaggregation, HCl acidification and dispersal on a slide; HF dissolution of cherts; thin sections
Calcareous nannofossils	Fine-grained clastics, oozes, marls	Disaggregation, concentration, and dispersal on slide
Foraminifera, tintinnids	Fine- to medium-grained clastics, carbonates	Disaggregation and picking

direction in which to cut the section. Some parts of the rock may contain more specimens than others, or the specimens may be oriented in a particular plane of the rock. If these can be determined in advance, a thin section can be cut in a direction to take maximal advantage of specimen concentrations or orientations. However, commonly the microfossils cannot be seen until after a section is examined under the microscope. Several thin sections, therefore, are usually made in different planes through the rock, each marked to identify the particular plane from which it was cut. Examination of these sections will usually indicate which planes in the rock will reveal the most or best preserved microfossils, and additional thin sections can be made in the same direction. For study of microfossils that are dispersed sparsely in the rock, a thicker section that just transmits light will be most useful. These sections are sometimes called thick sections because they exceed the thickness of standard petrographic thin sections. Sections that are wider and longer than the standard sections are useful in finding microfossils because fewer sections need be cut. In addition, these bigger thin sections commonly provide data on the relationships of the microfossils to one another.

Thin sections are made by first cutting a rock with a lapidary saw along the selected plane to make a slice a few milimeters thick. One side is then carefully polished on a lapidary wheel or ground-glass plate to which a fine abrasive powder has been added. The polished side is cemented to a glass slide, and the other side of the slice is ground to a selected thickness, polished, and a glass cover-slip may be cemented to it for microscopic examination. A variety of cements are availabe for mounting rock sections which can be obtained from any lapidary supply company. In some cases the thickness of the section is too great to permit viewing with a coverslip in place; then, leave the rock section uncovered and view it under a drop of water or a special immersion oil for use under the microscope. These oils are also available from lapidary or scientific supply houses.

Oriented thin sections

Some kinds of microfossils, especially foraminifera, may require study of sequential sections in order to make ontogenetic or morphologic observations on single specimens. In these cases the rock is cut and polished to reveal the specimen for study. It is then etched in 10% HCl or EDTA for limestones and 10% HF for cherts for a time short enough to produce a microrelief of the specimen, usually about 30 seconds or so. After washing the surface with plenty of water, the etched specimen is covered with a drop of acetone and a piece of acetate large enough to cover the specimen is pressed over it. After 10 minutes the acetate is pulled off the specimen and an impression of the specimen's cross-section is obtained on the acetate. Commonly, bits and pieces of the rock or fossil may adhere to the acetate peel but repetition of the process will produce clear peels. The rock section is then ground to a new plane and the process repeated. In this way, a series of sections of the specimen can be obtained that are available for long-term study, even though the specimen is destroyed in the process. The peels can be directly viewed in optical and scanning electron microscopes and the complete specimen reconstructed. When modern computer programs for morphology are applied to the sections, the entire specimen can be reconstructed, reoriented, and measured for whatever purpose.

Some isolated, single specimens larger than 0.5 mm or so that can be handled with forceps or tiny camel hair paintbrushes may require sections through particular parts to be useful. Thus, free specimens must be oriented so that a section can be made in the plane required for study. These are made by manipulating a single specimen under the microscope. In order to do this, a specimen is glued to a glass coverslip temporarily with any kind of cement and oriented so that the desired section is parallel to the glass. The specimen is covered with an epoxy or other resin and ground on a lapidary wheel or glass plate until the desired plane of observation is approached. At that position, the specimen may be further ground in a controlled way by using a glass slide that has been thoroughly ground on one side by grinding on a glass plate covered with a fine abrasive. The specimen, mounted on a glass slide, is placed under the microscope; the ground glass slide is stroked across the surface of the specimen while it is covered with water. The water allows the microscopist to observe the progress of sectioning while it is under way. Acetate peels, of course, can be made at any stage in this process too.

Free specimens

Most microfossils are best and most easily studied as specimens freed from their enclosing rock. Two ways are possible: for specimens in uncemented clastic rocks, the sample may be disaggregated, sieved, and the specimens picked from the residue; for cemented or precipitate rocks, acid dissolution of the sample may yield microfossils that are resistant to the acids.

Calcareous and agglutinated microfossils. These microfossils are made of materials that can be destroyed by acid treatment. They may be recovered only by nondestructive, disaggregation methods. Because calcareous nannofossils are so much smaller than foraminifera, tintinnids, and thecamebians, special concentrating techniques must be utilized to recover abundant specimens from some samples.

Most clastic rocks can be disaggregated by thoroughly drying the sample and then following one of the following procedures:

1 The dried sample is placed in a solution of dilute H_2O_2 and sodium pyrophosphate (15%), and gently heated.

2 The sample is placed in a solution of a surface-active agent (Quaternary O).

3 The sample is soaked in a cleaning solvent like kerosene overnight in a pot, the solvent is then poured off, the sample is covered with water to which a handful of sodium bicarbonate is added, and boiled.

When the sample has broken down as much as possible, it is sieved over a 180 or 200 mesh sieve to remove the clay and fine mud, and dried thoroughly again. The process is repeated until the sample is completely disaggregated and the microfossils are clean. The microfossils are then picked from the dried residues saved on the sieves with fine camel-hair brushes and mounted on standard cardboard micropaleontology slides for microscopic study.

Calcareous nannofossils are most abundant in fine-grained clastic rocks and deep-sea oozes. These rocks can be disaggregated as described above or by mashing a small piece about the size of a matchhead in distilled water. If the nannofossils are abundant, preparations may be made directly from the suspension of water and nannofossils. If the nannofossils are not abundant, the 2–25 μm size fraction, in which almost all the calcareous nannoplankton occur, may be concentrated by repeated gravity settling in a beaker, followed by decantation. The nannofossils may also be concentrated by repeated centrifugation at 3000 revolutions/minute for 45 seconds. Most of the material coarser than 25 μm will settle to the bottom of the centrifuge tube, the nannofossils will appear as a white layer near the top, and material finer than 2 μm will remain suspended and can be decanted. The white layer may be sampled directly with a spatula into a water suspension.

Strewn slides can be prepared by dispersing the water with suspended fossils on a glass coverslip and drying. The coverslip is then mounted on a slide with Canada balsam or similar mounting medium for viewing. Slides intended for long-term use should be heated for several hours to harden the medium completely. Light-microscope examination is best accomplished at magnifications between ×800 and ×2000 using a microscope equipped with cross polarizers and phase-contrast illumination. For scanning electron microscopy the suspension is dispersed and dried on a scanning electron microscope (SEM) stub or coverslip to be attached to the stub. The stub is coated in a sputter-coater with a thin layer (30–50 nm thick) of gold or gold–palladium for conductivity. The sample is then ready for scanning. Routine examination of an SEM stub is usually done at ×2000–3000, with ×5000–10 000 used for photography.

Siliceous microfossils. Processing techniques that concentrate and clean radiolarians, diatom frustules, silicoflagellates, ebridians, and actiniscans typically involve treatment in hydrogen peroxide to remove organic material and to disaggregate the sediments. This is followed by removal of carbonate with hydrochloric acid. Additional cleaning can be accomplished with nitric acid and/or sulfuric acid, and clays can be suspended and removed from the sample after treatment in a dilute solution of sodium pyrophosphate. Radiolarians may also be removed from cherts by dissolving the rock in 10% hydrofluoric acid (use extreme caution and proper laboratory care with HF).

Because silicoflagellates, ebridians, and actiniscans are smaller than most radiolarians and diatoms, the sample might be sieved to separate the larger microfossils. Sieves may be used in this process, but they are easily contaminated and must be thoroughly cleaned after each use. Strewn slides of the processed material are prepared using a mounting medium such as Hyrax with a relatively high refractive index (ca. 1.71), because the opaline silica is transparent and difficult to see in water or standard mounting media. Dried specimens or suspended material may be placed on stubs for viewing in the SEM.

Organic-walled microfossils. Organic-walled prokaryotes and protists (acritarchs, prasinophytes, and dinoflagellates) are found in chiefly marine sedimentary rocks including shale, siltstone, fine-grained sandstone, carbonate, and chert. Shales and other mudrocks are probably best, especially if the rocks are relatively dark in color (green to gray). Black shales, however, may be so organic-rich that the isolation of the microfossils from amorphous organic matter may be difficult or impossible; in addition, some black shales may be too thermally mature to yield well-preserved organic microfossils.

The microfossils are extracted from rocks by following palynologic procedures similar to those used in studies of pollen, including digestion in hydrochloric, hydrofluoric, and nitric acids. The goal is to separate the relatively light, acid-resistant microfossils from mineral particles and organic debris. Usually up to 100 g of sample are processed with 10% hydrochloric acid to remove carbonates, 10% hydrofluoric acid to remove silicates, a controlled oxidation step to remove organic debris without eroding the microfossils, and flotation in a heavy liquid solution (such as zinc chloride, specific gravity approximately 2) to remove mineral material. Abundant amorphous organic matter commonly obscures the microfossils so that additional concentration steps are required for satisfactory results. Humic substances may be solubilized in a dilute solution of potassium hydroxide or ammonium hydroxide. The resultant residues may be sieved into size fractions (e.g., <20 μm, 20–36 μm, and >36 μm), but this may compromise quantitative descriptions of the assemblage. The microfossils may be stained and mounted in a plastic- or glycerin-based medium on a glass microscope slide.

Palynologic processing is far from standardized; typically, a great deal of experimentation is necessary before optimal results are achieved. The chemicals used are extremely hazardous. Hydrofluoric acid can be fatal; the inhalation of the fumes is more dangerous than the burns from direct contact. All work with hazardous chemicals should be done in a fumehood with the handler wearing gloves, apron, and a faceshield.

Strewn slides prepared from the residues are studied using a light microscope, preferably equipped with differential interference contrast optics. Most organic-walled microfossils are translucent when viewed under transmitted light. This translucency allows the viewer to observe both sides of the specimen by focussing up and down. When focussing through the specimen, viewers must remember that they are looking at the outside of the upper surface, then at a cross-section of the specimen, and then at the inside of the lower surface. The wall features of the lower surface will appear to be right–left reversed. Specimens mounted in a flexible medium can be rolled to allow observation from different vantages.

Examination by light microscope should be complemented by study under the SEM. The SEM provides only detailed surface views of a specimen and internal features seen with light microscopy cannot be viewed. The following procedure has proven successful in preparing specimens for a combination of light microscopy and SEM. The microfossils are picked under a binocular microscope using an eyelash affixed by paraffin to a dissecting needle. The specimens are mounted on glass coverslips, sputter-coated with gold or gold–palladium, and studied using an SEM. Upon completion of SEM observation, the coverslips may be inverted and mounted permanently on slides so that the same specimen may be studied under the light microscope. This or a similar method should be employed routinely, especially in erecting new taxa, because light microscopy and SEM provide very different kinds of complementary information.

Supplementary reading

Barss MS, Williams GL. Palynology and nannofossil processing techniques. *Geol Surv Can Paper* 1973; 73:26–25. Describes methods of preparation used to recover acritarchs and dinoflagellates from marine rocks.

Colbath GK. A comparison of palynological extraction techniques using samples from the Silurian Bainbridge Formation, Missouri, USA. *Rev Palaeobot Palynol* 1985; 44:153–64.

Doher LI. Palynomorph preparation procedures currently used in the paleontology and stratigraphy laboratories, US Geological Survey. *US Geol Surv Circ* 1980; 830:1–29. Methods to prepare acritarchs, dinoflagellates, and other organic-walled microfossils.

Evitt WR. Some techniques for preparing, manipulating and mounting dinoflagellates. *J Micropalaeontol* 1984; 3:11–18.

Funkhouser JW, Evitt WR. Preparation techniques for acid-insoluble microfossils. *Micropaleontology* 1959; 5:369–75.

Gray P, ed. *Encyclopedia of Microscopy and Microtechnique*. New York: Van Nostrand Reinhold, 1973:638. An encyclopedia describing a wide variety of techniques and methods used in microscopic examination of various subjects, including many useful in studying fossil prokaryotes and protists.

Hodgkinson RL. Microfossil processing: a damage report. *Micropaleontology* 1991; 37:320–6. An assessment of damage done to microfossils by various preparation techniques.

Kummel B, Raup D. *Handbook of Paleontological Techniques*. San Francisco: WH Freeman, 1965:852. A compendium of articles dealing with collection and preparation of paleontologic materials, including many protistan fossils.

Lee JJ, Small EB, Lynn DH, Bovee EC. Some techniques for collecting, cultivating and observing protozoa. In: Lee JJ, Hutner SH, Bovee EC, eds. *An Illustrated Guide to the Protozoa*. Lawrence, Kansas: Society of Protozoologists, 1985:1–7. Methods for the collection, culture, and observation of freshwater and marine protists.

Leffingwell HA, Hodgkin N. Techniques for preparing fossil palynomorphs for study with the scanning and transmission electron microscopes. *Rev Palaeobot Palynol* 1971; 11:177–99.

Lipps JH. Microfossils. In: Gray P, ed. *Encyclopedia of Microscopy and Microtechnique*. New York: Van Nostrand Reinhold, 1973:308–12. Short descriptions of techniques for processing and studying fossil protists and other microfossils.

MacAdam RB. Preservation of organic-walled microfossil types studied by scanning electron microscopy. *J Paleontol* 1971; 45:140.

Traverse A. *Paleopalynology*. Boston: Unwin Hyman, 1988:600. The appendix contains detailed laboratory techniques for the study of acritarchs, dinoflagellates, and other organic-walled microfossils.

Vidal G. A palynological preparation method. *Palynology* 1988; 12:215–20. Description of a useful method of preparing marine rocks for acritarchs, dinoflagellates, and other organic-walled microfossils.

Glossary

Abapical. The side of a silicoflagellate basal ring that contains no skeletal elements; the side opposite the apical side.

Abyssal. Pertaining to the deepest parts of the oceans, below 4000 m, and the organisms, sediments, and other features therein.

Acantharian. A group of planktic protozoa, classified as a suborder of the radiolaria or a class in the Actinopodea, characterized by skeletons made of strontium sulfate. No fossil record.

Acanthomorph. An acritarch that has a spherical vesicle, lacks an inner body and crests, and has prominent simple or branching processes; "subgroup" Acanthomorphitae.

Accessory apertures. In foraminifera, openings in the test into cavities associated with the primary aperture other than the primary chambers themselves.

Acicular. Refers to needle-like crystals.

Acritarch. Organic-walled microfossil of unknown biologic affinities; literally, "of uncertain origin."

Adherent operculum. A dinoflagellate operculum that is not permanently attached to the cyst but that remains touching the cyst.

Adnate operculum. A dinoflagellate operculum incompletely surrounded by sutures, and thus permanently attached to the cyst.

Adsorption. The binding of substances to a phase boundary through physical forces. The reversible binding of dissolved calcium ions to the ion exchange resin in a water softener is an example of adsorption.

Aerobic. Organisms capable of living in the presence of free oxygen, or environments or conditions possessing free oxygen.

Agamont. The generation or individual that reproduces asexually, by a complex cell division (syn. schizont).

Agglutinated. Composed of foreign material bound together to form a skeletal covering, as in the tests of some foraminifera and thecamebians, and the lorica of some tintinnids.

Akinete. A resting spore of a cyanobacterium.

Ala. The equatorial flange found in the phycoma stage of some modern and fossil prasinophytes and in pteromorph acritarchs.

Aliphatic. Hydrocarbons containing straight or branched carbon chains.

Allopatric. Refers to a speciation process involving geographic isolation of the parent and daughter species.

Alternation of generations. The alternation of the sexual, or gamete-producing phase, with the asexual, or cell division phase, of a protist, particularly foraminifera. Usually involves the alternation of haploid and diploid generations.

Amino acids. Compounds that contain carboxylic acid groups ($-COOH$) and amino groups ($-NH_2$).

Amphiesma. Cell covering of a motile dinoflagellate, consisting of an outer membrane, a thin layer of cytoplasm, a layer of flattened vesicles, and a discontinuous layer of microtubules (not always present).

Amphiphiles. Compounds that have polar chemical groups that tend to dissolve in water as well as nonpolar groups that tend to dissolve in oil phases. Soaps, detergents, and most lipids are examples of amphiphiles.

Anaerobic. Organisms capable of living in the absence of free oxygen, or environments or conditions without free oxygen.

Antapex. Posterior end of a dinoflagellate.

Antapical plate. One or more plates that touch the antapex in dinoflagellates.

Anterior. Generally, the end of a protist cell that leads in the direction of movement.

Anterior intercalary plate. Plate lying between the apical and precingular series of plates in dinoflagellates.

Aperture. An opening to a shell, test, or similar skeletal structure through which an animal or protist may extend various structures such as pseudopodia, heads, feet, or filtering apparatuses.

Apex. The anterior end of a protist, or the highest part of a protist skeleton, as in silicoflagellates.

Apical. Referring to the apex.

Apical cell. The rapidly growing, sometimes substrate

penetrating cells at the tips of branches of particular cyanobacteria.

Apical closing plates. Very small plates within the apical series on a dinoflagellate; also called preapicals and canal plates.

Apical plate. Plate that touches the apex on a dinoflagellate.

Apical ring. A closed ring, usually in the shape of a polygon, on the apical structure of a silicoflagellate.

Apical structure. In silicoflagellates, the dome-shaped structure made up of skeletal elements that is attached to the basal ring.

Aragonite. An orthorhombic $CaCO_3$ mineral that composes the tests of some protists, the foraminiferal suborder Robertinina in particular, and the skeletons of many metazoans.

Archeopyle. The opening in the dinoflagellate cyst wall through which the cytoplasm emerges during excystment, forming along definite preformed lines that correspond to the boundaries of paraplates genetically determined and of primary importance in classification.

Areal. A structure, particularly an aperture of a foraminiferan, that is isolated on a single surface, i.e., not in contact with another surface of the test.

Armored dinoflagellates. Dinoflagellates possessing a rigid, tabulate layer below the outer membrane of the theca.

Aromatic hydrocarbons. Compounds that contain structures based on the benzene ring. Benzene, naphthalene, anthracene, and pyrene are all aromatic hydrocarbons.

Assemblage. Fossils or living organisms that occur together.

ATP. Adenosine triphosphate, a compound involved with metabolic energy transformation within cells.

Autapomorphy. A uniquely derived character state that diagnoses a terminal taxon.

Autocyst. Dinoflagellate cyst with a single wall layer.

Autophragm. Single-layered dinoflagellate cyst, or the inner wall or a two-layered cyst if the two layers are separated by numerous processes or other extensively distributed structures.

Autotroph. Organisms that obtain their energy from chemical oxidation–reduction reactions (chemoautotrophs) or from light through photosynthesis (photoautotroph).

Auxospore. The zygote stage in the life-cycle of diatoms that is formed from sexual fusion of male and female gametes; they produce diatom frustules of the maximum cell size.

Auxotroph. Photosynthetic alga that requires organic materials from an external supply.

Axopods. Straight, slender, commonly stiffened, ray-like pseudopods.

Bachmannocenid. A silicoflagellate skeleton, usually of the genus *Bachmannocena*, that consists of a basal ring without an apical structure.

Bacillariophyceae. The botanical class within the algal division Chrysophyta to which diatoms are assigned.

Bacterivore. An organism that eats chiefly bacteria.

Baculate. An acritarch surface sculpture characterized by small, nontapering, rod-like projections, which typically have blunt ends.

Banded iron formations. Sedimentary rocks, commonly laminated, with a greater than 15% oxidized iron mineral content; chiefly Precambrian in age.

Basal ring. A closed ring, forming the base of the silicoflagellate skeleton, often in the shape of a polygon, with spines at the corners.

Basal side. A section of a polygonal basal ring between the corners of a silicoflagellate skeleton.

Bathyal. Ocean depths between 200 and 2000 m.

Benthic. Organisms that live on the bottom of the sea or a lake.

Binary fission. Division of a cell into two more or less equal parts during asexual reproduction.

Biocoenosis. A group of organisms living together, or fossil organisms found in the same place where they lived.

Biofacies. Part of a stratigraphic unit that is distinguished by a distinctive fossil content.

Biogenous. Biologically derived, such as biologically derived sediment.

Biogeography. The distribution of organisms in space; the study of that distribution.

Bioherm. A buildup above the sea floor of a mound-like or lens-like structure by calcium carbonate secreting organisms, including prokaryotes, algae, foraminifera, and a variety of metazoans.

Biohorizon. A biostratigraphic surface marked by a change or of a distinct character; usually a biozone boundary.

Bioluminescence. The production by enzymatic reactions of a glowing, cold light by organisms.

Biomineralization. The production of minerals, commonly in skeletons but elsewhere as well, by organisms.

Biostratigraphy. Stratigraphy of rocks based on their fossil content.

Biostrome. A body of rock formed by organisms, that is more or less parallel to the bedding of the enclosing sedimentary rocks.

Bioturbation. The burrowing and disruption of sedimentary deposits by organisms.

Biozone. Generally, the smallest biostratigraphic unit that can be recognized for intercontinental or worldwide correlation.

Body fossil. Fossils of the body or body parts of an organism, for example, its skeleton, impression, mold, cast, or similar remains.

Boreal. Pertaining to the north; used in biology and paleontology to refer to northern biotic provinces.

Boundary stratotype. A designated point in a specific stratigraphic sequence that serves as the standard for the definition of a boundary for a stratigraphic unit.

Brackish. Water with a salinity between that of normal fresh water and normal sea water.

Bridge. A series of skeletal elements that attach to the strut on the apical structure of a silicoflagellate skeleton without forming an apical ring.

Budding. The asexual reproduction of an organism by the development and separation of one or more new individuals from the body of a parent.

C-axis. The vertical crystallographic axis used for reference in the description of crystals; in paleontology, most commonly used in describing the crystallographic orientation of calcite, a common mineral in the skeletons of organisms.

Calcareous nannofossils. A collective term for all the minute carbonate plates and rods that probably derived from algal protists.

Calcareous nannoplankton. Planktic algae that bear carbonate plates and rods at some time in their life-cycle.

Calcite compensation depth (CCD). The depth in the oceans at which calcite dissolves as rapidly as it is deposited.

Canal plates. In dinoflagellates, the very small plates within the apical series, also called preapicals and apical closing plates.

Cannibal. An organism that eats members of its own species.

Cannopilid. A silicoflagellate skeleton with a multiwindowed apical structure. Such skeletons commonly belong to the genus *Cannopilus* or *Distephanus*.

Carbonaceous chondrites. Stony meteorites with small pebbles (chondrules) embedded in their mineral matrix, and commonly with several percent of their mass as organic carbon compounds.

Carnivore. An organism that eats animals.

Carotenoid. A type of fat-soluble pigment characterized by a yellow or orange color.

Cavate cyst. A dinoflagellate cyst with widely separated wall layers not separated by numerous processes or other extensively distributed structures.

CCD. Abbreviation for the calcite compensation depth.

Cell. A fundamental unit of life, defined as a membrane-bound system of catalytic and replicating macromolecules, capable of using nutrients and energy to grow and reproduce.

Cellulose. A resistant carbohydrate used by organisms in cell construction.

CENOP. Abbreviation for the Cenozoic Paleoceanography (CENOP) multiuniversity research program that developed out of CLIMAP and that reconstructed the climate for portions of the Neogene.

Central capsule. A pseudochitinous or mucinoid membrane, characteristic of radiolarians, that separates the protoplasm into endoplasm (within) and ectoplasm (without).

Centrales. One of two orders of diatoms (Bacillariophyceae); characterized by radial symmetry of the valve.

Chamberlets. In foraminifera, divisions of a chamber into partitions separated by test walls.

Chemical fossil. Chemicals preserved in the rock that were originally manufactured by organisms.

Chemotactic. Movement of an organism in response to a chemical stimulus.

Chitin. A resistant nitrogen-based organic compound used by organisms in the construction of some invertebrate skeletons, inner linings of foraminiferal tests, and in some fungi.

Chitinozoan. An extinct group of chitinous, flask-shaped microfossils of uncertain but probably metazoan affinities.

Chloroplast. Membrane-bound organelles found in eukaryote cells that contain chlorophyll pigments and enzyme systems used in photosynthesis.

Chorate cyst. A dinoflagellate cyst in which the diameter of the inner body is considerably less than the overall diameter (with processes).

Cingular plate. In dinoflagellates, the plate that forms the cingulum.

Cingulum. In dinoflagellates, the depression occupied by the transverse flagellum that divides the cell into the anterior epitheca and posterior hypotheca.

Clade. In evolutionary biology, a group of lineages derived from the branching of a single evolutionary lineage. In ebridians, exterior connecting skeletal elements.

Cladistics. A junior synonym of phylogenetics.

Cladogenesis. The branching of a lineage to form a clade or divergent evolution caused by speciation.

CLIMAP. Abbreviation for the Climate Long Range Investigation Mapping and Planning (CLIMAP) multi-university research program that reconstructed Pleistocene climates.

Coccolithophore. A minute, marine planktic alga belonging to the chrysophytes with two flagella and a covering of miniscule calcareous plates.

Coccoliths. Calcite plates forming the exterior covering of coccolithophores.

Coccosphere. The more or less spherical cell covering of a coccolithophore, consisting of interlocking coccoliths.

Combination archeopyle. An archeopyle formed by the loss of plates from more than one plate series of a dinoflagellate.

Competition. The simultaneous attempt by a species or individual to acquire a resource utilized by another individual or species.

Compound operculum. A dinoflagellate operculum consisting of several pieces.

Consensus tree. A tree, or branching diagram showing a phylogenetic hypothesis, that incorporates the knowledge or perceived knowledge from two other trees.

Consumer. An heterotrophic organism that feeds on primary and/or secondary producers.

Continental shelf. The more or less flat part of the continental margin from the subtidal to the continental slope, generally at 200 m.

Convergence. A place where different types of water or different currents meet at the surface of the ocean.

Corbisemid. A silicoflagellate skeleton with a triangular basal outline, usually a member of the genus *Corbisema*.

Core. In oceanography and marine biology or geology, a cylindrical sample of the substrate, or sedimentary or other kind of rock, taken with tubular devices of various designs and power source.

Cosmopolitan. An organism that is widely distributed in a variety of provinces throughout the world.

Costate. An acritarch surface sculpture of elongate, parallel ridges; areas between ridges are relatively wide.

Cristae. Structures within mitochondria formed by membranes that separate enzymes involved in ATP generation; the several kinds of cristae may reflect different origins for the mitochondria.

Cristate. An acritarch surface sculpture characterized by low, flange-like crests that may anastomose.

Crossed nicols. In a polarizing microscope, the two prisms, one below the microscope stage and the other above it, that are oriented so that the transmission planes of each are at right angles, thus preventing the transmission of light unless a substance intervenes between them.

Cryptarch. An informal group comprising organic-walled, hollow vesicles or cellular bodies (which may be associated in pairs, clusters, or colonies) and aseptate filamentous microfossils; as defined, the "group" Cryptarcha includes the "subgroup" Sphaeromorphitae of the "group" Acritarcha.

Cryptopylome. A type of excystment structure in acritarchs similar to pylome, but not preformed and thus not visible on vesicle prior to excystment.

Cyanobacteria. A group of eubacteria characterized by their low state of cell differentiation with single or colonial cell arrangements commonly covered by a mucilagenous sheath. They are common prokaryote fossils dating back to 3.5 billion years ago. They are also known as blue–green algae, cyanophytes, or myxophytes.

Cyst. Nonmotile, resistant dormant stage in the life-cycle of many protists; in paleontology, the chief fossil remains of dinoflagellates, prasinophytes, and acritarchs. Any structure, formed of debris collected by a foraminifera, occupied by the organism during feeding, growth, or reproductive activities.

Cytoplasm. The contents of a cell, excluding the nucleus.

Cytoskeleton. The system of microtubules and microfilaments, or other secreted inorganic or organic material, found on, in, or below the surface of a protist that involves all or most of the body and that may provide rigidity to the organism's shape.

Deep Sea Drilling Project. A program run by the United States National Science Foundation to drill into the deep-sea sediments of the world for scientific information, superceded by the Ocean Drilling Program.

Detritivore. An organism that eats chiefly organic-rich sediment or pieces of organic detritus.

Diacromorph. An acritarch that has a spherical to ellipsoidal vesicle that is characterized by ornamented polar areas and an equatorial zone free of ornamentation; "subgroup" Diacromorphitae.

Diagenesis. All the physical, chemical, and biologic changes undergone by a sediment or skeleton of any organism after its initial deposition, exclusive of weathering and metamorphism.

Diatomite. A rock consisting largely of the opaline skeletons of diatoms. They may be of nonmarine or marine origin and they are commonly mined for commercial uses, including filters and mild abrasives.

Dictyochid. A silicoflagellate skeleton that has more than three basal sides and an apical structure that does not include an apical ring; commonly belonging to the genus *Dictyocha*.

Dimorphic. A species with two distinct morphologic forms.

Dinocyst. A dinoflagellate cyst.

Diploid. Nuclei with a double number of chromosomes occurring as a short or long stage in the life-cycle of most species of eukaryotes.

Disintegration archeopyle. An archeopyle in dinoflagellates in which the epicyst breaks up, first by splitting in the middorsal intercalary regions, then by loss of adjacent paraplates.

Dissolution. A process in which minerals pass into solution.

Distephanid. A silicoflagellate skeleton with a single apical ring, commonly belonging to the genus *Distephanus*.

Dithecate. A coccolithophore constructed of two layers of coccoliths; each layer may contain coccoliths of different shapes.

Divergence. In oceanography, the movement of waters away from one another at the sea surface; divergences are usually places of upwelling. In evolution, the change in two or more species of some character or characters away from one another.

Dorsal surface. The back or upper surface of most organisms; in most dinoflagellates, the surface opposite the insertion of the flagella; in dinocysts, the surface opposite the parasulcus.

DSDP. Abbreviation for Deep Sea Drilling Project.

Echinate. In acritarchs, a surface sculpture of small, tapering, thorn-like spines (or spinules).

Ectophragm. Outer layer of a dinoflagellate cyst with two or three wall layers, if this layer is separated from the next inner layer by numerous processes or other extensively distributed structures.

Ectosymbiont. A symbiont living on the external surface of another organism.

Embryo. In any organisms, the product of sexual reproduction while still in an early stage of development.

Emigration. The movement or migration of organisms to another geographic area.

Enantiomeric. Compounds, such as amino acids and carbohydrates with asymmetric carbon atoms surrounded by four different chemical groups, having two possible chemical structures (called enantiomers) that represent mirror images of each other. Living organisms use only one of the two possible enantiomers, common examples being L-amino acids and D-sugars. The L and D refer to optical properties, in which the plane of polarized light passing through solutions is rotated to the left (levorotatory) or to the right (dextrorotatory).

Endemic. An organism restricted to a particular geographic region or environment.

Endolithic. An organism living in rocks, usually by boring.

Endoperculum. A dinoflagellate opercular piece(s) that is from the endophragm.

Endophragm. The inner layer of a dinoflagellate cyst with two wall layers, if the outer layer is not separated from the next inner layer by numerous processes or other extensively distributed structures; the inner layer of a cyst with three wall layers. The inner wall in acritarchs that have a bilayered vesicle.

Endoplasmic reticulum. A system of intracellular membranes separating parts of the cell from the remaining cytoplasm and allowing the separation of substances and conditions within the cell.

Endospore. In prokaryotes, a spore utilizing part of the original cell; in dinoflagellates, the innermost cyst layer.

Endosymbiont. A symbiont living within a cell or tissue of another organism.

Enzymes. Proteins in cells that speed up (catalyze) chemical reactions.

Epeiric seas. A sea on the continental shelf or within a continental land mass. Used for ancient seaways, in particular.

Epicyst. Anterior part of a dinoflagellate cyst, above the cingulum.

Epifauna. The animals or protozoa that live on the surface of a substrate.

Epilithic. Refers to living, or life, on rock surfaces.

Epiphyton. Photosynthetic organisms living on or attached to a plant.

Epitheca. Anterior part of a dinoflagellate theca, above the cingulum.

Epityche. Type of excystment structure in acritarchs characterized by a curving slit that allows a flap of the vesicle to fold out.

Epivalve. The larger of the two valves of a diatom frustule.

Ester bond. A bond in which an acidic group of one compound is chemically linked with the hydroxyl group of another. The equivalent of a water molecule is lost, a process often referred to as a condensation reaction. The monomers of many important biochemical polymers, for example DNA and RNA, are linked together through ester bonds.

Estuary. The zone of mixing of fresh and sea water in the mouths of rivers, bays, or lagoons.

Euendolith. Organism that always live in rock borings, crevices, or between grains.

Eukaryotes. Organisms with well-organized intracellular structure, including a nucleus with chromosomes and various membrane-bound organelles.

Eurybathyal. Refers to a distribution across a wide depth range.

Eurytopic. Occurring in many different habitats. Tolerant of a variety of conditions.

Eutrophic. Waters characterized by having a high nutrient level, such as in upwelling parts of the ocean.

Evolute. Shells or tests in which the later chambers do not completely envelop earlier ones, leaving all or some of the chambers in earlier whorls visible.

Excystment structure. Escape opening on a cyst or phycoma of various algae through which a vegetative cell passes.

Extant. A living, as opposed to an extinct, organism.

Extramural. Protoplasm or other parts of a foraminiferan residing outside the test.

Eyespot. A light-sensitive, pigment-containing body in the cell of many phytoflagellates.

Fecal pellet. A small ovoid particle of excrement made from waste products by an organism.

Fission. The asexual reproduction of protists by cell division.

Flagellum. A relatively thin and long, protoplasmic extension from the cell of a protist or metazoan.

Flysch. Marine sedimentary rocks chiefly composed of calcareous shales, muds or marls, or facies that are characterized by graded, thin beds largely lacking fossils.

Foramen. In foraminifera, the opening between chambers representing either previous apertures or secondarily dissolved holes.

Form genus. A genus based on plant or algal parts with the same form or morphology.

Fossulate. In acritarchs, a surface sculpture characterized by burrow-like depressions or closely spaced, elevated granules separated by irregular grooves (usage not consistent).

Foveolate. In acritarchs, a surface sculpture of closely spaced pits; the diameter of the pits typically exceeds that of the areas between the pits.

Free operculum. A dinoflagellate operculum completely separated from the cyst by sutures.

Free-living. An organism that is free to move about during its life; not cemented or attached to a substrate, nor parasitic.

Frustule. The external skeleton of diatoms that is composed of opaline silica and consists of two overlapping halves, termed valves.

Functional morphology. The use or function of various parts of organisms; the study of such uses.

Fusiform. An elongate morphology resulting from a long axis much greater than the shorter axis.

Gametogenesis. The process of the formation and release of gametes.

Gamont. The generation or individual that produces gametes.

Gemmate. In acritarchs, a surface sculpture of small, spherical projections that are somewhat constricted at the base.

Genera. The plural of genus.

Genus. In systematic classification of organisms, the category above species and below family; always used as the first part of the scientific name of a species.

Geopetal. Any rock feature that indicates the top during its formation.

Golgi apparatus or body. A membranous organelle within the cell of a eukaryote, usually containing various products of cell synthesis.

Gonal processes. Processes that occur at the intersection of three or more paraplates in dinoflagellates.

Gonyaulacoid. A dinoflagellate with tabulation or paratabulation like modern *Gonyaulax* with six precingulars and six postcingulars; in Kofoidian notation, Opr, 5′, Oa, 6″, 6c, 6‴, 1p, 1⁗, 5s.

Grab sample. A sample obtained by any of a number of different kinds of devices that take, in a single motion, a portion of the uppermost substrate from an aquatic environment.

Grade. A level of evolutionary organization obtained by two or more taxa.

Gradual evolution. The evolution of new forms by a long continuous, and imperceptible, gradation from one form to another.

Granulate. In acritarchs, a surface sculpture of scattered granules (or grana).

Haploid. Nucleus with a single number of chromosomes occurring during a short or long stage in the life-cycle of eukaryotes.

Haptonema. A relatively thin and long, protoplasmic extension which may be straight or coiled and may serve for temporary attachment in the haptophycean algae.

Herbivore. An organism that eats chiefly plants or algae.

Herkomorph. An acritarch that has a spherical to polygonal vesicle, the surface of which is divided by crests into polygonal fields; "subgroup" Herkomorphitae. Most herkomorphs are probably prasinophytes.

Heterococcolith. A coccolith constructed of calcite crystals that differ in size and shape.

Heterocyst. An enlarged cyanobacterial cell, with thickened, two-layered wall and pore at the ends, that occurs singly or in sequence.

Heteromorphic. Processes on a single acritarch vesicle

that are not all similar in morphology.

Heterotroph. Organisms that obtain their energy by ingesting organic material, alive or dead; includes herbivores, carnivores, and scavengers.

HMS *Challenger*. The British oceanographic vessel that pioneered modern oceanography in an around-the-world expedition from 1872 to 1876.

Holococcolith. A coccolith constructed of calcite crystals that are essentially identical in size and shape.

Holoplanktic. An organism that lives its entire life-cycle in the plankton.

Homeomorphic. Inorganic crystals or biologic species that have superficial similarity to each other.

Homomorphic. A single acritarch vesicle that has processes that are all similar in morphology; also, the superficial resemblance in the morphology of different organic species or objects.

Homoplasy. The similarity of a character that is present in two taxa, but is absent in their common ancestor.

Hormogone. Segments of cyanobacterial filaments that are capable of movement and reproduction to form a new filament.

Hyaline wall. A wall type in foraminifera that is translucent and consists of interlocking, calcite granules arranged in one of two different crystallographic orientations.

Hydrocarbons. Compounds containing only hydrogen and carbon.

Hydrosphere. Generally, the liquid and frozen (ice) waters of the Earth—oceans, lakes, rivers, and ground water, as well as snow and ice.

Hypersaline. Waters with a salinity considerably greater than normal sea water.

Hypnozygote. A very thick-walled, nonmotile zygote; a zygote cyst.

Hypocyst. Posterior part of a dinoflagellate cyst, below the cingulum.

Hypotheca. Posterior part of a dinoflagellate theca, below the cingulum.

Hypovalve. The smaller of the two valves of a diatom frustule.

Hystrichospheres. A term, no longer used, for spiny cyst-like microfossils, many of which are dinoflagellate cysts, whereas others are now called acanthomorph acritarchs.

Ichthyolith. Small denticles, scales, and miscellaneous pieces of bony material presumed to be parts of fishes, sharks, or rays.

Immigration. The movement of organisms into another geographic region.

Infauna. The animals or protozoa that live beneath the surface of a substrate.

Ingroup. The study group or taxon in a phylogenetic analysis. The sister taxon of the ingroup is the outgroup.

Intergonal processes. Processes that occur along the paraplate boundaries of dinoflagellate cysts away from triple junctions.

International Code of Botanical Nomenclature. The rules for naming and recognizing the proper names of plant species, accepted by all botanists; used to govern the use of names in the algal protists.

International Code of Zoological Nomenclature. The rules for naming and recognizing the proper names of animal species, accepted by all zoologists; used also to govern the use of names in the protozoan protists.

Interstitial. Pertaining to the spaces between sand particles.

Intertidal zone. The zone between the lowest and highest tidal levels along the shore of the oceans.

Interval zone. Strata between two distinctive biostratigraphic horizons.

Intratabular features. Characters that emphasize the central parts of plates or paraplates of dinoflagellates.

Involute. Shells or tests in which later chambers completely envelop earlier ones.

Isobathyal. Of the same depth.

Isotope. One of two or more forms of the same chemical element but with a different number of neutrons in the nucleus.

Iterative evolution. Patterns of similar morphologic evolution repeated at different times in geologic time.

Kerogen. Poorly defined organic polymers found in coals and other organic substances that have undergone heating and aging.

Kofoidian notation. The standard shorthand for expressing dinoflagellate tabulation: ′ for apical plates, a for anterior intercalary, ″ for precingular, c for cingular, ‴ for postcingular, p for posterior intercalary, ⁗ for antapical, s for sulcal.

Labiate process. Tubes or openings through the valve wall of diatoms with an internal flattened tube or slit often surrounded by two lips. Labiate processes have been very useful in classifying diatoms.

Laevigate. In acritarchs, a smooth surface characterized by a lack of ornamentation.

Leiospheres. Simple spheromorph acritarchs that have a relatively smooth wall.

Lobulate. The lobate termination of acritarch processes.

Longitudinal flagellum. The posterior flagellum of a dinoflagellate.

Lorica (pl., **Loricae**). The test, shell, or external covering secreted or agglutinated from foreign material of a protozoan.

Macrofauna. Large animals visible with the naked eye.

Macromolecule. Organic molecules made of polymeric structures composed of smaller monomeric subunits. Examples include proteins, nucleic acids, and large carbohydrates like starch and cellulose.

Magnetotactic. Movement of organisms in response to a magnetic field.

Marine snow. Aggregates of living, fecal, and nonliving matter floating in or settling through the ocean waters.

Marker species. In biostratigraphy, species with distinctive biostratigraphic distributions.

Mass extinction. The extinction of a significant proportion of the extant species at a particular stratigraphic horizon.

Median split. A type of excystment structure in acritarchs resulting in the separation of the vesicle into two nearly equal parts.

Megalospheric test. A foraminiferal test with a large proloculus formed during fission during asexual reproduction.

Melanocyrillids. Complex vase-shaped Proterozoic acritarchs exhibiting a distinctive triangular opening (excystment structure?) typically surrounded by a hexagonal or triangular collar.

Membranelle. One of several compound ciliary structures arranged around the anterior end of a ciliate.

Meristem. Regions of cyanobacterial trichomes where rapid cell differentiation occurs.

Meroplanktic. An organism that lives part of its life-cycle in the plankton.

Mesocenid. A type of silicoflagellate skeleton. See bachmannocenid.

Mesophragm. The middle layer of a few dinoflagellate cysts with three wall layers in which the layers are not connected by supporting structures.

Metazoa. Heterotrophic, multicellular eukaryotes; animals.

Micrograngular wall. A wall type in foraminifera composed of tiny equidimensional granules of crystalline calcite.

Microgranulate. A surface of an acritarch or other protist skeleton that has a microscopic grain resolvable only in an electron microscope.

Microspheric test. A foraminiferal test with a small proloculus formed by the fusion of gametes during sexual reproduction.

Microtubule. Hollow, cylindrical structures found in the cytoplasm and nucleoplasm of protists; may be involved in the extension and contraction of protoplasm. Provides structure to various organelles.

Mitochondria. Conspicuous double-membraned organelles within cells functioning in cell respiration.

Monomers. The subunits of larger polymeric compounds. Examples of biologically important monomers include amino acids, nucleotides, and monsaccharides like glucose.

Monophyly. The condition of a taxon or clade when it is composed of the ancestral species and all of its descendants.

Morphometric. Refers to the measurement of morphologic characteristics on an organism.

Multilocular test. A foraminiferal test with more than one chamber.

Multinucleate. A cell containing many nuclei, as in certain asexual stages of foraminifera.

Mutualism. When two or more different species live together and benefit from the association.

Nannoflora. All the species of nannoplankton or nannofossils within a single environment or rock sample.

Nannoherbivore. An organism that eats chiefly nannoplankton.

Nannolith. One of a variety of miniscule calcareous plates, probably from a planktic alga.

Nannoplankton. Planktic organisms of very small size.

Nassellaria. Cone-shaped radiolarians (order Nassellarida) of the class Polycystinea.

Natural classification. A classification that reflects the evolutionary history of the taxon.

Naviculopsid. An elongate silicoflagellate, generally of the genus *Naviculopsis*, that has two sides of its skeleton bowed outwards and connected in the middle by a single cross-member (the bridge).

Nektic. Organism capable of swimming in the open waters.

Nematocyst. Ejectile organelle of unknown function in dinoflagellates. Also present in some metazoans, particularly cnidarians.

Neritic. The shallow-water area along the coastal region of the sea; those nearshore marine waters influenced by coastal conditions.

Netromorph. An acritarch that has an elongate to fusiform vesicle that may have processes or spines at one or both poles; "subgroup" Netromorphitae.

Niche. All the ecologic factors affecting or potentially affecting a single species.

Nontabular. A dinoflagellate cyst that has no indication of the plates on the theca.

Nucleic acids. Polymers of nucleotides responsible for the storage and utilization of genetic information in the cell. Examples include DNA (deoxyribonucleic acid) and RNA (ribonucleic acid).

Nucleolus. RNA-containing bodies in protists.

Nucleotides. The monomers of nucleic acids. Examples include AMP, TMP, GMP, and CMP, the monomers of DNA.

Nucleus. Eukaryote organelle central to sexual reproduction and containing the chromosomes in the cell.

Numerical taxonomy. The use of similarity coefficients and clustering techniques to arrange individuals into nested sets based on overall similarity.

Nutricline. A zone of rapid increase in dissolved nutrients at depth in the ocean.

Obligate. An organism that can grow only under certain conditions.

Ocean Drilling Program. See Deep Sea Drilling Project.

Oceanic. Dwelling in or from the open ocean.

Oceanic gyre. The current systems circulating in the northern and southern hemispheres of the world's oceans.

Ocellus. Complex, light-sensory apparatus in dinoflagellates.

Oligotrophic. Parts of the ocean or a lake characterized by a low nutrient level.

Omnivore. An organism that eats most organic material, whether it is animal, plant, or other.

Oncolite. A calcareous, concentrically layered sedimentary structure, usually spherical or oval but also irregularly shaped and less than 10 cm in diameter, formed by successive layers of cyanobacterial sheaths.

Oömorph. An acritarch that has an egg-shaped, polarized vesicle, which is relatively smooth on one end and heavily ornamented on the other; "subgroup" Oömorphitae.

Ooze. A deep-sea sediment dominated by the skeletons of some organism, and usually named for that organism, such as radiolarian ooze.

Operculum. A cover for a skeletal opening. In dinoflagellates, the cover over the archeopyle. In acritarchs, the plug-like object that fits within, and thus closes, the pylome; may be ornamented.

Opisthe. The posterior cell formed from regular binary fission in ciliates.

Opportunitistic. An organism that takes advantage of any opportunity, especially with regard to food preferences.

Optic axes. The direction in crystals along which there is no double refraction.

Optic extinction pattern. The dark pattern formed by areas of a rock or fossil skeleton perpendicular or parallel to the crossed nicols of a polarizing microscope that do not transmit light.

Organ genus. A generic name based on the morphology of an organ from a fossil, in particular the organ of a plant, such as a leaf, flower, stem, etc., or parts of algae.

Organelle. Structures within cells that perform separate functions.

Osmotrophe. An organism that absorbs its organic nutrients directly from sea water.

Outgroup. A taxon that shares a common ancestor with the group under study, and is typically used to determine primitive and derived character states in the study group. More than one outgroup can be used.

Overgrowth. The secondary deposition of skeletal material through later processes onto the primary skeleton of an organism.

Paleobathymetry. The depths of ancient marine deposits, usually inferred from the fossil content of the sedimentary rock.

Paleobiogeography. The distribution of fossil organisms; the study of such distributions.

Paleoceanography. The oceanography, particularly current patterns, temperatures, and vertical structure, of ancient oceans, inferred from paleontologic, sedimentologic, and/or geochemical data.

Paleodepth. The water depth of ancient marine deposits; usually determined by the assemblage of benthic microfossils contained in the rocks.

Palmate. A flat or broad structure or test.

Palynologist. A scientist who studies palynomorphs or pollen, whether fossil or modern.

Palynomorphs. Organic-walled, and typically acid-resistant, microfossils; includes spores and pollen grains of vascular plants, dinoflagellate cysts, prasinophytes, fungal spores, chitinozoans, scolecodonts, and acritarchs.

Paraphyly. A grouping that includes an ancestor and some, but not all, of its descendants.

Parasite. An organism that derives its food and other benefits from another organism without killing it but at its expense.

Parasutural. Features that emphasize boundaries between plates or paraplates in dinoflagellates and acritarchs.

Parsimony. The simplest or most economic explanation. In phylogenetics parsimony refers to trees constructed with the fewest number of character state changes.

Partial rupture. Type of excystment structure in acritarchs; the vesicle partially splits, leaving a gap.

Peduncle. A retractible protoplasmic extension that may play a part in attachment to substrate or host.

Pelitic. A description for a sedimentary rock composed chiefly of clay.

Pellicle. The external, nonmineralized covering of the

cell of a protist.

Penitabular. Features that lie approximately parallel to paraplate boundaries within individual paraplates in dinoflagellates.

Pennales. One of two orders of diatoms (Bacillariophyceae) that is basically characterized by bilateral symmetry of the valve.

Peptide. Amino acids, the monomers of proteins, link together through a specialized bond called the peptide bond. The resulting polymer is sometimes referred to as a peptide, particularly if it only contains a few amino acids. Larger peptides with 50 or more amino acids are usually called proteins.

Peridinioid. Tabulation or paratabulation in dinoflagellates like modern *Peridinium*, with seven precingular plates and five postcingular plates; in Kofoidian notation, 2pr, 4′, 3a, 7″, Xc, 5‴, Op, 2⁗, Xs.

Perioperculum. The opercular piece(s) that forms from the periphragm of acritarchs.

Periphyton. Photosynthetic organisms living on rocks.

Peritidal. The zone between the highest influence of the ocean to somewhat below the lowest tides.

Periphragm. Outer layer of a dinoflagellate cyst with two or three walls, if the next inner layer is not connected to it by supporting structures. Outer wall in acritarchs that have a bilayered vesicle.

Phaeodarian. A radiolarian belonging to the class Phaeodarea and which possesses hollow skeletal structures of an admixture of silica and organic matter.

Phenetics. The theory and practice of taxonomy based on the overall similarity of the individuals that make up the study group.

Photic. Refers to light; in ecology, particularly the lighted part of the water column.

Photoautotroph. Organisms that obtain their energy from the sunlight.

Photosystem I. A system of photosynthesis that utilizes long wavelengths of light.

Photosystem II. A system of photosynthesis that utilizes short wavelengths of light.

Phototactic. Movement in response to light.

Phototrophic. Refers to organisms that utilize photosynthesis for the production of their energy needs.

Phycologist. A specialist on algae.

Phycoma (pl. **phycomata**). Nonmotile stage in life-cycle of some prasinophytes; grows during its existence, and therefore should not be called a cyst.

Phyletic evolution. Evolution by gradual increments of change over long periods of time.

Phylogenetics. The theory and practice of taxonomy based on the work of Willi Hennig. Phylogenetists use character state changes to reconstruct the phylogenetic relationships among organisms, and then translate these hypotheses directly into classifications that reflect these relationships.

Phytoflagellate. A photosynthetic protist with a flagellum.

Phytoplankton. Mostly planktic, photosynthetic protists.

Picoplankton. Very tiny plankton (0.2–2.0 μm), usually prokaryotes.

Pike. Relatively small spines pointing abapically at the approximate midpoint of a basal side of a silicoflagellate.

Planispiral. A shell or test that coils in a single plane.

Planktic. Refers to organisms that float in open waters.

Plankton. Organisms that live in the open waters of the ocean or a lake, and that largely are moved by water currents rather than by self-locomotion.

Plasma membrane. The bounding membrane of the cell.

Plastid. A membrane-bound organelle in eukaryotic protist and plant cells that are usually involved in photosynthesis.

Plates. Thin cellulosic pieces comprising the theca of a dinoflagellate or acritarch.

Pleisomorphy. The primitive or preexisting state of a character.

Pleomorphic. A species with two distinct morphologic forms at different phases of the life-cycle.

Polarity. The sequence of states for a character from primitive to derived.

Polycystine. A radiolarian with an opaline skeleton belonging to the class Polycystinea, which contains the spumellarian and nassellarian subgroups.

Polygonomorph. An acritarch that has a vesicle with a polygonal outline, and typically simple processes; "subgroup" Polygonomorphitae.

Polymerization. The process by which monomers link together to form polymers.

Polyphyly. The condition of a taxon or grouping that contains members having different ancestors.

Porcelaneous wall. A wall type in foraminifera that is variably translucent to opaque, composed of minute laths of high-magnesium calcite in various arrangements.

Postcingular plate. In dinoflagellates, a plate that touches the cinculum posteriorly.

Posterior intercalary plates. In dinoflagellates, plates lying between the postcingular and antapical series of plates.

Prasinophytes. Phylogenetically primitive, primarily marine green algae; motile cells typically quadriflagellate and scale bearing; flagella also typically bear scales; life-cycle of some includes phycoma stage.

Preapical plates. In dinoflagellates, very small plates within the apical series, also called apical closing plates and canal plates.

Precingular plate. In dinoflagellates, a plate that touches the cingulum anteriorly.

Predation. The act of eating another animal or protozoan.

Primary aperture. In foraminifera, the principal opening into the interior of the test through which reticulopodia pass.

Primary producers. Photosynthetic organisms that are capable of converting light energy into organic matter.

Primary productivity. The amount of organic material synthesized by organisms using photosynthesis in a specified volume of water.

Prismatomorph. An acritarch that has a prismatic to polygonal vesicle, the edges of which may be produced into a flange; "subgroup" Prismatomorphitae. Most prismatomorphs are probably prasinophytes.

Problematica. Organisms, fossil or living, whose systematic placement and relationships cannot be determined.

Process. A projection (solid or hollow) from the surface of an acritarch vesicle or dinoflagellate cyst, typically more than 5 μm in length.

Proloculus. The initial chamber of a foraminiferal test.

Proter. The anterior cell formed from regular binary fission in ciliates.

Protozoologist. A scientist who studies protozoans.

Province. A broad region of the Earth which is characterized by particular associations of organisms.

Proximate. A dinoflagellate cyst that closely resembles the theca in which it formed in both size and shape.

Pseudopodia. Protoplasmic extensions from the body of a protist used in locomotion, feeding, and attachment.

Pseudopylome. In acritarchs, a thickening of the vesicle wall that resembles the pylome rim, but which typically occurs opposite the true pylome.

Pteromorph. An acritarch that has a spheroidal vesicle surrounded by an equatorial flange (ala); "subgroup" Pteromorphitae. Most pteromorphs are probably prasinophytes.

Punctate. A surface sculpture of scattered shallow pits.

Punctuated equilibrium. An evolutionary mode characterized by relatively long periods of morphologic stasis broken by sudden morphologic changes resulting from allopatric isolation of small ancestral populations.

Punctuated gradualism. A pattern of evolution through time exhibited particularly by some foraminifera, in which morphologic change progresses gradually but is marked by sudden rapid changes.

Pusule. A series of membrane-bound tubules of uncertain function.

Pylome. A type of excystment structure in acritarchs; rounded to subangular, preformed opening, closed by a plug-like operculum; in some cases bordered by thickened rim or collar, or by tubular neck.

Pyrenoid. Specially differentiated regions of the chloroplasts that are composed of proteinaceous material and that have a granular appearance; specific function is not known; they may be for storage or production of proteins, starch, or enzymes.

Pyrolytic. At sufficiently high temperatures, organic compounds begin to break down into smaller fragments. This is termed a pyrolytic process, from Greek words for "fire" and "breaking."

Ranked classification. Classifications that follow the Linnaean system of kingdom, order, class, family, genus, and species.

Rhizopodia. Anastomozing pseudopodia extending away from the body of a protist, commonly used for locomotion.

Rugulate. In acritarchs, a surface sculpture characterized by a finely wrinkled surface (usage not consistent).

Sabka. A supratidal environment, gradational between the land surface and the intertidal zone, characterized by evaporites, tidal floods, and eolian deposits.

Scavenger. An organism that searches for and eats dead organisms.

Schizont. See agamont.

Sea snow. See marine snow.

Secondary apertures. In foraminifera, secondary openings into the interior of the test; may be present with a primary aperture or may replace it.

Septum (pl., **septa)**. A partition between similar parts of an organism. In filamentous algae, a perpendicular structure crossing the filament. In foraminifera, the portion of the test wall between chambers.

Sheath. A structure surrounding a series of cells, usually in a linear fashion.

Simple operculum. In dinoflagellates and acritarchs, an operculum consisting of a single piece made up of one or several paraplates.

Sister group. A taxon that is the nearest genealogic relative of another.

Species. A kind of organism, ranked below the genus level. Among sexually reproducing organisms, a group of interbreeding populations reproductively isolated from other such groups. In prokaryotes, many protists, and most fossil organisms, a collection of specimens of similar morphology that is usually gradational across all specimens, and that can be reasonably assumed to have lived in spatial or temporal continuity.

Species richness. The number of species found in a particular place or sample.

Sphaeromorph. An acritarch that has a spherical to ellipsoidal vesicle and lacks processes; "subgroup" Sphaeromorphitae. Some sphaeromorphs are probably prasinophytes.

Spiny acritarchs. Process-bearing (acanthomorph) acritarchs.

Sporangia. Structures in which spores are produced.

Spore. A specialized asexually produced cell freed from the parent and eventually capable of producing a new organism.

Sporopollenin. A high molecular weight and acid-resistant polymer of carbon, hydrogen, and oxygen; probably composes the wall of most acritarchs and dinoflagellate cysts, as well as spores and pollen grains of vascular plants.

Spumellarian. Radiolarians with a morphology based on a sphere and belonging to the class Polycystinea.

Standing crop. The biomass of a specified population of one or more species existing at a particular time.

Statospore. A resting spore of diatoms that have two overlapping siliceous valves like the diatom frustule, but they lack a girdle band and the two valves are dissimilar. Statospores are most typical of planktic neritic diatoms.

Stenotopic. An organism with a narrow range of habitats or tolerances.

Sterane. A hydrocarbon, found in sedimentary rocks, derived from the sterols of eukaryotes.

Stromatolite. A layered sedimentary structure produced by the trapping and binding of sedimentary particles and precipitation of minerals by organisms, uniquely prokaryotes in the Archean and early Proterozoic, but possibly protists and metazoans in the late Proterozoic, and protists and metazoans in the Phanerozoic.

Strut. In silicoflagellates, a skeletal element that is attached to the basal ring and supports additional elements that with the struts comprise the apical structure.

Sulcal plates. Plates in the sulcal area of a dinoflagellate or an acritarch.

Sulcus. In dinoflagellates, the depression, roughly perpendicular to the cingulum and containing the flagellar pores.

Supplementary aperture. Apertures in addition to the primary aperture.

Surfactant. Any molecule that tends to accumulate at surfaces is called a surfactant. Examples include soaps and detergents, which produce monomolecular films at air–water interfaces.

Suspension feeder. An organism that captures food particles suspended in fluid.

Suture. The external expression on the surface of a test or shell of the internal chamber walls, or the line of junction between adjacent plates of a dinoflagellate.

Symbiont. An organism that derives benefits from and provides benefits to another organism, without harm to either one.

Symbiosis. A relationship between two or more organisms which benefits each one.

Symbiotrophe. An organism that derives its chief nutrients from algal symbionts.

Sympatric. Refers to a speciation process taking place without geographic isolation of the parent and daughter species.

Synapomorphy. A derived character state shared by two or more taxa.

Tabulation. Number and arrangement of plates of a dinoflagellate.

Taphonomy. The branch of paleontology concerned with the death, burial, and fossilization of organisms.

Tasmanitids. A group of distinctive thick-walled and pore-bearing fossil prasinophytes.

Temperate. Usually the mid latitudes that do not have extremes of temperature; a moderate climate or range of ocean temperatures.

Terrigenous. Land-derived, such as a sediment that comes from land.

Test. A shell or covering of a protozoan or animal that may be enclosed in protoplasm; in foraminifera and thecamebians, in particular, it may consist of organic materials, secreted minerals (calcite, aragonite, silica), or agglutinated particles.

Testate. An organism with a test.

Thanatocoenoses. The death assemblages of organisms.

Theca. A cell covering consisting of numerous rigid cellulosic plates lying beneath an outer flexible membrane, found in certain dinoflagellates.

Thermocline. A vertical zone of rapid increase in temperature in the ocean.

Thermohaline. Temperature and salinity, usually referring to vertical oceanic circulation that is caused by differences in temperature and salinity.

Thermophilic. Prokaryotes that live in hot waters, such as volcanic fumeroles.

Trace fossil. The traces of an ancient organism, such as a trackway, trail, burrow, stromatolite or other evidence of their former existence; preserved as sedimentary structures rather than impressions or parts of the actual organism.

Transverse flagellum. Flagellum that encircles the dinoflagellate cell, occupying the cingulum.

Trichome. A chain of prokaryote cells linked together commonly by a surrounding sheath.

Trichocyst. Ejectile organelle of unknown function in dinoflagellates and ciliates.

Trochospiral. A shell or test that coils in a spire.

Trochospiral suture. Excystment structure in acritarchs; spiral-shaped slit.

Trophic. Of or pertaining to nutrition.

Turbidite. A sediment or rock deposited from a turbidity current; typically with graded bedding.

Type species. The species on which a genus is based; the typical species for the genus.

Umbilicus. Space formed between the chambers of the last whorl of a trochospiral foraminiferal test.

Unarmored. Dinoflagellates that do not have a prominent tabulate layer below the outer membrane (although scanning electron microscope studies may reveal a very thin tabulate layer).

Unicellular. A single-celled organism.

Unilocular test. A foraminiferal test consisting of a single chamber.

Uniserial. A foraminiferal test with chambers arranged in a single row.

Unranked classification. A classification with taxa arranged into nested sets that reflect the branching patterns present in a cladogram. This grouping is typically done by indenting taxon names in a style similar to outlining. No rank is given any of the taxa that make up the hierarchy.

Upwelling. The movement of water from deeper parts of the ocean or a lake toward the surface, usually bringing nutrient-rich water to the surface and usually resulting in high productivity in surface waters.

Vacuole. A globular structure in the cytoplasm of a cell; may contain food particles, fats, oil, etc., or may function to expel material from the cell.

Vaterite. A hexagonal form of $CaCO_3$; rarely used by some organisms in their skeletons; unstable.

Vegetative reproduction. Asexual reproduction.

Ventral surface. Generally the lower surface of an organism, but defined in a variety of ways for protists lacking a clear lower surface.

Vesicle. The hollow body of an acritarch.

Water mass. A contiguous body of water in the ocean with similar temperature and salinity values, and with a similar history and biota.

Weight. The relative importance of a character. Weights may be either quantitative in the case of phylogenetic analyses or qualitative in the case of evolutionary taxonomy.

Whorl. A single coil of a shell or test.

Xanthidia. An abandoned term designating microfossils that resembled *Xanthidium*, a living desmid; many of these microfossils turned out to be dinoflagellate cysts.

Xanthophyll. A general term for a group of yellow, arotenoid pigments.

Xanthosome. Brown or yellow bodies contained in the cytoplasm of foraminifera; may be excretion products.

Zooplankton. Animal or nonphotosynthetic protist plankton.

Zooxanthellae. A general term for endosymbiotic dinoflagellates, from the genus *Zooxanthella*.

Zygote. The product of the fusion of gametes while still in a very early stage of development.

Index

Page references in *italics* refer to figures; page numbers in **bold** refer to tables.